Use this card to mask the answer column as you complete each page.

Use this card to mask the answer column as you complete each page.

Use this card to mask the answer column as you complete each page.

Chemical
Problem-Solving
by Dimensional Analysis

THIRD EDITION

Chemical
Problem-Solving
by Dimensional Analysis

A Self-Instructional Program

Arnold B. Loebel

Merritt College, Oakland, California

Houghton Mifflin Company · Boston

Dallas   Geneva, Illinois
Lawrenceville, New Jersey   Palo Alto

Printed in the U.S.A.
Library of Congress Catalog Card Number: 86-81913
ISBN: 0-395-35678-4

BCDEFGHIJ-A-8987

Contents

Preface   ix
To the Student   xi

Pretest	*1*	**Chapter 1**
Pretest Answers	*2*	Mathematics Review 1
Section A Exponential Notation	*2*	
Section B Logarithms	*18*	
Section C Significant Figures	*33*	
Problem Set	*43*	
Problem Set Answers	*45*	

Section A Conversion Factors	*47*	**Chapter 2**
Section B Solving Problems by		What is Dimensional
Dimensional Analysis	*49*	Analysis? 47
Problem Set	*53*	
Problem Set Answers	*55*	

Pretest	*57*	**Chapter 3**
Pretest Answers	*58*	The Metric System 57
Section A The Metric System	*58*	
Section B The British System of		
Weights and Measures	*67*	
Section C Temperature Conversions	*70*	
Problem Set	*72*	
Problem Set Answers	*74*	

Pretest	*77*	**Chapter 4**
Pretest Answers	*78*	Density, Specific Gravity, and
Section A Density	*78*	Percentage 77
Section B Specific Gravity	*81*	
Section C Percentage	*82*	
Problem Set	*88*	
Problem Set Answers	*91*	

Section A Atomic Weight	*95*	**Chapter 5**
Section B Molecular Weight	*96*	Mole: A Chemical Conversion
Section C The Mole	*97*	Factor 95
Problem Set	*104*	
Problem Set Answers	*105*	

Pretest	*111*	**Chapter 6**
Pretest Answers	*111*	Percentage Composition of
Section A Percentage Composition		Molecules 111
by Dimensional Analysis	*112*	
Section B A Shortcut Method	*114*	
Section C Percentage of a		
Complexed Molecule in a Compound	*114*	
Problem Set	*115*	
Problem Set Answers	*116*	

Pretest 117
Pretest Answers 117
Section A   The Empirical Formula 118
Section B   The Molecular Formula 119
Problem Set 121
Problem Set Answers 122

Chapter 7
The Simplest, or Empirical,
   Formula   117

Pretest 123
Pretest Answers 124
Section A   Balancing Simple Chemical
Equations by Inspection 125
Section B   Net Ionic Equations 129
Section C   Oxidation Numbers 137
Section D   What Examination of Oxidation
Numbers Tells You 140
Problem Set 143
Problem Set Answers 145

Chapter 8
The Balanced Chemical
   Equation   123

Pretest 147
Pretest Answers 147
Section A   Names and Symbols
of the Elements 148
Section B   Oxidation Number 149
Section C   Binary Compounds 153
Section D   Oxyacids 165
Section E   Polyatomic Ions from Oxyacids 173
Section F   Salts of the Oxyacids 179
Problem Set 184
Problem Set Answers 186

Chapter 9
Simple Inorganic
   Nomenclature   147

Pretest 189
Pretest Answers 189
Section A   Molar Stoichiometry 190
Section B   Weight Stoichiometry 192
Section C   Reaction-Controlling Component 194
Section D   Percent Yield 197
Problem Set 198
Problem Set Answers 200

Chapter 10
Stoichiometry   189

Pretest 205
Pretest Answers 206
Section A   Nonmetric Units 206
Section B   Impure Substances 210
Section C   Solutions and Density 213
Section D   Solution Concentrations
in Molarity 215
Section E   Miscellaneous Complications 218
Section F   Consecutive Reactions 222
Problem Set 224
Problem Set Answers 227

Chapter 11
Complicated Stoichiometry   205

Pretest 233
Pretest Answers 234
Section A   Gas Dimensions 235
Section B   Change-of-Conditions Problems 235

Chapter 12
Gas Laws   233

Section C   Determining an
Unknown Parameter   240
Section D   Determining Molecular Weight   243
Section E   Mixtures of Gases; Dalton's Law   244
Section F   Diffusion of Gases   246
Problem Set   247
Problem Set Answers   249

Pretest   255
Pretest Answers   256
Section A   Problems in which P, V, and T
Are Given   256
Section B   Problems in which One of the
Variables, P, V, or T, Is Missing   260
Section C   Problems Concerning Two Gases,
One with P, V, and T Given, and the Other
with Only Two Variables Given   263
Section D   Problems in which Both Gases
Are at the Same Temperature and Pressure   265
Problem Set   268
Problem Set Answers   271

Chapter 13
Stoichiometry Involving
Gases   255

Pretest   279
Pretest Answers   280
Section A   Molarity   280
Section B   Formality   286
Section C   Molality   287
Section D   Preparing Dilute Solutions
from Concentrated Ones   291
Section E   Mixtures of Solutions
of Different Molarity   295
Problem Set   296
Problem Set Answers   297

Chapter 14
Solution Concentration: Molarity,
Formality, Molality   279

Pretest   303
Pretest Answers   303
Section A   What pH Means   304
Section B   Calculating the pH from the
H^+ Ion Concentration   304
Section C   Determining the H^+ Ion
Concentration from the pH   307
Section D   Calculating the pOH from the
OH^- Ion Concentration   310
Section E   Determining the OH^- Ion
Concentration from the pOH   311
Section F   The Water Equilibrium   311
Section G   The Relationship Between
pH and pOH   313
Problem Set   315
Problem Set Answers   316

Chapter 15
pH: The Way Chemists Express
the Strength of Acidic and
Basic Solutions   303

Pretest   319
Pretest Answers   320
Section A   Problems with a Single Solution
of Known Molarity   320

Chapter 16
Stoichiometry Involving
Solutions   319

Section B   Standardization of Solutions   325
Section C   Problems with Two Solutions
Both in Molarity   327
Section D   Back-titration   329
Problem Set   333
Problem Set Answers   337

Section A   Colligative Properties   343
Section B   Determining K_F or K_B 346
Section C   Determining Molecular Weight   348
Section D   Determining Freezing and
Boiling Points of Solutions   350
Section E   Preparing Solutions with
Known Freezing and Boiling Points   352
Section F   What Happens when the Solute
Breaks up into Several Particles   354
Section G   Osmotic Pressure   358
Problem Set   360
Problem Set Answers   362

Chapter 17
Colligative Properties of
Solutions   343

Pretest   369
Pretest Answers   370
Section A   Specific Heat   370
Section B   Heat Involved in Changes of Phase   377
Section C   Heat Involved in Chemical
Reactions   381
Section D   Calorimetry   390
Problem Set   398
Problem Set Answers   401

Chapter 18
Calculations Involving
Heat   369

Four-Place Logarithms   410
Conversion Factors   412
International System of Units (SI)   415

Appendix I
Appendix II
Appendix III

Preface

The continuing demand for this totally programmed problem-solving book thirteen years after its first introduction has shown that there is, indeed, the need for such a text: one that covers only those types of mathematical problems relevant to the first semester of General Chemistry, pre-General Chemistry, or courses in chemistry geared to the allied health-science students. This third edition has been written to take into account the changes occurring in chemistry curricula and also to respond to the requests from various instructors using this text.

The basic format of the earlier editions has been retained. The book is divided into relatively short chapters, each of which covers one type of problem (or occasionally two related types). A Pretest precedes each chapter. Numerical answers to the Pretest problems and a scoring procedure follow immediately after. Students can grade themselves and use their score to determine whether they should go on to the Problem Set at the end of the chapter or whether they should work through certain or all sections of the chapter. All of the Pretest problems are used as examples in the body of the chapter. A Problem Set follows each chapter and all of the problems are completely worked out.

The Problem Sets in the third edition have been expanded so that there are now roughly one and one-half times as many worked-out problems. Also, the wording in the problems has been varied, forcing the students to avoid a rote treatment. The problems are now grouped so that the student will first solve problems identified as corresponding to specific sections in the chapter. Then, following these, is a final group of problems chosen randomly from all sections.

The concept of pH was introduced in the second edition of the text as an appendix, since the mathematical manipulations are not amenable to a dimensional-analysis solution. However, because of its importance, pH is now presented in a separate chapter (Chapter 15) immediately following the chapter on solution concentrations. As a consequence of this emphasis, the section in Chapter 1 on logarithms has been expanded and the use of scientific calculators is included. Then, in Chapter 15, the solutions to the pH problems are worked out using log tables (Appendix I) and enclosed in a box. The direct answer supplied by a calculator follows immediately. The number of significant figures allowed in the logarithmic calculations is emphasized continuously.

Inorganic nomenclature, a topic that doesn't fit into the category of mathematical problem-solving at all, has been expanded from what was an appendix in earlier editions to a chapter (Chapter 9) complete with Pretest and Problem Set. Anyone who teaches beginning chemistry knows the need for a complete and thorough treatment of this nettlesome topic. The chapter is written so that the instructor can assign it only in part or spaced out during the course of the semester (or quarter) as he or she wishes.

The balancing of chemical equations, another topic not falling into the category of mathematical problem-solving, has been made into a separate chapter (Chapter 8) in the third edition. It covers just balancing by inspection, with only a brief mention of redox reactions and a method for balancing them by counting the number of electrons gained and lost by changes in oxidation number. However, the use of net ionic equations is covered thoroughly and the method of predicting whether a metathesis reaction will or will not occur is emphasized.

The chapter on thermodynamics (Chapter 18) has been expanded but still does not cover entropy or free energy. These topics generally appear in more advanced courses than those to which this text is directed. Chapter 18 is restricted to enthalpies of temperature and phase changes and of chemical reactions. Hess's Law is introduced and the enthalpies of formation are used to calculate the enthalpies of reactions. Also, there is a detailed section on calorimetry.

I have completely eliminated the chapter on electrochemistry that appeared in earlier editions. Users of this text have indicated that, as with entropy and free energy, this is a topic not generally covered in their courses. Also, I have eliminated the use of equivalents and normality. These concepts, not permitted in the SI system, are surely facing the fate of the slide rule—a topic that was eliminated after the first edition. However, a note is included mentioning the existence of equivalents and normality and saying that the instructor will discuss them should he or she feel it is desirable.

A.B.L.

To the Student

You've probably never used a book designed the way this one is. It's called a self-instructional program. If it's going to do the job it's intended to do (and it really can), you've got to work with it in a special way. If you do, you'll find that in a short time you will be easily solving problems of such dazzling complexity that even reading them at this point can make your head spin. For example, after you've finished Chapter 16, you're going to zip through a problem that reads "How many milliliters of an H_2SO_4 solution whose density is 1.40 g/mL and that is 50 wt % H_2SO_4 will react with 35.0 mL of a 20 wt % solution of NaOH whose density is 0.90 g/mL?" You'll work that one just as easily as if you were figuring out the number of minutes in three hours. You're going to think somebody else is doing the work.

But—and here's where it is really in *your* lap—you have to *work* this book. You can't just read it like a normal text.

In the body of the material, on the inside columns of the pages, you'll see blanks, _____, in the middle of sentences or after questions. In the column on the outside, on the same line as the blank, you will find the word or number that fills the blank. The one thing you *must* do—and without this it won't work at all—is to write the response for each and every blank without looking at the response in the outside column. You can do this best by covering the column on the outside with the card located at the front of the book. Read the text material, and when you get to a blank, write the appropriate word or number either in the blank or on a piece of scratch paper. (You are going to need scratch paper anyway, since as you work through the book your responses will involve a progressive building up of the solution to the problems. You will have to keep a running solution in order to proceed from one blank to the next.) Then uncover the response for the blank and check the answer you wrote, to see that it is the same. It will always be the same. If, by some chance, it isn't, just go back a few blanks and try it over again until you're sure you understand.

You must work through the material and write down the responses for every blank. When you are working with a program, you're not taking a test or simply filling in blanks. You are learning *by doing*. The program works so that each response gives you enough understanding to make the next response, and so on and on and on.

The book is divided into 18 chapters, each of which covers pretty much a single type of problem. The order of topics is similar to that in most texts. However, your instructor (or text) may want you to use a modified sequence. No sweat. Any possible variation has been considered and if a switch is made you won't be without the background you need. Sometimes some very advanced—and difficult—material is presented. Your instructor may feel that you don't need to know this at this point in your chemical career. He or she will tell you to skip it. There is usually a notation in the text telling you that what's coming up may not be required.

At the beginning of each chapter is a short set of problems that cover the material in the chapter. Most are even used as examples in the chapter. This is called a *pretest*. If you think that you can handle these calculations, take the pretest. You're told how long it should take you to do the test. Immediately after the pretest there are answers to the problems and a method of scoring yourself. Your score will tell you either that you can skip the chapter and go immediately to the problem set at the end of the chapter, or that you had better work through certain sections of the chapter first.

Once you're ready, you can do the problem set at the end of the chapter. There are in some cases special instructions that tell you that you don't necessarily need to do certain problems (that's how hard they are). At other times, an asterisk (*) precedes the problem. This means that it's a hard one. Four or five problems even have *two* asterisks. (You can figure out what *that* means.) Right after the problem set, all the problems are completely worked out. It's a good idea to check the answer as soon as you have finished a problem. Then you can see how you're doing.

I count on the fact that all of you have a pocket calculator. What I don't know is whether your particular model is a *scientific* calculator and can get logarithms and roots. (You'll be using these functions in Chapters 1 and 15.) So what I did was to box in the calculations for those of you who don't have a scientific calculator. If you do have a scientific calculator, you can skip the boxed material and just check the answer at the very end.

I've taken great care to see that there are no mistakes in the text. But no matter how hard one tries, a slip or two always seems to creep in. You'll do me—and future students who use this book—a big favor if you take the time to write me and let me know where I blew it. As a matter of fact, I would like to hear from you anyhow—comments, criticisms, whatever. I promise to answer. My address is Merritt College, 12500 Campus Drive, Oakland, CA 94619.

Now it's time for you to get to work. See how you do on the pretest for Chapter 1.

Arnold Loebel

Mathematics Review

A word of advice: You are going to find a pocket calculator invaluable in doing all the problems in this book. A *scientific calculator*, with exponential notation and logarithmic capabilities (such as the inexpensive Texas Instrument TI-30 or TI-35 or the Sharp EL-506), is good for your purposes.

PRETEST

1. Express 56,000 in scientific notation (that is, as a number between 1 and 10 times 10 raised to a power).
2. Express 0.0031 in scientific notation.
3. Perform the operation $3.5 \times 10^{-6} \times 2.0 \times 10^{2}$.
4. Perform the operation $\dfrac{8.4 \times 10^{-4}}{2.0 \times 10^{2}}$.
5. Perform the operation $6.7 \times 10^{18} + 3.00 \times 10^{19}$.
6. Perform the operation $(3.0 \times 10^{-8})^{3}$.

Use either your calculator or the four-place table of logarithms in Appendix I to find the answers to the next six questions.

7. $\log 10^{3}$
8. $\log 55.5$
9. $\log 0.000558$
10. antilog -9.2291
11. $e^{0.25}$
12. $\sqrt[5]{35}$
13. Express 2.4321640 to three significant figures.
14. Perform the operation $6.0 \times 4.3 \times \dfrac{4.3271}{3}$.
15. Perform the operation $4.321 + 6.5$. Express your answer to the correct number of significant figures.

PRETEST ANSWERS

1. 5.6×10^4 2. 3.1×10^{-3} 3. 7.0×10^{-4}
4. 4.2×10^{-6}
5. 3.67×10^{19} or 36.7×10^{18}
6. 2.7×10^{-23} or 27×10^{-24}
7. 3   8. 1.7443   9. -3.2534
10. 5.90×10^{-10} 11. 1.284   12. 2.036
13. 2.43   14. $4. \times 10^1$ 15. 10.8

If you had all the answers right, you're good. You can skip Chapter 1 and do the problem set at the end immediately.

If you had all but one right in numbers 1 to 6, you can probably skip Section A in Chapter 1. If you missed two or more out of questions 1 to 6, work through Section A. If you got all of numbers 7 to 12 correct, you can skip Section B in Chapter 1. If you missed only number 11, look at the part of Section B on natural logarithms. If you missed only number 12, read the last part of Section B, Determining Powers and Roots. If you did numbers 13, 14, and 15 correctly, you can skip Section C.

The purpose of this chapter is to give you the tools you will need to get a number answer once you have set up a chemical problem. The chapter is divided into three sections that, though quite different, all have to do with handling numbers. Remember that the way to use this text is to cover the outside of the page with a card (see the card bound into the front of the book), or a piece of paper, so that you cannot see the answers that go in the blanks. Then you read the inside column. When you get to a _____, write down what goes in it. Then slide your card down the outside column to see whether you have written the correct word or number. We are going to start with what is called *exponential notation*.

SECTION A   Exponential Notation

There is a very convenient way of showing in mathematical shorthand that you have multiplied the same number repetitively. You write a small, raised number after the digit that is being multiplied by itself. This raised number is called an *exponent* or a *power*. For example,

$$2 \times 2 \times 2 = 2^3$$

You have multiplied three 2's together and the exponent is 3. You can also say that you have 2 to the third power. (Your calculator may give you answers in exponential notation. However, it's a good idea to read this section to see what you are really doing.)

Using the same system,

2^4 $2 \times 2 \times 2 \times 2 = $ _____

four; 4   You have multiplied _____ 2's together, and the exponent is __.

You can also say that you have 2 to the _____ power or write _____.

fourth; 2^4

Using the same system,

$$5 \times 5 \times 5 = \text{_____}$$

5^3

which is 5 to the _____ power. Similarly,

third

$$7.5 \times 7.5 \times 7.5 \times 7.5 \times 7.5 \times 7.5 = \text{_____}$$

7.5^6

You have multiplied _____ 7.5's together, so you can also say that you

six

have 7.5 to the _____ power. If you multiply two 2's together, which is

sixth

2 to the _____ power, you get 4. If you have only one 2, it

second

equals __. Of course, if you only have one of a number, you say that it is

2

to the _____ power, which you can write exponentially as _____. Any

first; 2^1

number raised to the first power is the number itself. Therefore,

$$6.6^1 = \text{\_\_\_\_}$$

6.6

In this case, there is only _____ 6.6 and you can say that it is raised to

one

the _____ power. If you had 10^1, this would be equal to _____, since

first; 10

any number to the _____ power is equal to itself. If you had 10 to the

first

second power, written exponentially as _____, this would mean that

10^2

you have multiplied _____ 10's together, 10×10, which equals

two

_____. If you have 10^3, which is 10 to the _____ power, you would

100; third

have multiplied _____ 10's together, $10 \times 10 \times 10$, which equals

three

_____. Let's look at the values of these powers of 10 and see if any

1000

pattern shows up.

$$10 = 10^1 = \qquad 10$$
$$10 \times 10 = 10^2 = \qquad 100$$
$$10 \times 10 \times 10 = 10^3 = \qquad 1000$$
$$10 \times 10 \times 10 \times 10 = 10^4 = 10,000$$

The number 10 contains _____ zero and is equal to 10 raised to the

one

_____ power. The number 100 contains _____ zeros and is equal to

first; two

10 raised to the _____ power. The number _____ contains

second; 1000

three zeros and is equal to 10 raised to the _____ power. If you wrote

third

the number 10,000 exponentially, you would have an exponent of __,
since there are four zeros. What about the number 1,000,000,000,000?

4

Written exponentially, it is _____.

10^{12}

Now you can see why we bother to write numbers exponentially. It is
a neat way of writing very large (or, as you will see, very small) numbers
in a small space.

Now think about how you could write the number 1 expressed as 10
to a power. The number 1 has _____ zeros, so you write this

no

10^0; zero

exponentially as _____, which is 10 to the _____ power. This may look strange. However, it does not mean that 10 is multiplied by zero, which would be equal to zero. It simply means that no 10's are multiplied together at all. *Any number to the zero power equals* 1.

Next let us express exponentially numbers that have digits other than 1 and 0 in them. Suppose that you have the number 400, which equals 4 × 100. You know from the above that you can write 100 exponen-

10^2

tially as _____. So instead of writing 4 × 100, you can write

10^2; 10^3

4 × _____. Using this method, you can write 6000. as 6 × _____,

5×10^4

and 50,000. as _____.

Actually, you are simply moving the decimal point. The first example shown was 400., which equals 4.00×10^2. The decimal point is moved two places to the left,

$$4\underset{2\ \ 1}{\text{.0 0}}\text{.}$$

Then you multiply by 10 raised to the second power. In the second example, 6000., you move the decimal point three places to the left,

$$6\underset{3\ 2\ 1}{\text{.0 0 0}}\text{.}$$

You get 6.000×10^3. The exponent of 10 is equal to the number of

places

_____ you move the decimal point to the left. Then all you do is write the number and count the number of places you move the decimal point. Put in the decimal at the new location and multiply by 10 raised to a power that equals the number of places moved. Try this one:

10^6

9,000,000. = 9.000000 × _____.

10^1

What about 470? This equals 47.0 × _____, but it also equals

10^2; 10^3

4.70 × _____. It even equals 0.470 × _____. You can put the decimal point anywhere you want. All you need do is count the number

decimal point

of places you move the _____ to the left and multiply by 10 raised to that power.

How do you decide to which place to move the decimal point? It is most convenient, when nothing else affects your decision, to move the decimal point far enough so that you have a number between 1 and 10 times 10 to the appropriate power. This is called *scientific notation*. (Your calculator may be programmed to do this for you.) Therefore 470 expressed in scientific notation is 4.70×10^2, because 4.70 (the digit part) is between 1 and 10. The number 3476 expressed in scientific nota-

3.476×10^3; 1 and 10

tion is _____ because 3.476 is between _____.

Now practice a bit to make sure that you have all of this straight. Write the following numbers in scientific notation.

43. = _____	4.3×10^1
7569. = _____	7.569×10^3
432,506.43 = _____	4.3250643×10^5
9.0 = _____	9.0

The last one is tricky. Actually it is 9.0×10^0. But 10^0 equals 1, and multiplying by 1 doesn't change anything. Thus you need not write 10^0. However, some feel that it is not true scientific notation unless you show the 10^0.

Now see if you can do it backward. When written as a number not in exponential notation, the number 6.54×10^5 is _____ .   654,000

If you got the correct answer, and feel fairly confident, skip the next section and go directly to the section on negative exponents. If you had trouble, read what is immediately following. (Incidentally, if your calculator gives you the number in exponential form, it can reverse the process.)

Consider $5. \times 10^2$. This is a short way of writing $5. \times 10 \times 10 = 5. \times 100 = 500$. The easy way of getting to 500 is to write $5. \times 10^2$ with many zeros after the 5: 5.0000000×10^2.

The exponent on the 10 tells you how many places the decimal was moved to the left. All you have to do is move the decimal back to the right that many places.

5 . 0 0 0 0 0
     1 2

The number becomes 500. Now take the number 3.5678×10^3. The 10^3 tells you that the decimal point has been moved _____ places to the   three

left. Therefore you move it back to the right _____ places. The digital,   three

nonexponential number equal to 3.5678×10^3, therefore, is _____ .   3567.8

What about 6.5×10^4? You move the decimal point back to the right

four places because the 10 is raised to the _____ power. Since   fourth

there are not that many numbers to move the decimal point, you can

add some _____, which don't change the value of the number. Thus   zeros

you would have 6.5000000×10^4. Moving the decimal point back to the

right _____ places gives you _____ .   four; 65,000.

6 . 5 0 0 0 . 0 0 0
     1 2 3 4

You should now be able to do the first example. When written as a

nonexponential number, 6.54×10^5 is _____ .   654,000.

Negative Exponents

There is another convenient trick with exponential notation. Suppose that you divide by a number multiplied repetitively,

$$\frac{5}{2 \times 2 \times 2}$$ which you could write exponentially as $\frac{5}{2^3}$

You can move the number written exponentially to the top (numerator) if you change its sign,

$$5 \times 2^{-3}$$

The minus sign simply shows that you are dividing by the number raised to that power. Therefore, using the same procedure,

5^{-4}

$$\frac{47}{5 \times 5 \times 5 \times 5} = \frac{47}{5^4} = 47 \times \underline{\hspace{1cm}}$$

or

$10^{-2}; 10^5$

$$\frac{2}{10^2} = 2 \times \underline{\hspace{1cm}} \quad \text{and} \quad \frac{7}{10^{-5}} = 7 \times \underline{\hspace{1cm}}$$

When you shift a number written exponentially from the bottom (the denominator) to the top (the numerator) of the fraction, you change the sign. In the last example, you had a −5 in the denominator and you changed the sign from − to + when you shifted it to the numerator. (It is customary to omit plus signs. They are understood.) Now do a few for practice.

$11 \times 10^{-4}; 5 \times 10^8; \frac{3 \times 10^{-2}}{2}$

$$\frac{11}{10^4} = \underline{\hspace{1cm}} \quad\quad \frac{5}{10^{-8}} = \underline{\hspace{1cm}} \quad\quad \frac{3}{2 \times 10^2} = \underline{\hspace{1cm}}$$

(Notice that you can shift only the numbers that are written exponentially.)

$\frac{6 \times 10^2 \times 10^{-7}}{5}$

$$\frac{6}{5 \times 10^{-2} \times 10^7} = \underline{\hspace{1cm}}$$

What about this?

$$4 \times 10^{-3} = \frac{4}{\underline{\hspace{0.5cm}}}$$

10^3

That's right. If you shift a number written exponentially from the top to

sign

the bottom, you change its _____. It works both ways.

The reason that we bother with this is that it allows you to express very small numbers in a convenient way. Consider $1. \times 10^{-2}$. This is really

$$\frac{1.}{10^2} = \frac{1.}{10 \times 10} = \frac{1.}{100}$$

Dividing this out, you get

0.01   So $1. \times 10^{-2} = 0.01$

The decimal point has been moved _____ places to the left in going

from 1. $\times$ 10^{-2} to .01. The power to which 10 is raised is _____, so
the number of places you move the decimal point is equal to the power

to which _____ is raised.

   We can put this another way: When you move a decimal point to the

right, the number of _____ that you move it gives you the

negative _____ of the 10 by which you must multiply.
Here is the sequence.

$$0.1 \quad = 1 \times 10^{-1}$$
$$0.01 \quad = 1 \times 10^{-2}$$
$$0.001 \quad = 1 \times 10^{-3}$$
$$0.0001 = 1 \times 10^{-4}$$

Following this pattern, 0.000000001 would equal 1. $\times$ _____, since

you had to move the decimal point _____ places to the right.

$$0 \underbrace{.0\ 0\ 0\ 0\ 0\ 0\ 0\ 0\ 1}_{1\ 2\ 3\ 4\ 5\ 6\ 7\ 8\ 9}$$

   There is a little memory trick to use in deciding whether the sign on
the power of 10 is minus or plus when you move the decimal point left
or right. When the number becomes **bigger**, you move the decimal to the
right and the power of 10 is negative. All the words have a **g**. Thus,

$$0.01 = 1. \times 10^{-2}$$

(The number 0.01 becomes 1, which is bigger; you move the decimal
point to the right, and the power is negative.)
   When the number becomes smaller, you move the decimal point to the
left and the power of 10 is plus. All words have an **l**.

$$100. = 1. \times 10^{2}$$

(The number 100 becomes 1, which is smaller; you move the decimal
point to the left, and the power is plus.)

   The number 0.00001 can be written as 1. $\times$ _____. Since 1. is

_____ than 0.00001, the sign on the power to which the 10 is

raised is _____. The decimal point was moved to the _____.

The number 1,000,000. can be written as 1. $\times$ _____. Since 1. is

_____ than 1,000,000, the sign on the power to which 10 is

raised is _____. The decimal point has been moved to the _____.

   Here is a little practice. Simply note whether the power to which the
10 has been raised is plus or negative.

---

two

-2

10

places

power (exponent)

10^{-9}

nine

10^{-5}

bigger

negative; right

10^{6}

smaller

plus; left

– | $0.0001 = 1. \times 10\underline{\hphantom{xx}}^4$ (bigger; negative; right)

+ | $1000. \quad = 1. \times 10\underline{\hphantom{xx}}^3$ (smaller; plus; left)

– | $0.1 \quad = 1. \times 10\underline{\hphantom{xx}}^1$ (bigger; negative; right)

+ | $10. \quad = 1. \times 10\underline{\hphantom{xx}}^1$ (smaller; plus; left)

So far we have looked at numbers containing only the digits 1 and 0. But just as you were able to count the number of places moved to the left in large numbers to find the power of 10, you can count places in small numbers when you move the decimal point to the right to make them bigger. For example,

$$0.02 = 2. \times 10^{-2}$$

two; right | You have moved the decimal point _____ places to the _____ and therefore have a –2 power.

$$0 \,.\, \underline{0\ 2}$$
$$1\ 2$$

Using the same procedure, to write the number 0.00035 in scientific

four; right | notation, you move the decimal point _____ places to the _____.

–4 | Therefore the power of 10 will be _____. The result is that 0.00035

3.5×10^{-4} | expressed in scientific notation is _____.

1; 10 | Recall that scientific notation is any number between __ and _____ times 10 to some power. If you move the decimal point all the way to

10^{-5} | the end so that you have 35 × _____, then 35 is larger than 10. If you

10^{-3} | stop the decimal point in front of the 3, you have 0.35 × _____, and 0.35 is less than 1.

To prove how well you can do this now, write the following numbers in scientific notation.

4.76×10^{-5} | $0.0000476 =$ _____

3×10^{-2} | $0.03 \quad\ =$ _____

3.567×10^1 | $35.67 \quad =$ _____

$4. \times 10^{-5}$ | $0.00004 \ =$ _____

4.86715×10^4 | $48671.5 \quad =$ _____

[You have probably observed that when a number starts after a decimal point, there is usually a zero to the left of the decimal point. For example, we write 0.25, not .25. This is only a convention, so that someone reading the number won't think that the decimal point is just a spot of dirt on the paper. To prevent any such doubts, a zero is put before the decimal. There are times when you have tables (see Appendix I) and all the numbers start after the decimal. Then the zero is left out.]

Let's see now if you can do these negative exponents backward. The

0.0000037 | digital number equal to 3.7×10^{-6} is _____. If you got this answer, you can skip the next section and go right to the section called Multiplying Numbers Expressed in Exponential Notation. If you didn't get the answer, or don't feel confident, work through the next short section.

If you have a number such as 6.25×10^{-3}, the negative sign on the exponent means that the decimal place must have been moved to the

_____ . The number of the exponent, 3, means that it has been moved   right

_____ places. Consequently, to get a decimal number you must move   three

the decimal point _____ places back to the _____ . Since there are no   three; left
numbers three places to the left, you put in some extra zeros

$$0\ 0.0\ 0\ 6\ .\ 2\ 5$$
$$3\ 2\ 1$$

and count off the number of places you need. The result, then, is that

the decimal number equal to 6.25×10^{-3} is _____ .   0.00625

Another example is 4.3615×10^{-8}, which has a negative 8 exponent on the 10. This means that the decimal point must have been moved

_____ places to the _____ to get the number. Therefore, to convert it   eight; right

back to decimal notation, you move the decimal point back _____   eight

places to the _____ . So you put some _____ in front of the number   left; zeros

_____   0000000004.3615

and count off the number of places you need. The result is that the

decimal number equal to 4.3615×10^{-8} is _____ .   0.000000043615

Now you can try the first example again. The decimal number that is

equal to 3.7×10^{-6} is _____ . See how easy it is?   0.0000037

Multiplying Numbers Expressed in Exponential Notation

Suppose that you want to multiply $10^2 \times 10^3$.

$$10^2 = 10 \times 10$$

$$10^3 = \text{_____}$$ $10 \times 10 \times 10$

Therefore

$$10^2 \times 10^3 = (10 \times 10) \times (10 \times 10 \times 10)$$
$$= 10 \times 10 \times 10 \times 10 \times 10$$

A much shorter way of writing five 10's multiplied together is _____ . 10^5
Therefore, $10^2 \times 10^3 = 10^5$.

Do you see any relationship between $10^2 \times 10^3$ and 10^5? All you

have done is _____ the exponents. To check this, suppose that you   add
want to multiply $10^4 \times 10^5$.

$10^4 = 10 \times 10 \times 10 \times 10$

$10 \times 10 \times 10 \times 10 \times 10$ | $10^5 =$ ________________

Thus $10^4 \times 10^5 = 10 \times 10 \times 10 \times 10 \times 10 \times 10 \times 10 \times 10 \times 10$.

A shorter way of writing nine 10's multiplied together gives you the final result, $10^4 \times 10^5 = 10^9$.

Once again you can see that, when you multiply a number raised to a

add | power by the same number also raised to a power, you just ______ the powers (exponents). You can do a few more yourself.

10^{13} | $10^{11} \times 10^2 =$ ______

10^4 | $10^3 \times 10 =$ ______ (The first power is understood)

10^{12} | $10^4 \times 10^8 =$ ______

If you have numbers other than 10's raised to a power, you multiply these separately. It is much easier if you group the digit numbers together and all the 10's at the end. For example,

$(3 \times 10^2) \times (4 \times 10^5)$

can be rewritten as $3 \times 4 \times 10^2 \times 10^5$.

Then you multiply the numbers as you would normally, and multiply the 10's to a power by adding the exponents.

$3 \times 4 \times 10^2 \times 10^5 = 12 \times 10^7$

Therefore, if you want to multiply $(5 \times 10^5) \times (6 \times 10^6)$, you can

$5 \times 6 \times 10^5 \times 10^6$ | rewrite this as ________________. Then you multiply the numbers normally and the 10's to a power by adding exponents to give the final

30×10^{11} | answer ____________. This answer is not in scientific notation, but you

3.0×10^1 | know that 30 can be written in scientific notation as ____________.

3.0×10^1 | Therefore 30×10^{11} is equal to ____________ $\times 10^{11}$. Multiplying the

3.0×10^{12} | powers of 10 by adding exponents gives ____________. Try a few more yourself.

8×10^{24} | $4 \times 10^{11} \times 2 \times 10^{13} =$ ____________

2.8×10^7 | $7 \times 10^2 \times 4 \times 10^4 =$ ____________ (in scientific notation)

3.6×10^{13} | $2 \times 10^4 \times 3 \times 10^6 \times 6 \times 10^2 =$ ________ (in scientific notation)

(It doesn't matter how many numbers you have multiplied together.)

What happens if the exponent is negative? Consider $10^3 \times 10^{-2}$. Recall that a negative exponent simply means that the exponential number was raised from the bottom of the fraction. You could rewrite this as

$$\frac{10^3}{10^2}$$

Breaking this down to 10's multiplied repetitively, you get

$$\frac{10 \times 10 \times 10}{10 \times 10}$$

Now you can see that two of the 10's in the denominator cancel two 10's in the numerator.

$$\frac{\cancel{10} \times \cancel{10} \times 10}{\cancel{10} \times \cancel{10}}$$

leaving 10^1. This is just what you should expect. You have added the exponents 3 and –2 algebraically to give 1. Try a few yourself.

$$10^{11} \times 10^{-6} = \underline{\quad}\;\underline{\quad} \qquad\qquad 10^5$$
$$10^{-9} \times 10^5 = \underline{\qquad} \qquad\qquad 10^{-4}$$
$$10^{-3} \times 10^{-2} = \underline{\qquad} \qquad\qquad 10^{-5}$$
$$10^{15} \times 10^{-4} \times 10^{-6} = \underline{\qquad} \qquad\qquad 10^5$$
$$10^{-27} \times 10^3 \times 10^{-1} \times 10^4 = \underline{\qquad} \qquad\qquad 10^{-21}$$

If you were able to do all the above, skip the next section and go on to Dividing Numbers Expressed in Exponential Notation. If you had any problem, read the rules for algebraic addition given next.

Algebraic Addition

If you add numbers of the same sign (either + or –), the answer is the sum of the numbers and the sign is the same as that of the numbers. If you add numbers of different signs, you subtract the smaller from the larger, and the sign is the same as the sign of the larger. If you have more than two numbers, you add all those of the same sign separately. Then subtract the smaller sum from the larger sum and give your answer the sign of the larger. Here are some examples.

$$2 + 6 = \underline{\quad} \qquad\qquad 8$$
$$-2 + (-6) = \underline{\qquad} \qquad\qquad -8$$
$$5 + (-3) = \underline{\quad} \qquad\qquad 2$$
$$3 + (-5) = \underline{\qquad} \qquad\qquad -2$$
$$6 + (-4) + (-3) = \underline{\qquad} \qquad\qquad -1$$
$$(-7) + (-4) + 3 + (-9) = \underline{\qquad} \qquad\qquad -17$$

Remember that a plus sign (+) is usually understood, so you need not write it. A minus sign (–) is always written.

Dividing Numbers Expressed in Exponential Notation

To divide two numbers, you set up a fraction with the number you are dividing on the top (numerator) and the number you are dividing it by on the bottom (denominator). Therefore, if you want to divide 18 by 3, you write this as the fraction \underline{\qquad}. $\qquad \dfrac{18}{3}$

Margin answers:

$\dfrac{10^5}{10^2}$

sign

$10^5 \times 10^{-2}$

adding

10^3

$10^1;\ 10^{-5}$

$10^{-4};\ 10^{12}$

$1\ (=10^0)$

10^{-3}

10^{10}

If you want to divide 10 raised to a power by another 10 raised to a power, you set up the same kind of fraction. For example, 10^5 divided by 10^2 can be written as the fraction _____. However, you know that you can shift a number raised to a power from the bottom of a fraction to the top simply by changing the _____ of the exponent. Doing this to $\dfrac{10^5}{10^2}$ gives _____. You now have a multiplication that you perform by algebraically _____ the exponents. The final answer for the division of 10^5 by 10^2 is therefore _____.

It is really very simple. Try a few.

$$\frac{10^7}{10^6} = \text{\_\_\_\_\_} \qquad\qquad \frac{10^4}{10^9} = \text{\_\_\_\_\_}$$

$$\frac{10^{-1}}{10^3} = \text{\_\_\_\_\_} \qquad\qquad \frac{10^7}{10^{-5}} = \text{\_\_\_\_\_}$$

$$\frac{10^4}{10^4} = \text{_____}$$

(For the last fraction, the number on the bottom equals the number on the top. They simply cancel each other out. If you remember this when you have a lot of exponential numbers on the top and bottom, you can get a solution more quickly.)

$$\frac{\cancel{10^{-6}} \times 10^2 \times \cancel{10^{-3}}}{\cancel{10^{-3}} \times 10^5 \times \cancel{10^{-6}}} = \frac{10^2}{10^5} = \text{\_\_\_\_\_}$$

You will also find that if you have more than one exponential number on the top or bottom, it may be faster if you algebraically add the exponents on the bottom and top separately and then shift the denominator to the numerator.

$$\frac{10^5 \times 10^{-3} \times 10^7}{10^4 \times 10^3 \times 10^{-8}} = \frac{10^9}{10^{-1}} = \text{\_\_\_\_\_}$$

When you are solving real problems, you will have numbers as well as 10's raised to various powers. You will find it much simpler if you re-write the fraction, grouping all the numbers together at the beginning and putting all the 10's at the end. Then you can handle the numbers in the normal fashion, making the multiplications and divisions and adding the exponents for the powers of 10. (If your calculator can express numbers in exponential notation, you don't have to bother with this. Just put in the number, then enter the exponent to which 10 is raised.) For example,

$$\frac{3 \times 10^3 \times 4 \times 10^{-5}}{2 \times 10^7 \times 6 \times 10^1} = \frac{3 \times 4}{2 \times 6} \times \frac{10^3 \times 10^{-5}}{10^7 \times 10^1} = 1 \times 10^{-10}$$

Now suppose that you have the operation

$$\frac{1 \times 10^2 \times 2 \times 10^{-6}}{2 \times 10^8 \times 3 \times 10^{-4}}$$

First you rewrite to group all the numbers together to give

_____. Then you solve the multiplication and division

of the numbers to give _____. And then you solve the powers of 10 to

give _____. When you carry out the calculation, you get _____.
But this isn't in scientific notation. You can change the 0.33 to 3.3 $\times$

_____, so the answer is now 3.3 $\times$ _____ $\times$ 10^{-8}. Or, finishing the

multiplication, you get _____. Try a few for practice.

$$\frac{3 \times 10^6 \times 8 \times 10^2}{6 \times 10^{-4}} = \underline{\hspace{2cm}}$$

(If you got 4×10^4 for your answer, you forgot to change the sign on
the 10^{-4} when you shifted it to the numerator.)

$$\frac{5 \times 10^{-3} \times 4 \times 10^{-5} \times 3 \times 10^1}{2 \times 10^{11} \times 2 \times 10^{-17}} = \underline{\hspace{1cm}}$$

(If you got 15×10^{-1}, you were right. But you didn't express your
answer in scientific notation. If you got 15×10^{-13}, you forgot to
change the sign of the exponent.)

Adding and Subtracting Numbers Expressed in Exponential Notation

(*Note*: If your calculator handles numbers in exponential notation, you
can skip this short section.)

   If you want to add or subtract numbers that are multiplied by 10 to a
power, you can do this only if the power of 10 is the same for all numbers. The reason for this will become clear if you write the exponential
numbers as ordinary numbers. Consider

$$\begin{array}{r} 4 \times 10^{-1} \\ + 3 \times 10^2 \\ \hline \end{array}$$

If you write 4×10^{-1} as a digital number, it equals _____. If you write

3×10^2 as a digital number, it equals _____. Then you add

$$\begin{array}{r} 0.4 \\ + 300. \\ \hline 300.4 \end{array}$$

which equals 3.004×10^2. You see, then, that you simply cannot say
$3 + 4$ is 7. But if the numbers are times the same power of 10, then they
can be added and the result multiplied by that power of 10. Consider

$$\begin{array}{r} 5 \times 10^{-1} \\ + 2 \times 10^{-1} \\ \hline \end{array}$$

If you write 5×10^{-1} as an ordinary number, it equals _____. If you

write 2×10^{-1} as an ordinary number, it equals _____. Then you add

---

(answer column)

$\dfrac{1 \times 2}{2 \times 3} \times \dfrac{10^2 \times 10^{-6}}{10^8 \times 10^{-4}}$

0.33

10^{-8}; 0.33×10^{-8}

10^{-1}; 10^{-1}

3.3×10^{-9}

4×10^{12}

1.5

0.4

300

0.5

0.2

$$0.5$$
$$+\ 0.2$$
$$0.7$$

which equals $7. \times 10^{-1}$.

Thus, to either add or subtract numbers that are times a power of 10, you simply convert all the numbers to the same power of 10. Remember that if you multiply by a 10 to a negative exponent, the number becomes bigger because the decimal point is moved to the right. When you multiply by 10 to a plus exponent, the number becomes smaller, since you have moved the decimal point to the left.

Suppose that you want to add

$$4. \times 10^1$$
$$+\ 5. \times 10^2$$

You can convert either $4. \times 10^1$ to something $\times 10^2$ or $5. \times 10^2$ to something $\times 10^1$.

Suppose that you multiply $4. \times 10^1$ by 10^1. Since this is a plus exponent, the number becomes smaller and the decimal point moves to the left. You then have

$$0.4 \times 10^1 \times 10^1 = 0.4 \times 10^2$$
which you can add to $\quad 5. \ \times 10^2$
to give $\quad 5.4 \times 10^2$

Or you could multiply $5. \times 10^2$ by 10^{-1}. Since this is a negative exponent, the number becomes bigger and the decimal point moves to the right. You then have

$$50. \times 10^2 \times 10^{-1} = 50. \times 10^1$$
which you can add to $\quad 4. \times 10^1$
to give $\quad 54. \times 10^1 = 5.4 \times 10^2$

Here's another example. Add

$$8.56 \times 10^{-3}$$
$$+\ 6.10 \times 10^{-1}$$

If you decide to convert everything to $\times 10^{-1}$, you must multiply 8.56×10^{-3} by _____. This exponent is _____ so you must move the decimal point _____ places to the _____. The number becomes _____, having a value of _____. The new 10 raised to a power will be $10^{-3} \times$ _____. Multiplying these, you have _____. You can then make the addition

$$\underline{\hspace{2cm}}$$
$$+\ 6.10 \times 10^{-1}$$
$$\underline{\hspace{2cm}}$$

Margin answers:
10^2; +2
two; left
smaller; 0.0856
10^2; 10^{-1}
0.0856×10^{-1}
6.1856×10^{-1}

Try a few.

$$6. \times 10^5 + 9. \times 10^3 = \underline{\hspace{3cm}}$$

$$4.1 \times 10^{-3} + 7. \times 10^{-2} = \underline{\hspace{3cm}}$$

$$9. \times 10^1 - 3. \times 10^2 = \underline{\hspace{3cm}}$$

(Remember the rules for algebraic addition.)

$$8.7 \times 10^2 + 7.1 \times 10^4 + 1.0 \times 10^1 = \underline{\hspace{4cm}}$$

Raising Numbers Expressed in Exponential Notation to a Power

If you want to multiply a number expressed in exponential notation by itself repetitively, you can use the same shorthand method used for ordinary numbers. You write a small, raised number following the number that is multiplied by itself telling how many times it is done. Thus you can write

$$2.5 \times 10^2 \times 2.5 \times 10^2 \times 2.5 \times 10^2 \qquad \text{as} \qquad (2.5 \times 10^2)^3$$

The raised 3 indicates that you have multiplied three 2.5×10^2's together. You always write parentheses before and after the number so that it is clear just what is being multiplied by itself.

If we group the numbers at the beginning and the 10's to a power at the end, we get $2.5 \times 2.5 \times 2.5 \times 10^2 \times 10^2 \times 10^2$. Of course,

$10^2 \times 10^2 \times 10^2 = \underline{\hspace{1cm}}$ by adding exponents when you multiply. But 10^2 multiplied together three times also can be written by the

shorthand notation as $\underline{\hspace{1cm}}$. Can you see any relationship between

$(10^2)^3$ and 10^6? Since 2 $\underline{\hspace{1cm}}$ 3 equals 6, when you raise a number

already to a power to another power, you simply $\underline{\hspace{2cm}}$ powers. Thus we can write $(10^2)^3 = 10^{(2 \times 3)}$.

There isn't much we can do with the three 2.5's multiplied together except write $(2.5)^3$ and multiply them out when we need a real number. Therefore,

$$(2.5 \times 10^2)^3 = (2.5)^3 \times 10^6 = 15.6 \times 10^6$$

Consider $(10^2)^4$. This can be written as

$$10^2 \underline{\hspace{1cm}} \quad \text{which is equal to} \quad 10 \underline{\hspace{1cm}}$$

You try a few.

$$(10^6)^2 = \underline{\hspace{1cm}}$$

$$(4 \times 10^3)^2 = 4^2 \times (10^3)^2 = \underline{\hspace{2cm}}$$
(Take the 4^2 separately as 4×4.)

$$(10^{-3})^2 = \underline{\hspace{1cm}}$$

(The sign convention for multiplication is that if you multiply two numbers of the same sign, the sign of the answer is $+$. If you multiply two numbers of different signs, the sign of the answer is $-$.)

Answer column:

6.09×10^5

7.41×10^{-2}

-2.1×10^2

718.8×10^2 or 7.188×10^4 or 7188×10^1

10^6

$(10^2)^3$

times

multiply

$\times 4; \ 8$

10^{12}

16×10^6 or 1.6×10^7

10^{-6}

8×10^{-24}	$(2 \times 10^{-8})^3 =$ _____ $(2^3 = 2 \times 2 \times 2 = 8)$
0.11×10^{-22} or 1.1×10^{-23}	$(3 \times 10^{11})^{-2} =$ _____ $\left(3^{-2} = \dfrac{1}{3^2} = \dfrac{1}{3 \times 3} = 0.11\right)$
0.0156×10^3 or 1.56×10^1	$(4 \times 10^{-1})^{-3} =$ _____ $\left(4^{-3} = \dfrac{1}{4^3} = \dfrac{1}{4 \times 4 \times 4} = 0.0156\right)$

Determining the Roots of Numbers Expressed in Exponential Notation

It is possible to reverse the entire process of multiplying the same number together a certain number of times. Instead of asking what that will give, you ask, "What is the number which, if I multiply it by itself a certain number of times, will result in a given number?" This is called getting the *root*. An example that you will find familiar is "What is the square root of 9?" "Square" is the common way of saying "to the second power." So the question really is "What number multiplied by itself two times gives 9?" The answer, of course, is 3, since $3^2 = 9$.

Another familiar example is "What is the square root of 16?" This is

two; 16   asking what number multiplied by itself _____ times equals _____.

4   The answer is __.

The common way of saying "to the third power" is "cube." Thus, if you are asked for the cube root, you are asked what number multiplied

three   by itself _____ times will give a certain value.

2   What is the cube root of 8? If you multiply three __'s together, you

2   get 8. So the cube root of 8 is __.

Sometimes you see roots written

$$\sqrt{} \;\; \text{(square root)} \qquad \sqrt[3]{} \;\; \text{(cube root)}$$

However, the most convenient way to write a root of a number is to write 1 over whatever root it is. Thus the square root of 4 could be

$8^{1/3}$ written $4^{1/2}$ and the cube root of 8 could be written as _____. (To save space, we often write $\frac{1}{2}$ as 1/2.) In the same way, the fifteenth root of

$12^{1/15}$ 12 could be written as _____. The reason that this is so convenient will become obvious. If you have $(2^4)^2$, you multiply the two exponents to

2^8 get _____. If you want to know the square root of 2^4, you write $(2^4)^{1/2}$. When you multiply $4 \times 1/2$, you get the new exponent, 2^2. Let's check to see whether this answer is correct. You want to see if 2^2 is the

two; Yes   number that, multiplied together _____ times, equals 2^4. Does it? ____.

Since this seems to work, let us find the cube root of 10^9 by the same method. This can be written as $(10^9)^{1/3}$. If you multiply 9 by 1/3, you

3; three   get __. So the cube root of 10^9 must be 10^3. Check whether _____

Yes 10^3's multiplied together equal 10^9. Do they? _____.

What is the fifth root of 10^{10}? This can be written as (10^{10})——.

Multiplying these exponents gives _____. Using the same methods, the

eleventh root of 10^{-33} would be _____.

You probably now see what you are really doing. To get a root of a number expressed exponentially, you divide the exponent by the root you want to take. This is identical with multiplying by 1 over the root.

Thus the twelfth root of 10^{48} would be 10——. You have simply divided

_____ by _____.

If you have a digital number times a 10 to a power, you take the root of each separately. This is the same as when you raised numbers to a power. For example, the square root of 4×10^4 is written as $(4 \times 10^4)^{1/2}$.

Then you take the 1/2 powers separately: $4^{1/2} \times (10^4)^{1/2}$.

This equals 2×10^2.

Here is another example. The cube root of 8×10^{12} should be written

as _____. When you take the 1/3 powers separately, you get

_____. The cube root (1/3 power) of 8 is 2, and 10^{12} to

the 1/3 power is _____. Thus the answer for the cube root of

8×10^{12} is _____.

The cube root of 2.7×10^{-2} should be written as _____.

When you take the 1/3 powers separately, you get _____. However, you have a problem here. When you try to take the cube root of 10^{-2}, you find that you cannot divide 3 into –2 and get a whole number. You have $10^{-2/3}$, which isn't very useful. To end up with a whole-number exponent, the original power to which the 10 is raised must be exactly divisible by the root that you want to take. This isn't an impossible problem, since you can change the exponent by shifting the decimal place of the number. Thus

$$2.7 \times 10^{-2} = 27 \times 10^{-3}$$

(You multiply by 10^{-1}, which is negative, and the decimal point moves

to the _____.) Now you have an exponent that is exactly divisible by 3. You can rewrite the separated roots as

$$27^{1/3} \times (10^{-3})^{1/3}$$

Since the cube root of 27 is 3, the final answer is _____ .

Try this one: $(1.6 \times 10^7)^{1/2}$. You can see immediately that the

exponent, 7, is not exactly divisible by __. So you must change the exponent. It is convenient to have your decimal number greater than 1,

so you can shift the decimal point to the _____ and get 16. But then

you must multiply by 10 to the _____ power. (Remember, you get a

bigger number when you shift the decimal point to the _____, and the

---

Answer column:

$1/5$

10^2

10^{-3}

4

$48; 12$

$(8 \times 10^{12})^{1/3}$

$8^{1/3} \times (10^{12})^{1/3}$

10^4

2×10^4

$(2.7 \times 10^{-2})^{1/3}$

$2.7^{1/3} \times (10^{-2})^{1/3}$

right

3×10^{-1}

2

right

-1

right

negative

4×10^3

$32^{1/5} \times (10^{-10})^{1/5}$

2×10^{-2}

exponent will be _____.) The rewritten number will then be

$$16^{1/2} \times (10^6)^{1/2}$$

which equals _____.

How would you rewrite $(3.2 \times 10^{-9})^{1/5}$ so that the exponent is exactly divisible? _____. If $2^5 = 32$, what is the solution? _____.

For the moment, we will not worry about how to get roots of decimal numbers. As a matter of fact, you will rarely have to worry about roots other than square and cube (1/2 and 1/3 powers), which you might be able to get from a calculator. Or, as you will see in the next section, you can easily determine any root using logarithms.

SECTION B   Logarithms

It is possible that your calculator is capable of giving you the *logarithms* of numbers, both to the base 10 (log) and to the base e (ln or *natural logarithm*). (We'll explain these terms shortly.) You simply punch in the number and then depress the log or ln key. The calculator may also give *antilogarithms*. Different models work differently. In some cases, you punch in the number and depress the *antilog* (or anlog or 10^x) key for base 10 logarithms or the antiln (or anln or e^x) key for natural logarithms. Other models have an INV or 2nd function key. You enter the number whose antilog you want. Press the INV key, followed by the log or ln key. Read the instructions for your particular model.

No matter what your calculator can do, it's a good idea for you to go through this section—at least the next four pages and the part on determining *powers and roots*. Then when your calculator spews out numbers, you will have some idea what they are all about. In fact, your calculator is simply an automatic log table with an exponent tacked on.

Logarithms are a shorter way to handle numbers written in exponential notation. Instead of writing a number raised to a power, you just write the power. When you do this, you say that you have written the logarithm. Since we almost always write the number 10 raised to a power, when you write the power to which 10 is raised, you say that you have written the *common logarithm* or the *logarithm to the base* 10, or simply the *logarithm*. There is one number other than 10 whose power is often written as a logarithm. This is a special number, like π. It is abbreviated e. When you write the power to which e is raised, you say that you have written the *natural logarithm* or the *logarithm to the base e* or the *Naperian logarithm*. We'll mention these natural logarithms at the end of this section. You may use them in your more advanced courses. But for the moment, let us consider common logarithms.

Logarithms have all the properties of exponents. Thus if you want to multiply two numbers, you add their logarithms. When you have a number raised to a power, you can simply multiply its logarithm by the power. If you want to express a very large or a very small number, writing its logarithm will save writing a lot of zeros. It is for these reasons that logarithms are very useful.

Logarithms of Numbers Containing Only One Digit Other Than Zeros

Since the common logarithm (or simply the logarithm) is the exponent to which 10 is raised, the logarithm of 10^2 is equal to 2. The abbreviation for logarithm to the base 10 (common logarithm) is log (or to be super-exact, $\log_{10}$). Thus you can write

$\log 10^2 = 2$ and $\log 10^{-1} = -1$

or $\log 10^5 = \underline{\ \ }$ or $\log 10^{-3} = \underline{\ \ \ \ \ }$ 5; -3

As you see, the values of the logarithms may be either positive or negative. However, as with powers, don't bother to write plus signs. They are understood.

   To get the logarithm of a number that is not written in exponential notation, you must convert it to exponential notation. Then you write the power to which 10 is raised. For example, to get log 1000, you first

write the 1000 as 10 to a power: $1000 = \underline{\ \ \ \ \ }$. Now to get the 10^3

logarithm, you write the _____ to which 10 is raised: power

$\log 1000 = \log 10^3 = \underline{\ \ }$.   3

   If you want to find log 0.00001, you first write the number in

_____ notation: $\log 0.00001 = \log \underline{\ \ \ \ \ }$. Now you can write   exponential; 10^{-5}

the logarithm simply by writing the _____ to which 10 is raised: power

$\log 0.00001 = \log 10^{-5} = \underline{\ \ \ \ \ }$.   -5

   You have probably noticed that the sign of the logarithm, + or -, depends on whether the number whose logarithm you are getting is larger or smaller than 1. To show this, the following table lists a series of numbers from very large to very small. Some are shown in exponential notation. Others have their logarithms given. Complete the table.

Number	Exponential Notation	Log	
1000	_____	3	10^3
100	10^2	_____	2
10	_____	1	10^1
1	10^0	0	
0.1	_____	-1	10^{-1}
0.01	10^{-2}	_____	-2
0.001	_____	-3	10^{-3}

You can see from this list that numbers greater than 1 have a positive logarithm (the + is understood) and numbers less than 1 have a

– (negative); zero

_____ logarithm. The logarithm of 1 itself is _____ , since any number to the zero power equals 1. If you have forgotten this, check back to page 4.

0

Since log 1 = __ and log 10 = 1, the logarithm of a number between 1 and 10 must be equal to something between 0 and 1. It is some decimal number. The logarithms of the whole numbers between 1 and 10 are shown in the following table. (*Note*: These values are from a four-place table. If you use a calculator, you will get more digits.)

Number	Logarithm	Number	Logarithm
1	0.000	6	0.778
2	0.301	7	0.845
3	0.477	8	0.903
4	0.602	9	0.954
5	0.699	10	1.000

There isn't a direct relationship between the logarithm and the number. Although 5 is halfway between 1 and 10, its logarithm is not halfway between 0 and 1. So you always need a table of logarithms (or a calculator) to find the logarithms of numbers between 1 and 10. In tables of logarithms, the column on the left always lists the numbers

10

between 1 and _____ whose logarithms you are looking for. The values for the logarithms that appear in columns in the table always have values

1

between 0 and __ .

So far, we have considered only numbers that can be written as 10 to some power. But what happens if you want the logarithm of a number like 50? You can't just write this as 10 to some power. What you do is

5×10^1

write the number in scientific notation. You write 50 as _____ .

10

In this way you have written it as some number between 1 and _____ times 10 to some power. And now you can take advantage of what happens when you multiply numbers written as 10 to a power. You simply add the powers. Since logarithms are the powers (exponents),

add

when you multiply two numbers, you can _____ their logarithms. In the example, 50 is 5×10^1 in scientific notation. So you can say

$$\log 50 = \log (5 \times 10^1) = \log 5 + \log 10^1$$

Then all you have to do is look up log 5 in a table (it is 0.699), and add it to log 10^1.

$$\log 50 = \log (5 \times 10^1) = \log 5 + \log 10^1 = 0.699 + 1 = 1.699$$

If your calculator gives logs, it does this addition for you. So you can skip the rest of this section, and go to page 24, where antilogs are discussed.

Here is another example. To find log 0.0003, you first write the

scientific; 3×10^{-4}

number in _____ notation. This gives _____ .

$\log 0.0003 = \log (3 \times 10^{-4})$

Since 3 is multiplied by 10^{-4}, you can _____ their logarithms. | add

$\log 0.0003 = \log (3 \times 10^{-4}) = \log 3 \_\_ \log 10^{-4}$ | +

Then you look up the value for $\log 3$ in a table (it is 0.477). You know

that $\log 10^{-4} =$ _____. Writing all the steps, you have | -4

$\log 0.0003 = \log (3 \times 10^{-4}) = \log 3 + \log 10^{-4} = 0.477 + ($ _____ $)$ | -4

You now add 0.477 to -4. This is an algebraic addition. You subtract the smaller number from the larger, and give the answer the sign of the larger. If you have forgotten this, check back to page 11.

$\log 0.0003 = 0.477 + (-4) =$ _____ | -3.523

Here is another example of the same type. If you want $\log 0.006$, you

start by writing 0.006 in _____. | scientific notation

$\log 0.006 = \log ($ _____ $)$ | 6×10^{-3}

Since you have the logarithm of two numbers that are multiplied

together, you can _____ their logarithms. | add

$\log 0.006 = \log (6 \times 10^{-3}) =$ _____ | $\log 6 + \log 10^{-3}$

You can find $\log 6$ by looking it up in a _____. You know that | table

$\log 10^{-3} =$ _____. So | -3

$\log 0.006 = \log (6 \times 10^{-3}) = \log 6 + \log 10^{-3} =$ _____ $+ (-3)$ | 0.778

Making the algebraic addition, you have

$\log 0.006 =$ _____ | -2.222

Notice that, since you had the logarithm of a number that is less than 1,

the sign was _____. | – (negative)

Here is one more for practice. Suppose that you want $\log 400,000$. Since the number is greater than 1, you know that the sign of the

logarithm is _____. First you write the number in scientific notation. | + (positive)

$\log 400,000 = \log ($ _____ $)$ | 4×10^{5}

Looking up $\log \_\_$ in the table, you find that it is _____. You | 4; 0.602

know that $\log 10^{5}$ is $\_\_$. So | 5

$\log 400,000 = \log (4 \times 10^{5}) = \log 4$ _____ | $+ \log 10^{5}$

$\qquad =$ _____ $+ \_\_ =$ _____ | 0.602; 5; 5.602

In this example, both the numbers you added had a + sign. So you

simply added them and gave your answer a $\_\_$ sign, which you don't | +

write because it is _____. Notice that since the number | understood

400,000 is greater than 1, its logarithm has a $\_\_$ sign. | +

Logarithms of Numbers Containing More Than One Digit Other Than Zeros

Suppose that you want to find the logarithm of 55. You write it in

5.5×10^1

scientific notation: _____. But the table of logarithms that we have been using contains only *whole* numbers between 1 and 10. How can you find log 5.5? You need a table that lists one digit after the decimal. Such tables do exist. Such a table would show that numbers between 5.0 and 6.0 have the following values.

Number	Logarithm	Number	Logarithm
5.0	0.6990	5.6	0.7482
5.1	0.7076	5.7	0.7559
5.2	0.7160	5.8	0.7634
5.3	0.7243	5.9	0.7709
5.4	0.7324	6.0	0.7782
5.5	0.7404		

0.7404

So, from this table, log 5.5 is _____. Now you can get the value of log 55.

5.5×10^1; 5.5

$$\log 55 = \log (\underline{\hspace{3cm}}) = \log \underline{\hspace{1.5cm}} + \log 10^1$$

0.7404; 1; 1.7404

$$= \underline{\hspace{1.5cm}} + \underline{\hspace{0.7cm}} = \underline{\hspace{1.5cm}}$$

A complete table for all numbers between 1.0 and 9.9 would be ten times longer than the one shown here. If you want to know log 5.55—a number with three digits in it—you could make a table with all the numbers from 1.00 to 9.99, which would be ten times longer still. So, to save space, these tables list the first two digits down the left side and the last digit across the top. Then instead of having one very long list, you have ten shorter lists side by side. For numbers that have 5.5 for the first two digits, such a table looks like this.

	0	1	2	3	4	5	6	7	8	9
5.5	.7404	.7412	.7419	.7427	.7435	.7443	.7451	.7459	.7466	.7474

0.7443

Now you can find log 5.55. It is _____. And if you want log 55.5, you go through the following steps.

5.55×10^1; 5.55

$$\log 55.5 = (\underline{\hspace{3cm}}) = \log \underline{\hspace{1.5cm}} + \log 10^1$$

0.7443; 1.7443

$$= \underline{\hspace{1.5cm}} + 1 = \underline{\hspace{1.5cm}}$$

Here is another example. To find log 0.000558, you start by writing it

scientific

in _____ notation:

5.58×10^{-4}

$$\log 0.000558 = \log (\underline{\hspace{3cm}})$$

Then you add the logarithms of the numbers that are multiplied

5.58; 10^{-4}

together: $\log 0.000558 = \log \underline{\hspace{1.5cm}} + \log \underline{\hspace{1.5cm}}$.

You can find log 5.58 in the table above. It is _____. You | 0.7466

know that log 10^{-4} equals _____. So you can write | –4

    log 0.000558 = log 5.58 + log 10^{-4}

           = _____ + (_____) = _____ | 0.7466; –4; –3.2534

If you didn't get the number shown, it means that you didn't subtract the smaller from the larger. If you didn't get a – sign, it means that you didn't give your answer the same sign as the larger.

The type of table that enables you to find logarithms of numbers containing three digits always has the values of the logarithms expressed to four places to the right of the decimal point. Therefore it is called a *four-place table of logarithms.* You can find a complete four-place table in Appendix I of this book. It covers two pages, even though the print is very small. The print is somewhat larger in the *Handbook of Chemistry and Physics* (Boca Raton, FL: The Chemical Rubber Company). It also covers two pages. Still more exact tables are available. There is a five-place table in the *Handbook of Chemistry* that is good for numbers like 5.555. It covers 20 pages! Since everyone now has a pocket calculator to do multiplications and divisions, you will have very little use for such an exact table. (A calculator gives you seven-place logs.)

To use a four-place table of logarithms to find the logarithm of a number between 1 and 10, go down the column on the left that contains the first two digits. When you reach the line that gives the correct first two digits, go across this line until you get to the column for your third digit. The number printed there is the logarithm you're looking for.

For example, if you want the logarithm of 5.61, go down the left-hand columns in the table in Appendix I until you reach the line with 5.6. Then go across this line until you reach the column under 1. The number printed there, .7490, is the logarithm of 5.61. (The zero before the decimal is omitted in the table to save space.)

Let's try one. Use the four-place table in Appendix I to get log 2.46.

First go down the left-hand side until you reach the line for the _____ | 2.4
values. Then go across this line until you reach the column for all third

digits that are __. The logarithm for 2.46 that you find there is _____. | 6; .3909

Here is another one. You want log 0.0080. Since the number is less

than 1, the sign on the log will be _____. You start by writing | – (negative)

the number in _____ notation. | scientific

    log 0.0080 = log (_____) | 8.0×10^{-3}

Since you have two numbers multiplied together, you can _____ their | add
logarithms.

    log 0.0080 = log (8.0×10^{-3}) = _____ + log 10^{-3} | log 8.0

You know that log 10^{-3} equals _____. Now you must find log 8.0 in | –3
the log table. You go down the left side until you reach the line for the

first two digits, _____. But what is the third digit? Since no third digit | 8.0

appears in 8.0, you must assume that it is zero. So the value for the

.9031

logarithm will be in the 0 column. It is _____. Now you can finish the problem.

.9031; −3

$$\log 0.0080 = \log 8.0 + \log 10^{-3} = \underline{\hspace{2cm}} + \underline{\hspace{1.5cm}}$$

−2.0969

$$= \underline{\hspace{2cm}}$$

To make sure that you understand this perfectly, check yourself on these examples.

0.0253	log 1.06	= _____
4.9279	log 84700	= _____
−1.5229	log 0.03	= _____
−2.6716	log 0.00213	= _____
−8.0482	log 0. 00000000895	= _____

If you had trouble with any of these, you should go through all the examples again.

Incidentally, you may sometimes find tables of logarithms that don't show the decimal point in the left-hand column. Instead of 5.6, they have just 56. But you should realize that there is a decimal point after the first digit.

Antilogarithms

If you know the value of a logarithm and you want to know what number it represents, you simply reverse the process of finding the logarithm of a number. You find the *antilogarithm*. The correct abbreviation for antilogarithm is *antilog*.

For example, suppose that you know that the value of the logarithm of a number N is 3: log N = 3. To find the value of N, you write antilog 3. = N.

3

You know that $\log 10^3 = \underline{\hspace{0.5cm}}$. So it is easy to see that antilog 3. = 10^3. This means that the number whose logarithm is 3 is 10^3.

10^{-1}

If you want antilog −1., you know that it is equal to _____. All you have done is write 10 to the power that the antilog asks for. This is all that you need do to get antilogarithms of numbers to the left of the decimal point. However, if you want an antilogarithm of a number that is less than 1, say 0.8545, use a table of logarithms. Look through the values in the table until you find the number. Then read off the value that would have given you this number. Looking through the four-place table in Appendix I, you see that log 7.15 is 0.8545. Therefore

7.15

antilog 0.8545 = _____.

This is an easy search, since the values for the logarithms become larger as you go farther down and farther to the right in the table. (If your calculator gives you antilogarithms, you can skip the rest of this section and go to page 29, Natural Logarithms.)

Here is another example. For antilog 0.6180, you look through the

table until you find this number. It is on the _____ line and in the   4.1

column that means the third digit is __. Thus antilog 0.6180 = _____.   5; 4.15

    If you want antilog 0.0354, you look through the logarithms in the

table and find that it is on the line that has _____ as the first two digits.   2.0

It is in the column that has __ as the third digit. Therefore   2

antilog 0.3054 = _____.   2.02

    You must have noticed that you have used two different ways of

getting the antilogarithms of numbers. The first was antilog 3. = _____. 10^3

The number whose antilogarithm you wanted was greater than 1 (to the

left of the decimal), and you just wrote it as the power to which 10 is

raised. The second was antilog 0.3053. You had to look it up in a table

of logarithms to find that it was 2.02. In this case, the number whose

antilogarithm you wanted was less than 1 (to the right of the decimal).

Thus, when you get antilogarithms of numbers, you write whatever is to

the left of the decimal as _____ to that power. Whatever is to the right   10

of the decimal, you _____ in a table of logarithms.   look up

    Now what do you do if the number whose antilogarithm you want has

numbers both to the left and right of the decimal? For example, if you

want antilog 6.776, you write 6.776 = 6. + 0.776. It is the sum of two

numbers, one to the left of the decimal and the other to the right. You

can write

       antilog 6.776 = antilog (6. + _____)   0.776

Just as you add logarithms of numbers that are multiplied, you multiply

numbers whose antilogarithms are added. You can write

       antilog 6.776 = antilog (6. + 0.776)

                = antilog 6. × antilog _____   0.776

Now you can get antilog 6. just by writing 10 raised to the _____   6 (sixth)

power. And you can get antilog 0.776 by _____ it up in a log table.   looking

       antilog 6.776 = antilog (6. + 0.776)

                = antilog 6. × antilog 0.776

                = _____ × 5.97 10^6

Usually when numbers are written in exponential notation, the decimal

number comes first and the 10 to a power comes last. So you would

write this as antilog 6.776 = 5.97×10^6.

    Here is another example to make sure you have the whole process.

You want antilog 3.356.

       antilog 3.356 = antilog (_____)   3. + 0.356

                = antilog 3. __ antilog 0.356   ×

10^3; look	You know antilog 3. = _____ . To get antilog 0.356, you must _____
2.27	it up in a table. When you do, you find it is _____ . So you can write
	antilog 3.356 = antilog (3. + 0.356)
2.27	$\qquad$ = antilog 3. $\times$ antilog 0.356 = $10^3 \times$ _____
2.27×10^3	Or, written in the correct order, _____ .
	Do one yourself to see whether you get the right answer.
1.04×10^8	antilog 8.017 = _____
	Now what do you do with antilogarithms of negative numbers? You
-3	can easily get the antilog (−3.). It is just 10 to the _____ power. But there are no negative numbers in the table of logarithms. You have no way of looking up antilog (−0.676). Thus you write the negative number whose antilogarithm you want as the algebraic sum of a negative number that is greater than 1 (to the left of the decimal) and a positive number that is less than 1 (to the right of the decimal). For example, you can write −2.6 as (−3. + 0.4). Take one number greater than the number to the left of the decimal as the negative number (−3 for −2 in this example). Then subtract what is to the right of the decimal to get the positive part (you subtract 0.6 in this example). Thus
−8.; 0.2	$\qquad$ −7.8 $\quad$ = _____ + _____
−7.; 0.55	or $\quad$ −6.45 $\quad$ = _____ + _____
−1.; 0.243	or $\quad$ −0.757 = _____ + _____
	Suppose that you want to get the antilog −0.757. Break it up into the
greater	sum of a negative (−) number that is _____ than 1 and a
less	positive (+) number that is _____ than 1: antilog −0.757 =
−1. + 0.243; multiply	antilog (_____). Since you _____ the anti- logarithms of sums of numbers, you can write
	$\qquad$ antilog −0.757 = antilog (−1. + 0.243)
	$\qquad\qquad$ = antilog −1. $\times$ antilog 0.243
10^{-1}	You know that antilog −1. equals _____ , but you must look up
1.75	antilog 0.243 in a log table. It is _____ . Thus antilog −0.757 =
1.75×10^{-1}	_____ .
	Here is another example. You want antilog −9.2291. You start by writing the number as the sum of a negative whole number and a positive number less than 1.
−10 + 0.7709	$\qquad$ antilog −9.2291 = antilog (_____)
multiply	You can _____ the antilogarithms of the numbers that are added.
antilog 0.7709	$\qquad$ antilog −9.2291 = antilog −10 $\times$ _____

You know that antilog – 10 equals _____ , but you must look up

10^{-10}

antilog 0.7709 in a log table. It is _____ . Therefore

5.90

    antilog – 9.2291 = _____

5.90×10^{-10}

It is important for you to be able to find the antilogarithms of negative numbers, since they always come up when you are determining the concentration of acids from their pH values. (The idea of pH is covered in Chapter 15.)

Here is one last problem that you will face when you are doing antilogarithms. More often than not, the number you look up in the log table will fall between two values in adjacent columns. For example, suppose that you are trying to get antilog 0.8473. You can find 0.8470. Then the next number is 0.8476. The number you want is between these two. Since the antilog 0.8470 is 7.03 and the antilog 0.8476 is 7.04, the antilogarithm of your number, 0.8473, must be somewhere between 7.03 and 7.04. But where? You have to guess, or as mathematicians say, *interpolate*. Although it isn't exactly right, we're going to assume that there is a direct relationship (over this short range) between the number and its logarithm. Thus 0.8473 is halfway between 0.8470 and 0.8476, so the antilogarithm of your number must be halfway between 7.03 and 7.04. Well, halfway between 7.03 and 7.04 is 7.035. Therefore you estimate that antilog 0.8476 = 7.035.

Let's look at the mechanics of this interpolation process. What you want to know is what part of the way between two adjacent logarithms your number is. In our example it was halfway, or 0.5. Then you added this part to the lower of the two numbers, which gave you the two adjacent logarithms. In this example, you added the 5 to 7.03 and got 7.035. We will do more examples to make this clearer.

First, consider the example we just did. The two adjacent logarithms were 0.8476 and 0.8470. You subtract the smaller from the larger:

0.8476 – 0.8470 = _____ .

0.0006

Then you subtract the smaller, 0.08470, from the number you want

to interpolate: 0.8473 – 0.8470 = _____ .

0.0003

Your number is thus 0.0003 out of the 0.0006 total difference. You can get a decimal number for this by dividing your difference by the total difference.

    $\dfrac{0.0003}{0.0006}$ = _____

0.5

So you put a 5 at the end of the value for the antilogarithm of the

smaller, 0.8470. Since the antilog 0.8470 is _____ , your number will

7.03

have as its antilogarithm the value _____ .

7.035

Here is another example, for more practice. You want antilog 0.4716. Looking in the log table, you find that the nearest you can get is

between _____ and _____ . The difference between these two adjacent logarithms is

0.4713; 0.4728

0.4713; 0.0015	0.4728 – _____ = _____
	You subtract the smaller of these two from your number.
0.4713; 0.0003	0.4716 – _____ = _____
	So your number is three parts away out of a total of 15 parts. You divide to get the decimal.
0.2	$$\frac{0.0003}{0.0015} = \underline{\hspace{2cm}}$$
2	Therefore you put a __ on the end of the antilogarithm of the smaller, 0.4713.
2.96	antilog 0.4713 = _____ (smaller logarithm)
2.962	antilog 0.4716 = _____ (your logarithm)
	Try one more. You want antilog 0.5760. Looking in the log table, you
0.5752	see that the nearest values you can get are _____ and
0.5763	_____ . To get the total differerence between these two adjacent
subtract	values, you _____ the smaller from the larger.
0.0011	0.5763 – 0.5752 = _____
	The difference between your value and the smaller of the adjacent values is
0.5760; 0.0008	_____ – 0.5752 = _____
	To get the parts away, you divide
0.0008	$$\underline{\hspace{3cm}} = 0.727$$
0.0011	_____
	Since this interpolation is only an approximation, you can drop everything except the first number, 7, and put this digit on the end of the
smaller	antilogarithm of the _____ of the two adjacent values.
	antilog 0.5752 = 3.76 (smaller logarithm)
3.767	antilog 0.5760 = _____ (your logarithm)
	Now try one without any help. You want antilog 0.2725. First you make the subtraction
0.2742; 0.2718; 0.0024	_____ – _____ = _____
	Then you make the subtraction
0.2725; 0.2718; 0.0007	_____ – _____ = _____
	Then you divide
0.0007	$$\underline{\hspace{3cm}} = 0.2916$$
0.0024	_____

Since 0.2916 is very close to 0.3, you put a ___ on the end of the

antilogarithm of _____ .

    antilog 0.2718 = _____   antilog 0.2725 = _____

Here are a few more to try by yourself. Work them out on a separate
sheet of paper, and compare your answers with those in the margin.

    antilog 0.1007 = _____

    antilog 0.44   = _____   (remember 0.44 = 0.4400)

    antilog 0.609   = _____   (remember 0.609 = 0.6090)

(In the last one, you might have thought that the last digit was 4. When
we get to significant figures, you will see why 5 is better.)

Now we are going to pull this all together. You will be finding the
antilogarithm of a negative number that has digits both to the left and
right of the decimal and that needs interpolation.

    antilog -6.43

First you must make the negative number the sum of a negative whole

number and a positive number that is _____ than 1.

    antilog -6.43 = antilog (_____)

When you have the antilogarithm of a sum, you can _____ the
antilogarithms of the things that are added:

    antilog -6.43 = antilog $(-7 + 0.57)$ = antilog -7 ___ antilog 0.57

You know that the antilog -7 is _____ , but you must look up 0.57 in

a _____ . Realize that 0.57 is the same as 0.5700. When you
make the interpolation, as you did in the preceding part, you find that

    antilog 0.5700 = _____

Therefore you can write antilog $-6.43 = 10^{-7} \times$ _____ .

    Or, writing this in the usual order, antilog -6.43 = _____ .

Natural Logarithms

Recall that at the beginning of this section on logarithms we said that
there was another number besides 10 whose exponent you expressed as
a logarithm. This number, which has the symbol e, has a value of
2.71828. Like π, it is not a whole number, so you could add more digits
at the end indefinitely. When you express logarithms "to the base e,"
you are expressing the exponent to which e is raised. These are called
natural logarithms, abbreviated ln (or if you are superexact, $\log_e$).

There are tables that list numbers with their natural logarithms. However, you rarely need them because it is very simple to get from a natural
logarithm to a common (to the base 10) logarithm. You simply multiply
the common logarithm by 2.303 to get the natural logarithm.

Margin answers:

3

0.2718

1.87; 1.873

1.261

2.754

4.065

less

$-7 + 0.57$

multiply

$\times$

10^{-7}

log table

3.715

3.715

3.715×10^{-7}

$\ln N = 2.303 \times \log N$

Suppose that you want to find $\ln 10^3$. You write $\ln 10^3 = 2.303 \times \log 10^3$.

3.

But you know that $\log 10^3 = \underline{\quad}$. Therefore

6.909

$$\ln 10^3 = 2.303 \times 3 = \underline{\qquad}$$

Suppose that you want $\ln 24.5$. You write

2.303; 2.45 × 10¹

$\ln 24.5 = \underline{\qquad} \times \log 24.5 = 2.303 \times \log (\underline{\qquad\qquad})$

2.45; 10¹

$\qquad = 2.303\,(\log \underline{\qquad} + \log \underline{\qquad})$

0.3892; 1

$\qquad = 2.303\,(\underline{\qquad\qquad} + \underline{\quad})$

1.3892; 3.1999

$\qquad = 2.303\,(\underline{\qquad\qquad}) = \underline{\qquad\qquad}$

It's just as easy as that.

More often you will have an equation with e raised to a power and need to know what number this is equal to. What you actually need is the antilogarithm. For example, you want to know what $e^{0.25}$ is equal to. Since the natural logarithm is the power to which e is raised, you can write $\ln N = 0.25$. You know, from what we said before, that $\ln N = 2.303 \times \log N$. So you can say

$$\ln N = 0.25 = 2.303 \times \log N \qquad \text{or just} \qquad 0.25 = 2.303 \times \log N$$

If you divide both sides of the equation by 2.303,

$$\frac{0.25}{2.303} = \frac{2.303}{2.303} \times \log N$$

You know that any number divided by itself equals 1, so you can write

$$\frac{0.25}{2.303} = \log N \qquad \text{or} \qquad 0.1086 = \log N$$

All you need do now is find the antilogarithm of 0.1086 and you will

1.284

know the value for N: antilog $0.1086 = N = \underline{\qquad}$.

Let's go through this once more, but you do the work. You want to know what number $e^{4.705}$ is equal to. You know that the power to

natural

which e is raised is equal to its $\underline{\qquad\qquad}$ logarithm. So you write

ln

$$\underline{\quad}\ N = 4.705$$

To get from ln to log, you write

2.303

$$\ln N = \underline{\qquad} \times \log N$$

4.705

So you can now write $\underline{\qquad} = 2.303 \times \log N$.

2.303

Dividing both sides by $\underline{\qquad}$, you get

2.0430;
2.303

$$\frac{4.705}{\underline{\quad}} = \log N \qquad \text{or} \qquad \underline{\qquad} = \log N$$

antilog

All you must do now is find the $\underline{\qquad\qquad}$ of 2.0430.

N = antilog 2.0430 = antilog (_____)

$\quad$ = antilog 2. __ antilog 0.0430 = _____ $\times 10^2$

$2 + 0.0430$

$\times$; 1.104

Uses of Common Logarithms

Before inexpensive pocket calculators were available, chemists used two tools to do their calculations: the slide rule and the table of logarithms. The slide rule is very fast for doing multiplications, divisions, squares, cubes, square roots, and cube roots. But the answer that you can read from a slide rule usually has only three digits. The chemist who needed greater precision could use a four- or even five-place table of logarithms. The logarithm changes multiplication into simple addition and division into subtraction. You can get powers and roots by multiplying or dividing the logarithms.

$\quad$ Nowadays even the simplest calculator does multiplications and divisions to an eight-digit answer, so hardly anyone bothers to use logarithms to multiply or divide. As for powers and roots, most scientific calculators can get the squares and square roots of numbers by a single key stroke, and they can get any power or any root of a number using two or three key strokes. Today, logarithms are used only in certain electrochemical and thermodynamic expressions covered in more advanced courses and in pH (the way chemists express acidity), which you will find covered in Chapter 15.

Determining Powers and Roots

(Your instructor may suggest that you skip this section.)

$\quad$ In Section A of this chapter, pages 15 to 18, you learned that when you raise 10 to a power itself to a power (that is, multiplied it by itself repetitively), you can simply multiply the powers. Thus $(10^5)^3 = 10^{15}$.

Or $(10^{-2})^3 = $ _____. When you want to get a root of 10 to a power, you simply divide the power by whatever the root is. Thus $\sqrt{10^6} = 10^3$

and $\sqrt[3]{10^9} = $ _____. In fact, you get these roots by multiplying the power to which the 10 is raised by $\frac{1}{root}$, which is the same as dividing by the root.

$\quad$ Now, the common logarithm of a number is the _____ to which 10 is raised. Therefore, if you want a number to a power, you just multiply the logarithm of the number by the power. If you want the root of a number, you simply divide the _____ of the number by the root. Then you have the logarithm of your answer. To get the answer itself you take the _____.

$\quad$ Here is an example. To find the value of $(2.4)^{14}$, you start by getting the logarithm of the number: log 2.5 = _____.

$\quad$ Since you want the number to the 14 power, you _____ the logarithm by 14.

10^{-6}

10^3

power

logarithm

antilogarithm

0.3979

multiply

14; 0.3979; 5.5706

$\log (2.5)^{14} =$ _____ $\times \log 2.5 = 14 \times$ _____ $=$ _____

(Your calculator may do all this for you.)

You know that the logarithm of your number is 5.5706. To get the

antilogarithm

number itself, you take the _____ of 5.5706.

antilog

$(2.5)^{14} =$ _____ 5.5706

If you get the antilogarithm, as you did earlier in this section, you find that

3.721 × 10⁵

antilog 5.5706 = _____ So $(2.5)^{14} = 3.721 \times 10^5$

This isn't exactly correct because of the method you used in interpolating. The correct answer is 3.7204×10^5. (Your calculator would give 3.720488×10^5.) However, when you get to Section C, on significant figures, you will see that 3.7×10^5 is as near to the answer as you are allowed to write.

Getting the roots of numbers using logarithms is the same as taking numbers to a power. However, instead of multiplying the logarithm by the power, you divide the logarithm by the root. Here is an example. You want to know the value of $\sqrt[3]{2.5}$. You start by getting the logarithm of the number.

0.3979

$\log 2.5 =$ _____

3

Then, to get the cube root, you divide the logarithm by __ .

0.1326
3;

$\log \sqrt[3]{2.5} = \dfrac{0.3979}{\underline{}} =$ _____

You now know what the logarithm of $\sqrt[3]{2.5}$ is. To get the value, you

antilogarithm

take the _____ .

1.357

antilog 0.1326 = _____

Once again there is a slight error due to the method of interpolation. (A calculator would give 1.357063.) It is trivial, since the rules of significant figures say that the best you can write is just 1.4. The use of significant figures is going to be a revelation!

When you take roots of numbers expressed in scientific notation, you have to be careful with your signs. Here is an example. If you want

logarithm

$\sqrt[4]{2.5 \times 10^{-3}}$, you first get the _____ .

2.5; 10⁻³

$\log 2.5 \times 10^{-3} = \log$ _____ $+ \log$ _____

0.3979; (−3); −2.6021

$=$ _____ $+$ _____ $=$ _____

divide; 4

For the fourth root, you must _____ the logarithm by __ .

−0.6505
4;

$\log \sqrt[4]{2.5 \times 10^{-3}} = \dfrac{-2.6021}{\underline{}} =$ _____

That was the place where you might have lost the minus sign.

To get the value itself, you must find the _____ of
-0.6505.

antilog -0.6505 = antilog (_____)

= antilog -1 ___ antilog 0.3495 = _____

Once again the rules of significant figures tell you that you may write
this only as 2.2×10^{-1}.

antilogarithm

-1 + 0.3495

X; 2.236×10^{-1}

Determining pH

You will find the most use for logarithms when you determine the pH of
acid and base solutions. This is because the p function means $-\log$.
There are other p functions, like pOH, pK_a, and pK_w. They mean
$-\log [H^+]$, $-\log [OH^-]$, $-\log K_a$, and $-\log K_w$. The pH is such an important topic that we cover it separately in Chapter 15, so we will not
discuss it here.

Uses of Natural Logarithms

Earlier in this section, we mentioned natural logarithms: finding the
value of e raised to a power, and finding the natural logarithm of a
number. These uses come about in chemistry because there are several
important relationships that have the general form

$$x = Ae^{-y/RT}$$

Depending on what you are interested in, x might be related to the speed
of a chemical reaction, or to how easily a liquid evaporates, or even to
the useful work that you can get out of a chemical reaction. And y might
be related to the heat that a reaction needs, or the heat it takes to
evaporate a liquid, or the relative concentrations of reactants and
products in a chemical reaction. The T is always the temperature, and R
is called the *gas constant* (see Chapter 12). The A is a constant that you
are usually given. Because these relationships come up in advanced
chemistry courses, we will not cover them here. You will learn about
natural logarithms when you reach that stage.

SECTION C   Significant Figures

As great as calculators are, they can lead unsuspecting students to write
some very silly answers to their calculations. The reason is that calculators can produce answers with at least eight digits. Some even give as
many as twelve digits in the answer. Often, most of these digits are
meaningless. When you write them down, you are just writing a lot of
numbers that you are not at all sure of.

Let's see how this comes about. Suppose that you want to measure the area of a piece of paper with a 12-inch ruler. Each inch is divided into sixteenths. The piece of paper is rectangular, so its area is equal to the height times the width. You measure the height.

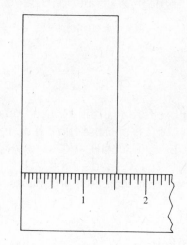

It looks as if it's about halfway between $1\frac{8}{16}$ and $1\frac{9}{16}$ inches. You guess that it is $1\frac{8.5}{16}$. A sixteenth of an inch, in decimal notation, is 0.0625 inch. Thus 8.5 sixteenths is 8.5 × 0.0625, or 0.53125 inch. So the height is 1.53125 inches.

Next you measure the width of the piece of paper.

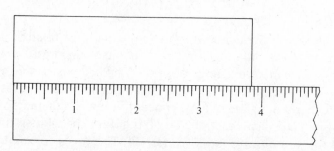

It looks as if it's a little more than $3\frac{13}{16}$ inches. You guess that it is two-tenths of the way between $\frac{13}{16}$ and $\frac{14}{16}$. So you call it $3\frac{13.2}{16}$ inches. Since one-sixteenth of an inch is 0.0625 inch, 13.2 sixteenths is 0.825 inch. So you guess that the width is 3.825 inches.

On your calculator, you get the area by multiplying 1.53125 by 3.825. With an eight-digit calculator, you get 5.8570313 square inches. But writing all these numbers is nonsense. As we will see, the last five are meaningless.

If you guessed a little wrong and said that the height was $1\frac{8.6}{16}$ instead of $1\frac{8.5}{16}$, you have a height of 1.5375 inches as a decimal number. Suppose that you also guessed wrong on the width. (It was really $3\frac{13.3}{16}$ and not $3\frac{13.2}{16}$.) In decimal notation this is 3.83125 inches. The area using these new values would be 5.8905469 square inches. Only the first two digits are the same as the area you got before, 5.8570312.

Now, let's go through the whole process again, but this time with the true height and width of $1\frac{8.4}{16}$ and $3\frac{13.1}{16}$ inches, respectively. Expressing these in decimal notation, you have 1.525 inches and 3.81875 inches. These dimensions give you an area of 5.8235938 square inches. Once again, only the first two digits in the area are the same as those you got before.

Here are all the numbers in a table, so that you can see what's going on.

Height, inches	Width, inches	Area, square inches
$1\frac{8.4}{16}$	$3\frac{13.1}{16}$	5.8235938
$1\frac{8.5}{16}$	$3\frac{13.2}{16}$	5.8570313
$1\frac{8.6}{16}$	$3\frac{13.3}{16}$	5.8905469

Just by making a very small error in guessing how far between the sixteenth markings on the ruler the edge of the paper was, you have areas that vary from 5.82. . . up to 5.89 . . . square inches. What is the sense in writing that the area is 5.8570312 square inches when you aren't really certain whether it is 5.82 . . . or 5.89 . . . square inches? Even the third digit is uncertain.

To take care of this problem, a convention has been adopted that says that you may write as many digits as you are sure of, and then one final digit that you are uncertain of. These digits are called *significant figures*. In this example, you can write the first two digits, 5.8, which are certain, and then one more digit, 6, that is not certain. The area of the paper is thus 5.86 square inches *expressed to three significant figures.*

Let's consider when and how you use this significant-figure convention. First you must realize that there are two different types of numbers. The first comes either from a definition or by counting. Examples are the 12 inches in a foot, the 60 minutes in an hour, or the 32 students (if you counted them) in a classroom. There is no uncertainty in this type of number. If you use it in a calculation, you do not consider it when you determine the allowed significant figures in your answer. You might think of it as containing an infinite number of significant figures.

The second type of number is one that you obtain by measuring something with a tool or an instrument—like a ruler, a stop watch, or a balance. With this type of number, the certainty depends on how good an instrument you used, how careful you were, and how well you can read the instrument. There is always a limit on the certainty of the number you have read. Therefore you must always obey the rules for significant figures.

The number of significant figures that a measured quantity has depends on two things: the instrument and the size of the measured object.

Suppose that you stand on a bathroom scale, and the arrow points to a number between 146 and 147.

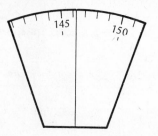

You are sure that you weigh 146 pounds plus some part of a pound, and you can guess that you are 0.3 pound over 146 pounds. You therefore express your weight as 146.3 pounds. The first three digits (1, 4, and 6)

sure; guessed

you are _____ of, and the last digit (3) you have _____. All these digits are significant. You can say that you have expressed your

four

weight to _____ significant figures.

Suppose that you weigh a grapefruit on the same scale, and see this:

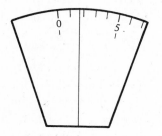

1

You are sure that the grapefruit weighs __ pound (lb). But it weighs

0.8 (my guess)

about _____ lb more than 1 lb. (If we didn't guess exactly the same on the last figure, this demonstrates the uncertainty.) To the correct number of significant figures, you can say that the grapefruit weighs

1.8; two

_____ lb. This number contains _____ significant figures, 1 (of which

sure; guess

you are _____) and 8 (which is a _____).

Try weighing a plum on the same bathroom scale. You see this:

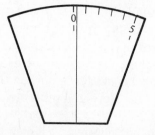

0.2

Its weight to the correct number of significant figures is _____. You must guess that it weighs 0.2 lb more than nothing. This number

one

contains _____ significant figure.

From these three weighings, you can see that the heavier the sample

you are weighing, the _____ the number of significant figures
you can get from the same measuring device (for example, the bathroom
scale).

greater (larger)

Now let us weigh the grapefruit on a kitchen scale. For convenience,
the pounds have been divided into tenths rather than ounces.

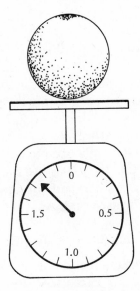

You are certain that the grapefruit weighs 1.7 lb and then a little more.
You can guess it is 0.05 lb more than 1.7 lb. Therefore, to the correct

number of significant figures, its weight is _____ lb. This number

1.75

contains _____ significant figures, while the previous weighing on the
bathroom scale gave only two significant figures.

three

Now weigh the plum on the same kitchen scale:

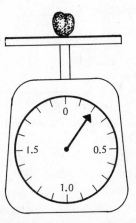

It weighs, as best you can read, exactly __ lb. You are guessing that it
does not weigh even a little more than 0.2 lb. So, to show that you have

.2

guessed a last place of zero, you write the weight as _____ lb. Now you

.20

two

have _____ significant figures in the weight, a 2 of which you are certain and a 0 that you have guessed. The zero is a significant figure.

Remember that when you weighed the plum on a bathroom scale, you had only one significant figure. Using a more sensitive measuring device, you can get more significant figures. You now know two ways of

larger sample

increasing the number of significant figures, (1) _____

more sensitive tool

and (2) _____.

When you wrote the weight of the plum as .20 lb, you could have put a zero in front of the decimal point, 0.20, to make sure that the decimal point isn't overlooked. This zero in front is not a significant figure. It is simply a place marker. As a matter of fact, all zeros written in front of a number are place markers and not significant figures. You could have written the weight in scientific notation as 2.0×10^{-1} lb. Here it is clear how many significant figures there are.

If you have a zero in the middle of a number, it is significant. For example, the number 0.000034010 has five significant figures. The first five zeros are simply place markers and not significant. But the zero between the 4 and the 1 *is* significant. And the zero at the end is significant.

four

Thus, in the number 0.03570, there are _____ significant figures.

zeros; zero

The _____ in front of the number don't count. The _____ at the end of the number is significant. You can see this clearly if you write the

3.570×10^{-2}

number in scientific notation, which gives the number _____.

four

The number 0.5670 has _____ significant figures because the zero

in front

_____ doesn't count. It is only a place marker. The zero

at the end

_____ does count because it is part of the measurement.

five

The number 0.00035001 has _____ significant figures. It can be

3.5001×10^{-4}

written in exponential notation as _____.

five

The number 5680.0 has _____ significant figures. In scientific

5.6800×10^3

notation, with all the significant figures, it is _____.

significant

All the zeros at the end are _____.

Unfortunately, there can be some doubt about whether final zeros are significant. If you haven't made the measurement yourself and are told that a beaker contains 500 grams (g) of water, it might contain between 400 and 600 g (one significant figure), between 490 and 510 g (two significant figures), or between 499 and 501 g (three significant figures). However, if you are especially careful and deal only with numbers expressed in scientific notation, this problem will not come up. The three possibilities would be written $5. \times 10^2$ g (one significant figure), 5.0×10^2 g (two significant figures), and 5.00×10^2 g (three significant figures). All zeros after the decimal point are significant.

Multiplication and Division

When numbers expressed to the correct number of significant figures are multiplied or divided, the answer can have no more figures than the number with the fewest significant figures.

   If you multiply 1. × 1.345, the 1. has _____ significant figure and

| | one |

the 1.345 has _____ significant figures. The product, if you express it to the correct number of significant figures, can have only one significant figure.

| | four |

   If you divide 35.3/6.8, the 35.3 has _____ significant figures. The

| | three |

6.8 has only _____ significant figures. The answer can then be expressed

| | two |

only to _____ significant figures. This is the fewest number of significant figures.

| | two |

   The result of (1.37 × 0.3)/(0.0356 × 1.5) can be expressed to only

_____ significant figure. The _____ has only one significant figure, and so .3 is the determining number.

| | one; 0.3 |

   We can check this rule for determining the number of significant figures allowed in multiplication and division by using the example we had at the beginning of this section. Recall that we calculated the area of a piece of paper from the measurements of its height and width. The height was measured to be $1\frac{8.5}{16}$ inches and the width was $3\frac{13.2}{16}$ inches. When we calculated the area, we showed that even as little a misreading as $\frac{0.1}{16}$ gave answers that varied from 5.82 to 5.89 square inches. The first two digits, 5.8, are certain, but the third is not. At that point, we said

that you could write the answer only as 5.86, which has _____ significant figures.

| | three |

   To check our rule for determining the number of significant figures, we must express the total number of sixteenths in $1\frac{8.5}{16}$ inches $\left(\frac{24.5}{16}\right)$

and in $3\frac{13.2}{16}$ inches $\left(\frac{61.2}{16}\right)$. Both 24.5 and 61.2 have _____ significant

| | three |

figures. Therefore the answer can have only _____ significant figures. And it does. As we said, any more figures are meaningless.

| | three |

Addition and Subtraction

When you add numbers expressed to the correct number of significant figures, you use a different set of rules for determining the correct number of significant figures allowed in the answer. In addition and subtraction, the location of the digits with respect to the decimal point controls the number of significant figures allowed. The answer may have more or fewer significant figures than the number that had the fewest significant figures. For example, if you add 0.5 + 9.74, the answer, expressed to the correct number of significant figures, is 10.2. The 10.2

has _____ significant figures, but the number 0.5 has only _____.

| | three; one |

   If you subtract 2.52 − 2.3, the answer, expressed to the correct

one

two

guessed

0.7

9.76

2.53

2.4

number of significant figures, is 0.2. The answer has only _____ significant figure, but the 2.3 has _____.

Let's see how this comes about. In the first example, 0.5 + 9.74, we'll assume that the last digit (it was _____) could have been wrong by 2. Thus the 0.5 could be as low as 0.3 or as high as _____. The 9.74 could be as low as 9.72 or as high as _____. We add the two low values and separately add the two high values.

$$\begin{array}{r} 0.3 \\ \underline{9.72} \\ 10.02 \end{array} \qquad \begin{array}{r} 0.7 \\ \underline{9.76} \\ 10.46 \end{array}$$

Only the first two digits, 1 and 0, are the same. You are uncertain of the third digit. Therefore the third digit is the first uncertain digit. It is the last that you can show in your significant figure. The answer is therefore 10.2 and the digit 2 is uncertain.

In the second example, the subtraction 2.52 − 2.3, let's assume that the last digit in each is wrong by only 1. Thus the 2.52 might be as low as 2.51 or as high as _____. The 2.3 might be as low as 2.2 or as high as _____. If we subtract the smaller possibility from the larger and the larger from the smaller, we have

$$\begin{array}{r} 2.53 \\ \underline{-2.2} \\ 0.33 \end{array} \qquad \begin{array}{r} 2.51 \\ \underline{-2.4} \\ 0.11 \end{array}$$

You can see that even the first digit is not certain. Thus all that you can write for your answer is 0.2.

To handle additions and subtractions, you write the numbers to be added or subtracted so that all the decimal points are lined up one above another. For the two examples, this would be

$$\begin{array}{r} 0.5 \\ \underline{9.74} \end{array} \qquad \begin{array}{r} 2.52 \\ \underline{-0.3} \end{array}$$

The last place that can appear in the answer is the column farthest to the right that has a digit in it for every number. You could draw a line down to show this.

$$\begin{array}{r} 0.5 \\ 9.7\,|\,4 \\ \underline{} \\ 10.2\,|\,4 \end{array} \qquad \begin{array}{r} 2.5\,|\,2 \\ -2.3 \\ \underline{} \\ 0.2\,|\,2 \end{array}$$

Anything to the right of the line cannot appear in your answer. The answers are 10.2 and 0.2.

Logarithms

Getting logarithms to the correct number of significant figures poses quite a problem. This is because the exponential nature of logarithms varies with the value of the number whose logarithms you are considering. For example, suppose that you have the number 1.03 but are uncertain about the last figure. It might be 1.02 or 1.04. The logarithms of the two uncertain possibilities are 0.0086 and 0.0170. These don't have any numbers in common. If, however, you had the number 9.03 and it might be as low as 9.02 or as high as 9.04, the two logarithms would be 0.9552 and 0.9562. In this case the first two numbers are the same.

The best we can do for logarithms is never to put more digits after the decimal in the logarithm than we had in our original number. Thus

$$\log 3.46 = 0.539 \quad not \ 0.5391 \quad and \quad \log 346 = 2.539.$$

(The 2 to the left of the decimal just shows that the number was 3.46×10^2.) If you wanted

$$\log 0.00346 = \log (3.46 \times \underline{\hspace{1cm}})$$

10^{-3}

you would have

$$\log (3.46 \times 10^{-3}) = 0.5391 \underline{\hspace{1cm}} = -2.4609$$

-3

(A calculator would give you -2.4609239.) But you are allowed only

_____ digits after the decimal. So you must write this as -2.461.

three

When you have antilogarithms, you may use as many significant figures as the number whose antilogarithm you get has after the decimal. Thus

$$\text{antilog } 0.598 = 3.96$$

not 3.963, which you would get by interpolation (or 3.96278). And

$$\text{antilog } 3.598 = 3.96 \times 10^3$$

(The 3 to the left of the decimal just tells you the power of 10.)

Rounding Off

Now that you can decide how many significant figures may appear in an answer, you have to know how to cut your answer down to the right number of digits. This process is called *rounding off*. It is very simple. There are only three rules, and they depend on the digits you want to remove.

(1) Consider only the digit immediately following the last digit you want to keep in your answer. If it is less than 5 (that is, it is 0, 1, 2, 3, or 4), discard all the digits you do not want. For example, round off 0.3752

to three significant figures. The third significant figure is __. (Recall that zeros preceding a number are not significant.) The digit following the 5

5

is __, which is less than 5. So you simply discard everything which

2

follows the 5, giving the answer _____.

0.375

Try another example. Round off 3.65748×10^3 to four significant

figures. The answer is _____. If you got the wrong answer, it is

3.657×10^3

possible that you didn't have the 7 as your fourth significant figure, which it is. Since the figure that follows the 7 is less than 5, you drop all the digits that follow the 7.

(2) If the digit following the last number you want to keep is greater than 5 (that is, if it is 6, 7, 8, or 9) or if it is 5 followed by any digit other than just zeros, you add 1 to the number you want to keep and discard all the digits you do not want.

For example, round off 45.253 to three significant figures. The third

2; 3   significant figure is __. The digits that follow it are 5 and __, which are

5   greater than just __, so you add 1 to the 2, giving 45.3

Here is another example. Round off 3.51866 to the fourth place after

3.3187   the decimal point. The answer is _____. Since the number

5   following the last digit you wanted to keep is greater than __, you add

1   __ to your last significant figure.

Try one more. If you round off 4.3256 to three significant figures, the

2; greater   third significant figure is the __. The digits that follow it are _____

add   than 5. In fact, they are 56. Thus you _____ 1 to the 2, giving the

4.33   answer _____.

(3) Only if the digit following the last figure you want to keep is exactly 5 (that is, 5 followed by nothing but zeros), do you add 1 to the last significant figure if it is an odd number and drop all the extra digits. If the last significant figure is an even number, discard all the unwanted digits.

For example, round off 3.635 to three significant figures. The last

3; add   number you want to keep is __. This is an odd number, so you _____

3.64   1 to it. The answer is _____.

If you round off 0.50450 to three significant figures, the answer is

0.504   _____. The reason you simply discard the numbers following the 4 is

even   that 4 is an _____ number.

If you round off 4.56050 to four significant figures, the last digit you

0   want to keep is __. It is considered an even number. Therefore the

4.560   answer is _____.

You have now reviewed all the mathematics you will need for the types of problems covered in this text. Next you will find a set of questions that covers all this material. Do this problem set and see if you can complete it in the suggested time.

PROBLEM SET

Time yourself while you do these problems. Do each section separately in the time noted. The answers will be found on the following pages. Do not check your answers until you have finished a complete section. A method of scoring yourself appears with the answers. You need not worry about significant figures in Sections A and B.

Section A (10 minutes)

1. Express 5650000. in scientific notation.

2. Express 0.0000565 in scientific notation.

3. Express 0.5×10^{-3} in scientific notation.

4. The number 55×10^{-8} equals _____ $\times 10^{-7}$.

5. Express 5.3×10^2 as an ordinary number.

6. Express 6.73×10^{-2} as an ordinary number.

7. Perform the operation $4 \times 10^2 \times 2 \times 10^{-4}$.

8. Perform the operation $3.5 \times 10^{-1} \times 2 \times 10^{-7}$.

9. Perform the operation $\dfrac{2 \times 10^{-2} \times 3 \times 10^1}{6 \times 10^{-5}}$

10. Perform the operation $(2 \times 10^{-4})^3$.

11. Perform the operation $\sqrt{2.5 \times 10^{-1}}$.

12. Perform the operation $0.5 \times 10^1 + 6 \times 10^{-1}$ and give the answer in scientific notation.

Section B (25 minutes using log tables or 15 minutes with a calculator)

Use a calculator or the four-place table of logarithms in Appendix I to determine the following.

13. $\log 10^7 =$

14. $\log 0.0001 =$

15. $\log 5.35 =$

16. $\log 2500 =$

17. $\log 0.00087 =$

18. antilog $3 =$

19. antilog $7.9571 =$

20. antilog $2.608 =$

21. antilog $-3.8456 =$

22. $e^{3.7} =$

23. $\ln 7.65 =$

24. $(2.6 \times 10^3)^9 =$

25. $\sqrt[3]{0.000042} =$

Section C (12 minutes)

26. How many significant figures are in the number 450.0?

27. How many significant figures are in the number 0.0032?

28. How many significant figures are in the number 0.12040?

29. Express the number 0.000352 to two significant figures.

30. Express the number 0.357 to two significant figures.

31. Express the number 0.305 to two significant figures.

32. Express the number 0.145 to two significant figures.

33. Perform the operation $\dfrac{3. \times 4.0 \times 3.254}{1.002 \times 2.05}$

    Express your answer to the correct number of significant figures.

34. Perform the operation $\dfrac{200.0 \times 20.0 \times 10.}{100.0}$

    Express your answer to the correct number of significant figures.

35. Perform the operation $41.56 + 72. + 7.3$. Express your answer to the correct number of significant figures.

36. What is the value of log 8.5 to the correct number of significant figures?

37. What is the antilog of 0.32 to the correct number of significant figures?

38. What is the value of antilog -8.198 to the correct number of significant figures?

PROBLEM SET ANSWERS

Section A

You should get at least ten correct answers out of twelve. If you do not, check your incorrect answers to see whether the error was just carelessness. If you really don't understand why your answer was incorrect, refer back to that part of the text which discusses the missed problems.

1. 5.650000×10^6
2. 5.65×10^{-5}
3. $5. \times 10^{-4}$
4. 5.5×10^{-7}
5. $530.$
6. 0.0673
7. $8. \times 10^{-2}$
8. $7. \times 10^{-8}$
9. $1. \times 10^4$
10. $8. \times 10^{-12}$
11. 5.0×10^{-1}
12. 5.6 (or just 6 to the correct number of significant figures)

Section B

You should get at least nine out of the first ten correct. If you didn't, review the first three parts of Section B. If your errors were in the last four of these, 19 through 22, you need review only the part on antilogarithms. You should be able to do both 23 and 24. If you miss either, redo the part of this section on natural logarithms. Questions 25 and 26 are covered in the section on the uses of common logarithms. If you get one of the two correct, that's good enough.

13. 7
14. −4
15. 0.7284
16. 3.3979
17. −3.0605
18. 10^3
19. 9.06×10^7
20. 4.055×10^2
21. 1.427×10^{-4}
22. 4.042×10^1 (4.0×10^1 to the correct number of significant figures)
23. 2.035
24. 5.430×10^{30} (5.4×10^{30} to the correct number of significant figures)
25. 3.476×10^{-2} (3.5×10^{-2} to the correct number of significant figures)

Section C

You should get at least nine correct answers out of the ten problems. If not, refer back to the text and redo the entire section.

26. four
27. two
28. five
29. 0.00035
30. 0.36
31. 0.30
32. 0.14
33. $2. \times 10^1$
34. 4.0×10^2 } (Notice how you can write these in scientific notation to the correct number of significant figures.)
35. 1.21×10^2
36. 0.93
37. 2.1
38. 6.34×10^{-9}

CHAPTER 2

What Is Dimensional Analysis?

There will be no pretest for this chapter because you have had no specific preparation for the material in it. Nevertheless, it is the most important chapter in the whole book, since it shows you the method you will use to solve virtually all of the chemical problems in this book.

Say someone asked you, "Do you have 5?" You would probably say, "Five what?" It might be 5 cents, or 5 fingers on a hand, or even 5 pounds of salt in 3 boxes. All numbers (with few exceptions) are quantities of something. But saying only the number (quantity) isn't enough. You have to say the quantity of whatever you are talking about. The "whatever" is called the *dimensions* (or sometimes *units*) of the number.

Let us return to the question "Do you have 5?" If the question asked were about money, it might be "Do you have 5 cents?" Cents is the dimension of the number 5. If it were about weight, it might be "Do you

have 5 pounds?" Pounds is the _____ of the number 5. If the   dimension
question were about length in yards, it would be "Do you have 5

_____ ?" Here yards is the dimension of the _____ 5.   yards; number

Thus, for a number to mean anything, it must have a _____   dimension
written after it, so that you know what you are talking about.

SECTION A   Conversion Factors

The other possibilities mentioned for what the 5 might be were 5 fingers per hand and 5 pounds of salt in 3 boxes. These bring us to what are called *conversion factors.* Let us start with a very simple example of a conversion factor.

We all know that there are 60 seconds in one minute. This really means that 60 seconds equal 1 minute. Mathematically, this is

60 sec = 1 min

If you divide both sides of an equality by the same thing, you still have an equality. Let's divide both sides by 1 min.

$$\frac{60 \text{ sec}}{1 \text{ min}} = \frac{1 \text{ min}}{1 \text{ min}}$$

But

$$\frac{1 \text{ min}}{1 \text{ min}} = 1 \qquad (\text{We say that they } cancel.)$$

Therefore

$$\frac{60 \text{ sec}}{1 \text{ min}} = 1$$

We call 60 sec/1 min a conversion factor. It equals 1. *All conversion factors equal* 1. The 1 is the same as the line in a fraction.

You are probably more used to saying, "There are 60 seconds per minute." *Per* means "divided by." The number 1 is usually not written, since it is understood.

What, then, is the conversion factor relating hours to days? Since there

24

24 hr

are _____ hours per day, you can write the conversion factor

$$\frac{\text{\_\_\_\_\_}}{\text{day}}$$

1

1

$\frac{100¢}{\$1}$

1

This conversion factor must equal __, since all conversion factors equal __.

The conversion factor relating cents (¢) to dollars ($) is _____ .

Since this is a conversion factor, it must equal __.

Let us go back to the first conversion factor, which came from 60 seconds equals 1 minute (60 sec = 1 min). What happens if you divide both sides of the equality by 60 sec?

$$\frac{60 \text{ sec}}{60 \text{ sec}} = \frac{1 \text{ min}}{60 \text{ sec}}$$

The 60 sec/60 sec cancels and is equal to 1. Therefore

$$1 = \frac{1 \text{ min}}{60 \text{ sec}}$$

Since 1 min/60 sec also equals 1, it must also be a conversion factor. The two conversion factors, then, are,

$$\frac{60 \text{ sec}}{\text{min}} \quad \text{and} \quad \frac{\text{min}}{60 \text{ sec}}$$

upside-down

They are the same, except that one is _____ compared to the other. The mathematical word for this is *inverted.* Thus any conversion factor can be inverted and it will still be a conversion factor. Sometimes we say that one is the *reciprocal* of the other.

$\frac{1 \text{ day}}{24 \text{ hr}}$

1

$\frac{12 \text{ in.}}{1 \text{ ft}}$

$\frac{1 \text{ ft}}{12 \text{ in.}}$; 1; equals

Similarly, the inverted form of 24 hr/day is _____ . It is still a conversion factor, and still equals __.

The two conversion factors relating feet to inches would be _____ and _____ . Both equal __, because 12 in. _____ 1 ft.

Many conversion factors have fixed definitions, such as, "There are 16 ounces per pound," which can be written as _____. Others are presented in a statement such as, "A worker makes $7.50 per hour," which you can write as _____, or as its reciprocal _____. You will quickly learn how to spot these factors in a problem.

$$\frac{16\ oz}{1\ lb}$$

$$\frac{\$7.50}{1\ hr}\ ,\ \frac{1\ hr}{\$7.50}$$

You must be careful when you write the dimensions of a number. The dimension must be explicit enough so that you cannot confuse it with the dimension of any other number. For example, if a problem says, "Onions cost 20¢ per pound," you can write the conversion factor _____. However, if in the same problem you have "Potatoes cost 39¢ per pound," then you must distinguish which vegetable you are talking about. Thus you write the complete conversion factor 20¢/1 lb onions for the onions; and then you write _____ for the potatoes. The dimensions are then pounds of onions and pounds of potatoes.

$$\frac{20¢}{1\ lb}$$

$$\frac{39¢}{1\ lb\ potatoes}$$

It has become customary to leave out the number 1 in conversion factors. We usually write 60 sec/min rather than 60 sec/1 min. The reason we can do this is that multiplying or dividing by 1 does not change the numerical value of what it is operating on. From this point on, therefore, all 1's in conversion factors will be omitted.

Conversion factors do not have to be for a single hour, a single day, a single foot, or a single pound. For example, if a problem states that "An earth-mover can dig a trench 500 feet long in 3 hours," the conversion factors are

$$\frac{500\ ft}{3\ hr} \quad or \quad \frac{3\ hr}{500\ ft}$$

SECTION B Solving Problems by Dimensional Analysis

Now we can get to the business at hand: how to solve a problem. It is really very simple. Practically all problems ask a question whose answer is a number *and its dimension.* Problems also give some information that is also a number *and its dimension.* You are going to consider the numbers and the dimensions separately. You multiply the information given by conversion factors so that you cancel all dimensions except the dimension you want in your answer. You can multiply by as many conversion factors as you wish, since all conversion factors equal __. And multiplying something by __ doesn't change its value.

1

1

Here is an example. How many seconds in 35 minutes? The question asked is "How many seconds?" Therefore, when you get an answer, it will have the dimension _____. The information given in the problem is 35 minutes.

seconds

You start by writing down the dimensions of the answer so that you can check yourself at the end:

? sec =

Then you supply the information given:

? sec = 35 min

You now want to multiply the 35 min by some conversion factor that has the dimension minutes in the denominator (on the bottom) to cancel the unwanted minutes in the information given, and that has the dimensions of the answer (seconds) in the numerator (on top). Try the conversion factor 60 sec/min.

$$? \text{ sec} = 35 \cancel{\text{min}} \times \frac{60 \text{ sec}}{\cancel{\text{min}}}$$

The unwanted dimension, minutes, cancels. The only remaining dimension is seconds.

? sec = 35 × 60 sec = 2100 sec

You know that this must be the correct answer because the only dimension left is the same as the dimension of the question.

Try another problem. How many hours are in 6 days? The question is

How many hours?

"_____" The dimensions of the answer will

| hours |

therefore be _____. So you write

| hours |

? _____ =

| 6 days |

The information given is _____, so you write

| 6 days |

? hr = _____

Now you want to multiply the information given by some conversion

| days |

factor that has the dimension _____ in the denominator (bottom) to

| days |

cancel the unwanted dimension _____ in the information given. This should lead toward the dimensions of the question asked. Such a conver-

| $\frac{24 \text{ hr}}{\text{day}}$ |

sion factor would be _____. If you do this, you have ? hr = 6 days ×

| $\frac{24 \text{ hr}}{\text{day}}$; days |

_____. The unwanted dimension, _____, cancels, leaving only the

| hours |

dimension _____, which is the dimensions of the question. This must be the answer. Its numerical value, once you have multiplied the numbers

| 144 hr |

together, is _____.

Multiplying by only one conversion factor usually will not get you all the way to the dimensions of the answer. However, since all conversion factors equal 1, you can multiply by as many as you need. Consider the problem, "How many seconds in 3 years?" The question asked is

| How many seconds?; 3 yr |

"_____" and the information given is _____. So you write

? sec = 3 yr

Now you need some conversion factor that has _____ in the denom- | years

inator, to cancel the _____ in the information given, and that will move | years
you in the direction of the dimensions of the question. What about
365 days per year?

$$? \text{ sec} = 3 \, \cancel{\text{yr}} \times \frac{365 \text{ days}}{\cancel{\text{yr}}}$$

The unwanted dimension years has been canceled, but now you have the
unwanted dimension days left. You need a conversion factor with days
in the denominator to cancel that. Try

$$? \text{ sec} = 3 \, \cancel{\text{yr}} \times \frac{365 \, \cancel{\text{days}}}{\cancel{\text{yr}}} \times \frac{24 \text{ hr}}{\cancel{\text{day}}}$$

The unwanted dimension now is _____. To move toward seconds, you | hours
might next multiply by the conversion factor _____. This gives | $\frac{60 \text{ min}}{\text{hr}}$

$$? \text{ sec} = 3 \, \cancel{\text{yr}} \times \frac{365 \, \cancel{\text{days}}}{\cancel{\text{yr}}} \times \frac{24 \, \cancel{\text{hr}}}{\cancel{\text{day}}} \times \frac{60 \text{ min}}{\cancel{\text{hr}}}$$

Finally, you must have a conversion factor with _____ in the | minutes
denominator to cancel out the unwanted dimension. This would be

_____. | $\frac{60 \text{ sec}}{\text{min}}$

$$? \text{ sec} = 3 \, \cancel{\text{yr}} \times \frac{365 \, \cancel{\text{days}}}{\cancel{\text{yr}}} \times \frac{24 \, \cancel{\text{hr}}}{\cancel{\text{day}}} \times \frac{60 \, \cancel{\text{min}}}{\cancel{\text{hr}}} \times \frac{60 \text{ sec}}{\cancel{\text{min}}}$$

The only dimension left is _____, which is the dimension of the | seconds
question. Thus the correct answer must be

$$? \text{ sec} = 3 \times 365 \times 24 \times 60 \times 60 \text{ sec}$$
$$= 94{,}608{,}000 \text{ sec} = 9.4608 \times 10^7 \text{ sec}$$

Sometimes you use a conversion factor in its inverted form. You
might know the conversion factor from feet to miles. Since there are

5280 feet per mile, the conversion factor is _____. In its in- | $\frac{5280 \text{ ft}}{\text{mi}}$

verted form, you would write _____. Now consider the problem | $\frac{\text{mi}}{5280 \text{ ft}}$
of how many miles are in 7920 ft. The question asked is

"_____" So the dimensions of your answer | How many miles?

must be _____. Since the information given is 7920 ft, you write ? mi | miles
= 7920 ft.

To cancel the unwanted dimension, feet, you need a conversion factor

with _____ in the denominator. This would be the inverted form | feet
shown above.

$$? \text{ mi} = 7920 \, \cancel{\text{ft}} \times \frac{\text{mi}}{5280 \, \cancel{\text{ft}}}$$

The unwanted dimension feet cancels, leaving only the dimension mile,
which is the dimension of the question and therefore must be the answer.

$$? \text{ mi} = 7920 \times \frac{\text{mi}}{5280} = 1.500 \text{ mi}$$

Often the problem itself has conversion factors in it. You may find it helpful at first to write down all the conversion factors before you start a problem.

Here is an example. What is the cost of three shirts, if a box containing 12 shirts costs $135? The conversion factors given in the problem are

$$\frac{12 \text{ shirts}}{\text{box}} \quad \text{and} \quad \frac{\$135}{\text{box}}$$

You must read the problem carefully to find these. It is stated that there are 12 shirts per box and the price is $135 per box.

The most important thing is to determine the dimensions of the answer to the question. In this problem the question asked is "What is the cost?" The dimensions of cost are dollars. You must reword the question as, "How many dollars?" Now you can write

? dollars =

3 shirts

The information given is _____, so you write

? dollars = 3 shirts

$$\frac{\text{box}}{12 \text{ shirts}}$$

The conversion factor that contains shirts as a dimension is _____. So you multiply by it.

$$? \text{ dollars} = 3 \,\cancel{\text{shirts}} \times \frac{\text{box}}{12 \,\cancel{\text{shirts}}}$$

cancel

$\text{box;} \dfrac{\$135}{\text{box}}$

Notice that it is used in the inverted form in order to _____ the dimension shirts in the information given. The unwanted dimension now is _____. So now you multiply by the conversion factor _____ to give

$$? \text{ dollars} = 3 \,\cancel{\text{shirts}} \times \frac{\cancel{\text{box}}}{12 \,\cancel{\text{shirts}}} \times \frac{\$135}{\cancel{\text{box}}}$$

The only dimension left is dollars, so your answer must be

$$? \text{ dollars} = 3 \times \frac{1}{12} \times \$135 = \$33.75$$

All the examples so far have been so simple that you could have solved them without dimensional analysis. Now here is a harder problem with some special points.

What is the gas consumption (in miles per gallon) of a car that uses 0.1 gal of gas in 100 sec when traveling 60 mi/hr?

The first thing to do is reword the question so that you know what the dimensions of the answer will be. "How many miles per gallon" can be reworded, "How many miles equal 1 gallon?" Thus the information given is 1 gal.

The conversion factors given in the problem are

$$\frac{0.1 \text{ gal}}{100 \text{ sec}} \quad \text{and} \quad \frac{60 \text{ mi}}{\text{hr}}$$

Starting with the dimensions of the answer and the information given, we have

$$? \text{ mi} = 1 \text{ gal} \times$$

The only conversion factor that you can write with gallons in the denominator is _____. So you multiply by this.

$\dfrac{100 \text{ sec}}{0.1 \text{ gal}}$

$$? \text{ mi} = 1\,\cancel{\text{gal}} \times \frac{100 \text{ sec}}{0.1\,\cancel{\text{gal}}}$$

The unwanted dimension now is _____. But the next conversion factor you have related to miles has hours as its other dimension. You must therefore convert from seconds to hours.

seconds

$$? \text{ mi} = 1\,\cancel{\text{gal}} \times \frac{100\,\cancel{\text{sec}}}{0.1\,\cancel{\text{gal}}} \times \frac{\cancel{\text{min}}}{60\,\cancel{\text{sec}}} \times \frac{\text{hr}}{60\,\cancel{\text{min}}}$$

Now the unwanted dimension is hours. You can multiply by the conversion factor given in the problem.

$$? \text{ mi} = 1\,\cancel{\text{gal}} \times \frac{100\,\cancel{\text{sec}}}{0.1\,\cancel{\text{gal}}} \times \frac{\cancel{\text{min}}}{60\,\cancel{\text{sec}}} \times \frac{\cancel{\text{hr}}}{60\,\cancel{\text{min}}} \times \frac{60 \text{ mi}}{\cancel{\text{hr}}}$$

$$? \text{ mi} = 1 \times \frac{100}{0.1} \times \frac{1}{60} \times \frac{1}{60} \times 60 \text{ mi} = 16.7 \text{ mi for 1 gallon}$$

As we proceed in this text, many example problems will be presented conversion factor by conversion factor. The blanks you will be filling in will be these conversion factors. However, it is a very good idea if, besides filling in the blanks, you also keep a running setup of the problem as we go along step by step, canceling the unwanted dimensions as you go.

PROBLEM SET

For these problems, set up the answer but don't bother to do all the multiplications and divisions. The answers follow directly. After you set up the answer, immediately check to see if you have done it correctly. What you want to know is whether you have mastered the method of dimensional analysis.

1. How many seconds in 800 minutes?

2. How many minutes in 5 years?

3. How many years in 500 days?

4. How many dozens of eggs are there in 3500 eggs?

5. How many miles in 12,000 yards? There are 5280 ft/mi and 3 ft/yd.

6. How many decades are there in 9 centuries? There are 10 years in a decade.

7. What is the cost of 6 onions if 3 onions weigh 1.5 lb and the price of onions is 20¢ per lb?

8. How many hours does it take to drive to Los Angeles from San Francisco at an average speed of 52 mi/hr? The distance between the two cities is 405 mi.

9. What is the cost of driving from San Francisco to Los Angeles (405 mi) if the cost of gasoline is $1.34/gal and the car gets 18 mi/gal of gasoline?

10. How many oranges are in a crate if the price of a crate is $1.60 and the price of oranges is $0.30/lb? On the average there are three oranges per pound.

11. The price of a ream of paper is $4.00. There are 500 sheets of paper in a ream. A sheet of paper weighs 0.500 oz. What is the price per pound of paper? (There are 16 oz in a pound.)

12. How many cars are there in a long freight train if it takes the entire train 2 min to pass a station as it travels 40 mi/hr? Each car is 50 ft long. There are 5280 ft in a mile.

13. A crate of eggs holds 27 cartons of eggs. Each carton contains a dozen eggs. It is found that the eggs in the crate weigh 19 lb. What is the average weight of an egg in ounces? (There are 16 ounces per pound.)

14. An 11-fluid-ounce bottle of cough medicine contains 6 fluid drams of terpenhydrate. How many milliliters (mL) of terpenhydrate are there in 14 cartons of bottles of cough medicine? A carton contains 24 bottles. (There are 0.125 fluid ounces per fluid dram and 29.5 mL per fluid ounce.)

15. How many French francs does it cost to drive a Renault from Paris to Bordeaux, a distance of 650 kilometers (km), if gasoline costs $2.25/gal and the Renault makes 12 kilometers per liter of gasoline? (Let's assume an exchange rate of 8.5 French francs per dollar. There are 4 quarts per gallon and 0.946 liter per quart.)

16. The contents of a truck containing 400 crates of avocados are correctly insured for $2000. How many avocados are in an average selection of 12 crates, if an average avocado weighs 0.4 pound and the value of avocados is $.30 per pound?

17. Radio waves travel at the speed of light, 3×10^{10} centimeters (cm) per second (sec). How many minutes did it take the radio signals from the Pioneer satellite passing Saturn to reach Earth when it was 7.6×10^8 mi from Earth? There are 10^{-2} meter (m) per centimeter (cm), 10^3 m per kilometer (km) and 1.6 km/mi.

PROBLEM SET ANSWERS

You should be able to do at least ten of these problems correctly. This chapter is so important that if you missed three or more problems, you had better go over the entire chapter again.

1. $? \text{ sec} = 800 \text{ min} \times \dfrac{60 \text{ sec}}{\text{min}} = 800 \times 60 \text{ sec}$

2. $? \text{ min} = 5 \text{ yr} \times \dfrac{365 \text{ days}}{\text{yr}} \times \dfrac{24 \text{ hr}}{\text{day}} \times \dfrac{60 \text{ min}}{\text{hr}} = 5 \times 365 \times 24 \times 60 \text{ min}$

3. $? \text{ yr} = 500 \text{ days} \times \dfrac{\text{yr}}{365 \text{ days}} = 500 \times \dfrac{1}{365} \text{ yr}$

4. $? \text{ doz} = 3500 \text{ eggs} \times \dfrac{\text{doz}}{12 \text{ eggs}} = 3500 \times \dfrac{1}{12} \text{ doz}$

5. $? \text{ mi} = 12{,}000 \text{ yd} \times \dfrac{3 \text{ ft}}{\text{yd}} \times \dfrac{\text{mi}}{5280 \text{ ft}} = 12{,}000 \times 3 \times \dfrac{1}{5280} \text{ mi}$

6. $? \text{ decades} = 9 \text{ centuries} \times \dfrac{100 \text{ yr}}{\text{century}} \times \dfrac{\text{decade}}{10 \text{ yr}} = 9 \times 100 \times \dfrac{1}{10} \text{ decades}$

7. $? \text{ cents} = 6 \text{ onions} \times \dfrac{1.5 \text{ lb}}{3 \text{ onions}} \times \dfrac{20\cancel{c}}{\text{lb}} = 6 \times \dfrac{1.5}{3} \times 20\cancel{c}$

8. $? \text{ hr} = 405 \text{ mi} \times \dfrac{\text{hr}}{52 \text{ mi}} = 405 \times \dfrac{1}{52} \text{ hr}$

9. $? \text{ dollars} = 405 \text{ mi} \times \dfrac{\text{gal}}{18 \text{ mi}} \times \dfrac{\$1.34}{\text{gal}} = 405 \times \dfrac{1}{18} \times \$1.34$

10. The question asked is "How many oranges are there in 1 crate?"

$? \text{ oranges} = 1 \text{ crate} \times \dfrac{\$1.60}{\text{crate}} \times \dfrac{\text{lb}}{\$0.30} \times \dfrac{3 \text{ oranges}}{\text{lb}}$

$= 1 \times 1.60 \times \dfrac{1}{0.30} \times 3 \text{ oranges}$

11. If you are having trouble, start by writing all the conversion factors given in the problem. Then you can simply select the one you need from the collection.

$? \text{ dollars} = 1 \text{ lb} \times \dfrac{16 \text{ oz}}{\text{lb}} \times \dfrac{1 \text{ sheet}}{0.500 \text{ oz}} \times \dfrac{\text{ream}}{500 \text{ sheets}} \times \dfrac{\$4.00}{\text{ream}}$

$= 1 \times 16 \times \dfrac{1}{0.500} \times \dfrac{1}{500} \times \$4.00$

12. $? \text{ cars} = 1 \text{ train} \times \dfrac{2 \text{ min}}{\text{train}} \times \dfrac{\text{hr}}{60 \text{ min}} \times \dfrac{40 \text{ mi}}{\text{hr}} \times \dfrac{5280 \text{ ft}}{\text{mi}} \times \dfrac{\text{car}}{50 \text{ ft}}$

$= 1 \times 2 \times \dfrac{1}{60} \times 40 \times 5280 \times \dfrac{1}{50} \text{ cars}$

13. $? \text{ ounces} = 1 \text{ egg} \times \dfrac{1 \text{ doz}}{12 \text{ eggs}} \times \dfrac{19 \text{ lb}}{27 \text{ doz}} \times \dfrac{16 \text{ oz}}{1 \text{ lb}}$

$= 1 \times \dfrac{1}{12} \times \dfrac{19}{27} \times 16 \text{ oz}$

14. ? milliliters = 14 ~~cartons~~ $\times \dfrac{24 \text{ \scriptsize bottles}}{1 \text{ \scriptsize carton}} \times \dfrac{6 \text{ \scriptsize fl dr}}{1 \text{ \scriptsize bottle}} \times \dfrac{0.125 \text{ \scriptsize fl oz}}{1 \text{ \scriptsize fl dr}} \times \dfrac{29.5 \text{ mL}}{1 \text{ \scriptsize fl oz}}$

= 14 × 24 × 6 × 0.125 × 29.5 mL

15. ? francs = 650 ~~km~~ $\times \dfrac{1 \text{ \scriptsize liter}}{12 \text{ \scriptsize km}} \times \dfrac{1 \text{ \scriptsize qt}}{0.946 \text{ \scriptsize liter}} \times \dfrac{1 \text{ \scriptsize gal}}{4 \text{ \scriptsize qt}} \times \dfrac{\$2.25}{\text{\scriptsize gal}} \times \dfrac{8.5 \text{ francs}}{\$}$

$= 650 \times \dfrac{1}{12} \times \dfrac{1}{0.946} \times \dfrac{1}{4} \times 2.25 \times 8.5$ francs

16. ? avocados = 12 ~~crates~~ $\times \dfrac{1 \text{ \scriptsize truck}}{400 \text{ \scriptsize crates}} \times \dfrac{\$2000}{\text{\scriptsize truck}} \times \dfrac{\text{\scriptsize lb}}{\$.30} \times \dfrac{\text{avocado}}{0.4 \text{ \scriptsize lb}}$

$= 12 \times \dfrac{1}{400} \times 2000 \times \dfrac{1}{.30} \times \dfrac{\text{avocado}}{0.4}$

17. ? minutes = 7.6 × 10⁸ ~~mi~~ $\times \dfrac{1.6 \text{ \scriptsize km}}{1 \text{ \scriptsize mi}} \times \dfrac{10^3 \text{ \scriptsize m}}{\text{\scriptsize km}} \times \dfrac{\text{\scriptsize cm}}{10^{-2} \text{ \scriptsize m}} \times \dfrac{\text{\scriptsize sec}}{3 \times 10^{10} \text{ \scriptsize cm}}$

$\times \dfrac{\text{min}}{60 \text{ \scriptsize sec}}$

$= 7.6 \times 10^8 \times 1.6 \times 10^3 \times \dfrac{1}{10^{-2}} \times \dfrac{1}{3 \times 10^{10}} \times \dfrac{\text{min}}{60}$

These last five problems were a little more difficult, since the conversion factors were hidden in the language of the problems. If you were able to do them, you are well on your way to being able to solve the types of chemical problems you are going to face in this book. If you were not able to do them don't worry. You are going to get a lot more practice.

CHAPTER 3

The Metric System

PRETEST

If you feel that you know the metric system of measurement, take this test. Try to work all the problems by dimensional analysis. A few conversion factors that you may not still have in your memory are given below. Many others that you should know are not. Try to complete this test in 45 minutes.

The answers follow immediately, with a method of scoring yourself.

Conversion Factors

$$\frac{2.54 \text{ cm}}{\text{in.}} \quad \frac{946 \text{ mL}}{\text{qt}} \quad \frac{454 \text{ g}}{\text{lb}} \quad \frac{4 \text{ qt}}{\text{gal}} \quad \frac{5280 \text{ ft}}{\text{mi}}$$

$$\frac{2 \text{ pt}}{\text{qt}} \quad \frac{16 \text{ fl oz}}{\text{pt}} \quad \frac{16 \text{ oz}}{\text{lb}} \quad \frac{2000 \text{ lb}}{\text{ton}} \quad \frac{1.61 \text{ km}}{\text{mi}}$$

1. How many milliliters are in 16 L? 2. How many yards are in a km?
3. How many μL are in 5 mL? 4. How many mg are in 3.00×10^{-4} ton?
5. How many angstroms are in 2×10^{-4} m?
6. How many nanometers are there in 3.5×10^{-2} m?
7. Let's assume that the price of milk in France is 3.6 francs/liter. What is the price in dollars for 1 qt? Use an exchange rate of 8.5 francs/dollar.
8. How long will it take a Porsche traveling at 120 km/hr to go 90 mi?
9. The speed of light is 3.0×10^{10} cm/sec. How many miles from the earth is the sun, if it takes light 6.0 min to cover the distance?
10. It takes 2.26 kilojoules (kJ) to vaporize 1.0 g of water at 100°C to steam. How many kilojoules does it take to vaporize 6.0 pt of water? (A liter of water weighs 0.96 kg at 100°C.)
11. How many dollars would it cost to drive a VW 5000 mi in Germany, where gasoline costs 0.90 DM/liter? The car averages 12 km/liter. (2.8 DM = $1.)
12. What is the temperature in °C corresponding to 65°F?
13. What is the temperature in °F corresponding to -14°C?

(These last two are not generally solved using dimensional analysis, but do so here for practice.)

PRETEST ANSWERS

If you got the correct answer to eight of the first ten problems, you are able to handle the metric system very well indeed. You can proceed immediately to the problem set at the end of Chapter 3. If, however, you were not able to do Questions 11 and 12, you should work the last section of Chapter 3, which gives the temperature conversions.

If you missed more than two of the first ten problems, you should work through all of Chapter 3.

1. 1.6×10^4 mL
2. 1.09×10^3 yd
3. 5×10^3 μL
4. 2.72×10^5 mg
5. 2×10^6 Å
6. 3.5×10^7 nm
7. $0.40
8. 1.2 hr
9. 6.7×10^7 mi
10. 6.2×10^3 kJ
11. $215
12. 18°C
13. 7°F

Don't be upset if you did poorly on the pretest. The problems were difficult. Now you are going to get a chance to go through the chapter and get more practice using dimensional analysis.

SECTION A   The Metric System

In this chapter we are going to examine the units of measurement used in all scientific work. These units are in what is called the *metric system*. Many countries use this system as their normal method of expressing measurements. As a matter of fact, only the United States and some other countries that used to be British colonies still use what is called the *English system* of weights and measures. (Britain itself has gone metric, so many people now call the former English system the *U.S. system*.) The metric system, as you will see, is vastly simpler than the English system. A Congressional committee is investigating what must be done to shift weights and measures to the metric system in the United States. Someday you will be saying, "What a hot day! It's 32 degrees in the shade," or, "My car is making good mileage—12 kilometers per liter of gas." At present, however, you will have to learn the metric system and also learn how to convert from the English system to the metric system.

There is a relatively new system of measurement, called the *SI system*, that is being used more and more often. Your textbook probably uses it for at least some dimensions. It is closely related to the metric system, and uses the same prefixes to indicate powers of 10. However, the base units are slightly different. The *calorie*, as a unit of heat, has been replaced by the *joule*. Also, in the SI system, temperatures are in *Kelvins* (K) rather than degrees Celsius (°C) or degrees Kelvin (°K). And the *mole* is symbolized by *mol*. The *angstrom* (Å) disappears completely in the SI system; instead, measurements of length at the micro level are given in *nanometers* (nm).

In this period of changeover from metric to SI, there is confusion as to which system is preferable. The SI system has many advantages in the interconvertibility of different types of measurement. However, most scientific literature still uses the metric system. In this book we use the metric system most of the time. But we present certain quantities in both metric and SI units (calories and joules, for example). And we use K and mol, rather than $°K$ and mole, as symbols for these dimensions.

You will find a more complete discussion of SI in Appendix III, along with the appropriate conversions from metric to SI units.

The beauty of the metric system is that all the conversion factors are simply 1.00 times 10 raised to some power. What is more, the name of the dimension tells you to what power 10 is raised. The following table shows how it works.

Prefix	Symbol	Meaning	Prefix	Symbol	Meaning
pico-	p	10^{-12}	deci-	d	10^{-1}
nano-	n	10^{-9}	kilo-	k	10^{3}
micro-	μ	10^{-6}	mega-	M	10^{6}
milli-	m	10^{-3}	giga-	G	10^{9}
centi-	c	10^{-2}	tetra-	T	10^{12}

Only about four of these prefixes are used commonly in introductory chemistry, so we are going to concentrate on them.

The basic unit of length in the metric and SI systems is the *meter.* A meter (m) is about a yard long, and it has 100 parts, each of which is called a *centimeter* (cm). Thus

$$1 \text{ cm} = 10^{-2} \text{ m}$$

The conversion factor and its reciprocal from this equality are

_____ and _____ .
$\dfrac{\text{cm}}{10^{-2}\text{ m}} ; \dfrac{10^{-2}\text{ m}}{\text{cm}}$

Try to calculate the number of centimeters (cm) in 4.5 meters (m).

The dimensions of the answer will be _____, so you write
centimeters

? _____ =
cm

The information given in the problem is _____, so you write
4.5 m

? cm = _____ .
4.5 m

The conversion factor that has _____ in the denominator to cancel
meters

the _____ in the information given is _____. You now write
meters; $\dfrac{\text{cm}}{10^{-2}\text{ m}}$

$$? \text{ cm} = 4.5 \text{ m} \times \frac{\text{cm}}{10^{-2}\text{ m}}$$

The only dimension left is _____, which is the dimension of
centimeters

the answer. The answer, then, must be

$$? \text{ cm} = 4.5 \times \frac{1}{10^{-2}} \text{ cm} = 4.5 \times 10^{2} \text{ cm}$$

The meter can also be divided into 1000 parts, each of which is called a *millimeter* (mm). Thus

$$1 \text{ mm} = 10^{-3} \text{ m}$$

$\dfrac{\text{mm}}{10^{-3} \text{ m}}, \dfrac{10^{-3} \text{ m}}{\text{mm}}$

The conversion factors from this equality are _____ and _____.

Try this more complicated problem. How many centimeters in 8×10^2 millimeters (mm)? The dimensions of the answer will be

centimeters; cm

_____, so you write ? _____ =.

8×10^2 mm

The information given is _____, so you write ? cm =

8×10^2 mm

_____.

The only conversion factor you have that has millimeters in the

$\dfrac{10^{-3} \text{ m}}{\text{mm}}$

denominator is _____, so you multiply by this factor.

$$? \text{ cm} = 8 \times 10^2 \text{ m̶m̶} \times \frac{10^{-3} \text{ m}}{\text{m̶m̶}}$$

This operation cancels the unwanted dimension millimeters, but leaves a

meters

new unwanted dimension, _____. A conversion factor that has

meters

the unwanted dimension, _____, in the denominator and is

$\dfrac{\text{cm}}{10^{-2} \text{ m}}$

related to the dimension of the answer to the question is _____.
Multiplying by this conversion factor gives

$$? \text{ cm} = 8 \times 10^2 \text{ m̶m̶} \times \frac{10^{-3} \text{ m̶}}{\text{m̶m̶}} \times \frac{\text{cm}}{10^{-2} \text{ m̶}}$$

$$= 8 \times 10^2 \times 10^{-3} \times \frac{1}{10^{-2}} \text{ cm} = 8 \times 10^1 \text{ cm}$$

It should now start to become apparent why the metric system is so easy to handle. All the conversions amount to simply multiplying or dividing by 10 raised to some exponent. If you are not very confident handling this exponential notation, refer back to Section A in Chapter 1.

The meter is divided into still smaller parts. If you divide a meter into 1 million parts, you have what used to be called a *micron*, symbolized by the Greek letter μ (mu). The micron is now called a *micrometer*, even though this looks like the name of a measuring device. Its symbol is μm. Thus

$$1 \text{ } \mu\text{m} = 10^{-6} \text{ m}$$

$\dfrac{\mu\text{m}}{10^{-6} \text{ m}}, \dfrac{10^{-6} \text{ m}}{\mu\text{m}}$

The conversion factors from this equality are _____ and _____.

The micrometer is one instance in which the SI notation has been pretty well accepted. No one says "micron" any longer.

In the metric system there is one unit of length whose name doesn't follow the systematic naming. It was chosen because it falls into the size range of atoms and the wavelength of x rays. It is the *angstrom* (A or Å) named in honor of the Swedish spectroscopist Anders Jonas Ångström (1814–1874), and

$$1 \text{ Å} = 10^{-8} \text{ cm}$$

You often see atomic sizes given in angstroms, but the tendency now (particularly in SI) is to use the *nanometer* (nm) or *picometer* (pm):

$$1 \text{ nm} = 10^{-9} \text{ m} \qquad 1 \text{ pm} = 10^{-12} \text{ m}$$

You might calculate the diameter of a Cl atom in picometers, knowing that it is 1.98 Å when expressed in angstroms. You ask, "How many picometers equal 1.98 angstroms?" The dimensions of the answer will be

_____, so you write ? _____ =.

The information given is _____, so you write ? pm = _____.
You can cancel the unwanted angstroms with the conversion factor

_____, which relates angstroms to centimeters:

$$? \text{ pm} = 1.98 \text{ Å} \times \underline{\hspace{3cm}}$$

The unwanted dimension is now _____, but the only conversion factor

you have for picometers contains the unit _____. You must

therefore convert the centimeter to _____. Thus the next step is

$$? \text{ pm} = 1.98 \text{ Å} \times \frac{10^{-8} \text{ cm}}{\text{Å}} \times \underline{\hspace{1.5cm}}$$

Now you can convert from the unwanted meters directly to the desired picometers.

$$? \text{ pm} = 1.98 \text{ Å} \times \frac{10^{-8} \text{ cm}}{\text{Å}} \times \frac{10^{-2} \text{ m}}{\text{cm}} \times \underline{\hspace{2cm}}$$

(Notice that you had to use the last conversion factor in its inverted form so that you could cancel the unwanted meters.) This gives you

$$? \text{ pm} = 1.98 \text{ Å} \times \frac{10^{-8} \text{ cm}}{\text{Å}} \times \frac{10^{-2} \text{ m}}{\text{cm}} \times \frac{\text{pm}}{10^{-12} \text{ m}}$$

All the dimensions cancel except for the desired picometers:

$$? \text{ pm} = 1.98 \times 10^{-8} \times 10^{-2} \times \frac{\text{pm}}{10^{-12}} = \underline{\hspace{3cm}} \text{ pm}$$

(If you came out with the wrong power of 10, check back to Section A of Chapter 1. It tells how to move 10 to a power from the denominator to the numerator.)

When you are dealing with very large lengths (which isn't too often in chemistry), the unit used is the *kilometer* (km). It is 1000 meters and the relationship is

$$1 \text{ km} = 10^3 \text{ m}$$

Here is a problem. How many kilometers in 6.3×10^{35} Å? The

question asked is "_____" and the information

given is _____. You therefore write ? km = 6.3×10^{35} Å.
A conversion factor that cancels out the unwanted dimension angstrom

is _____, so you write

---

picometers; pm

1.98 Å; 1.98 Å

$\dfrac{10^{-8} \text{ cm}}{\text{Å}}$
$\dfrac{10^{-8} \text{ cm}}{\text{Å}}$

cm

meter

meter

$\dfrac{10^{-2} \text{ m}}{\text{cm}}$

$\dfrac{\text{pm}}{10^{-12} \text{ m}}$

1.98×10^2

How many kilometers?

6.3×10^{35} Å

$\dfrac{10^{-8} \text{ cm}}{\text{Å}}$

$$? \text{ km} = 6.3 \times 10^{35} \, \cancel{\text{Å}} \times \frac{10^{-8} \text{ cm}}{\cancel{\text{Å}}}$$

$\dfrac{10^{-2} \text{ m}}{\text{cm}}$

Now you get to meters by multiplying by _____.

$\dfrac{10^{-2} \text{ m}}{\text{cm}}$

$$? \text{ km} = 6.3 \times 10^{35} \, \cancel{\text{Å}} \times \frac{10^{-8} \text{ cm}}{\cancel{\text{Å}}} \times \underline{\hspace{2cm}}$$

meters

$\dfrac{\text{km}}{10^3 \text{ m}}$

You can now get from the unwanted dimension _____ to kilometers, the dimensions of the answer, by multiplying by the conversion factor _____.

$$? \text{ km} = 6.3 \times 10^{35} \, \cancel{\text{Å}} \times \frac{10^{-10} \, \cancel{\text{m}}}{\cancel{\text{Å}}} \times \frac{\text{km}}{10^3 \, \cancel{\text{m}}}$$

$$= 6.3 \times 10^{35} \times 10^{-10} \times \frac{1}{10^3} \text{ km} = 6.3 \times 10^{22} \text{ km}$$

Now we can summarize these metric units of length in their conversion-factor form.

Picometer	$\dfrac{\text{pm}}{10^{-12} \text{ m}}$		Millimeter	$\dfrac{\text{mm}}{10^{-3} \text{ m}}$
Nanometer	$\dfrac{\text{nm}}{10^{-9} \text{ m}}$		Centimeter	$\dfrac{\text{cm}}{10^{-2} \text{ m}}$
Angstrom	$\dfrac{\text{Å}}{10^{-10} \text{ m}}$ or $\dfrac{\text{Å}}{10^{-8} \text{ cm}}$		Kilometer	$\dfrac{\text{km}}{10^3 \text{ m}}$
Micrometer	$\dfrac{\mu\text{m}}{10^{-6} \text{ m}}$			

The basic unit of volume in the metric system is the *liter*, which is a little more than 1 qt. The correct SI abbreviation for liter is a lower-case letter l. However, a lower-case l looks so much like the numeral 1 that for clarity we are using the capital letter L.

There are only two commonly used metric units of volume other than the liter itself. One is a *milliliter* (mL). From what you know about the prefix *milli-*, you know that

10^{-3}

$$1 \text{ mL} = \underline{\hspace{2cm}} \text{ L}$$

$\dfrac{\text{mL}}{10^{-3} \text{ L}}$
$\dfrac{10^{-3} \text{ L}}{\text{mL}}$

The conversion factors from this equality are _____ and

_____.

When the metric system was set up, it was done in such a way that 1 mL was the volume equivalent to a cube 1 cm on a side. That is,

$$1 \text{ mL} = 1 \text{ cm}^3$$

There was a very slight error made at that time, but for all your calculations you can consider this equality correct.

The other common unit of volume is the *microliter* (μL). From the prefix *micro-*, you know that

10^{-6}

$$1 \, \mu\text{L} = \underline{\hspace{2cm}} \text{ L}$$

Now we can summarize the metric units of volume in their conversion-factor form.

Milliliter $\dfrac{mL}{10^{-3}\ L}$ Microliter $\dfrac{\mu L}{10^{-6}\ L}$

(An SI unit for volume, 1 decimeter^3 = 1 liter, is not useful for chemists, so we will never see it in this book.)

Try a problem for practice. How many microliters in 4.7×10^{-2} mL?

The question asked is " _____ " and the information

given is _____. You therefore write

? _____ = _____

The dimension you want to cancel is _____. A conversion factor with this in the denominator is _____. Multiplying by this conversion factor leaves the unwanted dimension _____. The conversion factor that has this unwanted dimension in the denominator and that relates to the dimensions of the desired answer is _____.
Multiplying by this,

$$? \ \mu L = 4.7 \times 10^{-2} \ \cancel{mL} \times \frac{10^{-3}\ \cancel{L}}{\cancel{mL}} \times \frac{\mu L}{10^{-6}\ \cancel{L}}$$

leaves _____ as the only dimension. So the answer must be

$$? \ \mu L = 4.7 \times 10^{-2} \times 10^{-3} \times \frac{1}{10^{-6}} \ \mu L = 4.7 \times 10^{1} \ \mu L$$

Recall that we stated that 1 mL = 1 cm^3. The conversion factors that can be derived from this equality are _____ and _____. Actually, you can always express volumes as lengths cubed. Thus the SI system uses meter^3 as its prime unit of volume. In the English system of weights and measures, we talk about quarts and gallons, but also cubic yards and cubic feet.

In certain problems, you have to use both volume units based on liters and those based on some metric length cubed. The conversion factors relating milliliters to cubic centimeters enables you to go from one to the other. Here is a problem to illustrate this. How many milliliters are there in a cube that is 3.0 cm on each edge?

First you calculate the volume of the cube in cubic centimeters. The volume of a cube is its edge length raised to the third power:

Volume = $(3.0 \ cm)^3$

Both the number and its dimension must be cubed:

Volume = $(3.0)^3 \times (cm)^3 = 27 \ cm^3$

Margin notes:
How many μL?

4.7×10^{-2} mL

μL; 4.7×10^{-2} mL

milliliters

$\dfrac{10^{-3}\ L}{mL}$

liter

$\dfrac{\mu L}{10^{-6}\ L}$

microliters

$\dfrac{mL}{cm^3}; \dfrac{cm^3}{mL}$

How many milliliters?

Now you have the volume of the cube, which is the information given. The question asked is "_____" You therefore write

$$? \text{ mL} = 27 \text{ cm}^3$$

In order to cancel the dimensions of the information given and move toward the dimensions of the answer, you can multiply by the conver-

$\dfrac{\text{mL}}{\text{cm}^3}$

sion factor _____:

$$? \text{ mL} = 27 \, \cancel{\text{cm}^3} \times \frac{\text{mL}}{\cancel{\text{cm}^3}}$$

The only dimension left is the dimension of the answer, so

$$? \text{ mL} = 27 \text{ mL} \quad \text{must be the answer.}$$

Now find the volume in liters of a cube that is 2 Å on each edge. This problem may seem much harder, but it isn't. You simply use more conversion factors.

First you determine the volume in angstroms cubed (Å^3). The volume of a cube is the edge length cubed, so

$$\text{Volume} = (2 \text{ Å})^3 = 2^3 \text{ Å}^3 = 8 \text{ Å}^3$$

(Notice that both the number *and* the dimension of the number are cubed.) The question asked is "How many liters?" The information given has been calculated as 8 Å^3. Therefore you write

$$? \text{ L} = 8 \text{ Å}^3$$

$\dfrac{10^{-10} \text{ m}}{\text{Å}}$

You have no conversion factor with Å^3 as a dimension, but you do have one with Å, which is _____. If you cube this, it is still a conversion factor.

$$\left(\frac{10^{-10} \text{ m}}{\text{Å}}\right)^3 = \frac{10^{-30} \text{ m}^3}{\text{Å}^3}$$

(Notice that both the number and the dimensions are cubed.) Multiplying by this factor, you have

$$? \text{ L} = 8 \, \cancel{\text{Å}^3} \times \frac{10^{-30} \text{ m}^3}{\cancel{\text{Å}^3}}$$

cubic meters (m^3)

cubic meters

Now the unwanted dimension is _____, which you must multiply by some conversion factor containing _____ in the denominator. This will lead you toward the dimensions of the answer, liters. If you think about it, you will realize that you must work through the equality of milliliters and cubic centimeters. So you need some conversion factor containing cubic meters and cubic centimeters.

cubing

You can derive this factor by _____ the conversion factor relating meters to centimeters:

$$\left(\frac{\text{cm}}{10^{-2} \text{ m}}\right)^3 = \frac{\text{cm}^3}{10^{-6} \text{ m}^3}$$

Multiplying by this factor gives

$$? L = 8 \cancel{\text{Å}^3} \times \frac{10^{-30} \cancel{m^3}}{\cancel{\text{Å}^3}} \times \frac{cm^3}{10^{-6} \cancel{m^3}}$$

Now you can get from cubic centimeters to the volume dimension, milliliters, by multiplying by the converstion factor _____.

$$? L = 8 \cancel{\text{Å}^3} \times \frac{10^{-30} \cancel{m^3}}{\cancel{\text{Å}^3}} \times \frac{\cancel{cm^3}}{10^{-6} \cancel{m^3}} \times \frac{mL}{\cancel{cm^3}}$$

It is easy to convert from milliliters to liters.

$$? L = 8 \cancel{\text{Å}^3} \times \frac{10^{-30} \cancel{m^3}}{\cancel{\text{Å}^3}} \times \frac{\cancel{cm^3}}{10^{-6} \cancel{m^3}} \times \frac{\cancel{mL}}{\cancel{cm^3}} \times \frac{10^{-3} \text{ liter}}{\cancel{mL}}$$

The only dimension left is liters, so the answer must be

$$? L = 8 \times 10^{-30} \times \frac{1}{10^{-6}} \times 1 \times 10^{-3} \ L = 8 \times 10^{-27} \ L$$

If this seems labored, just remember that, to raise a quantity to a power, you treat the number and the dimension separately. You raise the number to the power and you also raise the dimension to the power.

For practice, find the conversion factor relating square millimeters and square meters. The conversion factor relating millimeters to meters is

_____. You must square both the numerator and the denominator.

$$\left(\frac{10^{-3} \text{ m}}{mm} \right)^2 = \frac{(10^{-3})^2 \ m^2}{mm^2} = \frac{10^{-6} \ m^2}{mm^2}$$

Check yourself on this one. The conversion factor relating cubic micrometers to cubic meters is _____.

The unit of weight in the metric system is the *gram* (g). Approximately 30 g equal an ounce. There are only three commonly used units of weight other than the gram itself. One is the *milligram* (mg):

I mg = _____ g

For very small quantities, there is the *microgram* (μg):

1 μg = _____ g

Finally, for larger weights, there is the *kilogram* (kg):

1 kg = _____ g

You can now see how simple the metric system of naming things is. The following table summarizes these units and gives their conversion factors.

| Microgram | $\frac{\mu g}{10^{-6} \ g}$ | Milligram | $\frac{mg}{10^{-3} \ g}$ | Kilogram | $\frac{kg}{10^3 \ g}$ |

Margin answers:
$\frac{mL}{cm^3}$

$\frac{10^{-3} \ m}{mm}$

$\frac{10^{-18} \ m^3}{\mu m^3}$

10^{-3}

10^{-6}

10^3

Do one problem for practice: How many kilograms in 10^{14} μg? Writing the dimensions of the answer and the dimensions of the information given, you have

kg; 10^{14} μg

? _____ = _____

Multiplying this by a conversion factor to remove the unwanted dimension, you have

$\dfrac{10^{-6} \text{ g}}{\mu\text{g}}$

? kg = 10^{14} μg $\times$ _____

grams

The unwanted dimension now is _____. You must relate this to

kilograms

_____, the dimensions of the answer. The conversion factor

$\dfrac{\text{kg}}{10^3 \text{ g}}$

that will do this is _____.

(Remember to keep writing down all the conversion factors on your own setup as we go along.) Multiplying by this factor, all the dimensions cancel except the dimension of the answer. So you have

$$? \text{ kg} = 10^{14} \ \cancel{\mu\text{g}} \times \frac{10^{-6} \ \cancel{\text{g}}}{\cancel{\mu\text{g}}} \times \frac{\text{kg}}{10^3 \ \cancel{\text{g}}} = 10^{14} \times 10^{-6} \times \frac{1}{10^3} \text{ kg} = 10^5 \text{ kg}$$

Go back to the pretest and do problems 1, 3, 5, and 6. See whether you get the correct answer. All the answers are expressed in scientific notation, which you should make a practice of doing.

Now you are going to do some easy problems using the metric system.

If 2.0 m of a sample of steel wire weigh 0.65 kg, how many milligrams

$\dfrac{0.65 \text{ kg}}{2.0 \text{ m}}$

do 10 mm weigh? The conversion factor in the problem is _____.

milligrams

The dimensions of the answer will be _____. The infor-

10 mm

mation given is _____. Setting this up, you have

? mg = 10 mm.

You now need a conversion factor with millimeters in the denominator that allows you to proceed toward the dimensions of the answer. Such a

$\dfrac{10^{-3} \text{ m}}{\text{mm}}$

factor might be _____.

$$? \text{ mg} = 10 \ \cancel{\text{mm}} \times \frac{10^{-3} \text{ m}}{\cancel{\text{mm}}}$$

So far, so good. You can now use the conversion factor shown in the problem.

$$? \text{ mg} = 10 \ \cancel{\text{mm}} \times \frac{10^{-3} \ \cancel{\text{m}}}{\cancel{\text{mm}}} \times \frac{0.65 \text{ kg}}{2.0 \ \cancel{\text{m}}}$$

To get from kilograms, the unwanted dimension, to milligrams, the dimensions of the answer, you simply multiply by the appropriate conversion factors.

$$? \text{ mg} = 10 \ \cancel{\text{mm}} \times \frac{10^{-3} \ \cancel{\text{m}}}{\cancel{\text{mm}}} \times \frac{0.65 \ \cancel{\text{kg}}}{2.0 \ \cancel{\text{m}}} \times \frac{10^3 \ \cancel{\text{g}}}{\cancel{\text{kg}}} \times \frac{\text{mg}}{10^{-3} \ \cancel{\text{g}}}$$

Therefore

$$? \text{ mg} = 10 \times 10^{-3} \times \frac{0.65}{2.0} \times 10^3 \times \frac{1}{10^{-3}} \text{ mg}$$

$$= 0.32 \times 10^4 \text{ mg} = 3.2 \times 10^3 \text{ mg}$$

Now do one yourself. A coal-burning locomotive uses 35 kg of coal to travel 3.0 km. How many grams of coal does it use to go 6.0 cm? The dimensions of the answer will be _____. The information given is _____. The conversion factor given in the problem is _____. Now perform your calculations. The final answer is _____. Did you forget to change the answer into scientific notation? If so, you got 70×10^{-2} g.

grams

$6.0 \text{ cm}; \dfrac{35 \text{ kg}}{3.0 \text{ km}}$

7.0×10^{-1} g

SECTION B   The English System (U.S. System) of Weights and Measures

Unhappily, we are stuck with a very clumsy and nonscientific system of measurement in the United States. It is very confusing to have 16 oz in an avoirdupois pound, 12 oz in a Troy pound (used only for precious metals), and 32 fl oz in a quart. From years of use, however, you probably remember most of the conversion factors in our system. To refresh your memory, and to have them all in one place, the most common ones are tabulated here.

Conversion Factors for English Units

Length	Volume	Weight
$\dfrac{12 \text{ in}}{\text{ft}}$	$\dfrac{16 \text{ fl oz}}{\text{pt}}$	$\dfrac{16 \text{ oz}}{\text{lb}}$
$\dfrac{3 \text{ ft}}{\text{yd}}$	$\dfrac{2 \text{ pt}}{\text{qt}}$	$\dfrac{2000 \text{ lb}}{\text{ton}}$
$\dfrac{5280 \text{ ft}}{\text{mi}}$	$\dfrac{4 \text{ qt}}{\text{gal}}$	

Although it isn't any harder to set up a problem in English (U.S.) units than in metric units, the arithmetic is much more laborious (although, with a calculator, it's less of a chore). For example, let's try this problem: How many miles in 6.0×10^7 in.? The question asked is "_____ _____" So the dimensions of the answer are _____. The information given is _____, so you write

$$? \text{ miles} = 6.0 \times 10^7 \text{ in.}$$

How many miles?

miles

6.0×10^7 in.

$\dfrac{\text{ft}}{12 \text{ in.}}$

The only conversion factor you have with inches is _____.

$$? \text{ miles} = 6.0 \times 10^7 \text{ in.} \times \frac{\text{ft}}{12 \text{ in.}}$$

foot

$\dfrac{\text{mile}}{5280 \text{ ft}}$

The unwanted dimension _____ can be canceled by multiplying by the conversion factor _____.

$$? \text{ miles} = 6.0 \times 10^7 \text{ in.} \times \frac{\text{ft}}{12 \text{ in.}} \times \frac{\text{mi}}{5280 \text{ ft}}$$

miles

The only dimension left is _____, so the answer must be

$$? \text{ miles} = 6.0 \times 10^7 \times \frac{1}{12} \times \frac{1}{5280} \text{ mi} = 9.5 \times 10^2 \text{ mi}$$

Converting from the English (U.S.) System to the Metric System

Most problems in introductory chemistry involve changing measurements given in English (U.S.) units to the equivalent measurements in metric units, or vice versa. The following table shows the conversion factors between the two systems.

Length	Volume	Weight
$\dfrac{2.54 \text{ cm}}{\text{in.}}$	$\dfrac{946 \text{ mL}}{\text{qt}}$	$\dfrac{454 \text{ g}}{\text{lb}}$
$\dfrac{1.61 \text{ km}}{\text{mi}}$		$\dfrac{\text{kg}}{2.2 \text{ lb}}$

With these conversion factors, it is very simple to convert any measurement in one system to the equivalent measurement in the other. Problem 2 in the pretest asks: How many yards are in a kilometer? Starting the problem as usual, you have

yd; 1 km

$$? \text{_____} = \text{_____}$$

$\dfrac{\text{mi}}{1.61 \text{ km}}$

You can multiply by the conversion factor _____ to cancel the unwanted dimension and, incidentally, to shift to English (U.S.) units.

miles

The unwanted dimension is now _____. You can get this to the wanted

$\dfrac{5280 \text{ ft}}{\text{mi}}$

$\dfrac{\text{yd}}{3 \text{ ft}}$; yards

dimension, yards, by multiplying by the two conversion factors _____ and _____. All the dimensions cancel except _____, so your answer must be

$$? \text{ yd} = 1 \times \frac{1}{1.61} \times 5280 \times \frac{1}{3} \text{ yd} = 1.09 \times 10^3 \text{ yd}$$

Look at Problem 4 in the pretest. It asks: How many milligrams are

How many milligrams?

there in 3×10^{-4} ton? The question asked is "_____" and the information given is _____. You therefore write

? _____ = _____

The only conversion factor you have that contains tons as a dimension is

_____. Multiplying by this factor, you have the unwanted di-

mension _____ left. Now, however, you have a conversion factor that
has pounds as one dimension and a metric unit as the other. This will
serve as your "bridge" (conversion factor) between the two systems. It

is _____. Multiplying by this "bridge," you now have as the un-

wanted dimension _____, which is easily converted to _____,

the dimensions of the answer, by using the conversion factor _____.
All the dimensions have now canceled except the dimensions of the
answer, so the answer must be

$$? \text{ mg} = 3 \times 10^{-4} \text{ ton} \times \frac{2000 \text{ lb}}{\text{ton}} \times \frac{454 \text{ g}}{\text{lb}} \times \frac{\text{mg}}{10^{-3} \text{ g}} = 2.72 \times 10^5 \text{ mg}$$

Try problem 8 in the pretest, and you will see how simple it is using
dimensional analysis. You want to find how long it will take a Porsche
doing 120 km/hr to go 90 mi. Do the problem yourself and then check
your setup with the following.

$$? \text{ hr} = 90 \text{ mi} \times \frac{1.61 \text{ km}}{\text{mi}} \times \frac{\text{hr}}{120 \text{ km}}$$

$$? \text{ hr} = 90 \times 1.61 \times \frac{1}{120} \text{ hr} = 1.2 \text{ hr}$$

In problem 8 there was just one conversion factor given: _____.
The only way problems can become more complicated is if more conversion factors are given in the question itself.

Problem 7 in the pretest says that the price of milk in France is 3.6
francs per liter and asks what the price is in dollars for 1 qt. (There are
8.5 francs per dollar.) There are two conversion factors in the question:

_____ and _____. Try to do the problem and see if
you get the correct answer, $.40. If you do not get this, check your
setup with the following:

$$? \text{ dollars} = 1 \text{ qt} \times \frac{946 \text{ mL}}{\text{qt}} \times \frac{10^{-3} \text{ L}}{\text{mL}} \times \frac{3.6 \text{ francs}}{\text{L}} \times \frac{\$1}{8.5 \text{ francs}}$$

Now do problems 9, 10, and 11 in the pretest. The setups are shown
below, but do not look at them until you have tried the problems
yourself.

Problem 9 $\quad ? \text{ mi} = 6 \text{ min} \times \frac{60 \text{ sec}}{\text{min}} \times \frac{3 \times 10^{10} \text{ cm}}{\text{sec}} \times \frac{10^{-2} \text{ m}}{\text{cm}}$

$$\times \frac{\text{km}}{10^3 \text{ m}} \times \frac{\text{mi}}{1.61 \text{ km}}$$

(margin notes)

3×10^{-4} ton

mg; 3×10^{-4} ton

$\dfrac{2000 \text{ lb}}{\text{ton}}$

pounds

$\dfrac{454 \text{ g}}{\text{lb}}$

grams; milligrams

$\dfrac{\text{mg}}{10^{-3} \text{ g}}$

$\dfrac{120 \text{ km}}{\text{hr}}$

$\dfrac{3.6 \text{ francs}}{\text{L}}$; $\dfrac{8.5 \text{ francs}}{\$}$

Problem 10 $? \text{kJ} = 6 \, \cancel{pt} \times \dfrac{\cancel{qt}}{2 \, \cancel{pt}} \times \dfrac{946 \, \cancel{mL}}{\cancel{qt}} \times \dfrac{10^{-3} \, \cancel{L}}{\cancel{mL}} \times \dfrac{0.96 \, \cancel{kg}}{\cancel{L}}$

$\times \dfrac{10^{3} \, \cancel{g}}{\cancel{kg}} \times \dfrac{2.26 \, \text{kJ}}{\cancel{g}}$

Problem 11 $? \, \$ = 5000 \, \cancel{mi} \times \dfrac{1.61 \, \cancel{km}}{\cancel{mi}} \times \dfrac{\cancel{L}}{12 \, \cancel{km}} \times \dfrac{0.90 \, \cancel{DM}}{\cancel{L}} \times \dfrac{\$}{2.8 \, \cancel{DM}}$

SECTION C   Temperature Conversions

In the metric system, temperatures are expressed in degrees Celsius, which can also be called degrees centigrade (°C). As you know, in the United States temperature is expressed in degrees Fahrenheit (°F). It is quite easy to convert a temperature from one system to the other. The mathematics are so simple that there is no reason to use the formal dimensional-analysis method to solve the problem.

In the English (U.S.) system, the temperature of freezing water is taken to be 32°F. The temperature of boiling water is assigned the value of 212°F. In the metric system, the temperature at which water freezes is taken to be 0°C, and the temperature at which it boils is assigned the value of 100°C.

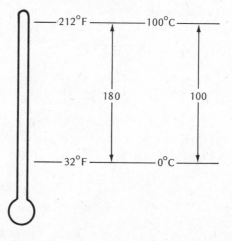

As you see, each °F is smaller than each °C. It takes 180 Fahrenheit degrees to equal only 100 centigrade (Celsius) degrees.

$$\frac{°F}{°C} = \frac{180}{100} \quad \text{or, simplifying,} \quad \frac{°F}{°C} = \frac{9}{5} \quad \text{or} \quad °F = \frac{9}{5} \, °C$$

When you convert from a temperature in one system to a temperature in the other, you use this conversion factor to find by how many degrees the temperature in which you are interested differs from the freezing point of water. A temperature of 55°C is 55 Celsius degrees away from the freezing point of water. Remember that a temperature of 55°F is only 23 Fahrenheit degrees away from the freezing point of water, since water freezes at 32°F.

Therefore, in making a conversion from one scale to the other, you always use the number of degrees away from this reference point, the freezing point of water.

Number of Fahrenheit degrees from reference point = temp. °F – 32

Number of Celsius degrees from reference point = temp. °C

To convert temperatures, use the relationship

$$\frac{\text{Fahrenheit degrees (°F)}}{\text{Celsius degrees (°C)}} = \frac{9}{5}$$

Substituting the degrees away from the freezing point of water, you have

$$\frac{\text{Temp. °F – 32}}{\text{Temp. °C}} = \frac{9}{5}$$

If you want to determine the temperature in °C, you use the form

$$\text{Temp. °C} = \left(\frac{5}{9}\right)(\text{temp. °F – 32})$$

And if you want to determine the temperature in °F, you use the form

$$\text{Temp. °F} = \left(\frac{9}{5}\right)(\text{temp. °C}) + 32$$

Look at problem 12 in the pretest. Find the temperature in °C corresponding to 65°F. Using the above form, you have

$$°C = \frac{5}{9}(°F – 32) = \frac{5}{9}(65 – 32) = \frac{5}{9}(33) = 18°C$$

Problem 13 asks: What is the temperature in °F corresponding to –14°C? Using the given form,

$$°F = \frac{9}{5}°C + 32 = \frac{9}{5}(-14) + 32 = -25 + 32 = 7°F$$

(If you have difficulties with algebraic addition of negative numbers, see the subsection on this topic in Section A in Chapter 1. Also, if you got 6.8°F for the last problem, you are not paying attention to the significant figures. Check the addition of significant figures in Section C, Chapter 1.)

There is one other temperature scale that you will use in gas-law and thermodynamic calculations. This is an absolute temperature, expressed in degrees Kelvin (°K). The size of the degree is identical with the size of a degree Celsius, but the freezing point of water is at 273°K. To get a temperature in °K, you simply add 273 to the temperature in °C. Thus the boiling point of water (100°C) would equal _____°K. We will return to the Kelvin scale in the chapter on gas laws.

[The only SI unit for temperature is the *Kelvin* (with no degree sign). Its symbol is just K. We will use both K and °K for absolute temperature in this text, but we'll use the SI K more often.]

You are now ready to do the problem set on the metric system.

373

PROBLEM SET

As you do these problems, immediately check your answer. Pages 74–76 give the setup and the numerical answers. The answers are expressed in scientific notation, to the correct number of significant figures. The entire set should take not more than 50 minutes to complete. The somewhat harder problems are marked with an asterisk (*).

1. How many millimeters are there in 3.8 m?

2. How many grams are there in 4.5×10^9 μg?

3. An 8.2 kg sample contains how many grams?

4. How many milliliters are in a 9.5×10^{16} μL sample?

5. Express 4×10^{11} Å in kilometers.

6. How many picometers in 7.4 μm?

7. A cube 8 cm on each side contains how many liters?

You may need the following conversion factors.

$\dfrac{2.54 \text{ cm}}{\text{in.}}$	$\dfrac{12 \text{ in.}}{\text{ft}}$	$\dfrac{454 \text{ g}}{\text{lb}}$	$\dfrac{16 \text{ oz}}{\text{lb}}$	$\dfrac{946 \text{ mL}}{\text{qt}}$
$\dfrac{1.61 \text{ km}}{\text{mi}}$	$\dfrac{5280 \text{ ft}}{\text{mi}}$	$\dfrac{2.2 \text{ lb}}{\text{kg}}$	$\dfrac{2000 \text{ lb}}{\text{ton}}$	$\dfrac{4 \text{ qt}}{\text{gal}}$

8. How many centimeters in 3.0 mi?

9. How many gallons in 43 L?

10. How many inches in 6.0 m?

11. How many ounces in 8.5 kg?

12. A square that is 2.5 m on a side has an area of how many square inches?

13. What is the cost of coal in dollars per ton if it costs $0.04/kg?

14. How many miles can you drive for $3.90 if the gas mileage of your car is 14 km/L of gas and the price is $1.34/gal?

15. Suppose that 6.0 L of mercury weigh 78 kg and cost $420. What is the price of mercury in dollars per pound?

16. A crane can pick up 3.0 tons of excavated earth in an hour. The crane operator's wages are $25 per hour. What is the cost of picking up 85 kg of excavated earth?

17. What would it cost to send a rocket to Mars, which is 7.9×10^7 km distant from Earth, if the average speed of the rocket is 1600 mi/min and the consumption of solid propellant, which costs $500/lb, averages 100 g every 1.5 sec?

18. What would it cost to inflate a weather balloon 14 ft in diameter with helium if the price of helium is $5.00/lb and 22.4 L of helium weigh 4.0 g? The volume of a sphere is

$$\text{Volume} = \frac{1}{6}\pi d^3$$

where d is the diameter and π (one of the few dimensionless numbers) equals 3.1416.

19. If you dug a hole through the earth to China, how many centuries would it be before you got there if you could dig at the rate of 4 mi per day? The diameter of the earth is 12,700 km.

20. A temperature of $-9°F$ is equivalent to what temperature in $°C$?

21. A temperature of $4500°C$ is equivalent to what temperature in $°F$?

22. A temperature of $-15°C$ is equivalent to what temperature in K?

23. A temperature of $80°F$ is equivalent to what temperature in K?

24. When the United States shifts to the metric system, a high fever of $103.0°F$ will be given as how many degrees Celsius?

*25. It requires 1 calorie (cal) to heat 1 g of water $1°C$. (*Hint:* This is a conversion factor with two units in the denominator, $1 \text{ cal}/[(g)(°C)]$.) How many calories are needed to heat 1.00 qt of water from $50°F$ to $140°F$? A gallon of water weighs 4.30 lb.

26. The smallest particle of an element is an atom. How many atoms of oxygen are there in 10.0 lb of oxygen? One atom weighs 2.68×10^{-23} g.

*27. Suppose that the oxygen atom is a sphere with a diameter of 1.22 Å. What is its volume in milliliters? Recall that the volume of a sphere is

$$\frac{1}{6}\pi d^3$$

where d is the diameter and π is 3.1416.

28. A piece of toast provides you with 50 kcal (kilocalories) of energy. How much energy, in joules (J), is in a gram of bread, if a one-pound loaf of bread contains 18 slices and there are 4.184 joules per calorie (cal)?

PROBLEM SET ANSWERS

Grade yourself on the problems in the following way: If you missed more than one problem in group 1–6, you had better redo Section A. If you missed three or more problems in 7–18, redo Section B. Problems 20–24 are covered in the short section on temperature, and you should get them all correct. If you did not, redo Section C.

Problems 25–28 are a little harder, involving a few units of heat.

1. $? \text{ mm} = 3.8 \text{ m} \times \dfrac{\text{mm}}{10^{-3} \text{ m}} = 3.8 \times 10^3 \text{ mm}$

2. $? \text{ g} = 4.5 \times 10^9 \text{ μg} \times \dfrac{10^{-6} \text{ g}}{\text{μg}} = 4.5 \times 10^3 \text{ g}$

3. $? \text{ g} = 8.2 \text{ kg} \times \dfrac{10^3 \text{ g}}{\text{kg}} = 8.2 \times 10^3 \text{ g}$

4. $? \text{ mL} = 9.5 \times 10^{16} \text{ μL} \times \dfrac{10^{-6} \text{ L}}{\text{μL}} \times \dfrac{\text{mL}}{10^{-3} \text{ L}} = 9.5 \times 10^{13} \text{ mL}$

5. $? \text{ km} = 4 \times 10^{11} \text{ Å} \times \dfrac{10^{-10} \text{ m}}{\text{Å}} \times \dfrac{\text{km}}{10^3 \text{ m}} = 4 \times 10^{-2} \text{ km}$

6. $? \text{ pm} = 7.4 \text{ μm} \times \dfrac{10^{-6} \text{ m}}{\text{μm}} \times \dfrac{\text{pm}}{10^{-12} \text{ m}} = 7.4 \times 10^6 \text{ pm}$

7. $? \text{ L} = 512 \text{ cm}^3 \times \dfrac{\text{mL}}{\text{cm}^3} \times \dfrac{10^{-3} \text{ L}}{\text{mL}} = 5.12 \times 10^{-1} \text{ L}$

8. $? \text{ cm} = 3.0 \text{ mi} \times \dfrac{1.61 \text{ km}}{\text{mi}} \times \dfrac{10^3 \text{ m}}{\text{km}} \times \dfrac{\text{cm}}{10^{-2} \text{ m}} = 4.8 \times 10^5 \text{ cm}$

9. $? \text{ gal} = 43 \text{ L} \times \dfrac{\text{mL}}{10^{-3} \text{ L}} \times \dfrac{\text{qt}}{946 \text{ mL}} \times \dfrac{\text{gal}}{4 \text{ qt}} = 11 \text{ gal}$

10. $? \text{ in.} = 6.0 \text{ m} \times \dfrac{\text{cm}}{10^{-2} \text{ m}} \times \dfrac{\text{in.}}{2.54 \text{ cm}} = 2.4 \times 10^2 \text{ in.}$

11. $? \text{ oz} = 8.5 \text{ kg} \times \dfrac{2.2 \text{ lb}}{\text{kg}} \times \dfrac{16 \text{ oz}}{\text{lb}} = 3.0 \times 10^2 \text{ oz}$

12. $? \text{ in}^2 = 6.25 \text{ m}^2 \times \dfrac{\text{cm}^2}{10^{-4} \text{ m}^2} \times \dfrac{\text{in}^2}{(2.54)^2 \text{ cm}^2} = 9.69 \times 10^3 \text{ in}^2$

13. $? \text{ dollars} = 1 \text{ ton} \times \dfrac{2000 \text{ lb}}{\text{ton}} \times \dfrac{454 \text{ g}}{\text{lb}} \times \dfrac{\text{kg}}{10^3 \text{ g}} \times \dfrac{\$0.04}{\text{kg}} = \$36.32$

If you are having difficulties, it would be a good idea, before you start, to write down all the conversion factors given in the problems.

14. $? \text{ mi} = \$3.90 \times \dfrac{\text{gal}}{\$1.34} \times \dfrac{4 \text{ qt}}{\text{gal}} \times \dfrac{946 \text{ mL}}{\text{qt}} \times \dfrac{10^{-3} \text{ L}}{\text{mL}} \times \dfrac{14 \text{ km}}{\text{L}} \times \dfrac{1 \text{ mile}}{1.61 \text{ km}}$

$= 9.6 \times 10^1 \text{ miles}$

15. ? dollars = $1 \text{ lb} \times \dfrac{454 \text{ g}}{\text{lb}} \times \dfrac{\text{kg}}{10^3 \text{ g}} \times \dfrac{\$420}{78 \text{ kg}} = \$2.44$

      = \$2.4 to the correct number of significant figures

16. ? dollars = $85 \text{ kg} \times \dfrac{2.2 \text{ lb}}{\text{kg}} \times \dfrac{\text{ton}}{2000 \text{ lb}} \times \dfrac{\text{hr}}{3.0 \text{ ton}} \times \dfrac{\$25}{\text{hr}} = \$0.78$

17. ? dollars = $7.9 \times 10^7 \text{ km} \times \dfrac{\text{mi}}{1.61 \text{ km}} \times \dfrac{\text{min}}{1600 \text{ mi}} \times \dfrac{60 \text{ sec}}{\text{min}} \times \dfrac{100 \text{ g}}{1.5 \text{ sec}}$

      $\times \dfrac{\text{lb}}{454 \text{ g}} \times \dfrac{\$500}{\text{lb}} = \$1.4 \times 10^8$ (140 million dollars!)

18. Volume = $\dfrac{1}{6}\pi(14 \text{ ft})^3 = 1.4 \times 10^3 \text{ ft}^3$

    ? dollars = $1.4 \times 10^3 \text{ ft}^3 \times \dfrac{(12)^3 \text{ in}^3}{\text{ft}^3} \times \dfrac{(2.54)^3 \text{ cm}^3}{\text{in}^3} \times \dfrac{\text{mL}}{\text{cm}^3}$

      $\times \dfrac{10^{-3} \text{ L}}{\text{mL}} \times \dfrac{4.0 \text{ g}}{22.4 \text{ L}} \times \dfrac{\text{lb}}{454 \text{ g}} \times \dfrac{\$5.00}{\text{lb}}$

      $= 1.4 \times 10^3 \times 1728 \times 16.39 \times 1 \times 10^{-3} \times \dfrac{4.0}{22.4} \times \dfrac{1}{454} \times \$5.00$

      $= \$7.8 \times 10^1$ (or \$78)

19. ? century = $12{,}700 \text{ km} \times \dfrac{\text{mi}}{1.61 \text{ km}} \times \dfrac{\text{day}}{4 \text{ mi}} \times \dfrac{\text{yr}}{365 \text{ days}} \times \dfrac{\text{century}}{100 \text{ yr}}$

      $= 5 \times 10^{-2}$ century

(This problem is so easy that it should have been a rest after doing problem 18. It is, however, at about the general level of difficulty of the problems you will face in chemistry.)

20. $°C = \dfrac{5}{9}(-9 - 32) = \dfrac{5}{9}(-41) = -23°C$

21. $°F = \dfrac{9}{5}(4500) + 32 = 8100 + 32 = 8132°F$

22. $K = -15 + 273 = 258 \text{ K}$

23. $°C = \dfrac{5}{9}(80 - 32) = \dfrac{5}{9}(48) = 27°C; \ K = 27 + 273 = 300 \text{ K}$

24. $°C = \dfrac{5}{9}(103.0 - 32) = 39.4°C$

25. You first determine how many degrees Celsius the water must be heated.

      $°C = \dfrac{5}{9}(140° - 32) = \dfrac{5}{9}(108) = 60.0°C$

      $°C = \dfrac{5}{9}(50 - 32) = \dfrac{5}{9}(18) = 10.0°C$

    $60°C - 10°C = 50.0°C$

    Then

$$? \text{ cal} = 1 \, \cancel{qt} \times \frac{1 \, \cancel{gal}}{4 \, \cancel{qt}} \times \frac{4.30 \, \cancel{lb}}{\cancel{gal}} \times \frac{454 \, \cancel{g}}{\cancel{lb}} \times \frac{1 \, \text{cal}}{\cancel{(g)(°C)}} \times 50.0°\cancel{C} = 2.44 \times 10^4 \text{ cal}$$

26. $? \text{ atoms} = 10 \, \cancel{lb} \times \frac{454 \, \cancel{g}}{\cancel{lb}} \times \frac{1 \text{ atom}}{2.68 \times 10^{-23} \, \cancel{g}} = 1.69 \times 10^{26} \text{ atoms}$

27. $\text{Volume} = \frac{1}{6}\pi d^3 = \frac{1}{6}(3.1416)(1.22 \, \text{Å})^3 = 0.951 \, \text{Å}^3$

$$? \text{ mL} = 0.951 \, \cancel{Å^3} \times \frac{(10^{-8})^3 \, \cancel{cm^3}}{\cancel{Å^3}} \times \frac{\text{mL}}{\cancel{cm^3}}$$

$$= 0.951 \times 10^{-24} \text{ mL} = 9.51 \times 10^{-25} \text{ mL}$$

(Notice how you go from 10^{-8} cm/Å to $(10^{-8})^3$ cm^3/Å^3. You cube each part separately.)

28. $? \text{ J} = 1.0 \text{ g}\,\cancel{bread} \times \frac{\text{lb}\,\cancel{bread}}{454 \text{ g}\,\cancel{bread}} \times \frac{18 \, \cancel{slices}}{\text{lb}\,\cancel{bread}} \times \frac{50 \, \cancel{kcal}}{\cancel{slice}} \times \frac{10^3 \, \cancel{cal}}{\cancel{kcal}} \times \frac{4.184 \text{ J}}{\cancel{cal}}$

$$= 8.3 \times 10^3 \text{ J}$$

CHAPTER 4

Density, Specific Gravity, and Percentage

In this chapter we will examine three conversion factors—density, specific gravity, and percentage. *Density* is a relationship between the mass of a substance and its volume. It has the dimensions of grams per milliliter or grams per liter (gases). *Specific gravity* works out to be the ratio of the density of a substance to the density of water, $d(X)/d(H_2O)$. The *percentage* is the parts of a substance per 100 total parts of the system in which the substance is contained. Using these conversions, try the pretest below. It should take no longer than 20 minutes.

PRETEST

1. What is the density of benzene in grams per milliliter if 5.5 L of benzene weigh 4.8 kg?
2. The density of air is 1.3 g/L. How many liters of air weigh 2.0 lb?
3. Given that the density of mercury is 13.6 g/mL, how many liters will 800 g of mercury occupy?
4. A cube of platinum that is 10 cm on each edge weighs 21 kg. What is the density of platinum?
5. What is the specific gravity of a solution of sulfuric acid, if 100 mL of solution weighs 150 g? The density of water at the temperature of interest is 0.996 g/mL.
6. A sample of calcite is 87.0% $CaCO_3$ by weight. How many grams of calcite are required to yield 125 g pure $CaCO_3$?
7. How many grams of iron are in 600 g iron ore if the iron ore is known to be 35.0% iron by weight?
8. A solution of HCl in water has a density of 1.1 g/mL. What is the weight percent of HCl in the solution if 150 mL of the solution contains 30 g HCl?
9. What is the percent purity of a sample of potassium permanganate if 21.0 g of the sample react with exactly 35.0 mL of a sodium oxalate solution? It is known that 10.0 mL of the sodium oxalate solution react with 5.70 g pure potassium permanganate.
10. How many grams of NaCl are in 500 mL of a solution that is 26.0 weight percent NaCl and has a density of 1.20 g/mL?

PRETEST ANSWERS

If you got nine out of the ten problems correct, you can go directly to the problem set at the end of this chapter. If you did problems 1 through 4 correctly, you can skip Section A, which is about density. If you got problem 5 correct and feel confident about specific gravity, you can skip Section B, which is on this topic. Problems 6 and 7 are concerned with simple density calculations. If you did those problems correctly, you know how to handle what is in Section C. Problems 8 through 10 are more complicated. They involve either percent purity or a combination of density and percentage. If you missed these, it would be a good idea to carefully read the end of Section C.

Many of these problems are worked out as examples in the body of the chapter.

1. 0.87 g/mL
2. 7.0×10^2 L
3. 5.88×10^{-2} L
4. 21 g/mL
5. 1.51
6. 144 g calcite
7. 210 g iron
8. 18%
9. 95%
10. 156 g NaCl

SECTION A Density

The density of a substance is defined as the mass per unit volume. Using the metric units of mass (grams) and volume (milliliters), we can say that the density is the number of grams that equal 1 mL. This equality is, of course, a conversion factor.

$$\text{Density} = \frac{\text{weight in grams}}{1 \text{ milliliter}}$$

$\dfrac{g}{mL}$

per

The dimensions of this factor will be _____ , or, as we usually say, density is the number of grams _____ milliliter.

We use the units g/mL for densities of liquids and solids. Very precise people use g/cm^3 for solids. However, the densities of gases are so small that it is more convenient to express them in grams per liter (g/L). Thus,

liter

for gases, the density is the weight in grams per _____ .

To determine the density of a substance, it is necessary to know only the weight of a known volume. For example, if a 20 mL sample of alcohol weighs 16 g, what is the density of alcohol? To be consistent, we can reword the question to ask how many grams 1 mL weighs. Now the

How many grams?

question asked is "_____" The information

1 mL

given is _____ . And in the problem, you are given the conversion factor that there are 20 mL per 16 g. Setting this up in the usual way, you have

g; mL; $\dfrac{16 \text{ g}}{20 \text{ mL}}$

? __ = _____ X _____

All dimensions except grams cancel, so the final answer is 0.80 g for
1 mL or 0.80 g/mL.

It is possible in this particular calculation to take a short cut. Since
density is always per 1.0 volume unit (milliliters or liters), you could say
that the question asked is "How many grams per milliliter?" So the dimensions of the answer will be grams per milliliter (g/mL).

$$? \frac{g}{mL} =$$

Then the conversion factor given in the problem is in the dimensions
of the answer, so you simply put it in:

$$? \frac{g}{mL} = \frac{16\ g}{20\ mL} = 0.80\ \frac{g}{mL}$$

The dimensions are exactly those of the answer.

Try one this way. What is the density of oxygen gas, if 11.2 L weigh
16.0 g? Since you want density as your answer, the dimensions of the
answer will be

$$? \underline{\hspace{2cm}} =$$

(*Note:* It is per liter because oxygen is a gas.) The conversion factor given

in the problem is then _____.

Putting this in, you see that you have exactly the dimensions of the
answer.

$$? \frac{g}{L} = \frac{16.0\ g}{11.2\ L} = 1.43\ \frac{g}{L}$$

This type of problem can become more complicated only when the
weight and/or volume are given in dimensions that must be converted to
grams and milliliters (or liters). However, since you are now expert at
converting metric and English (U.S.) dimensions, this should be very
easy.

Try one. What is the density in grams per milliliter of antifreeze, given

that 1.0 qt weighs 2.3 lb? The dimensions of the answer will be _____,

so you write ? _____ =

The conversion factor given in the problem is _____.
Therefore

$$? \frac{g}{mL} = \underline{\hspace{3cm}}$$

Now you must multiply by conversion factors that cancel the
unwanted dimensions and replace them with the dimensions of the
answer. The conversion factor relating pounds to grams is _____.

The conversion factor relating quarts to milliliters is _____. (If

---

$\frac{g}{L}$

$\frac{16.0\ g}{11.2\ L}$

$\frac{g}{mL}$

$\frac{g}{mL}$

$\frac{2.3\ lb}{qt}$

$\frac{2.3\ lb}{qt}$

$\frac{454\ g}{lb}$

$\frac{qt}{946\ mL}$

you have forgotten these conversion factors, check Section B, Chapter 3.) Multiplying by both these factors, you get

$$? \frac{g}{mL} = \frac{2.3 \, lb}{qt} \times \frac{454 \, g}{lb} \times \frac{qt}{946 \, mL} = 2.3 \times 454 \times \frac{1}{946} \frac{g}{mL}$$

This leaves only the dimensions of the answer, which is then

$$? \frac{g}{mL} = 1.1 \frac{g}{mL}$$

Incidentally, it is very common to write $\frac{g}{mL}$ as g/mL. This saves space, since it gets everything on one line.

Now we shall see how you use density as a conversion factor, which it really is. It converts weight to volume. Here is a problem. The density of carbon tetrachloride is 1.5 g/mL. What does 50 mL of carbon tetrachloride weigh? The question is: What does it weigh? So the dimensions

of the answer will be _____ . The information given is _____, and the conversion factor in the problem is the density, _____. Writing these in the usual way, you have

$$? \_\_ = _____ \times _____$$

The only dimension that does not cancel is _____, so the answer is

_____.

It is very important that, besides just filling in the blanks, you write down each conversion factor as you get it in your own setup. Only by doing this will you become expert at setting up these problems.

For a slightly more complicated example, let's do problem 2 in the pretest. The density of air is 1.3 g/L. How many liters of air weigh 2.0 lb?

The question asked is "_____" and so the dimensions of the answer will be _____. The information given is _____, and the conversion factor in the problem is the density, which is _____. Starting in the usual way, you write

$$? \_\_ = _____$$

The dimension of the information that you must cancel is _____. However, the dimension for weight in the conversion factor in the problem is _____. Thus, before you can use the density, you must convert _____ to _____. You know the conversion factor that will do this: _____.

The remaining unwanted dimension is _____, but this can be canceled using the density expressed so that the dimension _____ is in the denominator. You write the conversion factor as _____. Multiplying by

grams; 50 mL
$$\frac{1.5 \, g}{mL}$$

g; 50 mL; $\frac{1.5 \, g}{mL}$

grams

50 × 1.5 g = 75 g

How many liters?

liters

2 lb

1.3 g/L

L; 2.0 lb

pounds

grams

pounds; grams
$$\frac{454 \, g}{lb}$$

grams

grams

$$\frac{L}{1.3 \, g}$$

this factor, the only dimension left is _____, which is the dimension of the answer. Therefore, the answer must be

$$? \text{ L} = 2.0 \text{ lb} \times \frac{454 \text{ g}}{\text{lb}} \times \frac{\text{L}}{1.3 \text{ g}} = 698 \text{ L} = 7.0 \times 10^2 \text{ L}$$

Now try problems 1, 3, and 4 in the pretest. The setups are shown below, but don't look at them until you have done the problems.

Problem 1 $? \dfrac{\text{g}}{\text{mL}} = \dfrac{4.8 \text{ kg}}{5.5 \text{ L}} \times \dfrac{10^3 \text{ g}}{\text{kg}} \times \dfrac{10^{-3} \text{ L}}{\text{mL}} = 0.87 \text{ g/mL}$

Problem 3 $? \text{ L} = 800 \text{ g} \times \dfrac{\text{mL}}{13.6 \text{ g}} \times \dfrac{10^{-3} \text{ L}}{\text{mL}} = 5.88 \times 10^{-2} \text{ L}$

Problem 4   Volume $= (10)^3 \text{ cm}^3 = 10^3 \text{ mL}$

$$? \frac{\text{g}}{\text{mL}} = \frac{21 \text{ kg}}{10^3 \text{ mL}} \times \frac{10^3 \text{ g}}{\text{kg}} = 21 \text{ g/mL}$$

SECTION B   Specific Gravity

Sometimes another measurement of the mass–volume characteristics of a substance is used in preference to density. Such a measure is the *specific gravity* (sp gr), which is the ratio of the weight of the substance to the weight of water that occupies the same volume. Since a density is weight per milliliter, you can see that specific gravity also equals the ratio of the densities. So we can write

$$\text{Specific gravity} = \frac{\text{weight of sample}}{\text{weight of water occupying same volume}}$$

or

$$\text{Specific gravity} = \frac{\text{density of sample}}{\text{density of water}}$$

Water is generally the reference for solids and liquids. For gases the reference is usually to air.

It is often said that specific gravity is a dimensionless quantity, since it is g/g. However, this is not exactly true, since the dimensions are not simply grams; they are grams substance and grams water. It is the old story of pounds of potatoes and pounds of onions in the same problem.

As you probably know, the density of all substances varies with temperature. When you wish to convert from the density of a substance to its specific gravity, you must know at what temperature you have measured it and also what the density of water is at that temperature. The densities of water at various temperatures are tabulated on the following page to three decimal places. For greater precision, you must consult a handbook.

As you see from the table, the density of water is 1.0 g/mL expressed to two significant figures from a temperature of 0° to 40°C. If the table continued to 100°C, the value would still be 1.0 g/mL. Thus the specific

T (°C)	Density (g/mL)	T (°C)	Density (g/mL)
0°–12°	1.000	27°–30°	0.996
13°–18°	0.999	31°–33°	0.995
19°–23°	0.998	33°–36°	0.994
24°–26°	0.997	37°–39°	0.993

gravity would equal the density to two significant figures. If you require a greater number of significant figures, then you must perform the calculation.

Problem: What is the density of a sample if you know that its specific gravity is 1.13 at 25°C? This is so easy to solve that we need not use dimensional analysis. The relation is

$$\text{Specific gravity} = \frac{\text{density of sample}}{\underline{\hspace{3cm}}}$$

At 25°C the _____ is 0.997 g/mL. So you may write

$$1.13 = \frac{\text{density of sample}}{0.997}$$

Multiplying both sides by 0.997 (that is, cross-multiplying) gives

$$(0.997)(1.13) = 1.12661 = \frac{\cancel{(0.997)}(\text{density of sample})}{\cancel{0.997}} = \text{density of sample}$$

$$1.13 = \text{density of sample to three significant figures}$$

SECTION C   Percentage

Like density, percent (%) is a conversion factor. Recall the Latin prefixes used in naming the metric units. *Cent* means the number _____. So when you say that wine is 14% alcohol, this is a shorthand way of saying that there are 14 parts of alcohol per _____ parts of wine. As a fraction, we write

$$\frac{14 \text{ parts alcohol}}{100 \text{ parts wine}}$$

The percentage is the number of parts of a component in 100 parts of the total substance that contains the component.

The dimensions used to express the number of parts may vary, and you must be careful to note this. Thus a weight percent in the metric system is

$$\frac{\text{grams component}}{100 \text{ g total substance}}$$

Similarly, a volume percent is

$$\frac{\text{liters components}}{\underline{\hspace{3cm}}}$$

(margin answers)

density of water

density of water

100

100

100 L total substance

or also

milliliters component

_____

For weight percent, you could also have

_____
100 kg total substance

The unit in the numerator must be the same as the unit in the denominator. (There are some mixed-dimension percentages, but these are so unscientific and confusing that you should never use them.)

Here is a simple problem. What is the weight percent of salt in sea water if there are 5.0 g salt in 125 g sea water? You must reword the question to ask how many _____ of salt there are per _____ sea water. The dimensions of the answer will be _____, and the information given is _____. There is one conversion factor in the problem: _____. So you set the problem up in the usual way.

? _____ = _____

Multiplying this by the conversion factor given in the problem cancels the unwanted dimension, _____, and leaves only the dimensions of the answer, which must be

$$? \text{ g salt} = 100 \cancel{\text{ g sea water}} \times \frac{5.0 \text{ g salt}}{125 \cancel{\text{ g sea water}}} = 4.0 \text{ g salt}$$

You now know that there are 4.0 g salt per 100 g sea water, so the sea water must be 4.0% salt.

When you are doing percentage problems, the unit used in the numerator is always the same as that in the denominator. Therefore you must be very careful to specify what you are talking about. Thus you must write grams salt and grams sea water. It is not enough simply to write grams.

The problems can become a little more complicated when the dimensions of the conversion factor are not those of the question asked and the information given. This shouldn't be too much of a chore to convert. Consider the problem: What is the volume percent of fat in milk if 25 gal of milk contain 11 L of fat? First you reword the problem to ask how many milliliters of fat per _____ milk. The question asked is

"_____" The information now given is

_____. The conversion factor given in the problem is _____. Starting in the usual way, you write

? _____ = _____

However, the volume dimension given in the conversion factor for milk

---

Margin answers:

100 mL total substance

kilograms component

grams; 100 g

grams salt

100 g sea water
$$\frac{5.0 \text{ g salt}}{125 \text{ g sea water}}$$

g salt; 100 g sea water

grams sea water

100 mL

How many milliliters fat?

100 mL milk; $\dfrac{11 \text{ L fat}}{25 \text{ gal milk}}$

mL fat; 100 mL milk

gallons

is _____. So you must first convert the dimensions of the information given, 100 mL milk, to gallons. It will take two conversion factors to do this. You have

$$? \text{ mL fat} = 100 \text{ mL milk} \times \frac{\text{qt milk}}{946 \text{ mL milk}} \times \frac{\text{gal milk}}{4 \text{ qt milk}}$$

Now you can multiply by the conversion factor given in the problem.

$$? \text{ mL fat} = 100 \text{ mL milk} \times \frac{\text{qt milk}}{946 \text{ mL milk}} \times \frac{\text{gal milk}}{4 \text{ qt milk}} \times \frac{11 \text{ liters fat}}{25 \text{ gal milk}}$$

liters fat

milliliters fat

$$\frac{\text{mL fat}}{10^{-3} \text{ L fat}}$$

The unwanted dimension is now _____, and the dimensions of the answer will be _____. This is a very simple conversion. You multiply by the conversion factor _____.

$$? \text{ mL fat} = 100 \text{ mL milk}$$

$$\times \frac{\text{qt milk}}{946 \text{ mL milk}} \times \frac{\text{gal milk}}{4 \text{ qt milk}} \times \frac{11 \text{ L fat}}{25 \text{ gal milk}} \times \frac{\text{mL fat}}{10^{-3} \text{ L fat}}$$

$$= 12 \text{ mL fat in 100 mL milk} = 12\%$$

Although the last problem looks a little messy, it really isn't. You just keep multiplying by conversion factors to get from the dimensions of the information given, through the conversion factor given in the problem (the "bridge"), to the dimensions of the answer.

Look at problem 8 in the pretest. A solution of HCl in water has a density of 1.1 g/mL. What is the weight percent (wt %) of HCl in the solution if 150 mL of the solution contains 30 g HCl? First you reword

grams HCl; 100 g solution (soln)

the question to ask "How many _____ are in _____?"

There are two conversion factors given in the problem. The first is the density of the solution. The conversion factor is (writing it exactly, so

$$\frac{1.1 \text{ g soln}}{\text{mL soln}}$$

that you don't confuse grams solution with grams HCl) _____.

$$\frac{30 \text{ g HCl}}{150 \text{ mL soln}}$$

The second conversion factor is the "bridge." It is _____.

Now you start the problem in the usual way.

g HCl; 100 g soln

$$? \text{_____} = \text{_____}$$

grams soln

The unwanted dimension is _____, so you look at your conversion factors and find one that has this dimension in the denomi-

$$\frac{\text{mL soln}}{1.1 \text{ g soln}}$$

nator. The one that does is _____. Multiply. The

milliliters soln

$$\frac{30 \text{ g HCl}}{150 \text{ mL soln}}$$

unwanted dimension is now _____. So you can multiply

by the conversion factor _____. The only dimension remaining now is grams HCl, which is the dimension of the answer. Therefore the answer must be

$$? \text{ g HCl} = 100 \text{ g soln} \times \frac{\text{mL soln}}{1.1 \text{ g soln}} \times \frac{30 \text{ g HCl}}{150 \text{ mL soln}}$$

$$= 18 \text{ g HCl in 100 g soln} = 18\%$$

Notice that once again you must be very careful to write not only the unit (grams or milliliters), but also what it is a unit *of*. The density is grams per milliliter of *solution*, so you must write grams solution per milliliter solution. If you don't, in this case you will end up confusing grams HCl with grams solution.

Percent Purity

One percentage that is often of interest to chemists is the percent purity of a substance. It is almost always expressed as a weight percent, since volumes vary with temperature. Therefore, if you wish to know the percent purity of an impure sample, you must determine how many grams pure component are in 100 g impure sample. There is a danger of confusing the grams of pure substance with grams sample, because of the way the problem is often worded. For example, "What is the percent purity of an iron ore sample if . . .?" In such a case you must carefully distinguish between impure sample and pure iron. The best way to do this is always to refer to the total sample as *impure* (imp). You can re-word the question to ask how many grams iron are in 100 g imp.

Here is an example. What is the percent purity of a sample of gold ore given that 2.0 kg ore yield 90 g gold metal? The reworded question is

"How many _____ are in _____?"

grams gold; 100 g imp

The conversion factor ("bridge") given in the problem is _____.
Starting in the usual way, you write

$$\dfrac{90 \text{ g gold}}{2.0 \text{ kg imp}}$$

? _____ = _____

g gold; 100 g imp

The unwanted dimension, which must be canceled, is _____.
The dimension of weight of impure sample in the "bridge" is

grams impure

_____. Consequently, in order to be able to use the

kilograms impure

bridge, you must convert _____ to

grams impure

_____. This is easily done using the conversion factor

kilograms impure

_____ . Now you can multiply by the conversion factor. This

$$\dfrac{\text{kg imp}}{10^3 \text{ g imp}}$$

leaves, as the only dimension, _____, which is the dimension of the answer. Therefore the final answer will be

grams gold

$$? \text{ g gold} = 100\, \cancel{\text{g imp}} \times \frac{\cancel{\text{kg imp}}}{10^3\, \cancel{\text{g imp}}} \times \frac{90 \text{ g gold}}{2.0\, \cancel{\text{kg imp}}}$$

$$= 4.5 \text{ g gold in } 100 \text{ g imp} = 4.5\%$$

You must have noticed that the "bridge" (the conversion factor) has been mentioned more and more often. It will become increasingly important as you move on to problems that really involve chemistry. The "bridge" is a conversion factor that connects the substance about which the question is asked with the substance in the information given. In solving a problem, you always go from the information given, across the "bridge," to the substance in the answer.

Problem 9 in the pretest is a good example of a chemical-reaction "bridge." What is the percent purity of a sample of potassium permanganate, $KMnO_4$, if 21.0 g of the sample react with exactly 35.0 mL of a sodium oxalate, $Na_2C_2O_4$, solution? It is known that 10.0 mL of the sodium oxalate solution react with 5.70 g pure potassium permanganate.

You want to determine how many grams potassium permanganate are in 100 g impure sample. There are two conversion factors given in the problem. First, 21.0 g impure per 35.0 mL sodium oxalate solution,

$$\frac{21.0 \text{ g imp}}{35.0 \text{ mL sod. ox. soln}}$$

which, written as a fraction, is _____. The second is a "bridge," since it relates sodium oxalate solution to pure potassium

$$\frac{10.0 \text{ mL sod. ox. soln}}{5.70 \text{ g pot. perm.}}$$

permanganate. Written as a fraction, it is _____.

Starting in the usual way, you write

g pot. perm.; 100 g imp

? _____ = _____

grams impure

$$\frac{35.0 \text{ mL sod. ox. soln}}{21.0 \text{ g imp}}$$

The unwanted dimension, which must be canceled, is _____. As you see, the conversion factor _____ contains that dimension. Multiplying by this conversion factor leaves the unwanted

$$\frac{\text{mL sod. ox. soln}}{5.70 \text{ g pot. perm.}}$$

$$\frac{10.0 \text{ mL sod. ox. soln}}{\text{g pot. perm.}}$$

dimension _____. However, this appears in the "bridge."

So next you multiply by _____.

The only dimension left is _____, which is the dimension of the answer. Therefore, the answer must be

$$? \text{ g pot. perm.} = 100 \text{ g imp} \times \frac{35.0 \text{ mL sod. ox. soln}}{21.0 \text{ g imp}}$$

$$\times \frac{5.70 \text{ g pot. perm.}}{10.0 \text{ mL sod. ox. soln}}$$

$$= 95.0 \text{ g pot. perm. in 100 g imp} = 95.0\% \text{ pure}$$

Using Percentage as a Conversion Factor

In problems in which the percentage is given, you can use it as a conversion factor given in the problem. It is the "bridge" between pure component and impure substance.

Look at problem 6 in the pretest. A sample of calcite is 87.0% $CaCO_3$ by weight. How many grams of calcite are required to yield 125 g pure

How many grams of calcite?

125 g $CaCO_3$

$CaCO_3$? The question asked is "_____" The information given is _____. Starting the problem in the usual way, you write

? g calcite; 125 g $CaCO_3$

_____ = _____

$$\frac{100 \text{ g calcite}}{87.0 \text{ g } CaCO_3}$$

The percentage is the conversion factor given in the problem. Writing this as a fraction gives _____. (In this problem, the impure substance has a name, calcite.) Multiplying by this factor, all dimensions cancel

grams calcite

except _____. So the answer must be

$$? \text{ g calcite} = 125 \text{ g } \cancel{\text{CaCO}_3} \times \frac{100 \text{ g calcite}}{87.0 \text{ g } \cancel{\text{CaCO}_3}} = 144 \text{ g calcite}$$

Do problem 7 in the pretest. The setup is shown below, but try it before you check this setup.

$$? \text{ g iron} = 600 \text{ g ore} \times \frac{35.0 \text{ g iron}}{100 \text{ g ore}} = 210 \text{ g iron}$$

Problem 10 in the pretest is a perfect example of how carefully you must label all the dimensions. The problem asks how many grams of NaCl are in 500 mL of a solution that is 26.0 wt % NaCl and has a density of 1.20 g/mL. The dimensions of the answer will be

_____. The information given is _____. The conversion factors given in the problem are

_____ and _____

grams NaCl; 500 mL soln

$\dfrac{26.0 \text{ g NaCl}}{100 \text{ g soln}}$; $\dfrac{1.20 \text{ g soln}}{\text{mL soln}}$

Notice that the density is of the *solution*. Now set up the problem:

$$? \text{ g NaCl} = 500 \text{ mL soln} \times \frac{1.20 \text{ g soln}}{\text{mL soln}} \times \frac{26.0 \text{ g NaCl}}{100 \text{ g soln}}$$

$$= 500 \times 1.20 \times \frac{26.0}{100} \text{ g NaCl} = 156 \text{ g NaCl}$$

The Sum of the Percentages of All Components Equals 100

The percentage of a component equals the number of parts of the component in 100 parts of the total substance. Therefore, if you add all the percentages of all the components, they must equal 100. So if you have a solution made up of only NaCl in water and it is 15% by weight

of NaCl, the weight percent of water must be _____.

85%

Consequently, if you have a substance made up of two components and the percentage of one is given, you have two different conversion factors at your disposal. In the example just given, 15 wt % NaCl in water, the two conversion factors are

_____ and _____

$\dfrac{15 \text{ g NaCl}}{100 \text{ g soln}}$; $\dfrac{85 \text{ g water}}{100 \text{ g soln}}$

If you invert one of these conversion factors and multiply it by the other, you get a combined conversion factor that relates grams NaCl

to grams water. This is _____.

$\dfrac{15 \text{ g NaCl}}{85 \text{ g water}}$

Try one. How many grams of benzene are there in 5.0 g of a 40 wt % solution of naphthalene in benzene? The dimensions of the answer will

be _____. The information given is _____.
The conversion factor given in the problem as it is written is

grams benzene; 5.0 g soln

$\dfrac{40 \text{ g naphthalene}}{100 \text{ g soln}}$

_____.

You see that the conversion factor is about one component, naphthalene, and the dimensions of the answer are about the other component, benzene. You must therefore work out the equivalent

$\dfrac{60 \text{ g benzene}}{100 \text{ g soln}}$

100

conversion factor for benzene. It will be _____, which you get by subtracting 40 from _____.

Now the problem is very simple. Starting in the usual way, you write

g benzene; 5.0 g soln

$\dfrac{60 \text{ g benzene}}{100 \text{ g soln}}$

? _____ = _____

Then you multiply by the conversion factor _____ and all the unwanted dimensions cancel except the dimensions of the answer, which must be

$$? \text{ g benzene} = 5.0 \text{ g soln} \times \frac{60 \text{ g benzene}}{100 \text{ g soln}} = 3.0 \text{ g benzene}$$

PROBLEM SET

After doing each problem, check the answer pages, where you will find the setup as well as the final answer. If you miss a problem, go over the setup carefully to see where you went wrong. The first 20 problems are divided according to the section in the text that applies to them, and even subdivided into types of problems. In each such division the problems get more complicated as you go along. If you don't get a problem, don't go on to the next one until you understand how to do it. The last six problems are from all over, so to do them, you will have to know all the sections.

Section A

1. What is the density of gasoline, if 400 mL weigh 280 g?

2. What is the density of hydrogen gas, if 100 L weigh 8.93×10^{-3} kg?

3. What is the density of milk in g/mL if it is known that 1.00 qt weighs 1.00 kg?

4. What is the density of earth if 1.00 yd^3 weighs 22.0 lb? Express your answer in English (U.S.) units, lb/ft^3.

5. It is believed that dwarf stars may be made up of tightly packed hydrogen nuclei. Calculate the density of a hydrogen nucleus, in tons/cm^3, from the fact that the diameter of the spherical nucleus is 1.0×10^{-5} Å and 6.0×10^{23} nuclei weigh 1.0 g. The volume of a sphere is $1/6\pi d^3$, where d = diameter. (π equals 3.1416.)

6. The density of chloroform is 1.49 g/mL. A 100 g sample of chloroform occupies how many milliliters?

7. How many ounces does a 250 mL sample of sulfuric acid (density = 1.30 g/mL) weigh?

8. The density of brass is 8.0 g/mL. A cube of brass 2.0 in. on an edge will weigh how many grams?

9. The price of gold is $333 per Troy ounce. [This is the weight unit for precious metals; 1.00 Troy oz equals 1.097 avoirdupois (normal) oz.] If you had $1000, how large a cube of gold (in inches cubed) could you buy? The density of gold is 19.3 g/mL.

Section B

10. What is the specific gravity of a liquid whose density is 0.943 g/mL at 25°C? The density of water at this temperature is 0.997 g/mL.

11. What is the density of a metal whose specific gravity is 6.85 at 18°C? The density of water at this temperature is 0.999 g/mL.

12. A chunk of ore weighing 3.05 g when submerged in water displaces 0.642 mL of the water. The density of water at this temperature is 0.998 g/mL. What is the specific gravity of the ore?

Section C

13. What is the weight percent of gold in a sample of gold ore, if it was found that 1.00 g of the ore contains 45 mg of gold?

14. What is the volume percent of oxygen in air, if it is found that 150 L of air contain 48 g of oxygen? The density of oxygen at this temperature is 1.4 g/L.

15. What is the weight percent of sugar in a solution with a density of 1.10 g/mL, if it was found that 1.0 gal of the solution contains 2.0 lb of sugar?

16. What is the percent purity of an alcohol solution with a density of 0.84 g/mL, if 8.0 mL of the solution contain 2.0 mL of alcohol? The density of pure alcohol is 0.79 g/mL.

17. What is the percent purity of a sample of galena (lead sulfide ore), if it was found that 2.5 g liberate 220 mL of H_2S gas on reaction with HCl? It was found that 4.8 g pure lead sulfide liberate 0.45 L H_2S gas under the same conditions.

18. How many grams of tin are contained in a 2.0 lb sample of ore that is 85% pure tin?

19. How many cubic meters of gold-bearing rock that is 3.5 wt % gold must be processed to produce $5 worth of gold? The specific gravity at 10°C of the rock is 11. The price of gold is $304 per avoirdupois (normal) ounce.

20. A mining machine can dig out 15 yd³ of coal per minute and costs $35.00 per hour to operate. The density of the coal is 4.8 g/mL. The coal is 98% pure carbon. How much will it cost to dig out 5.0 tons of carbon? (*Hint:* Before you start, write down all the conversion factors in the problem.)

Mixed Group of Problems

21. A 2.50 lb sample of silver ore is analyzed and found to contain 140 g of pure silver. How would you express this in terms of the percent purity of the ore? Recall that there are 454 g/lb.

22. A sample of California wine states on the label that it is 14 volume percent alcohol. How many grams of the alcohol are in a 3.00 L jug of the wine? The density of alcohol is 0.780 g/mL and the density of the wine is 0.956 g/mL.

23. What is the specific gravity of a motor oil if its density is 0.875 g/mL at 25°C? The density of water at this temperature is 0.996 g/mL.

24. The density of air is found to be 1.21 g/L. How many pounds do three cubic feet (3 ft^3) of air weigh? There are 454 g/lb, 2.54 cm/in., and 12 in./ft.

25. A sample of pyrrhotite (a mineral containing iron and sulfur) weighs 6.37 g when submerged in water and displaces 1.40 mL of water. A 5.00 mL sample of water weighs 4.98 g. What is the specific gravity of the pyrrhotite?

26. Ethylene glycol, which is a chief ingredient in Dacron, is shipped to the manufacturers in 55 gal steel drums. How many pounds of ethylene glycol does one drum contain, if the specific gravity of ethylene glycol is 1.1 and a gallon of water weighs 8.34 lb?

PROBLEM SET ANSWERS

1. $? \dfrac{g}{mL} = \dfrac{280\ g}{400\ mL} = 0.700\ \dfrac{g}{mL}$

2. $? \dfrac{g}{L} = \dfrac{8.93 \times 10^{-3}\ \cancel{kg}}{100\ L} \times \dfrac{10^3\ g}{\cancel{kg}} = 8.93 \times 10^{-2}\ \dfrac{g}{L}$ (per liter because it is a gas)

3. $? \dfrac{g}{mL} = \dfrac{1.00\ \cancel{kg}}{1.00\ \cancel{qt}} \times \dfrac{10^3\ g}{\cancel{kg}} \times \dfrac{\cancel{qt}}{946\ mL} = 1.06\ \dfrac{g}{mL}$

4. $? \dfrac{lb}{ft^3} = \dfrac{22.0\ lb}{1.00\ \cancel{yd^3}} \times \dfrac{\cancel{yd^3}}{(3)^3\ ft^3} = 0.815\ \dfrac{lb}{ft^3}$

5. Volume $= \dfrac{\pi}{6}(1.0 \times 10^{-5}\ Å)^3 = 5.2 \times 10^{-16}\ Å^3$ per nucleus

   Rewording the question to ask "How many tons equal 1 cm³," you get

   $? \text{ tons} = \cancel{cm^3} \times \dfrac{\cancel{Å^3}}{(10^{-8})^3\ \cancel{cm^3}} \times \dfrac{\cancel{nucleus}}{5.2 \times 10^{-16}\ \cancel{Å^3}} \times \dfrac{1.0\ \cancel{g}}{6.0 \times 10^{23}\ \cancel{nuclei}}$

   $\times \dfrac{\cancel{lb}}{454\ \cancel{g}} \times \dfrac{ton}{2000\ \cancel{lb}}$

   $= 1 \times \dfrac{1}{10^{-24}} \times \dfrac{1}{5.2 \times 10^{-16}} \times \dfrac{1.0}{6.0 \times 10^{23}} \times \dfrac{1}{454} \times \dfrac{ton}{2000}$

   $= 3.5 \times 10^9\ \text{tons/cm}^3$

   (If you are able to do this last problem, you are doing well!)

6. $? mL = 100\ \cancel{g} \times \dfrac{mL}{1.49\ \cancel{g}} = 67.1\ mL$

7. $? oz = 250\ \cancel{mL} \times \dfrac{1.30\ \cancel{g}}{\cancel{mL}} \times \dfrac{\cancel{lb}}{454\ \cancel{g}} \times \dfrac{16\ oz}{\cancel{lb}} = 11.5\ oz$

8. Volume $= (2.0)^3\ in^3 = 8.0\ in^3$

   $? g = 8.0\ \cancel{in^3} \times \dfrac{(2.54)^3\ \cancel{cm^3}}{\cancel{in^3}} \times \dfrac{\cancel{mL}}{\cancel{cm^3}} \times \dfrac{8.0\ g}{\cancel{mL}} = 1.0 \times 10^3\ g$

9. $? in^3 = \cancel{\$1000} \times \dfrac{\cancel{oz\ Troy}}{\cancel{\$333}} \times \dfrac{1.097\ \cancel{oz}}{\cancel{oz\ Troy}} \times \dfrac{\cancel{lb}}{16\ \cancel{oz}} \times \dfrac{454\ \cancel{g}}{\cancel{lb}} \times \dfrac{\cancel{mL}}{19.3\ \cancel{g}} \times \dfrac{\cancel{cm^3}}{\cancel{mL}}$

   $\times \dfrac{in^3}{(2.54)^3\ \cancel{cm^3}} = 0.296\ in^3$

   (If you were able to do this one, you are doing *very* well!)

10. Specific gravity $= \dfrac{0.943\ g/mL}{0.997\ g/mL} = 0.946$

11. $6.85 = \dfrac{density}{0.999}$; $6.84\ g/mL = density$

12. Density ore $\dfrac{g}{mL} = \dfrac{3.05 \text{ g}}{0.642 \text{ mL}} = 4.75$ g/mL

(The volume of displaced water equals the volume of the submerged sample.)

$\text{sp gr} = \dfrac{4.75 \text{ g/mL}}{0.998 \text{ g/mL}} = 4.76$

13. ? g gold $= 100 \text{ g imp} \times \dfrac{45 \text{ mg gold}}{1.00 \text{ g imp}} \times \dfrac{10^{-3} \text{ g gold}}{\text{mg gold}} = 4.5$ g gold in 100 g imp

$= 4.5\%$

14. ? L oxygen $= 100 \text{ L air} \times \dfrac{48 \text{ g oxygen}}{150 \text{ L air}} \times \dfrac{\text{L oxygen}}{1.4 \text{ g oxygen}}$

$= 23$ L oxygen in 100 L air $= 23\%$

15. ? g sugar $= 100 \text{ g soln} \times \dfrac{\text{mL soln}}{1.10 \text{ g soln}} \times \dfrac{\text{qt soln}}{946 \text{ mL soln}} \times \dfrac{\text{gal soln}}{4 \text{ qt soln}}$

$\times \dfrac{2.0 \text{ lb sugar}}{\text{gal soln}} \times \dfrac{454 \text{ g sugar}}{\text{lb sugar}}$

$= 100 \times \dfrac{1}{1.10} \times \dfrac{1}{946} \times \dfrac{1}{4} \times 2.0 \times 454$ g sugar

$= 22$ g sugar in 100 g soln $= 22\%$

16. ? g alcohol $= 100 \text{ g soln} \times \dfrac{\text{mL soln}}{0.84 \text{ g soln}} \times \dfrac{2.0 \text{ mL alcohol}}{8.0 \text{ mL soln}} \times \dfrac{0.79 \text{ g alcohol}}{\text{mL alcohol}}$

$= 23.5$ g alcohol in 100 g soln $= 24\%$

(Be sure to watch your significant figures.)

17. ? g lead sulfide $= 100 \text{ g ore} \times \dfrac{220 \text{ mL H}_2\text{S}}{2.5 \text{ g ore}} \times \dfrac{10^{-3} \text{ L H}_2\text{S}}{\text{mL H}_2\text{S}}$

$\times \dfrac{4.8 \text{ g lead sulfide}}{0.45 \text{ L H}_2\text{S}} = 94$ g lead sulfide in 100 g ore $= 94\%$

18. ? g tin $= 2.0 \text{ lb ore} \times \dfrac{454 \text{ g ore}}{\text{lb ore}} \times \dfrac{85 \text{ g tin}}{100 \text{ g ore}} = 772$ g tin $= 7.7 \times 10^2$ g tin

19. ? m^3 $= \$5 \times \dfrac{\text{oz gold}}{\$304} \times \dfrac{\text{lb gold}}{16 \text{ oz gold}} \times \dfrac{454 \text{ g gold}}{\text{lb gold}} \times \dfrac{100 \text{ g ore}}{3.5 \text{ g gold}} \times \dfrac{\text{mL ore}}{11 \text{ g ore}}$

$\times \dfrac{\text{cm}^3 \text{ ore}}{\text{mL ore}} \times \dfrac{(10^{-2})^3 \text{ m}^3 \text{ ore}}{\text{cm}^3 \text{ ore}}$

$= 5 \times \dfrac{1}{304} \times \dfrac{1}{16} \times 454 \times \dfrac{100}{3.5} \times \dfrac{1}{11} \times 1 \times 10^{-6} \text{ m}^3$ ore

$= 1.2 \times 10^{-6}$ m^3 ore

(Notice that the specific gravity is equal to the density to two significant figures.)

20. $? \$ = 5 \text{ tons carbon} \times \dfrac{100 \text{ tons coal}}{98 \text{ tons carbon}} \times \dfrac{2000 \text{ lb coal}}{\text{ton coal}} \times \dfrac{454 \text{ g coal}}{\text{lb coal}}$

$\times \dfrac{\text{mL coal}}{4.8 \text{ g coal}} \times \dfrac{\text{cm}^3}{\text{mL coal}} \times \dfrac{\text{in}^3}{(2.54)^3 \text{ cm}^3} \times \dfrac{\text{yd}^3}{(36)^3 \text{ in}^3} \times \dfrac{\text{min}}{15 \text{ yd}^3}$

$\times \dfrac{\text{hr}}{60 \text{ min}} \times \dfrac{\$35.00}{\text{hr}}$

$= 5 \times \dfrac{100}{98} \times 2000 \times 454 \times \dfrac{1}{4.8} \times 1 \times \dfrac{1}{16.4} \times \dfrac{1}{4.65 \times 10^4}$

$\times \dfrac{1}{15} \times \dfrac{1}{60} \times \$35.00$

$= \$4.9 \times 10^{-2} = \$0.05$

(Hats off to you if you were able to drag your way through this last problem. If you didn't quite make it, that's okay, since you will never face any chemical problem with 10 conversion factors.)

21. $? \text{ g silver} = 100 \text{ g ore} \times \dfrac{\text{lb ore}}{454 \text{ g ore}} \times \dfrac{140 \text{ g silver}}{2.50 \text{ lb ore}} = 100 \times \dfrac{1}{454} \times \dfrac{140 \text{ g silver}}{2.50}$

$= 12.3 \text{ g silver in } 100 \text{ g ore} = 12.3\%$

22. $? \text{ g alcohol} = 3.00 \text{ L wine} \times \dfrac{14 \text{ L alcohol}}{100 \text{ L wine}} \times \dfrac{\text{mL alcohol}}{10^{-3} \text{ L alcohol}}$

$\times \dfrac{0.780 \text{ g alcohol}}{\text{mL alcohol}}$

$= 3.00 \times \dfrac{14}{100} \times \dfrac{1}{10^{-3}} \times 0.780 \text{ g alcohol} = 3.3 \times 10^2 \text{ g alcohol}$

[In your calculation, you did not need the density of wine, a conversion factor given in the problem. Don't assume that *all* the information you are given is useful. Also note that there are only two significant figures allowed (14% by volume), and so the answer must be expressed exponentially.]

23. $\text{Specific gravity} = \dfrac{0.875 \text{ g/mL} \quad \text{(oil)}}{0.996 \text{ g/mL} \quad \text{(water)}} = 0.879$

24. $? \text{ lb air} = 3 \text{ ft}^3 \times \dfrac{(12)^3 \text{ in}^3}{\text{ft}^3} \times \dfrac{(2.54)^3 \text{ cm}^3}{\text{in}^3} \times \dfrac{\text{mL}}{\text{cm}^3} \times \dfrac{10^{-3} \text{ L}}{\text{mL}} \times \dfrac{1.21 \text{ g air}}{\text{L}}$

$\times \dfrac{\text{lb air}}{454 \text{ g air}}$

$= 3 \times (12)^3 \times (2.54)^3 \times 1 \times 10^{-3} \times 1.21 \times \dfrac{\text{lb air}}{454}$

$= 2.26 \times 10^{-1} \text{ lb air}$

25. Specific gravity is defined as the weight of a substance divided by the weight of an equivalent volume of water. Therefore you must first find the weight of an equivalent (the amount displaced) volume of water.

$$? \text{ g water} = 1.40 \text{ mL water} \times \frac{4.98 \text{ g water}}{5.00 \text{ mL water}} = 1.39 \text{ g water}$$

$$\text{Specific gravity} = \frac{6.37 \text{ g mineral}}{1.39 \text{ g water}} = 4.58$$

26. Once again, specific gravity is defined as the weight of a substance divided by the weight of an equivalent volume of water. Thus you must first find the weight of 55 gal water.

$$? \text{ lb} = 55 \text{ gal water} \times \frac{8.34 \text{ lb}}{\text{gal water}} = 459 \text{ lb water}$$

Then you have

$$\text{Specific gravity} = \frac{\text{weight 55 gal ethylene glycol}}{\text{weight 55 gal water}}$$

Putting in values, you obtain

$$1.1 = \frac{\text{weight 55 gal ethylene glycol}}{459 \text{ lb water}}$$

Cross-multiplying gives

$(1.1)(459 \text{ lb}) = 505 \text{ lb} = \text{weight ethylene glycol}$

or, to the correct number of significant figures, since 55 gal has just two,

$5.0 \times 10^2 \text{ lb} = \text{weight ethylene glycol}$

CHAPTER 5

The Mole: A Chemical Conversion Factor

There will be no pretest for this short chapter. All that is contained in it is the introduction of two new conversion factors. One of these will be absolutely vital to you and will soon become second nature.

SECTION A   Atomic Weight

The size and mass of an atom are so small that it isn't possible to experimentally determine an absolute mass. However, it is extremely simple to determine masses relative to each other. Therefore what chemists have done is to assign an arbitrary number to one specific atom (most recently, the C^{12} isotope was chosen and assigned the number 12.00000) and use this as a measuring stick for all other atoms. Thus, if a magnesium atom weighs twice as much as the C^{12} isotope, the

atomic weight of magnesium is _____, that is, _____ times

_____.

| 24.00000; 2 |
| 12.00000 |

    Since it was found that a titanium atom was four times as heavy as the

C^{12} isotope, the atomic weight (at wt) of titanium is _____. It is as simple as that.

48.00000

    You will find the atomic weights of the common elements in a table inside the front cover of this text. The values are given only to the first place after the decimal, since you will not require more precision than that for the work here. However, much more precise values are known for most elements, and can be found either in other textbooks or in a handbook of chemistry and physics.

    You may also find in some texts a unit of mass for the atomic weights. It is called *atomic mass unit* (amu), or even u. This term will not be useful in this text, and so we won't consider it.

SECTION B   Molecular Weight

As you know, compounds are made up of combinations of atoms chemically bound together. Since the mass of the compound equals the sum of the masses of the atoms that make it up, all you need do is add up all the atomic weights. You will then have the molecular weight (MW). This, too, has no units (unless you labeled the atomic weights in amu). Some compounds are made up of molecules and others of ions. But we will always use the term *molecular weight.*

Chemists have a shorthand method of writing compounds. These are called *chemical formulas.* They write the symbols of the atoms that make up the compound, and put a *subscript* (below the line) number after the symbol of each atom to tell how many of that atom are present in the molecule. If there is only one of a given atom, you don't bother writing the 1. It is understood.

Consider the molecule H_2SO_4, sulfuric acid. It is made up of ___ H atoms, ___ S atom, and ___ O atoms. (The Atomic Weights Table at the beginning of the text lists the symbols as well as the names of the elements.) Thus, to get the molecular weight of sulfuric acid, you add the following.

$$H = 2 \times 1.0 = 2.0$$
$$S = 1 \times 32.1 = 32.1$$
$$O = 4 \times 16.0 = 64.0$$
$$98.1$$

To be certain that you have grasped this calculation, determine the molecular weight of $Na_4P_2O_7$, sodium pyrophosphate, using the values from the table at the beginning of this text.

$$Na = \underline{} \times 23.0 = \underline{}$$
$$P = \underline{} \times 31.0 = \underline{}$$
$$O = 7 \times \underline{} = \underline{}$$
$$\underline{}$$

When a group of atoms (called a *radical* or a *polyatomic ion*) is present in a compound more than once, it is customary to enclose the entire group in parentheses and write a subscript number to show how many of the polyatomic ions are present. Thus, in the molecule $Ca(NO_3)_2$, there are 2 (NO_3^-) polyatomic ions. The molecule has a total of ___ Ca, ___ N, and ___ O atoms.

Consider the molecule $Al_2(SO_4)_3$. It has ___ S atoms and ___ O atoms.

When one molecule is attached to another in a complex molecule, rather than use parentheses and a subscript number, you put a dot in the middle of the line between the two parts, and a normal number in front of the complexed molecule. Thus $Na_2P_4O_7 \cdot 10H_2O$ (sodium pyrophosphate decahydrate) has 10 water molecules attached. The total molecule has 20 H atoms and ___ O atoms.

Margin answers

2

1; 4

4; 92.0

2; 62.0

16.0; 112.0

266.0

1; 2

6

3; 12

17

SECTION C   The Mole

Since atomic weights and molecular weights do not have any dimension, two units called the *gram atomic weight* and the *gram molecular weight* were set up so that they would have the same number of grams as the number of the atomic weight and the molecular weight, respectively. Thus, if the atomic weight of O is 16.0, then its gram atomic weight is 16.0 g. If the molecular weight of H_2O is 18.0, then its gram molecular

weight is _____. (These expressions seem a little stilted now, so we won't use them. The most correct name is *molar mass*, which means exactly "the mass of a mole." But we are going to stick with the common expression, molecular weight.)

18.0 g

It has been found that it requires 6.02×10^{23} atoms of any element to weigh the atomic weight in grams. It also requires 6.02×10^{23} mole-

cules to weigh the _____ in grams. This number is called *Avogadro's number*.

molecular weight

We are now going to define a new unit called the *mole*. In the same way that a million equals 10^6 or a billion equals 10^9, a mole equals 6.02×10^{23}. The symbol for mole is mol.

We know that a given number of atoms weighs the atomic weight in

grams. This same number of molecules weighs the _____ in grams. Thus, we immediately have four conversion factors.

molecular weight

$$\frac{6.02 \times 10^{23} \text{ atoms}}{\text{mole}} \qquad \frac{6.02 \times 10^{23} \text{ molecules}}{\text{mole}}$$

$$\frac{\text{Atomic weight in grams}}{\text{mole}} \qquad \frac{\text{Molecular weight in grams}}{\text{mole}}$$

As a chemist, you will have very few opportunities to count numbers of atoms or molecules. The first two conversion factors have very limited uses. However, the second two are going to become second nature to you. When you specify molecular weight, you will give it the dimensions

of grams per mole. Thus H_2O has a molecular weight of 18.0 _____,

$\dfrac{g}{mol}$

and O has an atomic weight of 16.0 _____.

$\dfrac{g}{mol}$

Here are a few examples to practice on. There are 23.0 g of Na per

_____ of Na. You can write this as a conversion factor, _____.

mole; $\dfrac{23.0 \text{ g Na}}{\text{mol Na}}$

There are 6.02×10^{23} atoms of Na per _____ Na. You can write this as a conversion factor

mole

$$\frac{6.02 \times 10^{23} \text{ atoms Na}}{\text{mol Na}}$$

There are 16.0 g of oxygen in a mole of O atoms, but there are 32.0 g of oxygen in a mole of O_2 molecules. This is going to be a problem, so you must be especially careful to write down *complete* dimensions. Thus you write

$$\frac{16.0 \text{ g O}}{\text{mol O}} \quad \text{and} \quad \frac{32.0 \text{ g O}_2}{\rule{2cm}{0.4pt}}$$

mol O₂

Formerly a mole of atoms was called a *gram-atom* to eliminate the problem of whether you were referring to atoms or molecules. We will call everything *mole*, but be very certain to specify what constitutes the mole.

Some chemists even limit the term *mole* only to compounds that exist as molecules. For example, NaCl, which in solution or even in the solid or liquid states exists as Na^+ and Cl^- ions, could not be expressed in moles. The term *gram-formula weights* is used to make this distinction. We will not be so fussy, and will call everything a mole.

There are two types of possible problems concerning moles. The first isn't concerned with weight, either in the question asked or in the information given, and only numbers of atoms or molecules are mentioned. For these problems you will use the conversion factors

$$\frac{6.02 \times 10^{23} \text{ atoms}}{\text{mole atoms}} \quad \text{and} \quad \frac{6.02 \times 10^{23} \text{ molecules}}{\rule{2cm}{0.4pt}}$$

mole molecules

Look at a typical problem. How many S atoms in 4.000 mol of S? The

atoms S

dimensions of the answer will be _____. The information given

4.000 mol S

is _____. Starting in the usual way, you write

atoms S; 4.000 mol S

? _____ = _____

moles

The conversion factor having _____ in the denominator and heading

$$\frac{6.02 \times 10^{23} \text{ atoms S}}{\text{mol S}}$$

toward the dimensions of the answer is _____. Multi-plying by this factor cancels the unwanted dimension.

$$? \text{ atoms S} = 4.000 \text{ mol S} \times \frac{6.02 \times 10^{23} \text{ atoms S}}{\text{mol S}}$$

$$= 24.08 \times 10^{23} \text{ atoms S} = 2.41 \times 10^{24} \text{ atoms S}$$

Here is another example. You have 2.107×10^{24} molecules of H_2O. How many moles of H_2O do you have? Starting in the usual way, you write

mol H₂O;
2.107 × 10²⁴ molecules H₂O

? _____ = _____

You then multiply by the conversion factor:

$$\frac{\text{mol H}_2\text{O}}{6.02 \times 10^{23} \text{ molecules H}_2\text{O}}$$

_____

The unwanted dimension cancels, leaving

$$? \text{ mol H}_2\text{O} = 2.107 \times 10^{24} \text{ molecules H}_2\text{O}$$

3.50

$$\times \frac{\text{mol H}_2\text{O}}{6.02 \times 10^{23} \text{ molecules H}_2\text{O}} = \rule{1cm}{0.4pt} \text{ mol H}_2\text{O}$$

Here is a slightly different problem. How many atoms of N are there in 5.4 mol of N_2? There is a conversion factor that you can get from the formula for N_2. It relates the number of N atoms to N_2 molecules. You can therefore write

$$\frac{2 \text{ N atoms}}{\underline{\hspace{3cm}}}$$

1 molecule N_2

Now you can start the problem in the usual way.

$$? \underline{\hspace{3cm}} = \underline{\hspace{3cm}}$$

N atoms; 5.4 mol N_2

You convert to molecules of N_2 by multiplying by the conversion factor

$$\underline{\hspace{5cm}}$$

$\dfrac{6.02 \times 10^{23} \text{ molecules } N_2}{\text{mol } N_2}$

molecules N_2

The unwanted dimension left is _____. You can convert this to the dimensions of the question asked by multiplying by the conversion factor you found in the formula

$$\underline{\hspace{4cm}}$$

$\dfrac{2 \text{ N atoms}}{1 \text{ molecule } N_2}$

The final setup will then be

$$? \text{ N atoms} = 5.4 \;\cancel{\text{mol } N_2} \times \frac{6.02 \times 10^{23} \;\cancel{\text{molecules } N_2}}{\cancel{\text{mol } N_2}} \times \frac{2 \text{ N atoms}}{\cancel{\text{molecule } N_2}}$$

$$= \underline{\hspace{4cm}}$$

6.5×10^{24} N atoms

Try this problem. How many H atoms are in 4.2 mol of $(NH_4)_2SO_4$? The conversion factor that you can get from the formula relating H atoms to the molecule is

____ H atoms/molecule $(NH_4)_2SO_4$

8

Now work the problem through. The answer you get is _____.

2.0×10^{25} atoms H

Notice in the above problems that there is no mention of weights. Consequently, you did not use a conversion factor concerning weight, either the atomic or molecular weight. Now here is a problem with weight in it. Suppose that you are asked how many grams 6.0 mol of H atoms weigh. The dimensions of the answer will be _____, so you must now use the conversion factor that includes atomic weight.

grams H

The information given is _____. Setting up the problem, you write

6.0 mol H

$$? \underline{\hspace{2cm}} = \underline{\hspace{3cm}}$$

g H; 6.0 mol H

The conversion factor that will cancel the unwanted dimension and give you the dimension of the answer is _____, the atomic weight of H. Your answer will therefore be

$\dfrac{1.0 \text{ g H}}{\text{mol H}}$

$$? \text{ g H} = 6.0 \;\cancel{\text{mol H}} \times \frac{1.0 \text{ g H}}{\cancel{\text{mol H}}} = 6.0 \text{ g H}$$

If you are asked how many moles of H_2SO_4 are in 49.05 grams of H_2SO_4, you begin in the same way.

mol H_2SO_4; 49.05 g H_2SO_4

$$\dfrac{\text{mol } H_2SO_4}{98.1 \text{ g } H_2SO_4}$$

5.00×10^{-1}

molecular

141 g H_3AsO_4

mol H_3AsO_4

Avogadro's number

g H_3AsO_4; 0.50 mol H_3AsO_4

$\dfrac{141.9 \text{ g } H_3AsO_4}{\text{mol } H_3AsO_4}$; 71

g acetic acid

molecular

60.0

mol acetic acid

mol acetic acid;
75 mL acetic acid

? _____ = _____

The conversion factor by which you must multiply is

_____

This is the inverted form of the molecular weight of H_2SO_4. This operation cancels the unwanted dimension of the information given and leaves only the dimensions of the answer.

$$? \text{ mol } H_2SO_4 = 49.05 \text{ g } \cancel{H_2SO_4} \times \dfrac{\text{mol } H_2SO_4}{98.1 \text{ g } \cancel{H_2SO_4}}$$

$$= \text{_____ mol } H_2SO_4$$

The conversion factor that relates the molecular or atomic weight to moles of a substance is going to be one of your most often-used tools, so it will be useful to do more problems of this type.

How many grams of H_3AsO_4 will you need to have 0.50 mol of H_3AsO_4? Since the problem deals with weight, you will need the conversion factor you get from the _____ weight. It is

_____

Nothing in the problem mentions the number of atoms or molecules, so you will have no use for a conversion factor containing _____, 6.02×10^{23}. Don't try to use it. It will only confuse the picture. Starting in the usual way, then, you write

? _____ = _____

You can get to the dimensions of the question asked by multiplying by the molecular weight.

$$? \text{ g } H_3AsO_4 = 0.50 \text{ mol } H_3AsO_4 \times \text{_____} = \text{\_\_\_\_\_ g } H_3AsO_4$$

Here is a more complicated problem. How many moles of acetic acid, $C_2H_4O_2$, are there in 75.0 mL of acetic acid? The density of acetic acid is 1.05 g/mL. This problem doesn't deal directly with weight. It gives information about the volume. However, if you have the density of a compound, you can easily convert volume to weight. So there are two conversion factors that you will need. First the density:

$$\dfrac{1.05 \text{ _____}}{\text{mL acetic acid}}$$

And then the _____ weight:

$$\text{_____ g acetic acid}$$

_____

Starting in the usual way, then, you write

? _____ = _____

Now you simply choose a conversion factor that has _____ in the denominator in order to cancel the unwanted dimension in the information given. This would be

_____

Multiplying by this factor gives you

$$? \text{ mol acetic acid} = 75 \cancel{\text{ mL acetic acid}} \times \frac{1.05 \text{ g acetic acid}}{\cancel{\text{mL acetic acid}}}$$

If you now multiply by the molecular _____ the conversion factor that is in its reciprocal form, the only dimension left is the dimension of the question asked:

$$? \text{ mol acetic acid} = 75 \cancel{\text{ mL acetic acid}} \times \frac{1.05 \cancel{\text{ g acetic acid}}}{\cancel{\text{mL acetic acid}}}$$

$$\times \frac{\text{mol acetic acid}}{60.0 \cancel{\text{ g acetic acid}}}$$

$$= 1.3 \text{ mol acetic acid}$$

Here is a problem that you can try on your own. There will be just a few hints to get you started. You want to find how many grams of a 7.50 wt % solution of NaCl you need in order to have 0.85 mol of NaCl. The conversion factor given in the problem is found in the statement "7.50 wt % solution of NaCl." It is

_____

(If you have forgotten this, you can check it in Section C of Chapter 4.) Since this problem deals with the weight of NaCl (indirectly), you will

need the _____ conversion factor. It is

58.5 g NaCl
─────────
_____

The tricky part of this problem is asking the right question. It is "How

many grams of _____ ?" You must watch this carefully when you solve problems yourself.

Now you are ready to set up the problem and work it. If you got 663 g soln as your answer, you're doing well. If you found 1.1 or 3.7 g as your answer, you failed to label your dimensions correctly so that they cancel. This setup should be

$$? \text{ g soln} = 0.85 \cancel{\text{ mol NaCl}} \times \frac{58.5 \cancel{\text{ g NaCl}}}{\cancel{\text{mol NaCl}}} \times \frac{100 \text{ g soln}}{7.50 \cancel{\text{ g NaCl}}}$$

This type of problem concerning the number of moles and some property that can be related to the weight of the compound can become more complex only if the weight must be found through various conversion factors. However, you can always use as your "bridge" the

---

milliliters acetic acid

$\dfrac{1.05 \text{ g acetic acid}}{\text{mL acetic acid}}$

weight

$\dfrac{7.50 \text{ g NaCl}}{100 \text{ g soln}}$

molecular weight

mol NaCl

solution

molecular	conversion factor that is the _____ weight, expressed in its
grams	correct units, _____ per mole. You may, of course, use it in its reciprocal (inverted) form.

It is possible to have a problem involving both the weight of a compound and the number of atoms or molecules. In these cases you must use both conversion factors, the molecular weight and the one containing Avogadro's number. Suppose you are asked, for example, how many molecules of CO_2 are present in 3.2 g of CO_2. You start the problem in the usual way.

molecules CO_2 ; 3.2 g CO_2

$$? \underline{\hspace{3cm}} = \underline{\hspace{2cm}}$$

The only conversion factor you know that contains grams CO_2 is the

molecular	_____ weight. Multiplying by this, you have
$\dfrac{mol\ CO_2}{44.0\ g\ CO_2}$	? molecules CO_2 = 3.2 g CO_2 X _____

cancel	Notice that you used it in its inverted form in order to _____ the unwanted dimension in the information given. The unwanted dimension
mol CO_2	now left is _____. You can convert this to the dimension of the question asked simply by multiplying by
$\dfrac{6.02 \times 10^{23}\ \text{molecules}\ CO_2}{\text{mol}\ CO_2}$	_____

The final setup will then be

$$? \text{ molecules } CO_2 = 3.2\ \cancel{g\ CO_2} \times \frac{\cancel{mol\ CO_2}}{44.0\ \cancel{g\ CO_2}}$$

$$\times \frac{6.02 \times 10^{23}\ \text{molecules}\ CO_2}{\cancel{mol\ CO_2}}$$

$$= 4.4 \times 10^{22}\ \text{molecules}\ CO_2$$

Here is a more involved example of this type of problem. The density of iron is 7.8 g/mL. How many iron atoms are contained in a cube of pure iron 2.0 cm on an edge? First you must calculate the volume of Fe (the symbol for iron).

$$\text{Volume} = (2.0\ \text{cm})^3 = 8.0\ \text{cm}^3$$

Now the question can be reworded as, "How many Fe atoms are in 8.0 cm^3?" Starting in the usual way, you have

atoms Fe; 8.0 cm^3

$$? \underline{\hspace{2.5cm}} = \underline{\hspace{2cm}}$$

You know the conversion factor, 1.0 mL/cm^3, and also the conversion factor given in the problem, 7.8 g Fe/mL Fe, which is the density. Multiplying by these two conversion factors now leaves the unwanted

grams Fe	dimension _____. You can cancel this unwanted dimension by
$\dfrac{mol\ Fe}{55.8\ g\ Fe}$	multiplying by the conversion factor _____. This now leaves
mol Fe	the unwanted dimension _____, which can be converted to the dimensions of the answer by multiplying by the conversion factor

$$\frac{6.02 \times 10^{23} \text{ atoms Fe}}{\text{mol Fe}}$$

Now the only dimension left is the dimension of the answer, so the answer must be

$$? \text{ atoms Fe} = 8.0 \, \cancel{\text{cm}^3} \times \frac{\cancel{\text{mL}}}{\cancel{\text{cm}^3}} \times \frac{7.8 \, \cancel{\text{g Fe}}}{\cancel{\text{mL}}} \times \frac{\cancel{\text{mol Fe}}}{55.8 \, \cancel{\text{g Fe}}}$$

$$\times \frac{6.02 \times 10^{23} \text{ atoms Fe}}{\cancel{\text{mol Fe}}}$$

$$= 6.7 \times 10^{23} \text{ atoms Fe}$$

(Remember that you should be writing the conversion factors down in your own setup as you go along.)

The only way in which problems can become *still* more complicated is to have more conversion factors given in the problem. Here is such an example.

The price of an HCl solution is $6.00 for 2.0 kg. How much would 6.0 mol of HCl cost if the solution contains 310 g HCl per liter and its density is 1.1 g/mL? The first thing to do is to write down all the conversion factors given in the problem.

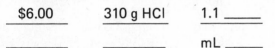

g soln

2 kg soln; L soln; soln

It is *vital* that you write the dimensions in such a way as to distinguish between grams solution and grams HCl. Since the problem concerns only weight and moles, you will need the conversion factor relating weight of

HCl to moles HCl. This is the molecular weight, _____.

$$\frac{36.5 \text{ g HCl}}{\text{mol HCl}}$$

Now start in the usual way by writing

$$? \, \underline{\hspace{2em}} = \underline{\hspace{6em}}$$

dollars; 6.0 mol HCl

The problem is now very simple. You keep multiplying by conversion factors, canceling unwanted dimensions until you finally arrive at the dimensions of the answer. Do it yourself and check your results against the answer below.

$$? \$ = 6.0 \, \cancel{\text{mol HCl}} \times \frac{36.5 \, \cancel{\text{g HCl}}}{\cancel{\text{mol HCl}}} \times \frac{\cancel{\text{L soln}}}{310 \, \cancel{\text{g HCl}}} \times \frac{\cancel{\text{mL soln}}}{10^{-3} \, \cancel{\text{L soln}}}$$

$$\times \frac{1.1 \, \cancel{\text{g soln}}}{\cancel{\text{mL soln}}} \times \frac{\cancel{\text{kg soln}}}{10^3 \, \cancel{\text{g soln}}} \times \frac{\$6.00}{2.0 \, \cancel{\text{kg soln}}}$$

$$= \$2.3 \quad \text{(to the correct number of significant figures)}$$

You are now ready to do the problem set.

PROBLEM SET

Molecular Weight

1. Using the Table of Atomic Weights at the beginning of the text, calculate the molecular weight for the following compounds.

   (a) HBR   (b) $CaCl_2$ (c) H_3PO_4

   (d) Na_2SO_4 (e) $(NH_4)_3AsO_3$ (f) $KF \cdot 2H_2O$

   (g) $Fe_3(Fe(CN)_6)_2$ (h) $H_3PW_{12}O_{40} \cdot 14H_2O$

Moles and Numbers of Atoms or Molecules

2. How many atoms of C are there in 8.0 mol of C?

3. A sample containing 3.01×10^{24} molecules of water contains how many moles of water?

4. How many O atoms are there in 3.0 mol of P_2O_5?

5. How many molecules of sugar are there in 150 mL of a solution that contains 3.5 mol of sugar in 12.0 L of solution?

6. A solution of alcohol in water is prepared by mixing 3.61×10^{22} molecules of alcohol with enough water to have 750 mL of solution. The density of the solution is 0.92 g/mL. How many moles of alcohol are present in 160 g of the solution?

Moles and a Weight-Related Property

7. How many moles of HCl are present in a sample of HCl weighing 272 g?

8. A sample of $KHSO_3$ contains 1.7 mol of $KHSO_3$. How many grams does it weigh?

9. How many grams of water must you take to have 2.5 mol of water?

10. The density of ethyl alcohol, C_2H_5OH, is 0.79 g/mL. How many milliliters of ethyl alcohol contain 0.35 mol C_2H_5OH?

11. What is the price per mole for NaCl if 100 g NaCl cost $.05?

12. How many moles of $CaCO_3$ are present in a 450-g sample of limestone which is 94% $CaCO_3$?

13. A solution was prepared by dissolving 0.25 mol of $K_2Cr_2O_7$ in enough water to prepare 3.0 L of solution. How many grams of $K_2Cr_2O_7$ are there in 125 mL of the solution?

14. A solution was prepared by dissolving 33.5 g of $Na_2C_2O_4$ in enough water to give 1.50 L of solution. The density of the solution was 1.08 g/mL. Then 0.125 mol of $Na_2C_2O_4$ is contained in how many grams of solution?

15. What is the edge length of a cube of cobalt, Co, that costs $3.50, if the price of Co is $7.00 per mole? The density of Co is 8.9 g/mL.

Weight and Number of Molecules and Atoms

16. How many atoms of Na are there in 92.0 g Na?

17. How many grams do 3.01×10^{24} molecules of H_2SO_4 weigh?

18. The density of Br_2 is 3.10 g/mL. How many milliliters of Br_2 must you take to have 2.11×10^{22} molecules of Br_2?

19. A 45.0 g sample of $(NH_4)_3AsO_3$ contains how many H atoms?

20. How many atoms, regardless of type, are there in 196 g H_2SO_4? (Be careful. There are 7 atoms of all types per molecule.)

21. What is the price for a million (1.0×10^6) atoms of Pt, given that a cube 1.0 in. on an edge costs $550? The density of Pt is 22.0 g/mL.

22. How many atoms of O are there in a cube of calcite ore that is 92% $CaCO_3$ and that measures 0.51 cm on the edge? The density of calcite is 4.26 g/mL.

Miscellaneous Problems

23. If you rub a piece of gold, Au, vigorously, you can rub off 2.00 mol of Au atoms in a century. How many minutes are needed to rub off 1.50 μg of gold?

24. Which of the following samples weighs the most: a sample containing 0.082 mol of H_2O or a sample containing 3.01×10^{21} molecules of H_2SO_4?

25. A sample of calcite ore is 94% $CaCO_3$. If you took 3.0 g of the calcite, how many molecules of $CaCO_3$ would you have?

26. How many pounds does a mole of mercury, Hg, weigh?

27. How many total atoms, regardless of type, are there in 11 mol of K_2HAsO_4?

28. Which of the following samples contains the most molecules: 0.500 mol of NaCl, 0.500 mL of C_2H_5OH (density = 0.789 g/mL), or 0.500 g of H_2O?

PROBLEM SET ANSWERS

You should have done all parts of problem 1 correctly. If you didn't, check your addition. If the error was not there, check the value for the atomic weights you used. If you still cannot find your error, go back and redo Section B in the chapter. (The last two problems concern very rare compounds.)

The rest of the problems were divided into groups depending on whether they concerned numbers of molecules or atoms, weight of compound, or both. The problems become progressively more complicated as you go through each group. You should be able to do a minimum of three in the first group, seven in the second group, and five in the third group.

1. (a) 80.9   (b) 111.1   (c) 98.0   (d) 142.1
   (e) 176.9   (f) 94.1   (g) 591.0   (h) 3132.8   (W is tungsten)

2. $? \text{ atoms} = 8.0 \, \cancel{\text{mol C}} \times \dfrac{6.02 \times 10^{23} \text{ atoms C}}{\cancel{\text{mol C}}} = 4.8 \times 10^{24} \text{ atoms C}$

3. $? \text{ mol } H_2O = 3.01 \times 10^{24} \, \cancel{\text{molecules } H_2O} \times \dfrac{\text{mol } H_2O}{6.02 \times 10^{23} \, \cancel{\text{molecules } H_2O}}$

$\qquad = 5.00 \text{ mol } H_2O$

4. $? \text{ atoms O} = 3.0 \, \cancel{\text{mol } P_2O_5} \times \dfrac{6.02 \times 10^{23} \, \cancel{\text{molecules } P_2O_5}}{\cancel{\text{mol } P_2O_5}}$

$\qquad \times \dfrac{5 \text{ atoms O}}{\cancel{\text{molecule } P_2O_5}}$ (Notice that the 5 atoms O per molecule of P_2O_5 is a counted number and doesn't affect the significant figures.)

$\qquad = 9.0 \times 10^{24} \text{ atoms O}$

5. $? \text{ molecules sugar} = 150 \, \cancel{\text{mL soln}} \times \dfrac{10^{-3} \, \cancel{\text{L soln}}}{\cancel{\text{mL soln}}} \times \dfrac{3.5 \, \cancel{\text{mol sugar}}}{12.0 \, \cancel{\text{L soln}}}$

$\qquad \times \dfrac{6.02 \times 10^{23} \text{ molecules sugar}}{\cancel{\text{mol sugar}}}$

$\qquad = 2.6 \times 10^{22} \text{ molecules sugar}$

6. $? \text{ mol alcohol} = 160 \, \cancel{\text{g soln}} \times \dfrac{\cancel{\text{mL soln}}}{0.92 \, \cancel{\text{g soln}}}$

$\qquad \times \dfrac{3.61 \times 10^{22} \, \cancel{\text{molecules alcohol}}}{750 \, \cancel{\text{mL soln}}}$

$\qquad \times \dfrac{\text{mol alcohol}}{6.02 \times 10^{23} \, \cancel{\text{molecules alcohol}}}$

$\qquad = 1.39 \times 10^{-2} \text{ mol alcohol}$

7. $? \text{ mol HCl} = 272 \, \cancel{\text{g HCl}} \times \dfrac{\text{mol HCl}}{36.5 \, \cancel{\text{g HCl}}} = 7.45 \text{ mol HCl}$

8. $? \text{ g } KHSO_3 = 1.7 \, \cancel{\text{mol } KHSO_3} \times \dfrac{120.2 \text{ g } KHSO_3}{\cancel{\text{mol } KHSO_3}} = 2.0 \times 10^2 \text{ g } KHSO_3$

9. $? \text{ g } H_2O = 2.5 \, \cancel{\text{mol } H_2O} \times \dfrac{18.0 \text{ g } H_2O}{\cancel{\text{mol } H_2O}} = 4.5 \times 10^1 \text{ g } H_2O$

(Notice in the last two problems the use of scientific notation to express the answers to the correct number of significant figures.)

10. $? \text{ mL ethyl alcohol} = 0.35 \, \cancel{\text{mol ethyl alcohol}} \times \dfrac{46.0 \, \cancel{\text{g ethyl alcohol}}}{\cancel{\text{mol ethyl alcohol}}}$

$\qquad \times \dfrac{\text{mL ethyl alcohol}}{0.79 \, \cancel{\text{g ethyl alcohol}}}$

$\qquad = 2.0 \times 10^1 \text{ mL ethyl alcohol}$

11. $? \text{ dollars} = 1.00 \, \cancel{\text{mol NaCl}} \times \dfrac{58.5 \, \cancel{\text{g NaCl}}}{\cancel{\text{mol NaCl}}} \times \dfrac{\$.05}{100 \, \cancel{\text{g NaCl}}} = \$.0292$

12. $? \text{ mol CaCO}_3 = 450 \, \cancel{\text{g imp}} \times \dfrac{94 \, \cancel{\text{g CaCO}_3}}{100 \, \cancel{\text{g imp}}} \times \dfrac{\text{mol CaCO}_3}{100.1 \, \cancel{\text{g CaCO}_3}}$

$= 423 \text{ mol CaCO}_3$

(If you have forgotten how to handle percent purity, check back to Section C of Chapter 4.)

13. $? \text{ g K}_2\text{Cr}_2\text{O}_7 = 125 \, \cancel{\text{mL soln}} \times \dfrac{10^{-3} \, \cancel{\text{L soln}}}{\cancel{\text{mL soln}}}$

$\times \dfrac{0.25 \, \cancel{\text{mol K}_2\text{Cr}_2\text{O}_7}}{3.0 \, \cancel{\text{L soln}}} \times \dfrac{294.2 \text{ g K}_2\text{Cr}_2\text{O}_7}{\cancel{\text{mol K}_2\text{Cr}_2\text{O}_7}}$

$= 3.1 \text{ g K}_2\text{Cr}_2\text{O}_7$

14. $? \text{ g soln} = 0.125 \, \cancel{\text{mol Na}_2\text{C}_2\text{O}_4} \times \dfrac{134.0 \, \cancel{\text{g Na}_2\text{C}_2\text{O}_4}}{\cancel{\text{mol Na}_2\text{C}_2\text{O}_4}} \times \dfrac{1.50 \, \cancel{\text{L soln}}}{33.5 \, \cancel{\text{g Na}_2\text{C}_2\text{O}_4}}$

$\times \dfrac{\cancel{\text{mL soln}}}{10^{-3} \, \cancel{\text{L soln}}} \times \dfrac{1.08 \text{ g soln}}{\cancel{\text{mL soln}}}$

$= 8.10 \times 10^2 \text{ g soln}$

15. First you must determine the volume of the cube in cubic centimeters.

$? \text{ cm}^3 = \cancel{\$3.50} \times \dfrac{\cancel{\text{mol Co}}}{\cancel{\$7.00}} \times \dfrac{58.9 \, \cancel{\text{g Co}}}{\cancel{\text{mol Co}}} \times \dfrac{\cancel{\text{mL Co}}}{8.9 \, \cancel{\text{g Co}}} \times \dfrac{\text{cm}^3}{\cancel{\text{mL}}} = 3.3 \text{ cm}^3$

You can then find the edge length by taking the cube root of 3.3 cm^3.

Edge length cm $= (3.3 \text{ cm}^3)^{1/3} = 1.5 \text{ cm}$

(If your calculator does not find roots, refer to Section B in Chapter 1 to see how to get them using logarithms.)

16. $? \text{ atoms Na} = 92.0 \, \cancel{\text{g Na}} \times \dfrac{\cancel{\text{mol Na}}}{23.0 \, \cancel{\text{g Na}}} \times \dfrac{6.02 \times 10^{23} \text{ atoms Na}}{\cancel{\text{mol Na}}}$

$= 2.41 \times 10^{24} \text{ atoms Na}$

17. $? \text{ g H}_2\text{SO}_4 = 3.01 \times 10^{24} \, \cancel{\text{molecules H}_2\text{SO}_4}$

$\times \dfrac{\cancel{\text{mol H}_2\text{SO}_4}}{6.02 \times 10^{23} \, \cancel{\text{molecules H}_2\text{SO}_4}} \times \dfrac{98.1 \text{ g H}_2\text{SO}_4}{\cancel{\text{mol H}_2\text{SO}_4}}$

$= 4.91 \times 10^2 \text{ g H}_2\text{SO}_4$

18. $? \text{ mL Br}_2 = 2.11 \times 10^{22} \text{ molecules Br}_2 \times \dfrac{\text{mol Br}_2}{6.02 \times 10^{23} \text{ molecules Br}_2}$

$\times \dfrac{159.8 \text{ g Br}_2}{\text{mol Br}_2} \times \dfrac{\text{mL Br}_2}{3.10 \text{ g Br}_2}$

$= 1.81 \text{ mL Br}_2$

19. $? \text{ atoms H} = 45.0 \, \cancel{\text{g (NH}_4)_3\text{AsO}_3} \times \dfrac{\cancel{\text{mol (NH}_4)_3\text{AsO}_3}}{176.9 \, \cancel{\text{g (NH}_4)_3\text{AsO}_3}}$

$\times \dfrac{6.02 \times 10^{23} \, \cancel{\text{molecules (NH}_4)_3\text{AsO}_3}}{\cancel{\text{mol (NH}_4)_3\text{AsO}_3}} \times \dfrac{12 \text{ atoms H}}{\cancel{\text{molecule (NH}_4)_3\text{AsO}_3}}$

$= 1.84 \times 10^{24} \text{ atoms H}$

20. $\text{? atoms} = 196 \, \cancel{\text{g H}_2\text{SO}_4} \times \dfrac{\cancel{\text{mol H}_2\text{SO}_4}}{98.1 \, \cancel{\text{g H}_2\text{SO}_4}}$

    $\times \dfrac{6.02 \times 10^{23} \, \cancel{\text{molecules H}_2\text{SO}_4}}{\cancel{\text{mol H}_2\text{SO}_4}} \times \dfrac{7 \text{ atoms}}{\cancel{\text{molecule H}_2\text{SO}_4}}$

    $= 8.42 \times 10^{24} \text{ atoms}$

21. The first thing to do is to find the volume of the cube in cm^3.

    $$\text{cm}^3 = (1.0)^3 \, \cancel{\text{in}^3} \times \dfrac{(2.54)^3 \, \text{cm}^3}{\cancel{\text{in}^3}} = 16.4 \text{ cm}^3$$

    The rest of the problem is straightforward.

    $\text{? dollars} = 1.0 \times 10^6 \, \cancel{\text{Pt atoms}} \times \dfrac{\cancel{\text{mol Pt}}}{6.02 \times 10^{23} \, \cancel{\text{Pt atoms}}} \times \dfrac{195.1 \, \cancel{\text{g Pt}}}{\cancel{\text{mol Pt}}}$

    $\times \dfrac{\cancel{\text{mL Pt}}}{22.0 \, \cancel{\text{g Pt}}} \times \dfrac{\cancel{\text{cm}^3 \text{ Pt}}}{\cancel{\text{mL Pt}}} \times \dfrac{\$550}{16.4 \, \cancel{\text{cm}^3 \text{ Pt}}}$

    $= \$4.9 \times 10^{-16}$

    (In decimal form, this is $0.00000000000000049, which seems to be a good bargain. Calculate how much this number, 1.0×10^6, of Pt atoms will weigh.)

    $\text{? g Pt} = 1.0 \times 10^6 \, \cancel{\text{atoms Pt}} \times \dfrac{\cancel{\text{mol Pt}}}{6.02 \times 10^{23} \, \cancel{\text{atoms Pt}}} \times \dfrac{195.1 \text{ g Pt}}{\cancel{\text{mol Pt}}}$

    $= 3.2 \times 10^{-16} \text{ g Pt}$

    (This is roughly 1,000,000,000,000th as much as the dot on this i weighs. Maybe it wasn't such a bargain.)

22. Once again you must calculate the volume of the sample.

    $$\text{cm}^3 = (0.51)^3 \, \text{cm}^3 = 0.13 \text{ cm}^3$$

    Since $1 \text{ cm}^3 = 1 \text{ mL}$, you can write Volume $= 0.13 \text{ mL}$.

    $\text{? atoms O} = 0.13 \, \cancel{\text{mL imp}} \times \dfrac{4.26 \, \cancel{\text{g imp}}}{\cancel{\text{mL imp}}} \times \dfrac{92 \, \cancel{\text{g CaCO}_3}}{100 \, \cancel{\text{g imp}}} \times \dfrac{\cancel{\text{mol CaCO}_3}}{101.1 \, \cancel{\text{g CaCO}_3}}$

    $\times \dfrac{6.02 \times 10^{23} \, \cancel{\text{molecules CaCO}_3}}{\cancel{\text{mol CaCO}_3}} \times \dfrac{3 \text{ atoms O}}{\cancel{\text{molecule CaCO}_3}}$

    $= 9.1 \times 10^{21} \text{ atoms O}$

23. $\text{? min} = 1.50 \, \cancel{\mu\text{g Au}} \times \dfrac{10^{-6} \, \cancel{\text{g Au}}}{\cancel{\mu\text{g Au}}} \times \dfrac{\cancel{\text{mol Au}}}{197 \, \cancel{\text{g Au}}} \times \dfrac{\cancel{\text{century}}}{2.00 \, \cancel{\text{mol Au}}}$

    $\times \dfrac{100 \, \cancel{\text{yr}}}{\cancel{\text{century}}} \times \dfrac{365 \, \cancel{\text{days}}}{\cancel{\text{yr}}} \times \dfrac{24 \, \cancel{\text{hr}}}{\cancel{\text{day}}} \times \dfrac{60 \text{ min}}{\cancel{\text{hr}}}$

    $= 1.5 \times 10^{-6} \times \dfrac{1}{197} \times \dfrac{1}{2.00} \times 100 \times 365 \times 24 \times 60 \text{ min}$

    $= 0.200 \text{ min}$

24. You must calculate the weight of each sample.

$$? \text{ g } H_2O = 0.082 \cancel{\text{ mol } H_2O} \times \frac{18.0 \text{ g } H_2O}{\cancel{\text{mol } H_2O}}$$

$$= 0.082 \times 18.0 \text{ g } H_2O = 1.476 \text{ g } H_2O$$

or 1.5 g H_2O to the correct significant figures

$$? \text{ g } H_2SO_4 = 3.01 \times 10^{21} \cancel{\text{ molecules } H_2SO_4} \times \frac{\cancel{\text{mol } H_2SO_4}}{6.02 \times 10^{23} \cancel{\text{ molecules } H_2SO_4}}$$

$$\times \frac{98.1 \text{ g } H_2SO_4}{\cancel{\text{mol } H_2SO_4}}$$

$$= 3.01 \times 10^{21} \times \frac{1}{6.02 \times 10^{23}} \times 98.1 \text{ g } H_2SO_4 = 0.4905 \text{ g } H_2SO_4$$

or 0.490 g H_2SO_4 to the correct significant figures

The water sample weighs more. (If you had trouble with the powers of 10 in the second calculation, remember that you shift the 10 to a power from the bottom to the top and change its sign.)

25. $$? \text{ molecules } CaCO_3 = 3.0 \cancel{\text{ g calcite}} \times \frac{94 \cancel{\text{ g } CaCO_3}}{100 \cancel{\text{ g calcite}}} \times \frac{\cancel{\text{mol } CaCO_3}}{100.1 \cancel{\text{ g } CaCO_3}}$$

$$\times \frac{6.02 \times 10^{23} \text{ molecules } CaCO_3}{\cancel{\text{mol } CaCO_3}}$$

$$= 3.0 \times \frac{94}{100} \times \frac{1}{100.1} \times 6.02 \times 10^{23} \text{ molecules } CaCO_3$$

$$= 1.7 \times 10^{22} \text{ molecules } CaCO_3$$

26. $$? \text{ lb } Hg = \cancel{\text{mol } Hg} \times \frac{200.6 \cancel{\text{ g } Hg}}{\cancel{\text{mol } Hg}} \times \frac{\text{lb } Hg}{454 \cancel{\text{ g } Hg}}$$

$$= 1 \times 200.6 \times \frac{\text{lb } Hg}{454} = 0.442 \text{ lb } Hg$$

27. First you count the number of atoms in one molecule of K_2HAsO_4. There are 2 K atoms, 1 H atom, 1 As atom, and 4 O atoms—a total of 8 atoms per molecule.

$$? \text{ atoms} = 11 \cancel{\text{ mol } K_2HAsO_4} \times \frac{6.02 \times 10^{23} \cancel{\text{ molecules}}}{\cancel{\text{mol } K_2HAsO_4}} \times \frac{8 \text{ atoms}}{\text{molecule}}$$

$$= 11 \times 6.02 \times 10^{23} \times 8 \text{ atoms}$$

$$= 5.3 \times 10^{25} \text{ atoms (to the correct significant figures)}$$

(The number of atoms in a molecule is a counted number and has no bearing on the significant figures.)

28. You determine the number of molecules in each sample separately.

$$? \text{ molecule } NaCl = 0.500 \cancel{\text{ mol } NaCl} \times \frac{6.02 \times 10^{23} \text{ molecules } NaCl}{\cancel{\text{mol } NaCl}}$$

$$= 0.500 \times 6.02 \times 10^{23} \text{ molecules } NaCl$$

$$= 3.01 \times 10^{23} \text{ molecules } NaCl$$

$$? \text{ molecule } C_2H_5OH = 0.500 \text{ mL } \cancel{C_2H_5OH} \times \frac{0.789 \text{ g } \cancel{C_2H_5OH}}{\text{mL } \cancel{C_2H_5OH}}$$

$$\times \frac{\cancel{\text{mol } C_2H_5OH}}{46.0 \text{ g } \cancel{C_2H_5OH}} \times \frac{6.02 \times 10^{23} \text{ molecules } C_2H_5OH}{\cancel{\text{mol } C_2H_5OH}}$$

$$= 0.500 \times 0.789 \times \frac{1}{46.0} \times 6.02 \times 10^{23} \text{ molecules } C_2H_5OH$$

$$= 5.16 \times 10^{21} \text{ molecules } C_2H_5OH$$

$$? \text{ molecule } H_2O = 0.500 \text{ g } \cancel{H_2O} \times \frac{\cancel{\text{mol } H_2O}}{18.0 \text{ g } \cancel{H_2O}} \times \frac{6.02 \times 10^{23} \text{ molecules } H_2O}{\cancel{\text{mol } H_2O}}$$

$$= 0.500 \times \frac{1}{18} \times 6.02 \times 10^{23} \text{ molecules } H_2O$$

$$= 1.67 \times 10^{22} \text{ molecules } H_2O$$

The 0.500 mol of NaCl has the most molecules.

CHAPTER 6

Percentage Composition of Molecules

PRETEST

You should be able to complete this test in 15 minutes.

1. What is the weight percent N in NO?
2. What is the weight percent N in N_2O?
3. What is the weight percent N in NO_2?
4. What is the weight percent C in $Ca(CN)_2$?
5. What is the weight percent C in $C_{16}H_{26}O_4N_2S$ (penicillin)?
6. What is the weight percent S in $Na_2S_2O_3 \cdot 5H_2O$?
7. What is the weight percent N in $(NH_4)_2SO_4$?
8. What is the weight percent water (H_2O) in $Na_2SO_4 \cdot 10H_2O$?

PRETEST ANSWERS

You probably got all these right unless you were not careful in adding the atomic weights to get the molecular weights. The molecular weight is shown along with the answers in order to see if this is the problem.

If you got all the answers correct except perhaps for one careless mistake in the molecular weights, go directly to the problem set at the end of the chapter. If you missed several, you had better study the chapter.

1. 46.7% N	(MW = 30.0 g/mol)		2. 63.6% N	(MW = 44.0 g/mol)
3. 30.4% N	(MW = 46.0 g/mol)		4. 26.1% C	(MW = 92.1 g/mol)
5. 56.1% C	(MW = 342.1 g/mol)		6. 25.9% S	(MW = 248.2 g/mol)
7. 21.2% N	(MW = 132.1 g/mol)		8. 55.9% H_2O	(MW = 322.1 g/mol)

SECTION A Percentage Composition by Dimensional Analysis

Chemists often want to know how much of a given element is present in a compound. They generally express this as the weight percent of the element in the compound, which makes it easy to calculate how much of the element is in a sample of any size of that particular compound. Once you know the formula of the compound, calculating the percentage composition is very simple. We will start by looking at how you can solve these problems by dimensional analysis, and then you will see a short-cut method for doing the same thing.

First we will find the weight percent of N in NO. Rewording the problem to express the question in usable dimensions, you can ask how many

100

grams of N there are in _____ grams of NO. (If you have forgotten this method for asking questions about percentage, check Section C of Chapter 4.)

Starting in the usual way, you write

g N; 100 g NO

$$? \rule{2cm}{0.4pt} = \rule{3cm}{0.4pt}$$

grams NO

The dimension that must be canceled is _____. The only con-

molecular

version factor available is the _____ weight of NO. So you

inverted (or reciprocal)

multiply by this conversion factor written in its _____ form so that the unwanted dimension cancels.

$$? \text{ g N} = 100 \text{ g NO} \times \rule{3cm}{0.4pt}$$

$\dfrac{\text{mol NO}}{30.0 \text{ g NO}}$

NO

The unwanted dimension now is mole _____, but the question asked

N

concerns grams __. You will need a "bridge" conversion factor that will

NO

relate N to _____.

one

The formula of the molecule, NO, tells you that there is _____ atom of N in one molecule of NO. There will always be the same number of

atoms

N _____ as there are NO molecules. Thus, if you had 6.02×10^{23}

mole

(which is called a _____) of NO molecules, you would have

6.02×10^{23}; N

_____ N atoms (which we call a mole of __ atoms). This is a conversion factor and can be written

N

$$\frac{1 \text{ mol \_\_ atoms}}{1 \rule{1.5cm}{0.4pt} \text{ NO molecules}}$$

mol

You now have the "bridge" that relates N to NO.

Continuing the problem, you cancel

$$? \text{ g N} = 100 \; \cancel{\text{g NO}} \times \frac{\text{mol NO}}{30.0 \; \cancel{\text{g NO}}}$$

and then you multiply by the "bridge," inverted so as to cancel the unwanted dimension.

$$? \text{ g N} = 100 \text{ g NO} \times \frac{\text{mol NO}}{30.0 \text{ g NO}} \times \underline{\hspace{2cm}}$$

$\dfrac{1 \text{ mol N atoms}}{\text{mol NO}}$

The unwanted dimension now is _____, from which you can get directly to the dimensions of the question asked by multiplying

mol N atoms

by the _____ weight of N.

atomic

$$? \text{ g N} = 100 \text{ g NO} \times \frac{\text{mol NO}}{30.0 \text{ g NO}} \times \frac{\text{mol N atoms}}{\text{mol NO}} \times \underline{\hspace{2cm}}$$

$$= \underline{\hspace{1.5cm}} \text{ g N in 100 g NO} = 46.7\% \text{ N}$$

$\dfrac{14.0 \text{ g N}}{\text{mol N atoms}}$

46.7

(Be careful to write "mole N atoms," because when you are talking you say "nitrogen," which could be either N or N_2. Be sure to write "mol N" and "mol N_2," or "mol O" and "mol O_2," or "mol H" and "mol H_2," etc.)

Let's try another example. What is the percent N in $(NH_4)_2SO_4$?

You will need three conversion factors: the _____ weight of $(NH_4)_2SO_4$,

molecular

$$\frac{132.1 \text{ g } (NH_4)_2SO_4}{\underline{\hspace{1.5cm}} (NH_4)_2SO_4}$$

mol

the _____ weight of N,

atomic

$$\frac{14.0 \underline{\hspace{1cm}}}{\text{mol} \underline{\hspace{0.5cm}}}$$

g N
N

and the "bridge" which will be

$$\frac{\underline{\hspace{0.5cm}} \text{ mol N}}{\text{mol} \underline{\hspace{1.5cm}}}$$

2
$(NH_4)_2SO_4$

If you didn't realize that there were two N atoms in each $(NH_4)_2SO_4$ molecule, recall that the subscript 2 after the parenthesis means that the entire NH_4 polyatomic ion is present two times in the molecule. Check back to Section B in Chapter 5 if you are still vague about this.

You can now solve the problem of what the weight percent N is in $(NH_4)_2SO_4$. You start by writing

$$? \text{ g N} = \underline{\hspace{1.5cm}} \text{ g } (NH_4)_2SO_4$$

100

Then you multiply by the three conversion factors, so that all unwanted dimensions cancel, giving as your final setup

$$? \text{ g N} = 100 \text{ g } (NH_4)_2SO_4 \times \frac{\text{mol } (NH_4)_2SO_4}{132.1 \text{ g } (NH_4)_2SO_4}$$

$$\times \frac{2 \text{ mol N}}{\text{mol } (NH_4)_2SO_4} \times \frac{14.0 \text{ g N}}{\text{mol N}}$$

$$= 21.2 \text{ g N in 100 g } (NH_4)_2SO_4 \quad \text{or } 21.2\% \text{ N}$$

SECTION B A Short-cut Method

These problems are actually so simple that it is hardly worth the effort to use a dimensional-analysis setup to get the answer. The answer always results from simply multiplying the ratio of the atomic weight of the atom of interest to the molecular weight of the molecule times the number of atoms of interest in the molecule times 100.

$$\frac{\text{Atomic weight of atom of interest}}{\text{Molecular weight of molecule}} \times \text{number of atoms of interest}$$

$$\times\ 100 = \text{weight percent of atom of interest}$$

Use the short-cut method to determine the weight percent S in $Na_2S_2O_3$. The ratio of the atomic weight of S to the molecular weight of

158.2

the compound is 32.1/_____. The number of S atoms per molecule is

2

___. Then the weight percent S is

100; 40.6%

$$\frac{32.1}{158.2} \times 2 \times \underline{\hspace{1cm}} = \underline{\hspace{1cm}}$$

SECTION C Percentage of a Complexed Molecule in a Compound

As mentioned in Chapter 5, there is a type of compound that is made up not only of atoms bonded together, but also of several compounds bonded to each other. In general, these are called complexes and are represented in their formulas by a dot joining the compounds that are bonded. Thus $Na_2S_2O_3 \cdot 5H_2O$, sodium thiosulfate pentahydrate, is five water molecules bonded to a sodium thiosulfate molecule. Or $CoCl_2 \cdot 2NH_3$

two

is cobaltous chloride with _____ ammonia molecules bonded to it. Sometimes you may be asked to find the percentage of the bonded molecule in these compounds.

For example, what is the percent H_2O in $Na_2S_2O_3 \cdot 5H_2O$? First you reword the question to ask how many grams of H_2O are in 100 g $Na_2S_2O_3 \cdot 5H_2O$.

g H_2O; 100 g $Na_2S_2O_3 \cdot 5H_2O$

? _____ = _____

Once again, you know the molecular weight of the $Na_2S_2O_3 \cdot 5H_2O$; and since the "bridge" must relate moles of H_2O to moles of $Na_2S_2O_3 \cdot 5H_2O$, this factor will be

5 mol H_2O

$$\frac{\underline{\hspace{2cm}}}{\text{mol } Na_2S_2O_3 \cdot 5H_2O}$$

grams H_2O

Since the answer's dimensions are _____, you must have a conversion factor relating moles H_2O to g H_2O. This will be the

molecular

_____ weight of H_2O. Your setup will therefore be

$$? \text{g } H_2O = 100 \text{ g Na}_2S_2O_3\cdot 5H_2O \times \frac{\text{mol Na}_2S_2O_3\cdot 5H_2O}{248.2 \text{ g Na}_2S_2O_3\cdot 5H_2O}$$

$$\times \frac{5 \text{ mol } H_2O}{\text{mol Na}_2S_2O_3\cdot 5H_2O} \times \frac{18.0 \text{ g } H_2O}{\text{mol } H_2O}$$

$$= 100 \times \frac{1}{248.2} \times 5 \times 18.0 \text{ g } H_2O$$

$$= 36.2 \text{ g } H_2O \text{ or } 36.2\% \text{ } H_2O$$

The short-cut method for setting up this type of problem would be to multiply the ratio of the molecular weight of the complexed molecule of interest to the molecular weight of the whole compound times the number of complexed molecules times 100.

$$\frac{\text{Molecular weight of complexed molecule}}{\text{Molecular weight of the whole compound}}$$

$$\times \text{ number of complexed molecules} \times 100$$

$$= \text{weight percent of complexed molecule}$$

PROBLEM SET

1. What is the weight percent O in MgO?

2. What is the weight percent Fe in Fe_2O_3?

3. What is the weight percent Cl in $Mg(ClO_3)_2$?

4. What is the weight percent As in $Ca_3(AsO_4)_2$?

5. What is the weight percent O in N_2O_5?

6. What is the weight percent C in CH_3COONa?

7. What is the weight percent C in C_2H_5OH?

8. What is the weight percent K in K_3PO_4?

9. What is the weight percent S in H_2S?

10. What is the weight percent H in H_2S?

11. What relationship is there between the answers to Problems 9 and 10?

12. What is the weight percent Cl in HCl?

13. What is the weight percent O in SnO_2?

14. What is the weight percent O in P_4O_{10}?

15. What is the weight percent H in $C_5H_9O_4N$?

16. What is the weight percent P in P_4O_{10}? (Don't bother to calculate this in the usual way. Just use your answer to Problem 14.)

17. What is the weight percent NH_3 in $Cu(OH)_2\cdot 4NH_3$?

18. What is the percent water in $CH_3COONa\cdot 3H_2O$?

19. What is the percent PbO in $PbSO_4 \cdot PbO$?

20. What is the percent CO in $Fe(CO)_5$?

PROBLEM SET ANSWERS

If you answered 17 out of the 20 problems correctly, you are ready to proceed to Chapter 7. If you missed more than three, first check to see that you were using the correct molecular weight. If your errors were not in the molecular weights, you should go over the chapter again.

The molecular weights of the molecules are shown after each answer.

1. 39.7% O   (MW = 40.3 g/mol)   2. 69.9% Fe   (MW = 159.6 g/mol)

3. 37.1% Cl   (MW = 191.3 g/mol)   4. 37.6% As   (MW = 398.1 g/mol)

5. 74.1% O   (MW = 108.0 g/mol)   6. 29.3% C   (MW = 82.0 g/mol)

7. 52.2% C   (MW = 46.0 g/mol)   8. 55.3% K   (MW = 212.3 g/mol)

9. 94.1% S   (MW = 34.1 g/mol)   10. 5.92% H   (MW = 34.1 g/mol)

11. The sum of the weight percents is 100.

12. 97.3% Cl   (MW = 36.5 g/mol)   13. 21.2% O   (MW = 150.7 g/mol)

14. 56.3% O   (MW = 284.0 g/mol)   15. 6.12% H   (MW = 147.0 g/mol)

16. 100 – % O = % P, so 100 – 56.3 = 43.7% P

17. 41.1% NH_3 (MW = 165.5 g/mol)   18. 39.7% H_2O (MW = 136.0 g/mol)

19. 42.4% PbO (MW = 526.4 g/mol)   20. 71.5% CO   (MW = 195.8 g/mol)

CHAPTER 7

The Simplest, or Empirical, Formula

PRETEST

You should be able to finish this test in 20 minutes. Do all the problems and then check your answers.

Determine the empirical formulas for the following.

1. A compound containing 5.9% H and 94.1% O.

2. A compound containing 75% C and 25% H.

3. A compound containing 69.6% Fe and the remainder is O.

4. A compound containing 32.4% Na, 22.6% S, and 45.0% O.

5. A compound containing 28.2% N, 8.1% H, 20.8% P, and 42.9% O. It is known to contain the (NH_4^+) radical (polyatomic ion). Write your formula accordingly.

6. A compound containing 92.3% C and the remainder H.

7. Suppose that, by another technique, the molecular weight of the compound in question 6 is found to be 78.0 g/mol. What is the molecular formula?

PRETEST ANSWERS

If you got all seven correct, you can go directly to the problem set at the end of the chapter. If you missed question 7 only, read through Section B at the end of the chapter. If you missed more than one out of questions 1–6, you had better go through the entire chapter.

1. HO   2. CH_4 3. Fe_2O_3 4. Na_2SO_4 5. $(NH_4)_3PO_4$

6. CH   7. C_6H_6

The Percentage Composition Turned the Other Way Around

In the preceding chapter you saw how it is possible to calculate the weight percent of all the atoms if you know the formula for the compound. In this chapter we are going to turn the whole process around. You will determine the formula of a compound if you know the weight percentage of the atoms that make it up.

SECTION A   The Empirical Formula

The formula of a compound shows the number of each type of atom in that compound. However, the ratio of the numbers of atoms is the same whether you consider 1 molecule, 10 molecules, a mole of molecules, or any sample of the compound.

A *percentage composition* of a substance tells you the weight of each type of atom in a 100 g sample of the substance. From these weights you can determine the number of moles of each type of atom in a 100 g sample. You can then determine their ratio in the compound. As you will see a little later on, you will get a formula for a molecule that has the correct ratio of the atoms, but not necessarily the correct number. This is why this is referred to as the *empirical*, or simplest, formula.

For example, what is the empirical formula for a compound that is 52.9% Cl and 47.1% Cu? You know that the compound must have a formula in the form

$$Cu_xCl_y$$

where x = the number of Cu atoms per molecule and y = the number of Cl atoms per molecule. (To be absolutely correct, there are no true molecules present, but only ions held together by the attractions of their opposite charges. Nevertheless, for simplicity's sake, we will refer to these ionic compounds as molecules.)

From the percentage composition you can calculate the number of moles of each atom in a 100 g sample, and you know that the ratio in the 100 g sample will be the same as the ratio in the molecule. Let us make the calculation.

$$x = \text{mol Cu} = 47.1 \, \text{g Cu} \times \frac{\text{mol Cu}}{63.5 \, \text{g Cu}} = 0.742 \, \text{mol Cu}$$

35.5; 1.49

$$y = \text{mol Cl} = 52.9 \, \text{g Cl} \times \frac{\text{mol Cl}}{\underline{\quad} \, \text{g Cl}} = \underline{\quad} \, \text{mol Cl}$$

You now know the ratio of atoms in the formula.

1.49

$$Cu_{0.742}Cl\underline{\quad}$$

We never use anything but whole numbers of atoms in a formula, so you must reduce these values to whole numbers. The best way to do this is to start by dividing all the values by the smallest number, since this will give you 1, at least for this number. Thus

$Cu_{0.742/0.742}Cl_{1.49/0.742}$ or $Cu_1Cl\_\_$ | 2

You know that the ratio of Cu atoms to Cl atoms in the molecule is 1 to 2. However, you do not know that the real formula (molecular formula) is $CuCl_2$. It might be Cu_2Cl_4, or Cu_3Cl_6, or any combination with a 1-to-2 ratio.

We will try another one. What is the empirical formula for a compound that is 18.4% Al, 32.6% S, and the remainder O? First you know that the

sum of the percentages will always be _____. Therefore it is possible to determine the percent O simply by subtracting the sum of all the other

percentages from _____. This, then, will give you the percent O as

_____ %.

| 100

| 100
| 49.0

The formula for the compound must be in the form $Al_xS_yO_z$. You must now determine the number of moles of each element in your 100 g sample represented by the percentage. Therefore

$$x = \text{mol Al} = 18.4 \, \cancel{\text{g Al}} \times \frac{\text{mol Al}}{27.0 \, \cancel{\text{g Al}}} = 0.681 \text{ mol Al}$$

$$y = \text{mol S} = 32.6 \text{ g S} \times \underline{\hspace{2cm}} = \underline{\hspace{2cm}}$$

$$z = \text{mol O} = \underline{\hspace{2cm}} \times \frac{\text{mol O}}{16.0 \text{ g O}} = \underline{\hspace{2cm}}$$

| $\frac{\text{mol S}}{32.1 \text{ g S}}$; 1.02 mol S

| 49.0 g O; 3.06 mol O

The formula can now be written as Al____ S____ O____ .

To get a whole-number ratio, you can divide each of these by _____.

You will now have Al_1S____ O____ .

But these are not whole numbers. However, if you multiply each of

them by __, you will end up with $Al_2S_3O_9$, which is the simplest, or empirical, formula.

Sometimes you may have to look very hard for a number to multiply by to get whole numbers for all the atoms. However, in 90% of the formulas that you run across, it will be 2, 3, 4, or 5.

| 0.681; 1.02; 3.06

| 0.681

| 1.50; 4.50

| 2

SECTION B The Molecular Formula

As you saw in Section A, you can find the simplest, or empirical, formula simply from the weights of the various atoms in a sample of substance. In order to get the true, or molecular, formula, you need one more piece of information, the molecular weight of the molecule.

Let's look at Question 7 in the pretest. You have determined the empirical formula in Question 6 through the following calculation.

$$? \text{ mol C} = 92.3 \text{ g C} \times \underline{\hspace{2cm}} = \underline{\hspace{2cm}}$$

$$? \text{ mol H} = \underline{\hspace{2cm}} \times \frac{\text{mol H}}{1.0 \text{ g H}} = \underline{\hspace{2cm}}$$

| $\frac{\text{mol C}}{12.0 \text{ g C}}$; 7.70 mol C

| 7.7 g H; 7.7 mol H

The empirical formula is _____. In question 7 you are told that the molecular weight is 78.0 g/mol. If the molecular formula were the same

| CH

as the empirical formula, CH, then the molecular weight would be

13.0; one   _____, which you get by simply adding the atomic weight of _____

one   C to the atomic weight of _____ H. But it isn't. Therefore, the molecular formula must be some other combination of C and H in a 1-to-1 ratio that has a molecular weight of 78.0 g/mol.

How many CH's are needed to have a molecular weight of 78.0 g/mol? You divide the true molecular weight by the molecular weight of the

13.0   empirical formula, _____ g/mol.

$$\frac{78.0}{13.0} = 6$$

This means that the true molecule must be made up of six empirical

C_6H_6 formulas. So the true molecular formula is _____.

Notice that the formula was written C_6H_6 and not 6CH. By using the subscripts, you indicate that you have one molecule with six C's and six H's bonded together. Writing 6CH would mean that you have six separate molecules, each of which has one C bonded to one H.

You are now ready for the problem set.

PROBLEM SET

Determine the empirical formula for each compound whose percentage composition is shown below.

1. 77.7% Fe and 22.3% O

2. 43% C and 57% O

3. 78.3% B and 21.7% H

4. 40.3% K, 26.7% Cr, and 33.0% O

5. 21.8% Mg, 27.8% P, and 50.4% O

6. 32.0% C, 42.6% O, 18.7% N, and the remainder H

7. 31.4% C, 1.6% H, 41.9% Br, and 25.1% O

8. 31.9% K, 28.9% Cl, and the remainder O

9. 42.1% C, 51.5% O, and the remainder H

10. 52.8% Sn, 12.4% Fe, 16.0% C, and 18.8% N

Determine the true molecular formulas for the following compounds.

11. 94.1% O and the remainder H with a molecular weight of 34.0 g/mol.

12. 84.9% Hg and the remainder Cl, with a molecular weight of 472.2 g/mol.

13. 40.0% C, 53.4% O, and the remainder H with a molecular weight of 90.0 g/mol.

14. 12.26% N, 3.54% H, 28.1% S, and 56.1% O. The molecular weight is 228.2 g/mol. The formula is known to contain the NH_4^+ grouping. Write your formula accordingly.

15. 37.7% Ce, 28.4% Cl, and 33.9% water (H_2O). The molecular weight is 372.6 and the formula will be written $Ce_xCl_y \cdot zH_2O$.

16. 71.5% Hg, 5.0% N, 17.1% O, and 6.4% H_2O, with a molecular weight of 561.2 g/mol.

PROBLEM SET ANSWERS

If you got the correct answers to eight of the ten problems, you may go on to Chapter 8. If you missed problems 8, 9, and 10, redo Section B of the chapter. If you missed two or more of problems 1–7, redo the entire chapter. Problem 6 is particularly difficult.

The answers are shown below along with the first formula prior to adjusting the ratio to whole numbers.

1. FeO $(Fe_{1.39}O_{1.39})$

2. CO $(C_{3.6}O_{3.6})$

3. BH_3 $(B_{7.25}H_{21.5})$

4. K_2CrO_4 $(K_{1.03}Cr_{0.514}O_{2.06})$

5. $Mg_2P_2O_7$ $(Mg_{0.897}P_{0.897}O_{3.15})$

6. $C_2O_2NH_5$ $(C_{2.67}O_{2.67}N_{1.34}H_{6.70})$

7. $C_5H_3BrO_3$ $(C_{2.62}H_{1.58}Br_{0.524}O_{1.57})$

8. $KClO_3$ $(K_{0.816}Cl_{0.814}O_{2.45})$

9. $C_{12}H_{22}O_{11}$ $(C_{3.51}H_{6.40}O_{3.22})$ (First $\div$ by 3.22; then $\times$ 11)

10. $Sn_2FeC_6N_6$ $(Sn_{0.444}Fe_{0.222}C_{1.33}N_{1.34})$

11. H_2O_2 $\left[H_{5.9}O_{5.9} = HO \ (MW = 17.0) \ \dfrac{34.0}{17.0} = 2 \right]$

12. Hg_2Cl_2 $\left[Hg_{0.423}Cl_{0.423} = HgCl \ (MW = 236.1) \ \dfrac{472.2}{236.1} = 2 \right]$

13. $C_3H_6O_3$ $\left[C_{3.33}H_{6.6}O_{3.33} = CH_2O \ (MW = 30.0) \ \dfrac{90.0}{30.0} = 3 \right]$

14. $(NH_4)_2S_2O_8$ $\left[N_{0.876}H_{3.50}H_{0.875}S_{0.875}O_{3.51} = NH_4SO_4 \ (MW = 114.1) \right.$
    $\left. \dfrac{228.2}{114.1} = 2 \right]$

15. $CeCl_3 \cdot 7H_2O$ $\left[Ce_{0.269}Cl_{0.800}(H_2O)_{1.88} = CeCl_3 \cdot 7H_2O \ (MW = 372.6) \right.$
    $\left. \dfrac{372.6}{372.6} = 1 \right]$

16. $Hg_2N_2O_6 \cdot 2H_2O$ $\left[Hg_{0.356}N_{0.375}O_{1.07}(H_2O)_{0.356} = HgNO_3 \cdot H_2O \right.$
    $(MW = 280.6) \ \dfrac{561.2}{280.6} = 2 \Big]$

CHAPTER 8

The Balanced Chemical Equation

PRETEST

You may not have been exposed to balancing chemical equations, or if so, only to handling the very simplest kinds (similar to the first four problems on this pretest). Net ionic equations and the balancing of electron-transfer (redox) equations may be a mystery to you. This pretest will let you know. If you really can handle the material, you should be able to do all of the following problems in 30 minutes.

Section A

Balance the following equations by inspection. Do not leave any fractional coefficients.

1. $Ca(OH)_2 + HCl \neq CaCl_2 + H_2O$ 2. $Fe + O_2 \neq Fe_2O_3$

3. $Na_2O_2 + H_2O \neq NaOH + O_2$ 4. $PbS + O_2 \neq PbO + SO_2$

Section B

Write the net ionic equations for the following reactions.

5. $AgNO_3(aq)$ with $NaCl(aq)$ produces $NaNO_3(aq)$ and $AgCl(s)$

6. $NaOH(aq)$ and $HCl(aq)$ produces $NaCl(aq)$ and H_2O

7. Which of the following metathesis (double-replacement) reactions will "go" (actually give a reaction)?

    (a) $Na_2CO_3(aq)$ and $HCl(aq)$ (b) $AgNO_3(aq)$ and $HCl(aq)$

    (c) $NaOH(aq)$ and $KCl(aq)$ (d) $KOH(aq)$ and $NH_4I(aq)$

Section C

8. What is the oxidation number for the following?

    (a) Fe in Fe^{3+} (b) S in H_2SO_3

    (c) S in S_8 (elemental sulfur)   (d) Mn in MnO_4^-

    (e) C in CH_3COOH (Consider the average value per C atom.)

Section D

9. In the following reactions, which of the reactants, if any, is oxidized?

    (a) $Mg + 2HCl = MgCl_2 + H_2$

    (b) $C_2O_4{}^{2-} + AsO_4{}^{3-} + 2H^+ = 2CO_2 + AsO_3{}^{3-} + H_2O$

    (c) $SOCl_2 + 2H_2O = H_2SO_3 + 2HCl$

    (d) $NaClO + H_2S = S + NaCl + H_2O$

10. Write the oxidation half reaction for equation (b) in Question 9.

Section E

11. Balance the following equations. Use no fractional coefficients.

    (a) $KMnO_4 + Fe + HCl \neq MnCl_2 + FeCl_3 + KCl + H_2O$

    (b) $Na_2Cr_2O_7 + H_2C_2O_4 + HCl \neq CrCl_3 + CO_2 + H_2O + NaCl$

    (c) $NaCN + KMnO_4 + H_2O \neq NaCNO + MnO_2 + KOH$

PRETEST ANSWERS

If you didn't get the correct answers for all the problems, you can find most of them worked out in the chapter. The pretest notes the section in the chapter that covers the various questions, and you can guide your studying on the basis of this.

1. $Ca(OH)_2 + 2HCl = CaCl_2 + 2H_2O$ 2. $4Fe + 3O_2 = 2Fe_2O_3$

3. $2Na_2O_2 + 2H_2O = 4NaOH + O_2$ 4. $2PbS + 3O_2 = 2PbO + 2SO_2$

5. $Ag^+ + Cl^- = AgCl$ 6. $OH^- + H^+ = H_2O$

7. (a), (b), and (d) will "go."

8. (a) +3 (b) +4 (c) 0 (d) +7 (e) 0

9. (a) Mg (b) $C_2O_4{}^{2-}$ (c) none (d) H_2S

10. $C_2O_4{}^{2-} = 2CO_2 + 2e^-$

11. (a) $3KMnO_4 + 5Fe + 24HCl = 3MnCl_2 + 5FeCl_3 + 12H_2O + 3KCl$

    (b) $Na_2Cr_2O_7 + 3H_2C_2O_4 + 8HCl = 2CrCl_3 + 6CO_2 + 7H_2O + 2NaCl$

    (c) $2KMnO_4 + 3NaCN + H_2O = 3NaCNO + 2MnO_2 + 2KOH$

When one or more substances undergo a change to form different substances, we say that a *chemical reaction* has occurred. It is possible to express the reaction in words, using the names of the substances. However, it is much simpler to express the various substances as formulas. Thus the fact that, when hydrogen *reacts* with chlorine, it forms hydrogen chloride is best expressed as

$$H_2 + Cl_2 = 2HCl$$

This is the *balanced chemical equation*. Before you can do any calculations about the amounts of any of the substances, you must have just such a balanced equation. The rest of this chapter will be spent examining how to balance them.

SECTION A   Balancing Simple Chemical Equations by Inspection

A chemical equation is a real equality. The left side equals the right side. By this we mean that the number of atoms of an element on one side equals the number of atoms of the same element on the other. Also, if there are charges on any of the substances (that is, if there are ions), the algebraic sum of the charges on one side must equal that on the other. The way that you achieve these equalities is to adjust the number of molecules or ions on both sides.

In the example in the introduction, we had H_2 and Cl_2 reacting to produce HCl. If we had written $H_2 + Cl_2 = HCl$, the equation would not balance. There are two H atoms on the left and only one H atom on the right. The Cl atoms also do not balance. There are _____ Cl atoms on | two

the left, and only _____ Cl atom on the right. It is very easy to balance | one
this equation by simply writing that two HCl molecules are formed.

$$H_2 + Cl_2 = 2HCl$$

Now the equation is balanced, for there are two H atoms on both sides

and _____ atoms on both sides. In words, the equation now reads, | two Cl
"One hydrogen molecule reacts with one chlorine molecule to yield two hydrogen chloride molecules." Notice that you put the 2 in front of the HCl. This is the way that you show that you have two HCl molecules. If you had written H_2Cl_2, you would have changed the molecule into a new molecule containing two H atoms and two Cl atoms all bonded together instead of the correct molecule containing only one H and one Cl bonded together. These numbers written in front of the formula of the molecule are called *coefficients*. They indicate the number of molecules present.

Most equations can be balanced by just looking at them and adjusting the coefficients of each molecule. If an equation has one molecule with a complicated formula, it is best to start with that molecule, counting the number of each type atom in it and inserting coefficients on the other side to balance these atoms. Then you work back to balance the number of atoms you now have.

This sounds complex, but another example will clarify the whole picture. Balance the equation

$$Sn(OH)_4 + H_3PO_4 \neq Sn_3(PO_4)_4 + H_2O$$

$Sn_3(PO_4)_4$; 3

4; 16

The most complicated formula is _____. It contains __ Sn atoms, __ P atoms and ____ O atoms. (If you have forgotten that a subscript after a parenthesis means that there are that many times each of the atoms inside the parentheses, check back to Section B of Chapter 5.)

3

To get three Sn atoms on the left, you put a coefficient __ in front of the $Sn(OH)_4$ molecule.

$$3Sn(OH)_4 + H_3PO_4 \neq Sn_3(PO_4)_4 + H_2O$$

four

4

H_3PO_4

The most complicated molecule also contains _____ P atoms. To get this many P atoms on the left, you have to put a coefficient of __ in front of the _____ molecule on the left.

$$3Sn(OH)_4 + 4H_3PO_4 \neq Sn_3(PO_4)_4 + H_2O$$

24

12; 12

Next you balance the H atoms. There are a total of ____ H atoms on the left. This is the sum of the ____ in the $3Sn(OH)_4$ and ____ in the $4H_3PO_4$.

[You must be *very* careful when you count atoms. If something is in parentheses, multiply the subscript *outside* the parentheses by any subscript following the atom *inside* the parentheses. That gives you the total number of atoms per molecule. If there is a coefficient in front of the molecule, that is the number of molecules. So you multiply the number

45

24

12

of atoms per molecule by this coefficient. For example, there are ____ O atoms in $5Al_2(SO_3)_3$.] Anyhow, since there are ____ H atoms on the left, this must come from ____ H_2O molecules.

$$3Sn(OH)_4 + 4H_3PO_4 = Sn_3(PO_4)_4 + 12H_2O$$

12

16; 28

16

12; 28

Now only the O atoms remain to be balanced. There are ____ O atoms in $3Sn(OH)_4$ and ____ O atoms in $4H_3PO_4$. That's a total of ____ O atoms on the left side of the equation. There are ____ O atoms in $Sn_3(PO_4)_4$ and ____ O atoms in $12H_2O$. That's a total of ____ O atoms on the right side. Both sides have the same number. Therefore the O atoms are balanced.

The balanced equation is thus

$$3Sn(OH)_4 + 4H_3PO_4 = Sn_3(PO_4)_4 + 12H_2O$$

3

3; 28

You should always check to see whether an equation is really balanced by counting atoms on both sides. There are __ Sn atoms on the left and __ Sn atoms on the right. There are a total of ____ O atoms on the left and _____ O atoms on the right. There are _____ H atoms on the left and

_____ H atoms on the right. There are __ P atoms on the left and __ P

atoms on the right. There are no charges on either side of the equation.
Therefore, it *must be balanced.*

Balancing equations by inspection takes a little practice. After a while
you will teach yourself some shortcuts. For example, in the equation we
just balanced, the PO_4 polyatomic ion (*radical*) appears exactly the same
on both sides. You could consider it just one unit. You see four PO_4's in
the $Sn_3(PO_4)_4$ and therefore you need four H_3PO_4's to balance them.
Here is a similar example. Balance

$$Pb(CH_3COO)_2 + (NH_4)_2SO_4 \neq PbSO_4 + NH_4(CH_3COO)$$

Here there are three different radicals (polyatomic ions): SO_4,
CH_3COO, and NH_4. They appear the same on both sides of the equation.
(CH_3COO^- is the acetate ion.) So you look at a complicated molecule,

say $Pb(CH_3COO)_2$. It has _____ CH_3COO groups. Therefore you need a

__ as the coefficient for the product _____.

$$Pb(CH_3COO)_2 + (NH_4)_2SO_4 \neq PbSO_4 + \_\_ NH_4(CH_3COO)$$

This means that there are _____ NH_4 groups on the right. Well, there

are already _____ NH_4 radicals on the left. So these NH_4's are balanced.

Also, there is _____ Pb on the right and one on the left. These are

_____. All that remains to be checked are the SO_4 polyatomic

ions. There is _____ on the left and _____ on the right. So everything
is balanced! That is certainly much easier than counting individual atoms.

Sometimes an atom appears with an even subscript on one side of the
equation in some molecule and an odd subscript on the other. In the pretest, you had

$$Fe + O_2 \neq Fe_2O_3$$

If you tried to put a coefficient of two in front of the Fe,

$$2Fe + O_2 \neq Fe_2O_3$$

you would have to multiply the O_2 by 1½ to get O_3. But for most purposes, fractional coefficients are not used. The solution to this problem
is to multiply the Fe_2O_3 by two, which will give you an even number of
O atoms:

$$Fe + O_2 \neq 2Fe_2O_3$$

Now the complicated molecules have a total of four Fe atoms and

_____ O atoms. Therefore you have to put a __ in front of the Fe and a

__ in front of the O_2:

$$4Fe + 3O_2 = 2Fe_2O_3$$

28; 24
24; 4; 4
two
2; $NH_4(CH_3COO)$
2
two
two
one
balanced
one; one
six; 4
3

4	This is now balanced, since there are ___ Fe atoms on both the left and
6	the right and ___ O atoms on both left and right.

The last question in this group on the pretest was the same sort of problem:

$$PbS + O_2 \neq PbO + SO_2$$

There is an odd number of O atoms in PbO, but an even number in the

2	O_2. Therefore you put a coefficient of ___ in front of the PbO:

$$PbS + O_2 \neq 2PbO + SO_2$$

two	This results in _____ Pb atoms on the right. Therefore you must put a
2	coefficient of ___ in front of the PbS.

$$2PbS + O_2 \neq 2PbO + SO_2$$

two	Putting this last coefficient in resulted in _____ S on the left side and so
2	you must put a coefficient of ___ in front of the SO_2 on the right:

$$2PbS + O_2 \neq 2PbO + 2SO_2$$

The only thing remaining to be balanced is the O atoms. There are a total

six	of _____ O atoms on the right. Therefore you must put a coefficient of
3	___ in front of the O_2.

$$2PbS + 3\,O_2 = 2PbO + 2SO_2$$

Question 3 on the pretest is a little more complicated. The O atoms appear in two compounds on both the left and right side. One of the compounds has an odd number of O's and the other has an even number so we don't know what to multiply. However, the H atom has a subscript of 2 (even) in water and 1 (understood) in NaOH. So you can start balancing

$$Na_2O_2 + H_2O \neq NaOH + O_2$$

2; NaOH	by putting a ___ as coefficient for the _____ :

$$Na_2O_2 + H_2O \neq 2NaOH + O_2$$

two	There are now two Na on the left and _____ on the right. This looks very promising. But when you count the O atoms, there are a total of
three; four	_____ on the left and _____ on the right. This means that you would
$\frac{1}{2}$	have to have a coefficient of ___ for the O_2, but that is not allowed. The
two	only way to get a whole number is to multiply *all* the coefficients by ____.

$$(2)Na_2O_2 + (2)H_2O = (2)2NaOH + (2)\frac{1}{2}O_2$$

or

2; 2; 4; 1	___ Na_2O_2 + ___ H_2O = ___ NaOH + ___ O_2

Check to see that everything is balanced: There are _____ Na atoms on | four

both sides, _____ H atoms on both sides and a total of _____ O atoms | four: six
on each side.

SECTION B   Net Ionic Equations

In Section A, we more or less assumed that the balanced equation represented the ratio of molecules of reactants used to molecules of products that
result from the chemical reaction. Thus our first example in Section A was

$$H_2 + Cl_2 = 2HCl$$

We said that this meant that one molecule of H_2 reacts with one molecule
of Cl_2 to yield two molecules of HCl. This is quite true. However, consider
another reaction:

$$NaOH + HCl = NaCl + H_2O$$

When this reaction is carried out in a water solution, molecules are not involved. The reason is that the NaOH, HCl, and NaCl—when dissolved in
water—break up (*dissociate*) into positively and negatively charged frag-
ments called *ions*.
   [A word of caution. Compounds like NaCl and NaOH (except when they
are in gaseous form) always exist as ions, whether they are dissolved in
water or not. The water serves only to free the ions from their oppositely
charged neighbors, that is, to *dissociate*.]
   Thus NaOH(*aq*) [the (*aq*) means "in water solution"] is present as
separated Na^+ (a sodium ion) and OH^- (a hydroxide ion). The HCl(*aq*) is
present as an H^+ (a hydrogen ion, which is sometimes shown combined with
H_2O to give H_3O^+, hydronium ion) and a Cl^-, a chloride ion. [HCl is a
covalent compound, and not present as ions until dissolved in water.] The

NaCl(*aq*) is present as a sodium ion, Na^+, and a chloride ion, _____. | Cl^-
   If you rewrite the balanced equation we had before, showing the ions
that are present, you have

$$Na^+ + OH^- + H^+ + \text{_____} = Na^+ + Cl^- + H_2O$$ | Cl^-

For all intents and purposes, the water doesn't dissociate into ions.
   But notice that the Na^+ is present as a *reactant* (left side of the equation)
and also as a *product* (right side of the equation). That means that the Na^+
undergoes no chemical change. Therefore it need not appear in an equation
that is written to represent a chemical change. You can therefore eliminate
Na^+ from both sides of the equation:

$$\cancel{Na^+} + OH^- + H^+ + Cl^- = \cancel{Na^+} + Cl^- + H_2O$$

The same thing is true for the Cl^- ions. They appear both as reactants and
products, and can also be eliminated:

$$OH^- + H^+ + \cancel{Cl^-} = \cancel{Cl^-} + H_2O$$

This leaves just

$$OH^- + H^+ = H_2O$$

This is what is called the *net ionic equation*.

The Na^+ and Cl^- ions present in the solution both before and after the reaction takes place are called *spectator ions.* There are reasons why it is often helpful to write balanced chemical equations in their net ionic form. It makes it possible to tell whether a reaction will *go.* It is easier to balance. It shows the similarity of various reactions. *But,* when you are doing any kind of stoichiometric problem, you *cannot* use the net ionic equation. You must use the complete equation and the weights of the compounds, *including* the spectator ions.

Here is a net ionic equation for you to try. Consider the reaction of $AgNO_3(aq)$ with $NaCl(aq)$ to yield $NaNO_3(aq)$ and $AgCl(s)$. [The (s) means that the substance is an *insoluble solid*, and therefore not dissociated into ions. If the insoluble solid is a product, it is often called a *precipitate.*]

Start by writing the equation showing those substances that are dissociated as their separated ions:

NO_3^-; Cl^-; Na^+; NO_3^- $Ag^+ + $ _____ $ + Na^+ + $ _____ $ = $ _____ $ + $ _____ $ + AgCl(s)$

NO_3^-; Na^+ The _____ and _____ ions appear as both reactants and products and can be eliminated. This leaves, for the *net ionic equation*, just

Ag^+; Cl^-; $AgCl(s)$ _____ $ + $ _____ $ = $ _____

Here is another example. When $NH_4Cl(aq)$ reacts with $NaOH(aq)$, the products are $NaCl(aq)$ and $NH_3 \cdot H_2O(aq)$. [You may sometimes see $NH_3 \cdot H_2O$ written as "NH_4OH". There is no evidence that a molecule with the formula NH_4OH exists. So if you want to be correct, write it as hydrated ammonia, that is, $NH_3 \cdot H_2O$.] The $NH_3 \cdot H_2O$, even though it is dissolved in water, doesn't ionize into NH_4^+ and OH^- to any appreciable extent. Those compounds that hardly dissociate at all are called *weak electrolytes*. Compounds like $NaCl$, NH_4Cl, and $NaOH$ that dissociate almost completely are called *strong electrolytes*. When you write ionic equations, you never show weak electrolytes as separate ions.

So, for our example, you start by showing those substances that *are*

ions   dissociated as _____, and those that are *not* dissociated as molecules:

NH_4^+; OH^-; Na^+; Cl^- _____ $ + Cl^- + Na^+ + $ _____ $ = $ _____ $ + $ _____ $ + NH_3 \cdot H_2O$

Na^+; Cl^- Both the _____ and _____ ions appear on both sides of the equation. So you can remove them. This leaves, for the net ionic equation,

NH_4^+; $NH_3 \cdot H_2O$ _____ $ + OH^- = $ _____

In the reaction of $Na_2CO_3(aq)$ with $HCl(aq)$, the products are $NaCl(aq)$, H_2O, and $CO_2(g)$. [The (g) means that the compound is a gas

$Ag\,NO_3 + NaCl \longrightarrow \begin{array}{l} Na\,NO_3 \\ Ag\,Cl \end{array}$

that escapes from the system. All gases are molecular. That is, they *do not dissociate.*] The complete ionic equation, with all species shown as ions or molecules, is

$$2Na^+ + \underline{\hspace{1cm}} + H^+ + Cl^- \neq Na^+ + \underline{\hspace{1cm}} + H_2O + CO_2(g)$$

CO_3^{2-}; Cl^-

But this isn't balanced: There are two Na^+'s on the left side, but just one on the right. There is one H on the left, but two on the right. You must double the number of H^+ and Cl^- (that is, HCl's). You will then have

$$2Na^+ + CO_3^{2-} + \underline{\hspace{1cm}} + \underline{\hspace{1cm}} = 2Na^+ + 2Cl^- + H_2O + \underline{\hspace{1cm}}$$

$2H^+$; $2Cl^-$; $CO_2(g)$

Now you can see that there are two Na^+ ions and \underline{\hspace{2cm}} ions on each side. You can eliminate them. This leaves, as the net ionic equation,

two Cl^-

$$CO_3^{2-} + \underline{\hspace{1cm}} = \underline{\hspace{1cm}} + \underline{\hspace{1cm}}$$

$2H^+$; H_2O; $CO_2(g)$

Which Compounds Dissociate?

In order to write a net ionic equation, you must know which compounds dissociate when they are in aqueous solution and which do not. The first type of non-dissociating compound that we considered was a water-insoluble substance (AgCl was used in the example). Table 8.1 on page 132 gives some general rules for solubility. But, when in doubt, you can check in the *Handbook of Chemistry and Physics*, which lists the solubilities of various compounds.

The second class of compounds considered, although soluble in water, did not dissociate appreciably into ions. Recall that these compounds are called *weak electrolytes*. They may be either an acid (a compound that dissociates to give an H^+ ion), a base (a compound that produces an OH^- ion when it dissociates), or water itself. Table 8.2 gives the general rules for dissociation.

It isn't too difficult to decide which compounds are weak electrolytes. The generalizations are fairly straightforward and the rules quite simple. It is actually easiest to remember which compounds are *strong* electrolytes (there are fewer) and then realize that all others are weak ones. For example, there are only seven acids that dissociate strongly. If you can remember these, you'll know that all other acids are weak electrolytes. And to remember which bases are weak electrolytes, all you must remember is that any metallic hydroxide is a strong base *if it is soluble* in water. But from the solubility rules in Table 8.1, you would know that the soluble metallic hydroxides are only those of Group IA and Group IIA below Ca. Thus the hydroxide of any other element or group is a weak electrolyte.

The last type of compound considered that did *not* dissociate into ions was gases that escaped from the aqueous solution. The common gases (at room temperature) are H_2, O_2, N_2, Cl_2, CO, CO_2, NO, NO_2, N_2O, SO_2, NH_3, and H_2S.

Table 8.1 General Rules for Solubility for Some Compounds

Water-soluble Compounds (minimum 10 g compound dissolve per liter of water)

Compounds that contain the anion		Exceptions
NO_3^-	Nitrate	None
CH_3COO^-	Acetate	*CH_3COOAg
ClO_3^-	Chlorate	None
Cl^-	Chloride	$AgCl$, Hg_2Cl_2, $PbCl_2$
Br^-	Bromide	$AgBr$, Hg_2Br_2, *$PbBr_2$, *$HgBr_2$
I^-	Iodide	AgI, Hg_2I_2, PbI_2, HgI_2
SO_4^{2-}	Sulfate	Ag_2SO_4, $BaSO_4$, *$CaSO_4$, Hg_2SO_4, $PbSO_4$, $SrSO_4$

Compounds that contain the cation		Exceptions
Li^+, Na^+, K^+, Rb^+, Cs^+	Group IA metal ions	None
NH_4^+	Ammonium ion	None
H^+	Acids	Organic acids with more than five C atoms

Water-insoluble Compounds (less than 1 g compound dissolves per liter of water)

Compounds that contain the anion		Exceptions
S^{2-}	Sulfide	Sulfides of Group IA and Group IIA elements, $(NH_4)_2S$
OH^-	Hydroxide	Hydroxides of Group IA elements, NH_4OH, $Ba(OH)_2$, *$Ca(OH)_2$, $Sr(OH)_2$
CO_3^{2-}	Carbonate	Carbonates of Group IA elements, $(NH_4)_2CO_3$
SO_3^{2-}	Sulfite	Sulfites of Group IA elements, $(NH_4)_2SO_3$
PO_4^{3-}	Phosphate	Phosphates of Group IA elements, $(NH_4)_3PO_4$

*Slightly soluble: between 1 g and 10 g dissolve per liter of water

Table 8.2 General Rules for Dissociation in Water Solution

1. All soluble salts are strong electrolytes [exceptions: $HgCl_2$, $CdSO_4$, $(CH_3COO)_2Pb$].

2. The only acids that are strong electrolytes are $HClO_4$, $HClO_3$, HCl, HBr, HI, HNO_3, and the first H coming off H_2SO_4.

3. The soluble hydroxides of metals (Group IA and Ba^{2+}, Ca^{2+}, Sr^{2+} hydroxides) are strong electrolytes (bases).

4. All other compounds are weak electrolytes.

Using the Net Ionic Equation to Predict Whether a Double-Replacement Reaction Will Go

The fact that you can write a balanced equation for a chemical reaction is no assurance that the reaction will really occur. There must be some driving force in order for the reaction to proceed in the direction shown, and to yield the products written in the equation. If the products, once formed, immediately returned to being the reactants, no change would be observed.

There are some reactions that go in the forward direction to yield the products and at the same time go in the backward direction to use them up. When the rate of the forward reaction equals the rate of the backward reaction, the system is said to be in *equilibrium*. You will learn a great deal about this in more advanced chemistry courses.

However, suppose that you have some method of quickly removing the products from the reaction site so that they cannot return to being reactants. Then there is no way that the reaction can do anything but go in the forward direction, with reactants giving products. No equilibrium can be reached. The net ionic equation is a big help in seeing whether this is happening when you are considering what are called *double replacement* (or—more scientific word—*metathesis*) *reactions*. What occurs in these reactions is an exchange of ions. The general form would be

$$AB + CD \rightarrow AD + CB$$

Thus the reaction can be made to go if in some way you can remove either (or both) AD or CB, the products, from the reaction site. One way to do this, if the reaction is an ionic one, is to produce some compound that does not dissociate appreciably (a _____ electrolyte). For example, consider the reaction of KOH(*aq*) with NH$_4$I(*aq*). You write this as the ionic equation:

$$K^+ + OH^- + NH_4^+ + I^- = K^+ + I^- + \underline{\hspace{2cm}}$$

The _____ and _____ are spectator ions, so you eliminate them. The net ionic equation is

$$\underline{\hspace{1.5cm}} + \underline{\hspace{1.5cm}} = \underline{\hspace{2cm}}$$

The weak electrolyte, _____, that is formed is removed from the reaction. The reaction will _____ as written.

What about the reaction of AgNO$_3$(*aq*) with HCl(*aq*)? This produces the insoluble salt, _____. Thus the net ionic equation is

$$\underline{\hspace{1.5cm}} + \underline{\hspace{1.5cm}} = \underline{\hspace{2cm}}$$

The insoluble salt is not dissociated in the solution, and so it is effectively removed from the reaction. The reaction cannot go in the _____

weak

NH$_3$·H$_2$O

K$^+$; I$^-$

NH$_4^+$; OH$^-$; NH$_3$·H$_2$O

NH$_3$·H$_2$O

go

AgCl

Ag$^+$; Cl$^-$; AgCl(*s*)

backward

direction. It can only go as written. So another way to have a reaction go

insoluble

is to form an _____ compound.

Still another way to make a reaction go is to have one of the products be a gas. As the gas bubbles out, it is removed from the reaction. Even if it remains dissolved, gases are molecules, and so non-electrolytes. Question 7(a) on the pretest shows an example. We looked at this in the preceding section:

$$Na_2CO_3(aq) + 2HCl(aq) = 2NaCl(aq) + H_2O + CO_2(g)$$

Written out, as the ionic equation, showing all the species actually present, you have

H_2O; CO_2

$$2Na^+ + CO_3^{2-} + 2H^+ + 2Cl^- = 2Na^+ + 2Cl^- + \underline{\hspace{1cm}} + \underline{\hspace{1cm}}$$

The net ionic equation is

CO_3^{2-}; $2H^+$

$$\underline{\hspace{1cm}} + \underline{\hspace{1cm}} = H_2O + CO_2$$

The reaction will go as written because not only is the weak electrolyte

H_2O; gas

$\underline{\hspace{1cm}}$ formed, but also the $\underline{\hspace{1cm}}$, CO_2, escapes from the reaction.

The reaction

$$Na_2SO_3(aq) + 2HCl(aq) = 2NaCl(aq) + H_2O + SO_2(g)$$

gas

will go because the SO_2 is a $\underline{\hspace{1cm}}$ and leaves the reaction site.

Another way a double-replacement reaction can go is if a reactant does not dissociate, but the products do. An example is

$$Cd(OH)_2(s) + 4NH_3 \cdot H_2O = Cd(NH_3)_4(OH)_2(aq) + 4H_2O$$

[The $Cd(NH_3)_4(OH)_2$ dissociates into what is called the *complex ion*, $Cd(NH_3)_4^{2+}$, and two OH^- ions.] There are no spectator ions, and the net ionic equation is

$Cd(NH_3)_4^{2+}$; $2(OH^-)$

$$Cd(OH)_2(s) + 4NH_3 \cdot H_2O = \underline{\hspace{2cm}} + \underline{\hspace{1cm}} + 4H_2O$$

You started with no dissociated reactants and produced a dissociated product and water.

Whenever you can write an ionic equation, where all the ions are not spectators (can't be eliminated from both sides), the reaction will go.

How the Net Ionic Equation Shows the Similarity of Various Reactions

Recall the net ionic equation for the reaction of NaOH(aq) with HCl(aq) to produce NaCl(aq) and H_2O:

$$Na^+ + OH^- + H^+ + Cl^- = Na^+ + Cl^- + H_2O$$

or

H^+; H_2O

$$OH^- + \underline{\hspace{1cm}} = \underline{\hspace{1cm}}$$

Look at the net ionic equation for the reaction of $Ba(OH)_2(aq)$ with HCl(aq) to yield $BaCl_2(aq)$ and water:

$$Ba(OH)_2(aq) + 2HCl(aq) = BaCl_2(aq) + 2H_2O$$

Written to show the ions present, it is

$$Ba^{2+} + \underline{\hspace{1cm}} + 2H^+ + \underline{\hspace{1cm}} = 2Ba^{2+} + \underline{\hspace{1cm}} + 2H_2O$$

$2OH^-$; $2Cl^-$; $2Cl^-$

Eliminating all the spectator ions leaves the net ionic equation:

$$\underline{\hspace{1cm}} + 2H^+ = \underline{\hspace{1cm}}$$

$2OH^-$; $2H_2O$

When you divide all the coefficients by two to get the simplest ratio, you have

$$\underline{\hspace{1cm}} + \underline{\hspace{1cm}} = \underline{\hspace{1cm}}$$

OH^-; H^+; H_2O

This is identical to the net ionic equation for the reaction of NaOH(*aq*) with HCl(*aq*). Both are reactions of a strong base with a strong acid. Any reaction of a strong base with a strong acid has this same net ionic equation. We therefore classify them as the same type: *strong acid-base reactions.*
   Or consider the reaction of NH_4Cl with NaOH,

$$NH_4Cl(aq) + NaOH(aq) = NaCl(aq) + NH_3 \cdot H_2O(aq)$$

The $NH_3 \cdot H_2O$ is a weak electrolyte. So the net ionic equation is

$$\underline{\hspace{1cm}} + OH^- = \underline{\hspace{1cm}}$$

NH_4^+; $NH_3 \cdot H_2O$

But if you had NH_4NO_3 and KOH,

$$NH_4NO_3(aq) + KOH(aq) = \underline{\hspace{2cm}} + \underline{\hspace{2cm}}$$

$KNO_3(aq)$; $NH_3 \cdot H_2O$

Everything is ionized except the _____, and so the net ionic equation is

$NH_3 \cdot H_2O$

$$\underline{\hspace{1cm}} + OH^- = \underline{\hspace{2cm}}$$

NH_4^+; $NH_3 \cdot H_2O$

This is just the same as the preceding net ionic equation. The reaction of any salt that contains NH_4^+ with any *base* (a compound that contains OH^-) is the same:

$$NH_4^+ + OH^- = NH_3 \cdot H_2O$$

Or take the reaction we considered before between Na_2CO_3 and HCl:

$$Na_2CO_3(aq) + 2HCl(aq) = 2NaCl(aq) + CO_2(g) + H_2O$$

This had, as its net ionic equation,

$$\underline{\hspace{1cm}} + \underline{\hspace{1cm}} = CO_2 + H_2O$$

CO_3^{2-}; $2H^+$

Any compound containing CO_3^{2-} reacts with any acid (compound containing H^+) to yield _____ and _____. You will need _____ moles of H^+ for each mole of the CO_3^{2-} compound.

$CO_2(g)$; H_2O; two

   Consider the reaction of Na_2S with HCl:

$$Na_2S(aq) + 2HCl(aq) = 2NaCl(aq) + H_2S(g)$$

The net ionic equation is

S^{2-}; $2H^+$

$$\underline{\hspace{2cm}} + \underline{\hspace{2cm}} = H_2S(g)$$

(The $H_2S(g)$ is only slightly soluble and so it escapes.) Any *sulfide* (com-

H^+

pound containing S^{2-}) reacts with any *acid* (compound containing $\underline{\hspace{1.5cm}}$)

H_2S

to produce $\underline{\hspace{1.5cm}}$ gas.

How Net Ionic Equations Can Help You to Balance Equations

You can usually balance double-replacement (metathesis) reactions by inspection without much difficulty. But sometimes writing the net ionic reaction first (eliminating spectator ions) makes this still easier. Recall that earlier you had the example of a reaction that yielded a gas. It was $Na_2CO_3(aq)$ reacting with $HCl(aq)$ to produce $NaCl(aq)$, $CO_2(g)$, and H_2O. Simply writing down all the ions did not give a balanced equation. You had to use two HCl's. If you had started with the net ionic equation, you would have known immediately that this was necessary.

The way you start with the net ionic equation is to say that the Na^+ and Cl^- ions must be spectators because they appear in dissociated compounds on both sides of the equation. You therefore eliminate them. Then you write all the remaining ions and molecules:

$$CO_3{}^{2-} + H^+ \neq H_2O + CO_2$$

H^+

This makes it obvious that you are missing one $\underline{\hspace{1.5cm}}$ on the left side of the equation. The balanced net ionic equation is

$2H^+$

$$CO_3{}^{2-} + \underline{\hspace{1.5cm}} = H_2O + CO_2$$

two

If you have two H^+ ions, they must have come from $\underline{\hspace{1.5cm}}$ HCl's,

two

which means that you produce $\underline{\hspace{1.5cm}}$ NaCl's. The complete equation is

$$Na_2CO_3 + \underline{2}HCl = \underline{2}NaCl + H_2O + CO_2$$

Although it is usually easy to balance equations for metathesis reactions and simple combination or single-replacement reactions, there are some electron-transfer reactions (*redox* reactions) that are so complex that writing them in their net ionic form is necessary. For example, try to balance the reaction

$$\underline{\hspace{0.5cm}}KMnO_4(aq) + \underline{\hspace{0.5cm}}Fe(s) + \underline{\hspace{0.5cm}}HCl_2(aq)$$

$$= \underline{\hspace{0.5cm}}MnCl_2(aq) + \underline{\hspace{0.5cm}}FeCl_3(aq) + \underline{\hspace{0.5cm}}KCl(aq) + \underline{\hspace{0.5cm}} H_2O$$

How long would it take you to get the correct equation? It is:

$$3KMnO_4 + 5Fe + 24HCl = 3MnCl_2 + 5FeCl_3 + 3KCl + 12H_2O$$

The next sections will give you a start on learning to balance this type of equation. First write the net ionic equation, eliminating the spectator ions, K^+ and Cl^-. The net ionic equation is then

$$MnO_4^- + Fe + H^+ \neq Mn^{2+} + Fe^{3+} + H_2O$$

When you balance it, you get

$$3MnO_4^- + 5Fe + 24H^+ = 3Mn^{2+} + 5Fe^{3+} + 12H_2O$$

Once you have this, then you can put the spectator ions back in to get the complete equation.

Before you learn how to balance these redox reactions, you must know how to recognize them. You can do this most simply by determining whether there is a change in *oxidation number* as you go from reactants to products.

Your instructor may not feel that at this stage it is necessary for you to learn to balance really complicated redox reactions—those that cannot be balanced by simple inspection—however, the next section on Oxidation Numbers is vital for you when you get to Chapter 9, Simple Inorganic Nomenclature. Do not skip over it!

Incidentally, redox reactions, unlike double-replacement reactions, may proceed only in the forward direction even though no un-ionized product is produced. This story will have to wait for a later chemistry course.

SECTION C   Oxidation Numbers

The oxidation number is a rather arbitrary method of counting the number of electrons an atom uses in forming bonds. For our purposes, we can consider it only a bookkeeping method, based on the following three rules.

(1) The oxidation number of H is +1 (except in the case of metal hydrides, when it is −1).
(2) The oxidation number of any group of atoms is equal to the charge of the group. A group may be just a single atom.
(3) The sum of the oxidation numbers of all the atoms in a group equals the oxidation number of the group.

What, then, is the oxidation number of Cl in HCl? Since HCl has no charge, the oxidation number of the group HCl must therefore equal

__. The oxidation number of H is _____. Therefore the oxidation number of Cl must be _____, so that the sum of the oxidation numbers equals __.   0; +1

−1

0

Here is another example. What is the oxidation number of O in OH^-?

The charge on the group is _____. [Notice that the charge on a group is   −1
written as a *superscript* (above the line) following the group. It shows the number of charges and a (+) or (−). The number 1 need not be written. Sometimes the number of charges is indicated by repeating the + or −. For example, Fe^{++} is Fe^{2+} or even $SO_4^=$ is SO_4^{2-}.]

Now, back to OH^-. The sum of the oxidation numbers of the O and

-1; +1

the H must equal _____. The oxidation number of H is _____. Con-

-2

sequently, the oxidation number of the O must be _____. You have $+1 - 2 = -1$.

What is the oxidation number of iron metal, Fe? Here the group con-

0

sists of only one atom, Fe, and the charge on it is __. That is, it has no

0

charge. Consequently, the oxidation number of Fe is __.

What is the oxidation number of Fe^{3+}? Here the charge on the group

3+; +3

(one atom) is _____, so the oxidation number of Fe^{3+} must be _____.

You can see that when you have a group that contains just a single atom, the oxidation number of the atom equals the charge. We often talk about the *valence* of an ion. This is really electrovalence, and means the *charge on the ion*. The oxidation number of an ion containing only one atom equals the electrovalence. And any substance in its elemental state has an oxidation number of zero. Thus iron metal, Fe, has an oxidation

0

number of __. And O_2, because it is an *element* (that is, it contains only

0

one type of atom) has an oxidation number of __. Even H_2 has an oxi-

0

dation number of __. The fact that it is an elemental substance takes precedence over the rule saying that the oxidation number of H is +1.

-2

What is the oxidation number of O in H_2O? It must be _____, since

0

each H is +1 and the charge on the group is __. You therefore have $2(+1) - 2 = 0$.

The oxidation number of O is almost always -2. The few exceptions

-1

are the peroxides, like H_2O_2, in which its oxidation number is _____, and the superoxides, as in the compound KO_2, in which the oxidation

number is $-\frac{1}{2}$. The peroxides and superoxides occur only with metals of

Groups I and II and H. Thus, in almost all cases, you can assign an oxidation number of -2 to oxygen.

What is the oxidation number of Cl in HClO? The sum of the oxida-

0; 0

tion numbers must be __ since the charge on the group is __. The oxida-

+1

tion number of H is _____, and the oxidation number of O is the usual

-2; +1

_____. Consequently the oxidation number of the Cl must be _____. You have $+1 + 1 - 2 = 0$.

What is the oxidation number of S in the $SO_4{}^{2-}$ ion? The charge on

2-

the group is _____, so the sum of the oxidation numbers must be

-2; -2

_____. The oxidation number of one O is _____, so the total oxida-

-8

tion number for four O's will be _____. Consequently, the oxidation

+6

number for the S must be _____ in order to make the sum equal -2:

$$-8 + 6 = -2$$

What is the oxidation number of S in the $SO_3{}^{2-}$ ion? The sum of the oxidation numbers of the group must add up to _____. Since the group contains three O's, which have a combined oxidation number of _____, then the S must have an oxidation number of _____.

-2

-6

+4

Notice that S has an oxidation number of +6 in $SO_4{}^{2-}$ and +4 in $SO_3{}^{2-}$. Many elements show different oxidation numbers in different groups. Figure 9.1 (page 150) is a periodic table that shows most of the common oxidation numbers for the elements you are likely to encounter. Look at it now. Note that the maximum positive oxidation number that an element can show is, with only a few exceptions, the same as its group number in the Periodic Table. Also note that the elements in Groups IA, IIA, and IIIA in the periodic table can show an oxidation number of only +1, +2, +3 respectively (except boron, B). Elements in all other groups can show several different oxidation numbers. Even H, if it is combined with a very metallic element such as Na or Ca, has a second oxidation number, –1.

Since the use of oxidation numbers is really a bookkeeping trick to keep track of the transfer of electrons, you shouldn't feel uneasy about some strange values. For example, find the oxidation number of C in CH_2O (formaldehyde). There is no charge on the group, so the sum of the oxidation numbers must be __. Since each H has an oxidation number of _____, there is a total of _____ for the two H's present. The oxidation number for O is the usual _____, so the oxidation number for the C must be __.

0

+1; +2

-2

0

Zero isn't so strange, but there are even a few compounds in which elements have fractional oxidation numbers. Consider Fe_3O_4. The four oxygens have a total oxidation number of _____. The entire group has no charge, so the sum of the oxidation numbers must be ____. Consequently, three Fe's must have a total oxidation number of ____. That means that each Fe must have an oxidation number of _____.

-8

0

+8

$+\dfrac{8}{3}$

Now figure out the oxidation number for Mo in the compound Mo_2B. The B will show its –5 oxidation number. The oxidation number of Mo is _____.

$+\dfrac{5}{2}$

Incidentally, the usual method for noting the oxidation number is to write it underneath the atom. So that there is no confusion, in this text we also put it in a circle. Thus we could write

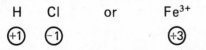

Recall that the electrovalence is written above the line and to the right. Also, we have put the sign *after* the number of charges on the electrovalence (for example, Fe^{3+}) but *before* the oxidation number.

(For example, for Fe^{3+}, the oxidation number is +3.) Also, although we don't show 1's for the single charge (H^+ or Cl^-), we do for the oxidation number

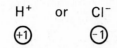

This is really nitpicking. You will find few chemists who are that fussy.

SECTION D What Examination of Oxidation Numbers Tells You

In Section C, we said that the oxidation number is only a bookkeeping trick. Nonetheless, you can get very real information by examining the oxidation number of an element both before and after it reacts. Consider the reaction of Mg(s) with two HCl(aq) to produce $MgCl_2$(aq) and H_2(g). In its ionic form, the reaction is

$$Mg(s) + 2H^+ + 2Cl^- = Mg^{2+} + 2Cl^- + H_2$$

The net ionic equation is

Mg; $2H^+$; Mg^{2+}; H_2 _____ + _____ = _____ + _____

0   The oxidation number of the element Mg is __. After it reacts to form

+2 Mg^{2+}, the oxidation number is _____. The oxidation number has increased (from 0 to +2), and we say that the Mg has been *oxidized*. It did this by giving up two of its electrons:

$$Mg \longrightarrow Mg^{2+} + 2 \text{ electrons}$$

The difference between the oxidation number of the Mg after reaction and its oxidation number before reaction is equal to the number of electrons that it gave up.

But where did these two electrons go? They had to be picked up by something else. The only thing available is the two positively charged H^+ ions.

$$2 \text{ electrons} + 2H^+ = H_2$$

+1   The oxidation number of the H^+ ion is _____ and that of the H in H_2 is

0   __. The oxidation number of H has decreased (from +1 to 0) and so we say it has been *reduced*. It did this by gaining an electron.

What has happened is that electrons have been transferred from Mg to H^+. This is called an *electron-transfer* reaction. In doing this, two H^+'s have been reduced to H_2 and the Mg has been oxidized to Mg^{2+}. So you can also call this a *reduction-oxidation* or *redox* reaction.

By determining whether a change in oxidation number occurs as you go from reactants to products, you can determine whether or not you are dealing with an electron-transfer (redox) reaction.

Now you try some. Consider the reaction

$$SOCl_2 + 2H_2O = H_2SO_3 + 2HCl$$

Is this an electron-transfer reaction? Check the oxidation numbers:

S O Cl$_2$ + H$_2$ O = H$_2$ S O$_3$ + H Cl

(+4)(-2) 2×(-1)   2×(+1)(-2)   2×(+1)(+4) 3×(-2)   (+1)(-1)

No atom changes its oxidation number as you go from reactants to products. The S is still +4, the O is –2, the Cl is –1, and the H is +1. Thus the reaction is *not* an electron-transfer reaction.

What about the reaction

$$H_2S + NaClO = S + NaCl + H_2O$$

Is this a redox reaction? You can find out by checking the

_____ of all the elements on both sides of the equa-	oxidation numbers
tion. The H in H$_2$S has an oxidation number of _____ and the H in the	+1
H$_2$O has an oxidation number of _____. So that's no change. The S in	+1
H$_2$S has an oxidation number of _____. But in the element S on the	–2
right side, it is __. Since the oxidation number increased, we know that	0
the S has been _____ by _____ electrons. The Na has	oxidized; losing
an oxidation number of _____ on both sides of the equation. But the	+1
Cl in NaClO has an oxidation number of _____, while in NaCl its oxi-	+1
dation number is _____. So the Cl has been _____ by gaining	–1; reduced
_____ electrons. The O atom is –2 on both sides of the equation. Since	two
the Cl is reduced and the S is oxidized, you have an _____	electron-transfer

or redox reaction.

Not only does the change in oxidation number tell you whether you have a redox reaction, but it also tells you how many electrons are gained or lost in the reaction. In the preceding example, you found that the S, in going from an oxidation number of –2 to an oxidation number

of 0, lost two electrons. The change in oxidation number was __. The	2
number of electrons lost was also __. At the same time the oxidation	2
number of the Cl went from _____ in NaClO to _____ in NaCl. The	+1; –1
difference between +1 and –1 is __. (If you have trouble with this, check the method for doing algebraic subtractions.) Therefore the Cl must be	2
gaining _____ electrons (the two electrons that the S lost).	two

Checking the number of electrons gained and lost leads to one of the methods for balancing redox equations. The reason is that, in the balanced equation, as many electrons must be gained by the substance

reduced; oxidized

that is _____ as are lost by the substance that is _____. In the preceding example, the S in H_2S lost two electrons and the Cl in NaClO gained two. Therefore one H_2S provides exactly the number of electrons one NaClO needs. But in our first example,

$$Mg + 2HCl = MgCl_2 + H_2$$

0

the Mg goes from an oxidation number of __ to an oxidation number of

+2; two

_____. This shows that it has lost _____ electrons. But the H^+ from

+1

the HCl goes from an oxidation number of _____ to an oxidation num-

0; one

ber of __ in H_2. It has gained just _____ electron. To use the two

two

electrons given off by the Mg, you have to have _____ H^+ ions. This

2

means that you need two HCl's. So there is a coefficient of __ before the HCl in the balanced equation.

At the end of Section B, you were challenged to balance by inspection,

$$\_KMnO_4 + \_Fe + \_HCl = \_MnCl_2 + \_FeCl_3 + \_KCl + \_H_2O$$

If you check the changes in oxidation numbers of the Mn and the Fe, you

+7

can get a good start. The oxidation number of Mn in $KMnO_4$ is _____

+2; five

and in $MnCl_2$ it is _____. This means that _____ electrons were

0

picked up. The Fe went from an oxidation number of __ in the elemental

+3; three

metal to _____ in $FeCl_3$. It has lost _____ electrons. Since each $KMnO_4$ picks up five electrons and each Fe gives up three, you need

fifteen

three $KMnO_4$'s to pick up a total of _____ electrons that are

five

released by _____ Fe atoms. You now know the coefficients of two of the substances in the balanced equation.

$$3KMnO_4 + 5Fe + ?HCl \neq ?MnCl_2 + ?FeCl_3 + ?KCl + ?H_2O$$

It is now not too difficult to continue the balancing. Since there are three $KMnO_4$'s, this means there are three K's, three Mn's, and twelve

3; 3; 12

O's. So there must be __ KCl's, __ $MnCl_2$'s, and _____ H_2O's. Since

5

there are five Fe's, this must produce __ $FeCl_3$.

$$3KMnO_4 + 5Fe + ?HCl = 3MnCl_2 + 5FeCl_3 + 3KCl + 12H_2O$$

The only thing remaining is the coefficient for the HCl. You count either

24

the number of H's or the number of Cl's on the right. Both equal _____,

24

and so the coefficient before the HCl must be _____.

$$3KMnO_4 + 5Fe + 24HCl = 3MnCl_2 + 5FeCl_3 + 3KCl + 12H_2O$$

This method of balancing electron-transfer reactions sometimes works quite well, but there are other times when it is less successful. There are several ways to balance very complicated redox equations and, if your instructor feels you need to do this at this point in your studies, he or she will show them to you.

PROBLEM SET

These problems are divided into groups for each of the sections in the chapter. Check your answers as soon as you have completed each group.

Section A
Balance the following equations by inspection, using no fractional coefficients.
1. $CaH_2 + H_2O \neq Ca(OH)_2 + H_2$
2. $NH_3 + O_2 \neq N_2 + H_2O$
3. $Al + Fe_3O_4 \neq Al_2O_3 + Fe$
4. $SiO_2 + C \neq SiC + CO$
5. $Cu + H_2SO_4 \neq CuSO_4 + SO_2 + H_2O$
6. $C_2H_5OH + O_2 \neq CO_2 + H_2O$
7. $HNO_3 + I_2 \neq HIO_3 + NO_2 + H_2O$

Section B
8. Write the net ionic equation for each of the following. Use the general rules for solubility (Table 8.1, page 132), dissociation (Table 8.2, page 132), and the gaseous substances (page 131) as your guide.

 (a) $H_2SO_4(aq) + NaOH(aq) = ?$

 (b) $H_2SO_4(aq) + BaCl_2(aq) = ?$

 (c) $H_2SO_4(aq) + Na_2S(aq) = ?$

 (d) $NaHSO_3(aq) + HCl(aq) = NaCl(aq) + SO_2(g) + H_2O$

 (e) $CH_3COONa(aq) + HCl(aq) = NaCl(aq) + CH_3COOH(aq)$ [The last substance named is acetic acid, and CH_3COONa is its sodium salt, sodium acetate]

 (f) $CaO(s) + 2HCl(aq) = CaCl_2(aq) + H_2O$

 (g) $BiCl_3(aq) + H_2O = BiOCl(s) + 2HCl(aq)$

 (h) $MnO_2(s) + 4HCl(aq) = MnCl_2(aq) + 2H_2O + Cl_2(g)$

9. Which of the following metathesis (double-replacement) reactions will go? Use the same guides as in Question 8.

 (a) $KCl(aq) + HNO_3(aq) = ?$

 (b) $NaClO(aq) + HCl(aq) = ?$

 (c) $Na_2CO_3(aq) + H_2SO_4(aq) = ?$

 (d) $Na_2CO_3(aq) + 2KOH(aq) = ?$

(e) $AgNO_3(aq) + CH_3COOH(aq)$ [acetic acid] $= ?$

(f) $H_2S(aq) + Pb(NO_3)_2(aq) = ?$

(g) $SO_3(g) + 2KOH(aq) = K_2SO_4(aq) + H_2O$

(h) $HCl(aq) + H_3PO_4(aq) = ?$

Section C

10. Give the oxidation number for the atom named in the compound shown.

   (a) Cl in Cl_2 (b) S in H_2SO_4 (c) S in $SO_3{}^{2-}$

   (d) S in $Na_2S_2O_3$ (e) C in CH_4 (f) C in CO_2

   *(g) C in CH_3COO^- (h) Fe in Fe (metal)   (i) Fe in FeO

   (j) As in $AsO_4{}^{3-}$ (k) H in AlH_3 (l) V in VCl_5

   (m) N in N_2O_3 (n) Cr in $Cr_2O_7{}^{2-}$ *(o) C in $H_2C_2O_4$

   *The average value for the two C atoms

Section D

11. How many electrons are gained or lost by the substance undergoing the change shown? Give this number for the substance, not just for single atoms.

   (a) $Fe^{3+} \rightarrow Fe$ (b) $C \rightarrow CO_2$ (c) $C_2O_4{}^{2-} \rightarrow 2CO_2$

   (d) $AsO_4{}^{3-} \rightarrow AsO_3{}^{3-}$ (e) $NaCl \rightarrow NaClO$ (f) $Cr_2O_7{}^{2-} \rightarrow 2Cr^{3+}$

Only attempt to balance the following equations if your instructor suggests it. They can be done by counting the electrons gained and lost and then by going back and forth, adjusting atoms on both sides of the equation.

12. $HNO_3 + H_2S \neq NO + S + H_2O$

13. $K_2Cr_2O_7 + BaCl_2 + H_2SO_4 \neq Cr_2(SO_4)_3 + Cl_2 + BaSO_4 + K_2SO_4 + H_2O$

14. $MnO_2 + FeSO_4 + H_2SO_4 \neq MnSO_4 + Fe_2(SO_4)_3 + H_2O$

15. $CH_2O + Ag_2O + OH^- \neq HCO_2{}^- + Ag + H_2O$

16. $CN^- + MnO_4{}^- + H_2O \neq CNO^- + MnO_2 + OH^- + H_2O$

17. $C_2H_4O + HNO_3 \neq C_2H_4O_2 + NO + H_2O$

18. $KMnO_4 + Na_2SO_3 + H_2O \neq MnO_2 + Na_2SO_4 + KOH$

PROBLEM SET ANSWERS

Section A

You should have been able to balance six out of the seven equations. If you weren't able to, go back over Section A once again.

1. $CaH_2 + 2H_2O = Ca(OH)_2 + 2H_2$ 2. $4NH_3 + 3O_2 = 2N_2 + 6H_2O$

3. $8Al + 3Fe_3O_4 = 4Al_2O_3 + 9Fe$ 4. $SiO_2 + 3C = SiC + 2CO$

5. $Cu + 2H_2SO_4 = CuSO_4 + SO_2 + 2H_2O$

6. $C_2H_5OH + 3O_2 = 2CO_2 + 3H_2O$

7. $I_2 + 10HNO_3 = 2HIO_3 + 10NO_2 + 4H_2O$

Section B

8. You should have been able to do at least six of these. If you couldn't, redo the first part of Section B.

(a) $H^+ + OH^- = H_2O$ (b) $SO_4{}^{2-} + Ba^{2+} = BaSO_4(s)$

(c) $2H^+ + S^{2-} = H_2S(g)$ (d) $HSO_3{}^- + H^+ = H_2O + SO_2(g)$

(e) $CH_3COO^- + H^+ = CH_3COOH$ (f) $CaO + 2H^+ = Ca^{2+} + H_2O$

(g) $Bi^{3+} + Cl^- + H_2O = BiOCl + 2H^+$

(h) $MnO_2 + 4H^+ + 2Cl^- = Mn^{2+} + Cl_2 + 2H_2O$

9. You should have been able to do all of these. The insoluble or weakly dissociated products are shown below. They are listed in Tables 8.1 and 8.2.

(a) Will not go   (b) Will go (HClO is a weak acid)

(c) Will go [$CO_2(g)$ and H_2O formed]   (d) Will not go

(e) Will go [$CH_3COOAg(s)$ formed]   (f) Will go [$PbS(s)$ formed]

(g) Will go (H_2O formed)   (h) Will not go

Section C

10. You should have been able to do all of these, but (k) might be a surprise.

(a) 0 (all elements zero)   (b) +6

(c) +4   (d) +2 (each S)

(e) −4 (H is +1 except with metals)   (f) +4

(g) 0 (one C is +3, the other −3)   (h) 0 (all elements zero)

(i) +2   (j) +5

(k) −1 (H with a metal)   (l) +5

(m) +3 (each N)   (n) +6 (each Cr)

(o) +3 (each C)

Section D

You should have been able to do at least five out of six correctly in Question 11. If you had difficulties, go back and reread the section.

11. The oxidation numbers before and after reaction are shown in parentheses.

(a) 3 electrons (0 to +3)   (b) 4 electrons (0 to +4)

(c) 2 electrons ($2 \times +3$ to $2 \times +4$)   (d) 2 electrons (+3 to +5)

(e) 2 electrons (−1 to +1)   (f) 6 electrons ($2 \times +6$ to $2 \times +3$)

You should have been able to balance at least five of Questions 12–18. Some pose problems with putting the spectator ions back into the balanced ionic equation. Below the balanced equations you will see the net ionic oxidation and reduction reactions that you write to get the number of electrons gained and lost (e^- means electron).

12. $2HNO_3 + 3H_2S = 2NO + 3S + 4H_2O$

$3e^- + NO_3^- \neq NO$

$S^{2-} - 2e^- \neq S$

13. $K_2Cr_2O_7 + 3BaCl_2 + 7H_2SO_4 = Cr_2(SO_4)_3 + 3Cl_2 + 3BaSO_4 + K_2SO_4 + 7H_2O$

$6e^- + Cr_2O_7^{2-} \neq 2Cr^{3+}$

$2Cl^- - 2e^- \neq Cl_2$

14. $MnO_2 + 2FeSO_4 + 2H_2SO_4 = MnSO_4 + Fe_2(SO_4)_3 + 2H_2O$

$2e^- + MnO_2 \neq Mn^{2+}$

$Fe^{2+} - 1e^- \neq Fe^{3+}$

15. $CH_2O + Ag_2O + OH^- = HCO_2^- + 2Ag + H_2O$

$2e^- + Ag_2O \neq 2Ag$

$CH_2O - 2e^- \neq HCO_2^-$

16. $3CN^- + 2MnO_4^- + 3H_2O = 3CNO^- + 2MnO_2 + 2OH^- + 2H_2O$

$3e^- + MnO_4^- \neq MnO_2$

$CN^- - 2e^- \neq CNO^-$

17. $3C_2H_4O + 2HNO_3 = 3C_2H_4O_2 + 2NO + H_2O$

$3e^- + NO_3^- \neq NO$

$C_2H_4O - 2e^- \neq C_2H_4O_2$

18. $2KMnO_4 + 3Na_2SO_3 + H_2O = 2MnO_2 + 3Na_2SO_4 + 2KOH$

$3e^- + MnO_4^- \neq MnO_2$

$SO_3^{2-} - 2e^- \neq SO_4^{2-}$

CHAPTER 9

Simple Inorganic Nomenclature

PRETEST

(You should be able to complete this test in about 15 minutes.)
Name the following compounds.

1. NaCl
2. CaH_2
3. NH_4CN
4. MnO_2
5. FeS
6. H_2SO_4 (as an acid)
7. $Co_3(PO_4)_2$
8. $NaHCO_3$
9. $KMnO_4$
10. HNO_2

Give the formulas for the following compounds.

11. Magnesium oxide
12. Titanium(IV) sulfide
13. Hydrogen peroxide
14. Zinc nitride
15. Ferric oxide
16. Silver nitrate
17. Chromium(III) phosphate
18. Sulfurous acid
19. Potassium perchlorate
20. Silicon tetrachloride

PRETEST ANSWERS

If you were able to get 17 right out of the 20, you can go directly to the
Problem Set at the end of the chapter. If you missed any more than that, you
had better work through the entire chapter.

1. Sodium chloride
2. Calcium hydride
3. Ammonium cyanide
4. Manganese(IV) oxide or manganese dioxide
5. Iron(II) sulfide or ferrous sulfide
6. Sulfuric acid
7. Cobalt(II) phosphate or cobaltous phosphate
8. Sodium hydrogen carbonate or sodium bicarbonate
9. Potassium permanganate
10. Nitrous acid

11. MgO   12. TiS_2 13. H_2O_2 14. Zn_3N_2
15. Fe_2O_3 16. $AgNO_3$ 17. $CrPO_4$ 18. H_2SO_3
19. $KClO_4$ 20. $SiCl_4$

Nomenclature—the naming of molecules—is the language of the chemist. When the chemist names a compound, he or she must describe it exactly and not leave any doubt as to what the formula is. One name describes just one molecule. It is important that all chemists agree on how to construct that one name. Unfortunately, there has not yet been a complete agreement on what system to use, so that some compounds have several possible names. Recently the International Union of Pure and Applied Chemistry (IUPAC) set up a commission to work out a system that all chemists would adopt. It is the best system worked out so far, but there are still some exceptions. You will no doubt still find older systems used in various texts and perhaps on the labels of chemicals in your labs. This brief discussion of nomenclature will give you enough information so that you will be able to name most of the compounds you see as formulas and also know what the formula is for most of the names you read, both in the IUPAC system and in the older systems.

SECTION A   Names and Symbols of
the Elements

Before you can start naming compounds whose formulas you are given, you must know the names of the atoms. By the same token, before you can write the formula for a compound whose name you are given, you must know the symbols of the elements it contains. There are about 106 known elements at the present time, and each has a name and a symbol. Inside the front cover of this book, you will find a list of 62 of the more common elements and their symbols. Most of the symbols are two letters. (Only the elements that have been known for a very long time have a single letter as their symbol.) The two letters come from the name and, if possible, are the first two letters of the name. Thus the symbol for the element calcium is Ca. The symbol for the element

Be; Co | beryllium is _____. The symbol for the element cobalt is _____. The
Os | symbol for the element osmium is _____.

Sometimes two or more elements have the same first two letters in their names. For example, magnesium and manganese. In order to distinguish between these two, the symbol is made up of the first and third letters. Thus the symbol for magnesium is Mg, and the symbol for

Mn | manganese is _____. The symbols for the two elements chlorine and
Cl; Cr | chromium would then be _____ and _____.

As we said before, some of the elements that have been known for a long time have only a single letter for their symbol. The symbol for

oxygen is O, for nitrogen is __, for fluorine is __, for hydrogen is H, for

iodine is __, for carbon is C, for sulfur is S, for phosphorus is P, and so forth.

N; F	
I	

Another group of elements are those whose symbols are derived from their Latin names. (Latin was the language of science in the old days. Even until recently medical doctors wrote their prescriptions in Latin.) Fortunately for us, there are only ten of these remaining. These are Fe, the symbol for iron (*ferrum*), K, the symbol for potassium (*kalium*), Pb, the symbol for lead (*plumbum*), Sn, the symbol for tin (*stannum*), Au, the symbol for gold (*aurum*), Ag, the symbol for silver (*argentum*), Cu, the symbol for copper (*cuprum*), Sb, the symbol for antimony (*stibium*), W, the symbol for tungsten (*wolfram*, which is a German, not a Latin, name), and Hg, the symbol for mercury (*hydragyrum*).

You will have to memorize these names and symbols. One of the best ways to do this is to prepare a set of cards for yourself. On one side, write the name of an element. On the other, write its symbol. Then you can randomly go through the stack of cards, read what is on one side and say or write what is on the other, checking the reverse to see whether you were correct. The 62 elements at the beginning of this text should be a sufficient number to learn, and they will cover almost all the formulas you will come in contact with.

SECTION B   Oxidation Number

The concept of oxidation number is covered thoroughly in Section C, Chapter 8. If you have not already worked through this section, now would be a good time to do it. Just to check yourself, see if you are able to do the following.

The oxidation number of the Cl in Cl⁻ is _____. The oxidation

number of Na in NaCl is _____. The oxidation number of Fe in $FeCl_2$

is _____, but the oxidation number of Fe in $FeCl_3$ is _____.

−1	
+1	
+2; +3	

The oxidation number of O in a compound is almost always _____.

Therefore the oxidation number of Mn in MnO_2 is _____, but when Mn

is present as in the ion Mn^{2+}, its oxidation number is _____.

−2	
+4	
+2	

The oxidation number of S in $SO_4{}^{2-}$ is _____. (The sum of all the oxidation numbers must be −2, since the ion has a 2− charge. The oxida-

tion number for __ O's is −8. Therefore the S must be +6.) The oxida-

tion number for S in H_2SO_3 is only _____.

+6	
4	
+4	

If you are able to do all of the above, you are sufficiently skilled in determining oxidation numbers. You will need this for naming compounds.

Figure 9.1 Some possible oxidation numbers of the common elements

IA	IIA	IIIB	IVB	VB	VIB	VIIB	VIII	VIII	VIII	IB	IIB	IIIA	IVA	VA	VIA	VIIA	O
H +1 (−1)																	
Li +1	Be +2											B +3 (−5)	C +4 (−4) +2 (0) (−2)	N +5 (+1) +4 −1 +3 (−2) (+2) −3	O −2 (−1) (−2)	F −1	
Na +1	Mg +2											Al +3	Si +4 (+2) −4	P +5 (+1) (+4)(−1) +3 (−2) (+2) −3	S +6 (+2) (+5)(+1) +4 −2 (+3)	Cl +7 −1 +5 +3 +1	
K +1	Ca +2	Sc +3	Ti +4 +3 +2	V +5 +4 +3 +2	Cr +6 +3* +2	Mn +7 +4 +3* +2	Fe +3* +2	Co +3* +2	Ni +2 (+3)	Cu +2* +1	Zn +2	Ga +3	Ge +4 +2 (−4)	As +5 +3 (−3)	Se +6 +4 (+2) −2	Br +5	
Rb +1	Sr +2									Ag +1			Sn +4* +2 (−4)	Sb +5 +3 (−3)	Te +6 +4	I +7 +5 +3 +1 −1	
Cs +1	Ba +2						Pt +4* +2			Au +3* +1	Hg +2* +1		Pb +4* +2	Bi (+5)* +3 (−3)			

*Indicates the "ic" oxidation number of metal ions
( )Indicates an uncommon oxidation state

Some common anions (arranged in order of group number)

Group III A
BO_3^{3-} Borate

Group IV A
CO_3^{2-} Carbonate
SiO_4^{4-} Silicate

Organic anions
CHO_2^- Formate
$C_2H_3O_2^-$ Acetate
$C_2O_4^{2-}$ Oxalate
$C_8H_4O_4^{2-}$ Phthalate

Group V A
NO_3^- Nitrate
NO_2^- Nitrite
PO_4^{3-} Phosphate
PO_3^{3-} Phosphite
AsO_4^{3-} Arsenate
AsO_3^{3-} Arsenite

Group VI A
SO_4^{2-} Sulfate
SO_3^{2-} Sulfite
$S_2O_3^{2-}$ Thiosulfate
$S_2O_8^{2-}$ Persulfate

Group VII A
ClO_4^- Perchlorate
ClO_3^- Chlorate
ClO_2^- Chlorite
ClO^- Hypochlorite
BrO_3^- Bromate
BrO^- Hypobromite
IO_4^- Periodate
IO_3^- Iodate
IO^- Hypoiodite

Group VI B
CrO_4^{2-} Chromate
$Cr_2O_7^{2-}$ Dichromate

Group VII B
MnO_4^- Permanganate

Miscellaneous
CN^- Cyanide
SCN^- Thiocyanate
O_2^{2-} Peroxide
OH^- Hydroxide

Figure 9.1 is an abbreviated periodic table containing the elements you will be seeing most often. We will use the traditional group numbers rather than the new and as yet rarely used sequence proposed by the IUPAC in 1983. The table shows the oxidation numbers each element can have in its usual compounds. There is a zigzag line running from the top of the periodic table at the element boron to the bottom of the table almost all the way to the right. This line forms a rough border between the elements that are considered metals (to the left) and the elements that are considered nonmetals (to the right). The metallic elements, when they form a compound, tend to give up electrons and become positive ions. The nonmetallic elements, when they form compounds with a metallic element, tend to pick up electrons and become

_____ ions. As you can see, about 80% of the elements are | negative
metallic.

The oxidation numbers for these metals are all positive. Some of the metals have only one possible oxidation number. This means that they can form only one type of ion. Thus the element Na has only a +1 oxi-

dation number. This means it can form only the _____ ion. The | Na^+
element Ca can have only a +2 oxidation number. Therefore it can

form only the _____ ion. | Ca^{2+}

However, some of the metals have several possible oxidation numbers. Fe, for example, as both a +2 and a +3 oxidation number. It can there-

fore form the two ions _____ and _____. The element Au can have | Fe^{2+}; Fe^{3+}
an Au^+ and an Au^{3+} ion, so it has the two possible oxidation numbers

_____ and _____. | +1; +3

The periodic table can be some help in figuring out what the possible oxidation numbers for an element are. The vertical columns in the table are referred to as *groups*, and each has a number. The long groups are called A groups and the shorter columns in the center of the periodic table are the B groups.

Look at the elements in Group IA. With the exception of H (which could also be put in Group VIIA), they can have only one possible oxi-

dation number, _____. So metals in Group IA can have only an oxida- | +1
tion number of +1. Look at the metals in Group IIA. The elements can

have only one _____ number, which is _____. The oxidation | oxidation; +2

number is the same as the _____ number. If you consider the metals | group

shown in Group IIIA, the only possible oxidation number is _____. | +3

You do not consider boron a metal because it is not to the _____ of | left
the zigzag line. Nonetheless, the only positive oxidation number it can show is +3, its group number. So, for A groups IA, IIA, and IIIA, the oxidation numbers of the metals are plus whatever the group number is. (For Group IIIA, some elements—not shown in Figure 9.1—do show a

+1 oxidation number.) Once you get to group numbers higher than IIIA, you cannot use this generalization.

Some of the B groups also follow this generalization. Group IIIB ele-

+3

ments can show an oxidation number of only _____ and Group IIB

+2

elements can only show an oxidation number of _____. (Hg appears to be an exception, since a +1 oxidation number is listed in the periodic table. This is because Hg forms a polyatomic ion Hg_2^{2+}, as well as its normal ion Hg^{2+}. This polyatomic ion has a calculated oxidation number of +1.) All the rest of the Group B elements, except Ag, can show several possible oxidation numbers and, consequently, several ions of different charge. When you are skilled in the reasons behind the location of the elements in the periodic table, you may be able to figure out just what positive charges are possible for an element. For the moment, however, you have to rely on memory to know what charges you have. Use the periodic table in Figure 9.1 as your guide.

What about the nonmetallic elements? The periodic table shows that they all have negative oxidation numbers, and—with the exception of C in Group IV, P and N in Group V, and O in Group VI—only one possible negative value. You can easily determine what this negative oxidation number is if you know the group number of the element. See if you can figure out how to do this.

Cl is in Group VIIA, and its negative oxidation number is –1. S is in Group VIA, and its negative oxidation number is –2. P is in Group VA, and its negative oxidation number is –3. You can determine the negative

8

oxidation number by subtracting the group number from __. Thus, the

–5

negative oxidation number for boron (B) in Group IIIA is _____. N in

–3

Group VA has a negative oxidation number of _____.

You get the negative oxidation number for *non*metallic elements when

negative

they gain electrons to form _____ ions. However—with the exception of O and F—all these nonmetals also show several positive oxidation numbers. These do not result from the losing of electrons in the sense that ions are formed. Rather they are the result of the nonmetals sharing electrons with other nonmetals. At the beginning of this section you worked out the oxidation number for S in SO_4^{2-}. It came out to be

+6

_____. This can be looked on as meaning that the S is sharing six electrons with the four O atoms. The O atom is more negative than the S atom. In the case of H_2SO_3, the oxidation number of the S was +4. This

sharing

can be seen as the S _____ four electrons with the three O atoms. In the compound $Na_2S_2O_3$, the oxidation number of one S atom

+2; two

is _____, so you can say that the S is sharing just _____ electrons. (This is an artifact of the system. Actually, the two S atoms are not identical in the compound.)

The highest possible value for the positive oxidation number of Cl in

Group VIIA is +7. The highest possible positive oxidation number for S in Group VIA is 6. And, as you would expect, the highest positive oxidation number for N in Group VA is _____. For C in Group IVB, it is

_____. You can now figure out what the highest positive oxidation

number is. It's the same as the _____.

The other possible positive oxidation numbers go down in steps of two. Therefore S in Group VIA has as its highest possible positive

oxidation number _____. Below that it can be +4, or it can be +2. Cl in

Group VIIA has +7 as its _____ possible positive oxidation

number, while the lower values it can show are +5, _____, and +1. P in

Group VA can show the values of _____ (the highest) and the lower

value of _____. It does not show a +1 value in any known compounds.

Silicon in Group IVA can show the two values, _____ and _____.

A few of the nonmetals are exceptional. N in Group VA can show a +4 and a +2 oxidation number in just two compounds. The oxidation number of C in organic molecules is strange, and can have almost any value. O normally has a −2 oxidation number and shows no positive values (because of the way the system of counting numbers was set up). However, it forms two polyatomic groups, O_2^{2-} (peroxide) and O_2^- (superoxide). These occur only very rarely and then only with very metallic elements or hydrogen. F shows no positive oxidation numbers at all.

All, as you can see, is not a bed of roses. Once again, you will have to rely on your memory to a certain extent. Nonetheless, the periodic table can still be a good guide for the majority of oxidation states.

+5
+4
group number
+6
highest
+3
+5
+3
+4; +2

SECTION C   Binary Compounds

When a molecule consists of only two different elements, we say that we

have a *binary* compound. Thus NaCl would be a _____ com-

pound since it contains only the _____ elements, Na and Cl. $MgCl_2$ is

also a binary compound, since it contains just two elements, Mg and ___.

However, H_2SO_4 is _____ a binary compound because it contains three

different _____, H, S, and O.

The binary compounds are the simplest to name. They can be made up of two elements, one of which is metallic and therefore *electropositive*,

since the metals have _____ oxidation numbers, and one that is

nonmetallic, that is, *electronegative*, and has a _____ oxidation

number. Or they can contain two nonmetallic elements, one of which is

binary
two
Cl
not
elements
positive
negative

more electropositive than the other. As you will see, there is a difference in how you name these two types of compounds. However, in both cases, the electropositive element is named first (and appears first in the formula). The electronegative element is named second, substituting "-ide" for its last syllable. So you can always spot a binary compound,

-ide

since its name always ends in "_____." Thus NaCl is called sodium

magnesium

chloride, $MgCl_2$ is _____ chloride, Na_2S is sodium

sulfide; calcium carbide

_____, and CaC_2 is called _____.
If the nonmetal name has a vowel at the end of the next-to-last syllable, you drop that too. Thus MgO is called magnesium oxide (not oxyide), BN is called boron nitride (not nitroide), and NaH is sodium

hydride; boron

_____ (not hydroide). The name for BP is _____ phosphide (not phosphoride).
In general, the metallic (electropositive) element is the element that is further to the left and lower in the periodic table. The electronegative

right

element is further to the _____ and higher in the periodic table. If both elements are more or less on the right in the periodic table, there is an order of increasing electronegative character which was agreed on for the purposes of nomenclature. It is the following.

B, Si, C, Sb, As, P, N, H, Te, Se, S, I, Br, Cl, O, F

Try your hand at naming some compounds.

calcium fluoride; aluminum hydride

CaF_2 _____ AlH_3 _____

magnesium oxide; hydrogen sulfide

MgO _____ H_2S _____

lithium nitride; cesium oxide

Li_3N _____ Cs_2O _____

zinc bromide; silicon carbide

$ZnBr_2$ _____ SiC _____

Incidentally, there are a few ions that are made up of several atoms so commonly found as a group that they are considered, insofar as nomenclature goes, as a single element. The only positive polyatomic ion like this is the ammonium ion, NH_4^+. It is considered, for nomenclature purposes only, to be like an element in Group IA. Thus the compound

chloride

NH_4Cl is named ammonium _____, just as if it were binary. The most common negative one that you will see is the hydroxyl ion, OH^-. When you use it in a binary compound, you drop the "-yl" and add

-ide; hydroxide

"_____." Thus NaOH is called sodium _____. The compound

hydroxide

$Ca(OH)_2$ is called calcium _____.
Another negative ion is CN^-, which is the *cyano* group. KCN is named

potassium cyanide

_____. The compound NH_4CN is named

ammonium cyanide

_____.

Much less common negative ions are $O_2{}^{2-}$ (the *peroxy* group), $N_3{}^-$ (the *azo* group), and $NH_2{}^-$ (the *amido* group). These form peroxides, azides, and amides, respectively.

All the binary compounds that we have considered so far have had the electropositive element chosen from those that have only one possible oxidation number. However, if the electropositive element has several possible oxidation numbers, you must name the compounds in such a manner that you know *which* of the possible oxidation states is present in the molecule.

Compounds of a Nonmetal with a Metal Having Several Oxidation States

It is at this point that you will have to distinguish between (a) binary compounds made up of a metallic element and a nonmetallic element and (b) binary compounds made up of two nonmetals. The first kind is *ionic*—that is, made up of positive and negative ions. The second kind is *covalently bonded*—that is, the atoms share electrons. We will consider the ionic binary compounds first.

The method the IUPAC proposed (*no* oxidation state indicated) was to use a Greek prefix to show the number of atoms of each element in the compound. However, chemists generally use another method for naming ionic compounds: the *Stock notation*, named after the German chemist, Albert Stock. With this system, the oxidation number of the metallic element is placed as a Roman numeral after the name of the element. Thus $FeCl_3$ is called iron(III) chloride. The oxidation number

of Fe is _____. And $FeCl_2$ is iron (____) chloride because the oxidation | +3; II

number of Fe is _____ in this compound. You would name the com- | +2
pound CoF_2 cobalt(II) fluoride and the compound CoF_3

_____. The compound SnO is called tin(____) | cobalt(III) fluoride; II

oxide, since the oxidation number of the Sn is _____. If you missed | +2
this, it was because you forgot that the oxidation number of O is always

_____ when it occurs alone. So the name of SnO_2 by the Stock system | −2

is _____. The two O's have a total oxidation number of | tin(IV) oxide

_____, so the Sn must have an oxidation number of _____. You | −4; +4
might have thought the $O_2{}^{2-}$ was a peroxide. It isn't. The peroxides only form with metals having just one oxidation number. How about the

compound $Cu(OH)_2$? The hydroxyl group is an ion with a _____ | 1−

charge. Therefore its oxidation number, as a group, must be _____. | −1
Since the molecule contains two of these hydroxide ions, the oxidation

number of the Cu must be _____, and the Stock name for the com- | +2

pound is copper(____) hydroxide. Here are a few compounds to give | II
you some practice.

manganese(II) chloride	$MnCl_2$	_____
titanium(IV) oxide	TiO_2	_____
iron(III) oxide	Fe_2O_3	_____
gold(I) sulfide	Au_2S	_____
gold(III) sulfide	Au_2S_3	_____

(The sulfide ion has a 2– charge, since it is a Group VIA element. Did you count correctly?) If you are having trouble, go back to Section C of Chapter 8.

One thing to be careful about is that it is considered *incorrect* to use Stock notation if the metallic element has only one oxidation number. Thus calling NaCl sodium (I) chloride is wrong. Sodium has *only* a

+1   _____ oxidation number, so you don't *have* to state it. Therefore you *don't*. Actually, the only reason to use any name that shows the oxidation number of the metallic element is to distinguish among several possible compounds of the same elements. There is no way of knowing

$FeCl_3$ whether iron chloride means $FeCl_2$ or _____. But zinc chloride could

$ZnCl_2$ only be _____, because zinc can have an oxidation number of only

+2   _____. You would say iron(II) chloride and iron(III) chloride, but *never* zinc(II) chloride. You'll need a lot of practice to remember which metallic elements have several different positive oxidation numbers. At first you'll have to use Figure 9.1 as a guide, but after awhile you'll start remembering them.

Here's some more practice for you. Name the following binary compounds, using Stock notation *when necessary*.

tin(IV) chloride; silver sulfide	$SnCl_4$ _____	Ag_2S _____
cobalt(III) hydroxide; lead(IV) oxide	$Co(OH)_3$ _____	PbO_2 _____
barium fluoride; iron(II) sulfide	BaF_2 _____	FeS _____
calcium hydride; potassium cyanide	CaH_2 _____	KCN _____
sodium peroxide; copper(I) oxide	Na_2O_2 _____	Cu_2O _____

[Did you remember the two negative ions, CN^- (cyano-) and O_2^{2-} (peroxy-)?]

There is one more way of naming compounds that have an electropositive element with several possible oxidation numbers. It is no longer recommended and is slowly disappearing, but you still find it in older texts and on reagent bottles in the lab. With this method, you name the metallic element (sometimes using its Latin name) and substitute "-ic" for the last syllable if the element is in its higher oxidation state or "-ous" if it is in its lower oxidation state. The compound $HgCl_2$ is therefore called mercuric chloride (the mercury is in the +2 or higher oxidation state). The compound Hg_2Cl_2 is called mercurous chloride (the

mercury is in the _____ oxidation state, which is its lower state).

+1

With this method of nomenclature, the compound $FeCl_3$ is called ferric chloride (the Latin name for iron is *ferrum*), and the compound

$FeCl_2$ is _____ chloride. The Fe in $FeCl_3$ has an oxidation

ferrous

number of _____, and in $FeCl_2$ it has an oxidation number of +2. The

+3

higher oxidation number is the "-ic" and the lower is the "_____."

-ous

The Latin name for copper is *cuprum*, so the name of the compound

CuO is cupric oxide, and for the compound Cu_2O it is _____

cuprous

oxide. The Latin name for tin is *stannum*. Using this method of naming

compounds, you say that $SnCl_4$ is _____ and $SnCl_2$ is

stannic chloride

stannous chloride. The Latin name for gold is *aurum*, so the names for

the two oxides, Au_2O and Au_2O_3, are _____ and

aurous oxide

_____, respectively.

auric oxide

The big problem with this method is remembering what the oxidation number is for the "-ic" ion of an element, and what the value is for the "-ous." The periodic table in Figure 9.1 has an asterisk after the "-ic" oxidation number. One clue is that if there is a +3 oxidation number, it is the "-ic" state. Thus Fe^{3+} is ferr*ic*, Co^{3+} is cobalt*ic*, and Mn^{3+} is mangan*ic*. But Sn only has the possible oxidation numbers of +4 and +2, so the +4 is the "-ic." And Cu has the values of only +2 and +1, so the +2 is the "-ic." You can see what trouble this system can cause.

Another problem arises when an element has more than two possible positive oxidation numbers. For example, Ti has the values +4, +3, and +2, and is in Group IVB. You might be tempted to call the +4 the "-ic"

form (although calling $TiCl_4$ _____ chloride is such a gross pun

titanic

that your instructor would never forgive you). However, you shouldn't.

Use the Stock notation. Then $TiCl_4$ is _____ chloride. Or you

titanium(IV)

could use the Greek-number system commonly used for binary compounds of two nonmetallic elements. With this system, you call $TiCl_4$ titanium *tetra*chloride (*tetra* is Greek for "four"). We will examine this method in the next part.

Compounds of Two Nonmetals

Suppose that you want to write the Stock-notation name for N_2O_5. You

determine that each of the N atoms has an oxidation number of _____,

+5

since the five O atoms have a total oxidation number of _____. You

−10

then call N_2O_5 _____ oxide.

nitrogen(V)

However, the method preferred by the IUPAC is to count the number of atoms present *in Greek*. Here are the Greek numerals up to eight.

mono	one	penta	five
di	two	*hexa*	six
tri	three	*hepta*	seven
tetra	four	*octa*	eight

pentoxide	Thus the preferred name for N_2O_5 is dinitrogen _____. (Notice that we omit the *a* on the "penta" to make the word easier to
di	pronounce.) Using this method, you would call N_2O_4 _____nitrogen
tetroxide; dichlorine; hept	_____ and Cl_2O_7 is _____ _____oxide.
	It is usual to omit "mono-"s for the first element. Thus the compound
dioxide; dioxide	SO_2 is called sulfur _____ and CO_2 is called carbon _____. You can even omit "mono-" for the second element: N_2O is dinitrogen oxide. But if there is any chance of confusion, use it. So you call CO
monoxide	carbon _____.
	Chemists tend to get a little lazy when it comes to naming compounds. Sometimes they omit the Greek count of the first element even when it is more than "mono." Our original example, P_2O_5, which we named
diphosphorus; pent	correctly as _____ _____oxide, is often called just phosphorus pentoxide. The excuse given is that if there were only one P atom in the compound, with five O atoms, the P would have to have an
+10	oxidation number of _____. But nothing ever has anything higher than
+5	+7, so therefore there must be two P atoms, each with a _____ oxida-
chlorine heptoxide	tion number. In the same way, Cl_2O_7 is called _____ _____.
	See if you can name these compounds. Give their correct name by the IUPAC notation.
carbon tetrachloride	CCl_4 _____
silicon hexafluoride	SiF_6 _____
diarsenic trioxide; arsenic trioxide	As_2O_3 _____ or _____
silicon dioxide	SiO_2 _____
diantimony trisulfide; antimony trisulfide	Sb_2S_3 _____ or _____
	These are really quite easy. Now let's go back a few paragraphs, to the place where we said that—even though the compound $TiCl_4$ was made up of a metallic element and a nonmetallic one—it was often named titanium tetrachloride, as if it were made up of two nonmetals. Consider the +4
MnO_2	oxidation state of Mn that has an oxide of the formula _____. You
dioxide	frequently see this compound called manganese _____, using the IUPAC "counting" method. And PbO_2 is usually referred to as
dioxide	lead _____.

At this point you are probably wondering what's the sense of having all these rules if there are so many exceptions. The problem is that initially there wasn't much system to the way compounds were named. (There was even sometimes a lack of system in deciding which element should appear first in a formula. The formula for ammonia, NH_3, has the more positive H atom placed at the end.) And old names die hard. It is doubtful that the day will come when H_2O is called dihydrogen oxide. The name *water* is here to stay.

Binary Compounds of Hydrogen

Hydrogen is in a unique situation with respect to whether it is electronegative or electropositive compared with another atom to which it is bonded. As you saw in Chapter 8, it can show either a +1 or a −1 oxidation number. When it combines with a Group VIA or Group VIIA element, it is definitely more positive. These groups are on the far

_____ of the periodic table. So gaseous HCl is called   right

_____ _____ and H_2S is called _____ sulfide.   hydrogen chloride; hydrogen
On the other hand, in the compound NaH, the H atom is less positive

than the Na. Thus the name of NaH is sodium _____. The com-   hydride

pound CaH_2 is called _____ _____ and AlH_3 is   calcium hydride

aluminum _____. Even Sn, in Group IVA, is positive enough   hydride
for its compound with hydrogen, SnH_4, to be called

_____ _____. These compounds in which H is the   tin(IV) hydride

negative end of the bond are always combinations of H with a _____   metallic

element. These appear to the _____ of the zigzag line in the periodic   left
table of Figure 9.1.
　But what about the elements that are neither much more positive nor much more negative than H? These are elements like B, C, N, Si, P, As, Sb, and Ge. These have trivial (nonsystematic) names that end either in "-ane" (for B, C, Si, and Ge) or "-ine" (for P, As, and Sb). Thus SiH_4 is called si*lane* (the last two syllables are dropped) and GeH_4 is named

_____. Boron forms many different compounds with hydrogen   germane
by linking up several B atoms. The simplest is B_2H_6, which is called diborane. The simplest compound of C and H is CH_4. It has the trivial name methane. But there are literally thousands of compounds made up of just carbon and hydrogen. They are called as a group *hydrocarbons*. Their nomenclature (luckily for us) falls into the province of organic chemistry and we are considering only inorganic compounds.
　The Group VA elements—P, As, and Sb—form the compounds PH_3,

phosph*ine*, AsH_3, _____, and _____, stibine. (The Latin name   arsine; SbH_3
for antimony, *stibbum*, is used for the stem.) The last member of Group

V, nitrogen, N, also forms a compound with hydrogen. It has the

NH₃   formula _____. Like water, it has a trivial name that is always used: *ammonia*.

At this stage of your chemical career, it's not too important that you know any of the last class of compounds *except* NH_3, called

ammonia; H₂O   _____ and _____, called water. But of course, you must know the names of compounds composed of very metallic elements with H and very nonmetallic elements with H.

In compounds of H with a very positive element (from Groups IA and IIA) the H atom actually takes an electron from these metallic elements and forms an H⁻ ion. However, when H atoms form compounds with a very negative element (from Groups VIA and VIIA), the H atoms don't give up an electron to form H⁺ ions. But if these compounds are dissolved in water, they do dissociate (split up) into an H⁺ and some negative (minus) ion. Whenever you add H⁺ ions to water, you say that you have an acid. Thus these compounds are named *acids* when they are present in water. There are not very many of them. You name them by writing "hydro-" for the hydrogen and then substitute "-ic" for the last syllable of the nonmetal's name. You follow this with "acid." Thus HCl dissolved in water is called hydrochloric acid, and HF is named

hydrofluoric acid; hydroselenic acid   _____. You call H_2Se _____,

acid   and for the compound HI dissolved in water you say hydriodic _____. (Notice that the *o* was dropped before the vowel *i* in iodic.) The thing to remember is that when you have an acid whose name starts with "hydr-"

H   it is a *binary* compound. It contains just __ and one other nonmetallic (Group VIA or VIIA) element.

Let's review what we have done so far. We have been dealing with

two   binary compounds. These contain just _____ different elements. You name the more electropositive element first and then the electronegative

-ide   element, substituting "_____" for the last syllable of its name. You can tell which is the electropositive (metallic) element, since it is located

left   further to the _____ and lower in the periodic table.

Only if the electropositive element has more than one possible positive oxidation state do you indicate which it is in the compound you are naming. There are three ways to do this. The first method is to write a

Roman   _____ numeral in parentheses after the name of the metal, the

number   number being equal to the oxidation _____. This system is

Stock   called the _____ system. It should be used only if your com-

metal   pound is made up of a _____ and a nonmetal.

If your compound is made up of two nonmetals (the elements that are

right; count   in the upper _____ in the periodic table), you then _____ the number of each type of atom in Greek. Generally, you omit the "mono"s and often you omit "di-" when it is the number of the first atom.

The old-fashioned method of naming a compound in which the electropositive element has several oxidation states is to name the metal

(usually in _____.) and substitute "_____" for the last syllable in its   Latin; -ic

name if it is in the higher oxidation state and "_____" if it is in the   -ous
lower oxidation state.

In all cases the electronegative (nonmetallic) element has "_____" as   -ide
its last syllable. Therefore any compound whose name ends in "-ide" is a

_____ compound and contains only _____ different elements. Aside   binary; two
from a few trivial names like water and ammonia, the only exception to
this statement is the case in which your binary compound contains H
and is dissolved in water to give an acid solution. These compounds are

named as acids. You write "hydro-" for the __ and then substitute "-ic"   H

for the last syllable of the nonmetal. You follow this with "_____" to   acid
show that it is an acid.

Now that you can name a compound whose formula is given, let's
reverse the process and find the formula for a compound whose name is
given.

Consider the compound sodium chloride, whose formula is _____.   NaCl

The oxidation number of Na is always _____, since it is in Group IA in   +1

the periodic table. The negative oxidation number of Cl is always _____, -1
since it is in Group VIIA. The sum of the oxidation numbers of a

molecule always equals __.   0

Consider the compound silver oxide. Since the last syllable of the

name is "-ide," you know that it must be a _____ compound.   binary
The only possible oxidation number for silver, Ag, is +1. The oxidation

number of O when it is named *oxide* is _____. In order for the sum of -2

the oxidation numbers for the molecule to be __, the formula must con-   0
tain two Ag's, the two +1's balancing the -2. The formula is therefore

_____. Ag_2O

How about aluminum sulfide? Aluminum is in Group IIIA. Its only

possible positive oxidation number is _____. Sulfur is in Group VIA.   +3

Its only negative oxidation number is _____ (8 minus the group -2
number). In order to have a formula that has a total oxidation number

of 0, you have to have __ Al's and __ S's. This way you have 2 × (+3)   2; 3

and 3 × (-2), so the formula is _____. Al_2S_3

See if you can work out the formula for calcium phosphide. Since the

last syllable is _____, you know that this is a binary compound.   "-ide"

+2	Calcium can have only one oxidation number: _____. (It is in Group
-3	IIA.) The usual negative oxidation number for P is _____, since it is in
VA	Group _____ in the periodic table. Therefore the formula for the
Ca_3P_2	molecule must be _____.
	Let's look at a few of the special ions that are considered just for nomenclature purposes as if they were single elements. Ammonium
NH_4Cl	chloride has the formula _____, since the NH_4^+ ion has an oxidation number of +1. (Recall that the oxidation number of a group of atoms
$(NH_4)_2S$	equals its charge.) Ammonium sulfide has the formula _____, since you need two of the NH_4^+ ions to balance the -2 oxidation number of S. Notice how the NH_4^+ is placed in parentheses and the subscript written after it. Recall the hydroxide ion, OH^-. The formula for
KOH	potassium hydroxide is _____ (the symbol for potassium is K). The formula for magnesium hydroxide is $Mg(OH)_2$, since the only possible
+2	oxidation number for Mg is _____. It is in Group IIA. You need two
-1	OH^- ions, which have an oxidation number of _____ each, to balance the +2 of the Mg. Notice once again that you put the group in parentheses.
cyanide	The CN^- ion is called _____. Since it has a 1- charge, it has
-1	an oxidation number of _____. The formula for zinc cyanide is
$Zn(CN)_2$	_____, since zinc can have only one oxidation number (+2).
	The peroxide group is O_2^{2-}. As a group, it has an oxidation number of
-2	_____, since its charge is 2-. Then the formula for hydrogen peroxide
H_2O_2	must be _____, since you need two H's to balance the -2 oxidation number of the peroxide ion. Calcium is in Group IIA and can have only a
+2; CaO_2	_____ oxidation number. The formula for calcium peroxide is _____, since only one Ca is needed to balance the -2 oxidation number of the
O_2^{2-}	_____ ion.
	Actually, it is harder getting the formula for compounds in which the more electropositive element has only one possible oxidation number than it is for compounds for which several oxidation states exist. In the latter case, the name tells you either what the oxidation state is or the number of atoms that are in the formula. For example, copper(I)
+1	chloride tells you that the oxidation number for the Cu is _____. The
CuCl	formula is _____ since the only possible oxidation number of Cl (Group
-1; $FeBr_3$	VIIA) is _____. Iron(III) bromide has the formula _____, since you
-1	need three Br's, each with an oxidation number of _____, to balance the +3 oxidation number of the Fe.
+4	Tin(IV) oxide has the formula SnO_2, since the Sn has _____ as its

oxidation number and you need two O's, each with a _____ oxidation number, to balance the oxidation number of Sn. In order to write a formula from the name chromium(III) oxide, you have to realize that you cannot take one and one-half oxygen atoms to get the needed −3 oxidation number. There is no way of having part of an atom. The only

way that you can write a formula is to have _____ Cr's. Then you can

balance the resulting +6 oxidation number with _____ O's. The

formula for the compound is _____.

See if you can work out a formula for manganese(II) arsenide. Arsenic

is in Group VA. Its only negative oxidation number is _____ (you

subtract 5 from __). From the name, you can tell that the oxidation

number of the Mn is _____. Since it is possible to write a formula with

only whole numbers of atoms, you take _____ Mn's, which have a total

oxidation number of _____, and _____ As's, which have a total oxi-

dation number of _____. Therefore the formula is Mn_3As_2.

There is a useful little trick to get the positive and negative oxidation numbers equal. It's similar to *cross multiplication*. You write the oxidation number of the second atom (or group of atoms) as the subscript of the first, and the oxidation number of the first atom (or group of atoms) as the subscript of the second. Suppose that you want the formula for copper(II) arsenide. The negative oxidation number of the As (it's in

Group VA) is _____. The name tells you that the oxidation number of

the Cu is _____:

Cu   As
(+2) (−3)

So you make the 3 a subscript for the Cu and the __ a subscript for the As:

Cu_3 As_2
(+2) (−3)

This gives you (+2 × 3) = +6   and   (−3 × 2) = −6. Therefore the sum

of the oxidation numbers equals __.

However, if both the subscripts can be divided by the same number, you have to simplify the formula. Thus the formula for tin(IV) oxide

Sn   O   is not Sn_2O_4
(+4) (−2)

Both the subscripts can be divided by _____. So you do it. The formula

is therefore _____.

−2
two
three
Cr_2O_3
−3
8
+2
three
+6; two
−6
−3
+2
2
0
two
SnO_2

+3	See if you can work out the formula for boron nitride. Since B is in Group IIIA, it has a _____ oxidation number. And N is in Group VA,
8	so its only negative oxidation number can be calculated as __ minus the
-3	group number, or _____. So the first formula you write is
3; 3	$$\underset{(+3)}{B} \quad \underset{(-3)}{N}$$
3	However, both subscripts are divisible by __, so the actual formula is
BN	_____.
	There are a few special ions you have to remember. The peroxide ion
$O_2{}^{2-}$	is _____ and the mercury(I) (mercurous) ion is $Hg_2{}^{2+}$. Thus the formula for mercury(I) chloride is
1; 2	$$\underset{(+2)}{Hg_2{}^{2+}} \quad \underset{(+1)}{Cl}$$
Hg_2Cl_2	Since subscript 1's are not written, the formula is _____.
	The formula for sodium peroxide is
2; 1	$$\underset{(+1)}{Na} \quad \underset{(2-)}{O_2{}^{2-}}$$
Na_2O_2	Omitting the subscript 1, you have _____.
	Writing the formula is simple if the name of the compound is given by
MnO_2	the counting method. The formula for manganese dioxide is _____. The
$TiCl_4$	formula for titanium tetrachloride is _____. The formula for uranium
UCl_5	pentachloride is _____, and the formula for octuranium trioxide is
eight	U_8O_3. (Recall that *octa* is Greek for _____, and that the final *a* is dropped before a vowel.) The formula for tetrasulfur tetranitride is
S_4N_4; Si_2I_6	_____. The formula for disilicon hexiodide is _____. The name literally *tells* you how many of each atom is present (*mono* usually being understood).
	The only problem you may have with these names is that occasionally the Greek number *di* for the first element may be omitted. Thus you could say chlorine heptoxide for Cl_2O_7. And the formula for phosphorus
P_2O_5	pentoxide is _____, not PO_5, which would mean that the oxidation
+10	number for P would be _____. The formula for arsenic trioxide is
As_2O_3	_____, since the "di-" in front of the arsenic is omitted.
	[Your instructor may suggest that you omit the next three paragraphs.]
	If your compound is named by the old-fashioned "-ic" and "-ous" system, you simply have to know what the oxidation number corre-

sponding to these is for the metallic element. The only way is either to have a reference like a textbook or the periodic table in Figure 9.1. After awhile, you'll simply remember some of the common ones like ferric [iron(III)], manganous [Mn(II)], stannic [Sn(IV)], and auric [Au(III)]. You can see the advantage of the other two systems.

Thus, if you are asked for the formula of ferric chloride, you have to

know that the _____ oxidation state of Fe is +3. Then you can | higher

write the formula as _____, by the same reasoning you would use if | $FeCl_3$
you were given the name iron(III) chloride for the same compound. You can work out the formula for stannous oxide if you know that the

_____ oxidation state of Sn is +2. The formula is _____. And if you | lower; SnO
know that the oxidation number corresponding to manganic is +3, you

can write the formula of manganic chloride as _____. | $MnCl_3$

Here is a short exercise so that you can check to see whether you can write the formulas for compounds whose names are given. Don't look at the answers until you have tried to do them yourself. Give the formulas for the following compounds.

potassium hydroxide _____	ammonium sulfide _____	KOH; $(NH_4)_2S$
tin(IV) phosphide _____	di-iodine tetroxide _____	Sn_3P_4; I_2O_4
arsenic pentoxide _____	aluminum hydride _____	As_2O_5; AlH_3
hydrofluoric acid _____	hydrogen cyanide _____	HF(*aq*); HCN

SECTION D Oxyacids

There is a group of acids that contain—besides H and a nonmetallic element—also oxygen. These are called *oxyacids*, or in IUPAC terminology, *oxo acids*. They are named by stating the name of the nonmetal and substituting "-ic" for its last syllable if it is at its highest oxidation state, and "-ous" for its last syllable if it is at the next-lower oxidation state. This is then followed by the word "acid." Thus the name for HNO_3 is nitric acid, and the name for HNO_2 is nitrous acid. The oxidation

number for the N in HNO_3 is _____, and the oxidation number for the | +5

N in HNO_2 is _____. | +3
You can figure out that +5 is the highest oxidation number for N if you have a periodic table. N is in Group VA, and the highest oxidation

number is _____ to the group number. | equal

The name for H_2SO_4 is _____. (In the case | sulfuric acid
of S and C, you keep the last syllable.) You can tell that this is an "-ic"

acid because the oxidation number of S in H_2SO_4 is _____. The name | +6

sulfurous acid	of H_2SO_3 is _____. The S in H_2SO_3 is at a
+4; lower	_____ oxidation number, which is the next _____ oxidation number to the +6.
	See if you can work out the name for H_2CO_3. The oxidation number
+4	for the C is _____. If you look in the periodic table, you see that C is
IVA	in Group _____. The highest oxidation number possible in Group IVA
+4; highest	is _____. Therefore the C in H_2CO_3 is at its _____ oxidation
carbonic acid	state, and the name of the acid is _____.
	The name for H_3PO_3 can be worked out if you calculate the oxidation
P; +3	number of the __, which works out to be _____. If you locate P in the periodic table, you'll see that it is in Group VA, in which the highest
+5	oxidation number possible is _____. But the oxidation number of the P
lower	in H_3PO_3 is two _____ than this. Thus the name of H_3PO_3 must be
phosphorous	_____ acid. (Incidentally, so that you don't confuse the acid with *phosphorus* the element, when you are speaking you shift the accent. You pronounce the acid phos-PHOR-ous and pronounce the element PHOS-phorus.)
	Some of the oxyacids that you will learn to name do not exist in the free state. However, as you'll see in the following section, their *salts* (combinations with metallic ions) do exist. It will be very helpful if you learn how to name these imaginary acids.
	So far you have learned how to name an oxyacid. When the non-
"-ic"	metallic element is in its highest oxidation state, you substitute _____ for the last syllable of the nonmetal's name. When it is in the next-lower
two	oxidation state (the oxidation number will be _____ less than for the
"-ous"	highest), you substitute _____ for the last syllable. What happens if there is an oxyacid in which the oxidation number of the nonmetal is still lower than the "-ous" acid? Chemists use the Greek word for "beneath," which is *hypo* (hypodermic means "beneath the skin") as the first part of the name. The acids are beneath the "-ous."
	Consider the three oxyacids H_3PO_4 (the oxidation number of P is
+5	_____), H_3PO_3 (the oxidation number of P is +3), and H_3PO_2 (the
+1	oxidation number of P is _____). The highest oxidation state possible
phosphoric	is +5, and so H_3PO_4 is called _____ acid. The oxidation
phosphorous	state two lower is +3, so H_3PO_3 is called _____ acid. The third acid, H_3PO_2, has an oxidation number two lower than the "-ous" acid, so it is named hypophosphorous acid.
	Try one more. When you have the acid $H_2N_2O_2$ and calculate the
+1	oxidation number of N, you find that it is _____. (Two N's are +2, so each N must be +1.) N is in Group VA, so its highest possible oxidation

number is +5. The next-lower state must be _____. Recall that the
oxidation numbers go down in steps of two. But in the acid $H_2N_2O_2$ the
oxidation number for N is only +1. Therefore its name must be

_____ acid.

    Actually, these hypo-ous oxidation states are very rare, except in the
nonmetals in Group VIIA: Cl, Br, and I. Here a further problem arises,
for there are *four* different positive oxidation states, +7, +5, +3, and +1.
Our "-ic," "-ous," and "hypo-ous" system allows for only three. Since
Br never shows a +7 oxidation number, chemists decided to consider the
+5 oxidation number the "-ic" (highest), the +3 (two lower) the

"_____," and the +1 the "hypo-ous." The +7 was then named "greater
than '-ic'" (in Greek, of course). This is *hyper*, but to avoid confusion it
was shortened to *per*. Thus the acid $HClO_4$, in which the oxidation

number of Cl is _____, is called *per*chloric acid. The acid HIO_4, in

which the I (iodine) has an oxidation number of _____, is named

_____ (pronounced "PER-eye-odic") acid. As a matter
of fact, the "per-ic" name is reserved for oxyacids in which the element
has an oxidation number of +7. Thus the name of the acid $HMnO_4$ is

_____ acid, since the oxidation number of the Mn is

_____. If you check the periodic table, you will see that Mn is in
Group VIIB. Thus Mn shows some oxidation states similar to those of
Group VIIA. The B Groups in the periodic table often have elements
that are similar to the elements in the A Group of the same number.
    Chromium, Cr, is in Group VIB, so you might expect it to have

oxyacids similar to those of elements in Group _____. The highest

oxidation number is _____, and thus the name for H_2CrO_4 is

_____ acid.
    Although the acid itself doesn't exist (though salts do), we can name

$H_2Cr_2O_7$. The oxidation number for one Cr is still _____. Therefore it
is still a chrom*ic* acid. However, to distinguish it from H_2CrO_4, you call
it dichromic acid. The "di-" tells you that there are two Cr's.
    Let's put all this into a table, so that you will have a ready reference
to the names for the oxidation numbers of elements in various groups.
(See Table 9.1.)

						+3
						hyponitrous
						-ous
						+7
						+7
						periodic
						permanganic
						+7
						VIA
						+6
						chromic
						+6

Table 9.1 Oxidation Number

Name	Group	IIIA	IVA	VA	VIA and B	VIIA and B
per_____ic						+7
_____ic		+3	+4	+5	+6	+5
_____ous			+2	+3	+4	+3
hypo_____ous				+1	+2	+1

You should now be able to name any oxyacid if you are given its formula, and if you look at a periodic table to see which group the non-metal is in. Try a few, to see how well you do.

In order to find the name for H_4SiO_4, you first locate _____ in the periodic table. It is in Group IVA. You know that the highest oxidation number must therefore be _____. Next you calculate the oxidation number of Si in H_4SiO_4. It is _____. Therefore the name must be _____ acid, since you are at the highest oxidation number.

In order to find the name of H_3AsO_3, you start by locating As in the _____ table. It is in Group VA. The highest oxidation state is _____. You then calculate the _____ number of the _____ in H_3AsO_3. It is +3. This is two below the value for the highest possible number, so the name is _____ acid.

To name H_3BO_3, you first find that B is in Group _____ in the periodic table. The highest oxidation number is _____. The oxidation number for B in H_3BO_3 is _____, so the name must be _____ acid, since you are at the highest oxidation number.

If you want the name for $HClO_2$, you first locate _____ in Group _____ in the periodic table. From this you know that the highest possible oxidation number is _____. The oxidation number for the Cl in $HClO_2$ is only _____. Recall that for Group VIIA the _____ oxidation number was taken to be the "-ic." The oxidation state of Cl in $HClO_2$ is _____ below the "-ic" value. So the name for $HClO_2$ must be _____ acid.

What about the name for HBrO? Br is in Group _____ also. The oxidation number for Br in HBrO is only _____. This is _____ lower than the "-ic" oxidation value for Group VIIA. Therefore the name must be _____ acid.

Getting the name of an oxyacid whose formula is given is simple if you can figure out the oxidation number of the nonmetallic element and have a periodic table so you can find its group. But some students prefer

Table 9.2 Common "-ic" Oxyacids

Boric acid	H_3BO_3	Sulfuric acid	H_2SO_4
Carbonic acid	H_2CO_3	Chromic acid	H_2CrO_4
Nitric acid	HNO_3	Chloric acid	$HClO_3$
Phosphoric acid	H_3PO_4	Bromic acid	$HBrO_3$
Arsenic acid	H_3AsO_4	Iodic acid	HIO_3

Si

+4

+4

silicic

periodic

+5; oxidation

As

arsenous

IIIA

+3

+3; boric

Cl

VIIA

+7

+3; +5

two

chlorous

VIIA

+1; four

hypobromous

to memorize. If that's your preferred method, Table 9.2 lists the "-ic" formulas of the common oxyacids you will encounter in first-year chemistry. The "-ous" form will have one less O atom.

If an oxyacid has two fewer O atoms than the "-ic" form, it will be a "hypo . . . ous" acid. If it has one more O atom, it will be a "per . . . ic" acid.

Table 9.3 shows some other acids you might want to commit to memory.

Let's see how good you are at naming oxyacids. Name the following acids.

Formula		Hint	Answer
H_3AsO_3	_____	(As is in Group VA)	Arsenous acid
$HClO_2$	_____	(Cl is in Group VIIA)	Chlorous acid
H_2CO_3	_____	(C is in Group IVA)	Carbonic acid
$HMnO_4$	_____	(Mn is in Group VIIB)	Permanganic acid
H_2SeO_4	_____	(Se is in Group VIA)	Selenic acid
HNO_2	_____	(N is in Group VA)	Nitrous acid
HIO	_____	(I is in Group VIIA)	Hypoiodous acid

Ortho-, Meta-, and Pyro- Acids

[Your instructor may suggest that you skip this next section. It is pretty specialized.]

It is possible to remove a water molecule from an oxyacid and still have an acid left *provided that the original oxyacid has at least three* H

atoms. For example, if you heat H_3PO_4, _____ acid, an   phosphoric
H_2O molecule will come off:

$$H_3PO_4 \rightarrow HPO_3 + H_2O$$

The oxidation number of the P in HPO_3 is still _____, so it's still an   +5
"-ic" acid. In order to show that the water has been removed, you write

"meta-" as a prefix to the name of the acid. Thus HPO_3 is _____phos-   meta
phoric acid.

Table 9.3 Some Special Oxyacids

*Dichromic acid	$H_2Cr_2O_7$
Permanganic acid	$HMnO_4$
Thiosulfuric acid	$H_2S_2O_3$
Acetic acid	$HC_2H_3O_2$
Oxalic acid	$H_2C_2O_4$

*Imaginary acid, but with salts

HBO$_2$	Removing a water molecule from boric acid, H$_3$BO$_3$, leaves _____,
metaboric	which is called _____ acid. Taking an H$_2$O from silicic acid,
H$_2$SiO$_3$; metasilicic acid	H$_4$SiO$_4$, leaves _____, which is called _____. (In fact, orthosilicic acid doesn't exist. Only the "meta-" form does.) In order to be super-correct, you should say that H$_3$PO$_4$ and H$_3$BO$_3$ and H$_4$SiO$_4$ did *not* have an H$_2$O molecule taken out. So you say that
ortho	these are *ortho-* forms of the acid. You call H$_3$PO$_4$ _____phosphoric
orthoboric	acid and H$_3$BO$_3$ _____ acid and H$_4$SiO$_4$
orthosilicic acid	_____. Usually chemists don't bother with the "ortho-" prefix. It is understood. It is even possible to remove one H$_2$O molecule from two oxyacid molecules and still have an acid left, even if the oxyacid has only two H atoms. In this process, the remains of the two oxyacids join to form a new molecule. For example,

$$2H_2SO_4 \rightarrow H_2S_2O_7 + H_2O$$

When this happens, you say that you have the *pyro-* form of the acid.

pyro	Thus H$_2$S$_2$O$_7$ is called _____sulfuric acid. If you remove one H$_2$O from
phosphoric	two molecules of H$_3$PO$_4$ (_____ acid), you have the acid
H$_4$P$_2$O$_7$; pyrophosphoric	with the formula _____, which is named _____ acid. Notice that the oxidation number of the P does not change. Removing two H atoms (a total oxidation number of +2) and one O atom (a −2 oxidation number) has no effect on the oxidation number of the P. Actually the nonexistent acid we called dichromic acid, H$_2$Cr$_2$O$_7$, is actually pyrochromic acid. It is produced by removing one H$_2$O from
chromic; H$_2$CrO$_4$	two _____ acid molecules: _____. You can tell when you have a "meta-" form of the acid only if you know how many H atoms the "ortho-" (that is the usual form) has. The
two	"meta-" has _____ less. You can usually spot the "pyro-" form, since it
two	has _____ of the nonmetal atoms. However, there are exceptions. [For example, H$_2$S$_2$O$_3$ is thiosulfuric acid (see Table 9.3). Sulfuric acid has simply exchanged one of the O atoms for an S atom.] But all "pyro-" forms have either five or seven O atoms.

Writing Formulas for Oxyacids

Getting the name of an oxyacid when you are given its formula is relatively easy. However, the reverse of this procedure is much more of a problem. If you are very sure of the Lewis-dot diagramming of molecules, you could figure out the formula. Or, as you will see, there are some semi-logical ways of doing it. However, it may be easier for you to simply memorize the formulas of the acids you will meet in first-year chemistry. Table 9.4 includes most of these acids (many appeared in the previous tables). Some of them (those marked with asterisks) are very unstable, or do not exist except as their salts. Table 9.4 gives them in the order of the groups in which the nonmetal element is found.

Now, if you would prefer to figure out the formulas, the way you can do it is this. First you must know the number of H atoms the acid contains. You can get this by subtracting the group number of the non-metallic element from 8. Thus the nonmetal in sulfuric acid is S, which is in Group VIA. It must therefore have (8 – 6) H atoms. That is, it has

_____ H atoms. Phosphorous acid has P, which is in Group VA.   two

Therefore it must have (8 – __) = _____ H atoms. And permanganic   5; three

acid contains Mn in Group VIIB, so it has _____ H atom.   one

There are three exceptions to this rule: B, C, and N oxyacids have two less H atoms than you would calculate. Thus carbonic acid (C is in Group

IVA) has two fewer H atoms than the _____ you would expect by   four

subtracting the group number from 8. Carbonic acid has _____ H   two

atoms. Nitric acid would then have only _____ H atom (you would   one

have calculated _____ by subtracting Group VA from 8).   three

Once you know the number of H atoms, you need to know how many O atoms the acid contains. You can do this if you know the oxidation number of the nonmetallic element. The name of the acid—whether it is "-ic," "-ous," "hypo . . . ous," or "per . . . ic"—and the group number of the element tell you the oxidation number. For example, if you have a Group VIA element, the maximum oxidation

number possible is _____. If the name of the acid ends in "-ic," this   +6
means that you are at the highest possible oxidation number,

Table 9.4 Common Oxyacids

Boric acid	H_3BO_3	Perchloric acid	$HClO_4$
Carbonic acid	H_2CO_3	Chloric acid	$HClO_3$
*Silicic acid	H_4SiO_4	*Chlorous acid	$HClO_2$
Nitric acid	HNO_3	*Hypochlorous acid	$HClO$
Nitrous acid	HNO_2	Bromic acid	$HBrO_3$
Phosphoric acid	H_3PO_4	Bromous acid	$HBrO_2$
Phosphorous acid	H_3PO_3	*Hypobromous acid	$HBrO$
Arsenic acid	H_3AsO_4	Periodic acid	HIO_4
Arsenious acid	H_3AsO_3	Iodic acid	HIO_3
Sulfuric acid	H_2SO_4	*Hypoiodous acid	HIO
Sulfurous acid	H_2SO_3	*Permanganic acid	$HMnO_4$
*Thiosulfuric acid	$H_2S_2O_3$	Acetic acid	$HC_2H_3O_2$
Chromic acid	H_2CrO_4	Oxalic acid	$H_2C_2O_4$
*Dichromic acid	$H_2Cr_2O_7$		

*Very unstable, or exists only in water solution or as salts.

+6

+7

+1

+3; two

_____. If the nonmetal is in Group VIIA and the acid's name is "hypo . . . ous," you know that the highest possible oxidation number is _____ and the "hypo . . . ous" is three steps (of two oxidation numbers) below this. So it is _____. If you have a Group VA element's "-ous" acid, the oxidation number of the nonmetal is _____. It is _____ less than the highest possible number for Group VA.

See if you can figure out the oxidation number for the nonmetallic element in the following acids.

+7

+4

+3

+5

Permanganic (Mn is Group VIIB)   _____

Silicic (Si is in Group IVA)   _____

Iodous (I is in Group VIIB)   _____

Nitric (N is in Group VA)   _____

Once you know the number of H atoms and the oxidation number of the nonmetal, it is easy to calculate the number of O atoms. Their total oxidation number must balance the sum of the positive oxidation numbers of the H atoms and the nonmetallic element. (We are assuming that our acids contain only a single nonmetal atom. If you have a "pyro-" form of the acid, you determine the "ortho-" formula and then subtract one H_2O molecule from two "ortho-" acids. Incidentally, if you have a "meta-" acid, the method we gave to get the number of H atoms is off by two. You determine the "ortho-" acid formula and subtract one H_2O from one acid molecule.)

Here is an example. Suppose that you want to find the formula for

two

sulfuric acid. First, since S is in Group VIA, the acid has _____ H atoms.

$$H_2 \ S \ O_?$$

+6

+2

Since it's an "-ic" acid, the oxidation number of the S is _____. The two H atoms have a total oxidation number of _____.

$$H_2 \ S \ O_?$$
(+2) (+6)

+8

Thus you have a total _____ positive oxidation number that must be balanced by the negative oxidation numbers of O atoms. Each O has a

-2; four

_____ oxidation number, so you need _____ O atoms to give you the -8 you need. The formula is therefore

$$H_2 \ S \ O_4$$
(+2) (+6) (-8)

Try another one: What is the formula for hypochlorous acid? Since Cl is in Group VIIA, the oxyacid has _____ H atom,

 H Cl O$_?$

And since it is a "hypo . . . ous" acid, the oxidation number of the Cl is

_____. The single H atom also has an oxidation number of _____.

 H Cl O$_?$

 (+1) (+1)

So the total positive oxidation number is _____. You need just _____
O atom (with a -2 oxidation number) to balance this.

 H Cl O

 (+1) (+1) (−2)

Therefore the formula for hypochlorous acid is _____.
 We'll do one more. Suppose that you want the formula for nitrous
acid. Since N is in Group VA, you would expect its oxyacid to have

_____ H atoms, *but* it is one of the exceptions. It has only _____.
Then, since N is in Group VA and this is an "-ous" acid, the oxidation

number of the N is _____.

 H N O$_?$

 (+1) (+3)

It takes _____ O atoms to balance the total $+4$ oxidation number of

the H and the N. Therefore the formula is _____.
 To see how well you can do this, try to give the formulas for the
following oxyacids. You are going to have to locate the elements in the
periodic table yourself.

Permanganic acid   _____   Silicic acid   _____

Iodous acid   _____   Nitric acid   _____

Chromic acid   _____

Answers column:
one

+1; +1

+2; one

$HClO$

three; one

+3

two

HNO_2

$HMnO_4$; H_4SiO_4

HIO_2; HNO_3

H_2CrO_4

SECTION E Polyatomic Ions from Oxyacids

All the oxyacids can lose one or more of their H atoms as H^+ ions. This
leaves behind a negatively charged ion (an *anion*). Anions can form compounds with positive ions (*cations*) from a metallic element. These compounds are called *salts*. The ammonium cation (the NH_4^+) acts like a
metallic cation.
 Names of salts are formed from the name of the cation plus the name
of the anion. You already know, from our discussion of binary com-

pounds, how to name cations. But how are anions named?

Here's how. The stem of the name of the anion is the same as the stem of the acid. But if the anion comes from an "-ic" (highest oxidation state) acid, you substitute the suffix "-ate" for the "-ic." Thus the anion from sulfuric acid is sulfate and the anion from phosphoric acid is

phosphate

_____. If the anion comes from the "-ous" (lower oxidation state) acid, you substitute the suffix "-ite" for the "-ous." The

nitrite

anion from nitrous acid is called _____ and that from chlorous

chlorite

acid is _____. If there is a "hypo-" or "per-" prefix in the acid's name, you just retain it. Thus the anion from hypoiodous acid would be hypoiodite. The anion from permanganic acid would be

permanganate

_____.

If you have the anion $AsO_4{}^{3-}$ and you want its name, you check the

As; +5

oxidation number of the _____ atom. It is _____. Since As is in

+5

Group VA, in which the highest possible oxidation number is _____,

arsenate

then the anion must be called _____. It is at the highest oxidation number.

But what is the name of the $SO_3{}^{2-}$ ion? The oxidation number of the

+4

S is _____. But S is in Group VIA, which has a highest oxidation num-

+6

ber of _____. This is only +4. Therefore the anion must be called

sulfite; sulfurous

_____. It would have come from _____ acid.

Try a few. What are the names of the following?

periodate; carbonate

$IO_4{}^-$ _____ $CO_3{}^{2-}$ _____

phosphite; nitrate

$PO_3{}^{3-}$ _____ $NO_3{}^-$ _____

borate; hypochlorite

$BO_3{}^{3-}$ _____ ClO^- _____

If an oxyacid has more than one H atom, it need not lose them all when it forms its anion. For example, sulfuric acid might lose just one H^+ ion: $H_2SO_4 \rightarrow H^+ + HSO_4{}^-$. The anion is named to show the remaining H atom as *hydrogen sulfate*. The anion $HCO_3{}^-$ is called

hydrogen

_____ carbonate and the name for $HSO_3{}^-$ is

hydrogen sulfite

_____. The oxidation number of the nonmetal is the same whether you have carbonate or hydrogen carbonate, sulfate or hydrogen sulfate, or sulfite or hydrogen sulfite.

[There are two other ways of showing that an H atom remains with the anion. The old-fashioned (and drugstore) method is to say "bi-" instead of hydrogen. Hydrogen carbonate is then *bicarbonate* and

bisulfite

hydrogen sulfite is _____. The other way is to say "acid" instead of hydrogen, since the anion has an H atom that it can lose as an H^+. This method is more usual with organic acids. Thus

$HC_8H_4O_4^-$, hydrogen phthalate, is often called *acid phthalate*.]

If the oxyacid has three H atoms, you can split off one or two of them and still leave H atoms on the anion. For example,

$$H_3PO_4 \rightarrow H^+ + H_2PO_4^- \qquad H_3PO_4 \rightarrow 2H^+ + HPO_4^{2-}$$

To distinguish between these two anions, you count (in Greek) the number of H atoms left on the anion. So $H_2PO_4^-$ is called *dihydrogen*

phosphate and HPO_4^{2-} is called _____hydrogen phosphate. (You have to say the "mono-" to make sure that it is understood that there aren't two H atoms.) You use "mono-" and "di-" only when two different H-containing anions exist. That means that the acid has to have more than two H atoms. To say that HCO_3^- is monohydrogen carbonate is incorrect, in the same way that saying sodium(I) chloride is incorrect. There is only one anion containing H and CO_3^{2-}. By the same reasoning, you never say sodium(I) chloride. There is only one compound that contains Na and Cl.

mono

Constructing the formula for one of these polyatomic ions when you are given its name is a little cumbersome. It requires that you use the same sort of reasoning you used to construct the formula of the oxyacid. Once again, if you prefer, you can memorize the formulas of the more common anions (and also their charges) in the "-ate" state. Then you can figure out that the "-ite" form has one less O atom. So if you remember

that SO_4^{2-} is sulfate, then _____ must be the formula for sulfite. Table 9.5 shows the most common anions in their "-ate" form.

SO_3^{2-}

You can find a much more complete list below the periodic table in Figure 9.1 (page 150).

If you want to work out the formulas, here is how to do it. Suppose that you want the formula for arsenite. First you determine its charge. You do this in the same way you determine the number of H atoms on

the acids. Recall that the number of H atoms equals __ minus the group

8

number (except for B, C, and N, which have _____ less than what you would calculate). Each H that is lost in the form of an H^+ gives the remaining anion one additional minus charge. Since arsenic (As) is in

two

Group VA, the oxyacid has _____ H atoms. The loss of these gives the

three

anion a _____ charge. You can start by writing

3−

$$As \; O_?{}^{3-}$$

Table 9.5 Some Common "-ate" Complex Ions

BO_3^{3-}	Borate	SO_4^{2-}	Sulfate	IO_3^-	Iodate
CO_3^{2-}	Carbonate	CrO_4^{2-}	Chromate	BrO_3^-	Bromate
NO_3^-	Nitrate	$Cr_2O_7^{2-}$	Dichromate	MnO_4^-	Permanganate
PO_4^{3-}	Phosphate	ClO_3^-	Chlorate		(there is no -ate)

Now to determine the number of O atoms, you use the name of the anion to tell you the nonmetal's oxidation number. We are interested in

two

arsen*ite*. The "-ite" tells you that the oxidation number is _____ less than the highest possible value. Since As is in Group VA, the highest

+5; +3

possible value is _____. Thus the As in arsenite must have a _____ oxidation number.

As O$_?^{3-}$
(+3)

Since the sum of the oxidation numbers of the As and the O must equal

−2

the charge, 3−, and each O present has an oxidation number of _____,

three; −6

you need _____ O atoms to give a total of _____ for the oxidation number that will leave the 3− for the charge.

As O$_3^{3-}$
(+3) (−6) (+3 − 6 = −3)

So the formula for arsenite is AsO$_3^{3-}$.

    Try another. Find the formula for chromate. First you find chromium (Cr) in the periodic table. It's in Group VIB. You would then expect it

two

to be similar to a Group VI nonmetal. So its acid would have _____ H atoms. (Remember, 8 minus the group number.) If it loses these, the

2−

charge on the polyatomic ion is _____.

Cr O$_?^{2-}$

Since the name ends in "-ate" and Cr is in Group VIB, its oxidation

+6

number must be _____, the highest possible value.

Cr O$_?^{2-}$
(+6)

In order to have a −2 left over when you add the oxidation numbers of

four; −2

the Cr and the O, you need _____ O atoms, each having a _____ oxidation number.

Cr O$_4^{2-}$
(+6) (−8) (+6 − 8 = −2)

Thus the formula for chromate is CrO$_4^{2-}$.

    When you have "hypo-" or "per-" anions, you determine the oxidation number of the nonmetal the same way you did for the oxyacids.

+7

Thus the oxidation number for iodine (I) in periodate is _____ ("per-"

+7

always means a _____). In hypochlorite, the oxidation number of the

+1

Cl is _____. An easy way to remember this is to count down: +7 is per . . . ate; +5 is -ate; +3 is -ite, and +1 is hypo . . . ite.

If the anion has a hydrogen in its name, you include this in your figuring out of the oxidation number. However, the charge will differ from what you expect it to be when you subtract the group number from 8. Each hydrogen that remains decreases the negative charge of the anion by _____. Thus the dihydrogen phosphate complex ion does not

| one |

have the expected _____ charge for P, a Group VA element. It is

| 3- |

_____ less, or _____. So you write:

| two; 1- |

    H_2 P $O_?^-$ (You don't write the "1" for 1- or 1+ charges.)

Since it is phosph*ate* and P is in Group VA, its oxidation number is

_____. The two H atoms have an oxidation number of _____.

| +5; +2 |

Therefore

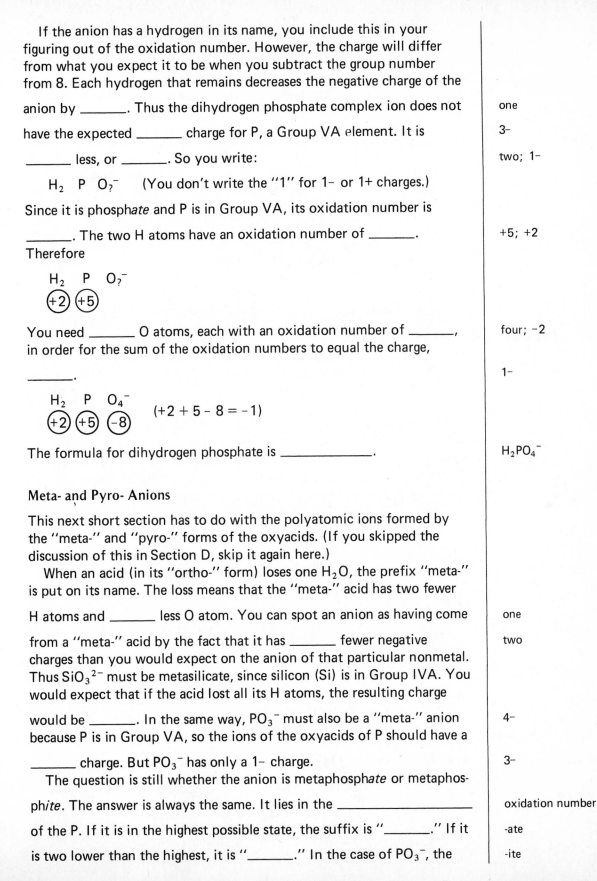

$$H_2 \quad P \quad O_?^-$$
$$\;(+2)\;(+5)$$

You need _____ O atoms, each with an oxidation number of _____, in order for the sum of the oxidation numbers to equal the charge,

| four; -2 |

_____.

| 1- |

$$H_2 \quad P \quad O_4^-$$
$$(+2)\;(+5)\;(-8) \quad (+2 + 5 - 8 = -1)$$

The formula for dihydrogen phosphate is _____.

| $H_2PO_4^-$ |

Meta- and Pyro- Anions

This next short section has to do with the polyatomic ions formed by the "meta-" and "pyro-" forms of the oxyacids. (If you skipped the discussion of this in Section D, skip it again here.)

    When an acid (in its "ortho-" form) loses one H_2O, the prefix "meta-" is put on its name. The loss means that the "meta-" acid has two fewer

H atoms and _____ less O atom. You can spot an anion as having come

| one |

from a "meta-" acid by the fact that it has _____ fewer negative charges than you would expect on the anion of that particular nonmetal. Thus SiO_3^{2-} must be metasilicate, since silicon (Si) is in Group IVA. You would expect that if the acid lost all its H atoms, the resulting charge

| two |

would be _____. In the same way, PO_3^- must also be a "meta-" anion because P is in Group VA, so the ions of the oxyacids of P should have a

| 4- |

_____ charge. But PO_3^- has only a 1- charge.

| 3- |

    The question is still whether the anion is metaphosph*ate* or metaphos-

ph*ite*. The answer is always the same. It lies in the _____

| oxidation number |

of the P. If it is in the highest possible state, the suffix is "_____." If it

| -ate |

is two lower than the highest, it is "_____." In the case of PO_3^-, the

| -ite |

+5	oxidation number of the P is _____. Since P is in Group VA, the name
metaphosphate	of PO_3^- is _____.
+4	In the first example, SiO_3^{2-}, the oxidation number of the Si is _____. Si is in Group IVA. Therefore the name we used before was correct.
metasilicate	SiO_3^{2-} is called _____.
	The way that you can spot a polyatomic ion as being a "pyro-" form is: (1) There are two of the nonmetal elements in the formula, and (2) since the "pyro-" form results from splitting out one H_2O from two oxyacids, there is an odd number of O atoms. As a matter of fact, there will be either five or seven O atoms. No oxyacid exists that could produce a "pyro-" form with any other number.
	So if you see $P_2O_5^{4-}$, you know that this is a "pyro-" form. But is it pyrophosph*ate* or pyrophosph*ite*? You know where the answer lies: in
oxidation number	the _____ of the P. In this case, the oxidation
+3	number of each P is _____. But P is in Group VA, so the ion must be
pyrophosphite	_____.
	Is the $S_2O_3^{2-}$ ion a "pyro-" form? It has the two S atoms that the "pyro-" form would have. Does it have enough O atoms? Three O atoms would indicate that either two of the "ortho-" forms had four O atoms or each had two O atoms. But, as we said, inorganic acids that have only two O atoms do not produce a "pyro-" form. Actually, $S_2O_3^{2-}$ is the result of substituting an S atom for an O atom in SO_4^{2-} (sulfate). Since both S and O are in Group VIA, this can happen. To indicate that there has been substitution, you use the Greek word for sulfur, "thio-." So the name is *thio*sulfate. In the same way, when S is substituted for the O in
thio	OCN^- (called *cyanate*), you have SCN^-. You call it _____cyanate.
	Let's consider $S_2O_5^{2-}$. Is this a "pyro-" form? It has the two S atoms you need. It also has five O atoms, which is a possible number for the "pyro-" form. (The "ortho-" acids have three each. You take one away when you remove an H_2O.) Is it pyrosulf*ate* or pyrosulf*ite*? Well, the
+4	oxidation number of the S in $S_2O_5^{2-}$ is _____. So the name of the
pyrosulfite	anion must be _____.
	To construct the formula for a "meta-" or "pyro-" ion from its name: (1) Determine the formula for the "meta-" or "pyro-" *acid*. (2) Take off the relevant number of H atoms. (3) Give the ion a negative charge equal to the number of H atoms removed. Thus, if you want the formula for
one	metaborate, you get the formula for metaboric acid by removing _____
H_3BO_3	H_2O from ("ortho-") boric acid. The formula for boric acid is _____.
HBO_2	The formula for metaboric acid is then _____. Removing the one H
BO_2^-	atom in the acid leaves the metaborate ion, _____.
	If you want the formula for pyrosulfate, you get the formula for

pyrosulfuric acid first. It would be _____ sulfuric acids minus

_____ _____. The formula for sulfuric acid is _____. So two mole-

cules of it would be $H_4S_2O_8$. Removing one _____ leaves _____.

The anion that forms when the H atoms leave the acid is _____, pyrosulfate.

two

one H_2O; H_2SO_4

H_2O; $H_2S_2O_7$

$S_2O_7^{2-}$

Now here's a good workout in naming the anions for the oxyacids. They are arranged in an order that will make it simpler for you.

CO_3^{2-} _____	HCO_3^- _____	Carbonate; Hydrogen carbonate
SO_4^{2-} _____	HSO_4^- _____	Sulfate; Hydrogen sulfate
SO_3^{2-} _____	HSO_3^- _____	Sulfite; Hydrogen sulfite
NO_3^- _____	NO_2^- _____	Nitrate; Nitrite
PO_4^{3-} _____	HPO_4^{2-} _____	Phosphate; Monohydrogen phosphate
$H_2PO_4^-$ _____	PO_3^{3-} _____	Dihydrogen; Phosphite phosphate
HPO_3^{2-} _____	$H_2PO_3^-$ _____	Monohydrogen; Dihydrogen phosphite phosphite
ClO_4^- _____	MnO_4^- _____	Perchlorate; Permanganate
ClO_3^- _____	BrO_2^- _____	Chlorate; Bromite
IO^- _____	CrO_4^{2-} _____	Hypoiodite; Chromate
$Cr_2O_7^{2-}$ _____		Dichromate

You may have noticed the two organic acids that were included in Table 9.4 in Section D. They were acetic acid, $HC_2H_3O_2$ (organic chemists write this as CH_3COOH), and oxalic acid, $H_2C_2O_4$. These form the anions acetate, $C_2H_3O_2^-$, and oxalate, $C_2O_4^{-2}$. There are no lower oxidation states for them.

For practice in finding the formula for an anion when you have its name, cover the formulas themselves and use the names in the answer column as your questions.

SECTION F   Salts of the Oxyacids

As we said earlier, a salt is composed of a cation (generally a metallic one) and some anion. If the anion comes from an oxyacid, it contains a

nonmetallic element and _____. This gives you a polyatomic ion. To name the salt, you write the name of the cation, followed by the

oxygen

name of the anion. If the cation has only one possible charge (oxidation number), you simply use the name of the element. If it has several possible charges (oxidation numbers), you use the Stock notation (or, if you dare, the "-ic"/"-ous" notation). Some examples will make this clear.

The compound $NaNO_3$ is called sodium nitrate. The metallic element is

Na; 1+	___. It is in Group IA, and can have only a ___ charge (oxidation number).
nitrate	The NO_3^- anion is called _____.
carbonate	The compound $CaCO_3$ is called calcium _____. The
2+	cation, Ca^{2+}, is in Group IIA, and can have only a _____ charge (oxidation number). The CO_3^{2-} ion is called carbonate.
aluminum phosphate	The compound $AlPO_4$ is called _____. The
Al; 3+	cation _____ is in Group IIIA, and can show only a _____ charge.
phosphate	The name of the anion is _____.
dihydrogen phosphite	The compound KH_2PO_3 is potassium _____, since K can have only a 1+ charge. The name of the anion is
dihydrogen phosphite	_____.
calcium nitrate	The name of the compound $Ca(NO_3)_2$ is _____.
2+	Since Ca is in Group IIA, it can have only a cation with a _____ charge. Notice that there are two NO_3^- anions in the formula, which is shown by
2	placing parentheses around the NO_3^- and writing a subscript ___ after the
1-	parentheses. Two of the anions, each with a charge of _____, were
2+	needed to balance the charge of the Ca^{2+} cation, which is _____.
sodium sulfate	Consider the compound Na_2SO_4. This is called _____. The reason that two Na^+ cations are required in the formula is that you
2+; 2-	need a _____ charge to balance the _____ charge of the SO_4^{2-} anion. The formula for aluminum sulfate is $Al_2(SO_4)_3$. Since the charge on
3+; 2-	the Al^{3+} cation is _____ and the charge on the SO_4^{2-} anion is _____,
6+	you need two Al^{3+} cations, which have a total charge of _____, to
6-	balance three SO_4^{2-} anions, which have a total charge of _____. The
0	charge of the whole molecule must equal ___.

So far we have examined salts whose cation has only one possible oxidation number. This is the simplest case. Consider the two salts Cu_2CO_3 and $CuCO_3$. They cannot both be called copper carbonate, so we must somehow note that the Cu is at a different oxidation number (charge) in the compounds. The Stock notation does this very well. You simply determine the oxidation number (charge) of the Cu cation in the compound. To do this, you have to remember that the CO_3^{2-} anion has a

2-   _____ charge. Then, if you look at the first compound, Cu_2CO_3, you see that it takes _____ Cu's to balance the _____ charge of the $CO_3{}^{2-}$ | two; 2–

anion. Therefore one Cu must have a charge of _____. Then you can | 1+

name the compound copper(__) carbonate. | I

In the second compound, $CuCO_3$, only one Cu is needed to balance

the _____ charge of the $CO_3{}^{2-}$ anion. Therefore the charge on the Cu | 2–

must be _____. This compound can then be named _____ | 2+; copper(II)
carbonate.

Let's try another example like this. There are two compounds: $SnSO_4$
and $Sn(SO_4)_2$. First you have to remember that the charge on the $SO_4{}^{2-}$

anion is _____. Then you see that in the first compound, $SnSO_4$, only | 2–

_____ sulfate anion is needed to balance the charge of one Sn. The | one

charge (oxidation number) on the Sn must be _____, so the name is | 2+

_____ sulfate. | tin(II)

In the second compound, $Sn(SO_4)_2$, _____ sulfate anions balance | two
the charge of one Sn. The charge (oxidation number) on the Sn must be

_____. The name of the compound is therefore _____. | 4+; tin(IV) sulfate

How about $Fe_3(PO_4)_2$ and $FePO_4$? You have to remember that the

phosphate anion has a _____ charge. Then, examining the first formula, | 3–

$Fe_3(PO_4)_2$, you see that you have _____ $PO_4{}^{3-}$ anions. This is a total | two

negative charge of _____, which exactly balances _____ Fe cations. | 6–; three

Therefore each Fe cation must have a charge of _____. The name of | 2+

the compound is _____. | iron(II) phosphate

In the second compound, $FePO_4$, only one Fe is needed to balance

the _____ charge of one $PO_4{}^{3-}$ anion. The charge on this one Fe then | 3–

must be _____ and the name of the compound is | 3+

_____. | iron(III) phosphate

To make sure you can handle this, practice on the following.

AgNO₃ _____ $Fe(HCO_3)_3$ _____ | Silver ; Iron(III) hydro-
nitrate   gen carbonate

$Ca_3(PO_3)_2$ _____ $Pb(C_2H_3O_2)_4$ _____ | Calcium ; Lead(IV)
phosphite   acetate

$Co_2(SO_4)_3$ _____ $Na_2C_2O_4$ _____ | Cobalt(III) ; Sodium
sulfate   oxalate

$CuCr_2O_7$ _____ $(NH_4)_3PO_4$ _____ | Copper(II) ; Ammonium
dichromate   phosphate

This may seem difficult, but—if you know the names and charges of
the polyatomic anions formed from oxyacids, and if you know whether
the metallic cations have more than one possible oxidation number
(charge)—with some practice you should be able to master these.

Chemical nomenclature is like any language. You simply have to learn the vocabulary.

Now we will try and reverse the whole process. That is, we'll write the formula for a salt when given its name. Once again you must know the formula and charge of the various polyatomic ions formed from oxyacids. The oxidation number (charge) of the cation is given in the name if there are several possible values. But if there is only one, you have to know it. The group number of the element tells you. (This was the same story with binary compounds.)

The formula for potassium nitrate is KNO_3. To arrive at this, you have

1+

NO_3^-

to know that the only possible charge of potassium is _____ and that the formula for the nitrate anion is _____. What about the formula for sodium sulfate? First you must know that the only possible charge on

1+

SO_4^{2-}; 2−; two

Na_2SO_4

the sodium cation is _____ and that the formula for the sulfate anion is _____, which has a _____ charge. You thus need _____ sodium cations to balance this 2− charge. The formula must be _____.

To get the formula for calcium chlorate, you must first know that the

2+

only possible charge on calcium is _____. Then you must remember the formula and charge for the chlorate anion. It is ClO_3^-. You need

two; 2+

$Ca(ClO_3)_2$

_____ ClO_3^- anions to balance the _____ charge of the Ca^{2+} cation. Thus the formula is _____.

Here is another example to really tax your memory. What is the formula of ammonium dichromate? First you must recall the formula

NH_4^+

$Cr_2O_7^{2-}$

two

2−; $(NH_4)_2Cr_2O_7$

and charge of the ammonium ion. It is _____. Next you must know the formula and charge for the dichromate anion. It is _____. Now you can set up the formula. You need _____ NH_4^+ cations to balance the _____ charge on $Cr_2O_7^{2-}$. The formula must be _____.

Let's see how you handle a compound like aluminum sulfate. The

3+

SO_4^{2-}

aluminum cation always has a charge of _____. The formula for the sulfate anion is _____, and it has a 2− charge. In order to get a formula which has as many positive charges as negative charges you have to have

two; $Al_2(SO_4)_3$

_____ Al^{3+} cations and three SO_4^{2-} anions. The formula is _____. You may prefer to set the formulas up so that this is easily seen. In this example you would write the ions first. Then you would see that you must take two Al^{3+} to balance three $(SO_4)^{2-}$.

First step: Al^{3+} $(SO_4)^{2-}$

Second step: Al^{3+} $(SO_4)^{2-}$ or $Al_2(SO_4)_3$

If you have a compound containing a cation with several possible charges, life is easier, since the charge on the cation is given in the name. Thus the compound cobalt(II) phosphate has cobalt cations with a

_____ charge. You must remember only that the formula and charge | 2+

for the phosphate anion is _____. Now you can balance the positive | PO_4^{3-}

and negative charges by taking _____ Co^{2+} cations and two PO_4^{3-} | three

anions. The formula is _____. | $Co_3(PO_4)_2$

$$Co\,\overset{2+}{\bigcirc}\,(PO_4)\,\overset{3-}{\bigcirc}$$
$$\underset{3}{} \longleftarrow \longrightarrow \underset{2}{}$$

Try another one. You want the formula for tin(IV) sulfate. You know

from the name that the cation is _____. You must remember the | Sn^{4+}

formula for the sulfate ion. It is _____. If you use the cross-multiplying | SO_4^{2-}
method,

$$Sn\,\overset{4+}{\bigcirc}\,(SO_4)\,\overset{2-}{\bigcirc}$$
$$\underset{2}{} \longleftarrow \longrightarrow \underset{4}{}$$

you get _____ as the formula. But in ionic compounds you | $Sn_2(SO_4)_4$
want to show only the simplest ratio of the ions. Both subscripts can be

divided by _____. Therefore the simplest ratio would be _____. | two; $Sn(SO_4)_2$
This is the correct formula. (You used this same technique to arrive at
the formulas for binary compounds.)
    When you run into the cation expressed in the "-ic"/"-ous" form, you
have to know the oxidation number (charge) that these endings refer to.

Thus ferr*ic* means Fe^{3+} and ferr*ous* means _____. Stann*ic* means | Fe^{2+}

_____ and stann*ous* means Sn^{2+}. | Sn^{4+}

    All these salts are ionic compounds. The "mono-, di-, tri-" counting
method is hardly ever used, though you might see it with +4 oxidation

numbers. Thus lead tetra-acetate, $Pb(C_2H_3O_2)$ __ , is sometimes used | 4

instead of the Stock notation _____ acetate. | lead(IV)

    If you want some practice before you do the problem set, redo the
practice group in which you named the salts when given the formulas.
But this time cover the formulas and use the names given in the answer
column as your questions.

PROBLEM SET

These compounds and ions are divided into groups corresponding to the sections in the chapter. If you run into difficulties, check back to the relevant section. The compounds and ions that fall into categories your instructor may have suggested you skip are marked with an asterisk. (Some of these are compounds of hydrogen and the "meta-" and "pyro-" forms of the oxyacids.) If you can work your way through this very long nomenclature problem set, you've got matters well in hand.

Section A

Write the symbol for the following elements.

1. calcium
2. sodium
3. iodine
4. boron
5. antimony
6. lead
7. tin
8. chlorine
9. cobalt
10. manganese
11. nitrogen
12. iron
13. arsenic
14. silicon
15. potassium

Section B

Give the oxidation number of the underlined element in the following (the value for a single atom). Make sure you indicate whether it is + or −.

16. $\underline{Fe}^{3+}$
17. $H\underline{Cl}O$
18. $\underline{S}O_4{}^{2-}$
19. $K\underline{Mn}O_4$
20. $\underline{Cl}^-$
21. $H_3\underline{As}O_3$
22. $Na\underline{H}$
23. $\underline{H}Cl$
24. $\underline{Cr}_2O_7{}^{2-}$
25. $Na_2\underline{Si}O_3$

Section C

Name the following compounds.

26. KF
27. ZnO
28. Ag_2S
29. NaOH
30. H_2O_2
31. NH_4CN
32. CuCl
33. Fe_2S_3
34. CoO
35. $HgCl_2$
36. Hg_2I_2
37. NaH
38. HCl(g) (gas)
39. HF(aq) (in water)
40. NH_3
*41. PH_3
*42. CH_4
*43. SbH_3

Give the formulas for the following compounds.

44. ammonium hydroxide
45. scandium bromide
46. calcium sulfide
47. potassium cyanide
48. sodium peroxide
49. silver fluoride
50. lithium nitride
51. iron(III) arsenide
52. chromium(III) oxide
53. copper(I) chloride

54. lead(IV) oxide

55. stannic iodide [†]

56. ferrous oxide [†]

57. mercuric oxide[†]

[†]If your instructor had you learn the "-ic/-ous" nomenclature

Name the following compounds. (Use the Greek count of atoms.)

58. N_2O_4

59. SO_2

60. Cl_2O_7

61. $TiCl_4$

62. CO_2

63. PbO_2

64. CCl_4

65. BF_3

66. SiF_6

67. PCl_5

68. CO

Give the formulas for the following compounds.

69. dinitrogen trioxide

70. phosphorus pentoxide

71. manganese dioxide

72. silicon tetrachloride

73. diboron trisulfide

74. sulfur trioxide

*75. arsine

Section D

Name the following oxyacids.

76. H_2SO_4

77. HNO_2

78. H_3AsO_3

79. H_2CrO_4

80. $HClO$

81. HIO_4

82. $HMnO_4$

83. H_3PO_3

84. $HC_2H_3O_2$

85. $HBrO_3$

86. H_4SiO_4

87. $H_2Cr_2O_7$

*88. HPO_3

*89. $H_4As_2O_7$

*90. $H_2S_2O_3$

Give the formulas for the following acids. (*Note*: Some are unstable or don't exist.)

91. nitric acid

92. phosphorous acid

93. dichromic acid

94. selenic acid

95. carbonic acid

96. chlorous acid

97. sulfurous acid

98. boric acid

99. perchloric acid

*100. metaphosphoric acid

*101. pyrosulfuric acid

*102. thiocyanic acid

Section E

Name the following anions.

103. NO_3^-

104. PO_3^{3-}

105. SO_4^{2-}

106. MnO_4^-

107. $Cr_2O_7^{2-}$

108. ClO_2^-

109. SeO_4^{2-}

110. CO_3^{2-}

111. SiO_4^{4-}

112. IO_4^-

113. $C_2H_3O_2^-$

114. OCN^-

*115. $P_2O_7^{4-}$

*116. BO_2^-

117. HSO_3^-

118. HPO_4^{2-}

119. HCO_3^-

120. $H_2BO_3^-$

Give the formulas for the following anions. Show the charge.

121. carbonate
122. sulfite
123. nitrate
124. chlorite
125. permanganate
126. acetate
127. phosphite
128. borate
129. oxalate
130. iodate
*131. metasilicate
*132. pyroarsenate
133. monohydrogen arsenate
134. hydrogen carbonate
135. hydrogen sulfate

Section F

Name the following compounds.

136. Na_2SO_3
137. $Ca_3(PO_4)_2$
138. $AgNO_3$
139. $Al(ClO_3)_3$
140. $(NH_4)_2C_2O_4$
141. $CuSO_4$
142. $Mn_3(PO_3)_2$
143. $Ca(HCO_3)_2$
144. $Fe(ClO_4)_2$
145. $CrSO_4$
146. NaH_2PO_4
147. $Fe(HCO_3)_3$
*148. KPO_3
*149. $Na_2S_2O_3$
*150. $Mn_2P_2O_7$

Write the formulas for the following compounds.

151. potassium permanganate
152. sodium dichromate
153. ferric oxalate
154. copper(II) nitrate
155. cobalt(II) silicate
156. tin(II) phosphate
157. sodium hydrogen carbonate
158. silver bromate
159. potassium arsenite
160. mercury(II) acetate
161. lead(II) perchlorate
162. zinc phosphate
163. calcium nitrite
164. barium sulfate
165. ammonium dichromate
*166. manganese(II) pyrophosphate
*167. lead(II) thiosulfate
*168. sodium metasilicate

PROBLEM SET ANSWERS

Section A

1. Ca	2. Na	3. I	4. B	5. Sb
6. Pb	7. Sn	8. Cl	9. Co	10. Mn
11. N	12. Fe	13. As	14. Si	15. K

Section B

16. +3	17. +1	18. +6	19. +7	20. −1
21. +3	22. −1	23. +1	24. +6	25. +4

Section C

26. potassium fluoride
27. zinc oxide
28. silver sulfide
29. sodium hydroxide
30. hydrogen peroxide
31. ammonium cyanide
32. copper(I) chloride (or cuprous chloride)
33. iron(III) sulfide (or ferric sulfide)
34. cobalt(II) oxide (or cobaltous oxide)
35. mercury(II) chloride (or mercuric chloride)
36. mercury(I) iodide (or mercurous iodide)
37. sodium hydride
38. hydrogen chloride
39. hydrofluoric acid
40. ammonia
41. phosphine
42. methane
43. stibine

44. NH_4OH
45. $ScBr_3$
46. CaS
47. KCN
48. Na_2O_2
49. AgF
50. Li_3N
51. $FeAs$
52. Cr_2O_3
53. $CuCl$
54. PbO_2
55. SnI_4
56. FeO
57. HgO

58. dinitrogen tetroxide
59. sulfur dioxide
60. (di)chlorine heptoxide
61. titanium tetrachloride
62. carbon dioxide
63. lead dioxide
64. carbon tetrachloride
65. boron trifluoride
66. silicon hexafluoride
67. phosphorus pentachloride
68. carbon monoxide

69. N_2O_3
70. P_2O_5
71. MnO_2
72. $SiCl_4$
73. B_2S_3
74. SO_3
75. AsH_3

Section D

76. sulfuric acid
77. nitrous acid
78. arsenious acid
79. chromic acid
80. hypochlorous acid
81. periodic acid
82. permanganic acid
83. phosphorous acid
84. acetic acid
85. bromic acid
86. silicic acid
87. dichromic acid
88. metaphosphoric acid
89. pyroarsenic acid
90. thiosulfuric acid

91. HNO_3
92. H_3PO_3
93. $H_2Cr_2O_7$
94. H_2SeO_4
95. H_2CO_3
96. $HClO_2$
97. H_2SO_3
98. H_3BO_3
99. $HClO_4$
100. HPO_3
101. $H_2S_2O_7$
102. $HCNS$

Section E

103. nitrate
104. phosphite
105. sulfate
106. permanganate
107. dichromate
108. chlorite
109. selenate
110. carbonate
111. silicate
112. periodate
113. acetate
114. cyanate
115. pyrophosphate
116. metaborate
117. hydrogen sulfite
118. monohydrogen phosphate
119. hydrogen carbonate
120. dihydrogen borate

121. CO_3^{2-}
122. SO_3^{2-}
123. NO_3^-
124. ClO_2^-
125. MnO_4^-
126. $C_2H_3O_2^-$
127. PO_3^{3-}
128. BO_3^{3-}
129. $C_2O_4^{2-}$
130. IO_3^-
131. SiO_3^{2-}
132. $As_2O_7^{4-}$
133. $HAsO_4^{2-}$
134. HCO_3^-
135. HSO_4^-

Section F

136. sodium sulfite
137. calcium phosphate
138. silver nitrate
139. aluminum chlorate
140. ammonium oxalate
141. copper(II) sulfate (or cupric sulfate)
142. manganese(II) phosphite (or manganous phosphite)
143. calcium hydrogen carbonate
144. iron(II) perchlorate (or ferrous perchlorate)
145. chromium(II) sulfate (or chromous sulfate)
146. sodium dihydrogen phosphate
147. iron(III) hydrogen carbonate
148. potassium metaphosphate
149. sodium thiosulfate
150. manganese(II) pyrophosphate

151. $KMnO_4$
152. $Na_2Cr_2O_7$
153. $Fe_2(C_2O_4)_3$
154. $Cu(NO_3)_2$
155. Co_2SiO_4
156. $Sn_3(PO_4)_2$
157. $NaHCO_3$
158. $AgBrO_3$
159. K_3AsO_3
160. $Hg(C_2H_3O_2)_2$
161. $Pb(ClO_4)_2$
162. $Zn_3(PO_4)_2$
163. $Ca(NO_2)_2$
164. $BaSO_4$
165. $(NH_4)_2Cr_2O_7$
166. $Mn_2P_2O_7$
167. PbS_2O_3
168. Na_2SiO_3

CHAPTER 10

Stoichiometry

PRETEST

If you are able to do simple stoichiometric problems, you should be able to complete this test in about 20 minutes. All the equations in the problems are balanced.

1. How many moles of $KClO_3$ are needed to prepare 7.1 mol O_2 according to the reaction
   $2KClO_3 = 2KCl + 3O_2$

2. How many moles of NaOH are required to react with 0.50 mol H_3PO_4 according to the reaction
   $H_3PO_4 + 3NaOH = Na_3PO_4 + 3H_2O$

3. How many grams of $KMnO_4$ are needed to react completely with 100g Fe according to
   $3KMnO_4 + 5Fe + 24HCl = 5FeCl_3 + 3MnCl_2 + 3KCl + 12H_2O$

4. How many grams of $Al_2(SO_4)_3$ are produced by the complete reaction of 27.0 g $Al(OH)_3$ from the reaction
   $2Al(OH)_3 + 3H_2SO_4 = Al_2(SO_4)_3 + 6H_2O$

5. What weight of Cu metal can be obtained by the most efficient method from 454g $CuSO_4 \cdot 5H_2O$? (You must figure out the bridge without an equation.)

6. How many moles of Al_2O_3 can be produced from a mixture of 2.0 mol Fe_2O_3 and 2.0 mol Al according to the reaction
   $Fe_2O_3 + 2Al = Al_2O_3 + 2Fe$

7. It was found that when 1.27 g of Cu reacted with an excess of S, 1.80 g of CuS were formed. What is the percent yield of this reaction?
   $Cu + S = CuS$

PRETEST ANSWERS

If you were able to do all the problems correctly, go straight to the problem set at the end of the chapter, where the problems are more difficult. If you missed only problems 1 and 2, you will find this material covered in Section A. Problems 3, 4, and 5 are covered in Section B, and Section C covers what is in problem 6. Problem 7 appears in Section D.

1. 4.7 mol $KClO_3$ 2. 1.5 mol NaOH
3. 170 g $KMnO_4$ 4. 59.2 g $Al_2(SO_4)_3$
5. 116 g Cu (You may assume that 1 mol of $CuSO_4 \cdot 5H_2O$ can yield a maximum of 1 mol of Cu.)
6. 1.0 mol Al_2O_3 7. 94.2%

Weight Relationships from Chemical Reactions

You are now going to learn how to do some basic chemical problems, involving such questions as, "How much of one substance reacts with how much of another substance to yield how much of a product?" This is called *stoichiometry*, and it is harder to spell the word than to do the problems. We will start with the "how much" expressed first as moles, then as weight, though in later chapters the "how much" may also be expressed as volumes of gases or of solutions.

SECTION A   Molar Stoichiometry

Before you can do any stoichiometry problem, you must have a balanced equation. Let's consider what the equation tells you. Consider the very first example in Chapter 9:

$$H_2 + Cl_2 = 2HCl$$

First, this equation says that 1 molecule of H_2 reacts with 1 molecule of Cl_2 to produce 2 molecules of HCl. However, if you had 10 molecules of

10; 20	H_2 they would react with _____ molecules of Cl_2 to form _____ molecules of HCl. Or if you had 6.02×10^{23} molecules of H_2, they
6.02×10^{23}; 1.20×10^{24}	would react with _____ molecules of Cl_2 to form _____
mole	molecules of HCl. Of course, 6.02×10^{23} molecules is called a _____ (abbreviated mol). So you can say that 1 mol of H_2 reacts with
1 mol; 2 mol	_____ of Cl_2 to form _____ of HCl. Many chemists find this the most practical way to express an equation, since from it you can get some very useful bridge-type conversion factors. Since there is 1 mol of
1	H_2 per __ mole of Cl_2, you can write the conversion factor

$$\frac{1 \text{ mol } H_2}{1 \text{ mol } Cl_2}$$

2 mol	And since there is 1 mol of H_2 per _____ of HCl, you can write the

conversion factor _____. The conversion factor that

you could write to relate Cl_2 to HCl would then be _____.

For practice, consider the reaction between HCl and Na_2CO_3, which will yield NaCl, CO_2, and H_2O. If you want to do any stoichiometric

calculations, you must first _____ the equation

$$HCl + Na_2CO_3 \neq H_2O + NaCl + CO_2$$

When you balance the equation, you get a coefficient of __ in front of

the HCl and a coefficient of __ in front of the NaCl. All the other

coefficients are __. From the balanced equation, you can get 10 conversion factors (not counting inverted forms). What is the conversion

factor relating moles of HCl to moles of CO_2? _____

The conversion factor relating NaCl to Na_2CO_3 is _____.

When you do a problem, you must select the appropriate conversion factor as your bridge. The reason these conversion factors are considered bridges is that they allow you to go from one substance to a different substance. Stoichiometric problems always ask a question about one substance and give some information about a different substance. In order to solve them, you go from the information given, across the bridge, to the substance asked for in the question.

Another example shows this clearly. According to the balanced equation shown below, how many moles of Br_2 are produced when 0.8 mol of KBr react?

$$KClO_3 + 6KBr + 3H_2SO_4 = KCl + 3Br_2 + 3H_2O + 3K_2SO_4$$

The question asked is: "How many moles _____?" The information

given is _____. You need a bridge relating _____ to _____, since these are the substances in the question and in the information given. Such a bridge would be the conversion factor from the balanced equation

$$\frac{\rule{2cm}{0.4pt}}{6 \text{ mol KBr}}$$

Starting the problem in the usual way, you have

? _____ = _____

If you now multiply by the bridge, the unwanted dimension of the

information given, _____, cancels. This leaves as the only

dimension _____. This is the dimension of the answer, so the answer must be

$$? \text{ mol } Br_2 = 0.8 \cancel{\text{ mol KBr}} \times \frac{3 \text{mol } Br_2}{6 \cancel{\text{ mol KBr}}} = 0.4 \text{ mol } Br_2$$

Margin answers:

$$\frac{1 \text{ mol } H_2}{2 \text{ mol HCl}}$$
$$\frac{1 \text{ mol } Cl_2}{2 \text{ mol HCl}}$$

balance

2

2

1

$$\frac{2 \text{ mol HCl}}{1 \text{ mol } CO_2}$$
$$\frac{2 \text{ mol NaCl}}{1 \text{ mol } Na_2 CO_3}$$

Br_2

0.8 mol KBr; Br_2 ; KBr

3 mol Br_2

mol Br_2 ; 0.8 mol KBr

mol KBr

mol Br_2

Try another. How many moles of H_2SO_4 are needed to react with 3.7 mol of $KClO_3$ according to the reaction shown above? The dimensions

mol H_2SO_4

of the answer will be _____. The information given is

3.7 mol $KClO_3$

_____. Therefore,

3.7 mol $KClO_3$

? mol H_2SO_4 = _____

The conversion factor by which you must multiply is the one that

H_2SO_4 ; $KClO_3$; $\dfrac{3 \text{ mol } H_2SO_4}{\text{mol } KClO_3}$

relates _____ to _____. It is _____.

mol $KClO_3$

The unwanted dimension, _____, now cancels, leaving as the

mol H_2SO_4

only dimension _____, which is the dimension of the answer. Therefore,

$$? \text{ mol } H_2SO_4 = 3.7 \text{ mol } \cancel{KClO_3} \times \frac{3 \text{ mol } H_2SO_4}{1 \text{ mol } \cancel{KClO_3}} = 11.1 \text{ mol } H_2SO_4$$

SECTION B   Weight Stoichiometry

So far we have considered only the number of moles of one substance that will be involved with a number of moles of another. However, in a laboratory you usually determine the exact amount of a substance on a balance and the information is obtained as *weight.* In metric units, this means that you will have grams as your dimensions. Therefore, you will have to convert our stoichiometric calculations from moles to grams. This is very easy, since you will always have available a conversion factor

molecular

that will do this: the _____ weight.

Let us look at an example. How many grams of HCl react with 25.0 g $Ca(OH)_2$ according to the balanced equation

$$2HCl + Ca(OH)_2 = CaCl_2 + 2H_2O$$

How many grams of HCl?

The question asked is "_____." The infor-

25.0 g $Ca(OH)_2$

mation given is _____. The bridge relating HCl to

$\dfrac{2 \text{ mol HCl}}{\text{mol } Ca(OH)_2}$

$Ca(OH)_2$ is _____. Starting in the usual way, then, you have

g HCl; 25.0 g $Ca(OH)_2$

? _____ = _____

(Be sure to write all this down in your own setup.)

Next you cancel the unwanted dimension and move toward a dimension in the bridge. Since the dimensions in the bridge are always moles,

moles

you convert grams $Ca(OH)_2$ to _____ $Ca(OH)_2$. The reciprocal of the molecular weight does exactly this. You therefore multiply by

$\dfrac{\text{mol } Ca(OH)_2}{74.1 \text{ g } Ca(OH)_2}$

_____. You now have

$$? \text{ g HCl} = 25.0 \text{ g } \cancel{Ca(OH)_2} \times \frac{\text{mol } Ca(OH)_2}{74.1 \text{ g } \cancel{Ca(OH)_2}}$$

Now you can use your bridge and get from $Ca(OH)_2$ to HCl. Multiplying gives you

$$? \text{ g HCl} = 25.0 \text{ g } \cancel{Ca(OH)_2} \times \frac{\text{mol } Ca(OH)_2}{74.1 \text{ g } \cancel{Ca(OH)_2}} \times \underline{\qquad\qquad}$$

The unwanted dimension is now _____, which can easily be converted to the dimension in the question by multiplying by the

_____ weight of HCl.

$$? \text{ g HCl} = 25.0 \text{ g } \cancel{Ca(OH)_2} \times \frac{\cancel{\text{mol } Ca(OH)_2}}{74.1 \text{ g } \cancel{Ca(OH)_2}}$$

$$\times \frac{2 \text{ mol HCl}}{\cancel{\text{mol } Ca(OH)_2}} \times \underline{\qquad\qquad}$$

The only dimension remaining is the dimension of the question asked. Therefore your answer must be

$$? \text{ g HCl} = 25.0 \text{ g } \cancel{Ca(OH)_2} \times \frac{\cancel{\text{mol } Ca(OH)_2}}{74.1 \text{ g } \cancel{Ca(OH)_2}}$$

$$\times \frac{2 \cancel{\text{mol HCl}}}{\cancel{\text{mol } Ca(OH)_2}} \times \frac{36.5 \text{ g HCl}}{\cancel{\text{mol HCl}}} = 24.6 \text{ g HCl}$$

As you see, these problems are not much harder than molar stoichiometry problems. You simply use the molecular weights as conversion factors to get from weight in grams to moles. You must always go to a dimension in moles, since the bridge from the balanced equation is always expressed in moles.

Try another problem. How many grams of NaOH are needed when 2.00 kg H_3PO_4 are used in the reaction.

$$H_3PO_4 + NaOH \neq Na_3PO_4 + H_2O$$

The first thing you do in order to solve any stoichiometry problem is to

_____ the equation. This can be done by putting a __ in front

of the NaOH and a __ in front of the H_2O. Since the problem asks about

NaOH and gives information about _____, the bridge conversion factor will contain both these substances. The bridge will be

_____.

Starting the problem in the usual way, you have

$$? \underline{\qquad\qquad} = \underline{\qquad\qquad\qquad}$$

The unwanted dimension in the information given is _____, but molecular weights are expressed in grams. The conversion is simple.

You multiply by _____. You can get to the dimensions in the bridge, moles H_3PO_4, by multiplying by the conversion factor from the molecular weight.

Right-margin answers:

$$\frac{2 \text{ mol HCl}}{\text{mol } Ca(OH)_2}$$

mol HCl

molecular

$$\frac{36.5 \text{ g HCl}}{\text{mol HCl}}$$

balance; 3

3

H_3PO_4

$$\frac{3 \text{ mol NaOH}}{\text{mol } H_3PO_4}$$

g NaOH; 2.00 kg H_3PO_4

kilograms H_3PO_4

$$\frac{10^3 \text{ g } H_3PO_4}{\text{kg } H_3PO_4}$$

$$\frac{\text{mol } H_3PO_4}{98.0 \text{ g } H_3PO_4}$$

$$? \text{ g NaOH} = 2.00 \text{ kg } H_3PO_4 \times \frac{10^3 \text{ g } H_3PO_4}{\text{kg } H_3PO_4} \times \underline{\hspace{3cm}}$$

Using the bridge, you can move out of the H_3PO_4 units and into the

NaOH units. So you multiply by _____

$$\frac{3 \text{ mol NaOH}}{\text{mol } H_3PO_4}$$

$$? \text{ g NaOH} = 2.00 \text{ kg } H_3PO_4 \times \frac{10^3 \text{ g } H_3PO_4}{\text{kg } H_3PO_4}$$

$$\times \frac{\text{mol } H_3PO_4}{98.0 \text{ g } H_3PO_4} \times \frac{3 \text{ mol NaOH}}{\text{mol } H_3PO_4}$$

moles NaOH

grams NaOH

molecular weight

$$\frac{40.0 \text{ g NaOH}}{\text{mol NaOH}}$$

The unwanted dimension left is _____, but the dimensions of

the answer are _____. It is simple to convert from moles to

weight by using the conversion factor from the _____.

You therefore multiply by _____.

The only dimension left is the dimension of the answer, which must be

$$? \text{ g NaOH} = 2.00 \text{ kg } H_3PO_4 \times \frac{10^3 \text{ g } H_3PO_4}{\text{kg } H_3PO_4} \times \frac{\text{mol } H_3PO_4}{98.0 \text{ g } H_3PO_4}$$

$$\times \frac{3 \text{ mol NaOH}}{\text{mol } H_3PO_4} \times \frac{40.0 \text{ g NaOH}}{\text{mol NaOH}} = 2.45 \times 10^3 \text{ g NaOH}$$

SECTION C   Reaction-Controlling Component

So far we have considered only cases in which one reacting component is
present in a specified amount and any others are assumed to be present
in sufficient amount to give a complete reaction. What would happen if
you were given specified amounts of several reactants? Let's take a look
at our first example,

$$H_2 + Cl_2 = 2HCl$$

This equation says that 1 mol of H_2 reacts with 1 mol of Cl_2. What
would happen if we started with 1 mol of H_2 but only 0.75 mol of Cl_2?

0.75

The 0.75 mol of Cl_2 could react only with _____ mol of H_2. The
excess H_2 would simply remain unchanged. We say that the Cl_2 *controls*
the reaction because it is the *limiting* (smaller) *amount*. This is like
saying that a chain is only as strong as its weakest link. It controls the
amount of HCl that is formed. Any stoichiometric calculations that you
make concerning the HCl must therefore be based on the amount of Cl_2.

1.5

Consequently, in the above example, you could form only _____ mol
HCl.

Consider another reaction: $2Na + Cl_2 = 2NaCl$. This balanced equation

2; 1; 2

says that __ mol of Na react with __ mol of Cl_2 to form __ mol of NaCl.

Thus, if you had only 1 mol of Na, you would need $\frac{1}{2}$ mol of Cl. If you
are going to compare moles Cl_2 with moles Na to determine the limiting

amount that controls the amount of NaCl formed, you must divide the moles of Na given in the problem by 2 before you compare it with the number of moles of Cl_2 given in the problem. The smaller of the two is the limiting amount.

Consider the reaction $4Fe + 3O_2 = 2Fe_2O_3$. If you are given amounts of Fe and O_2 in the problem and you want to determine which amount is limiting and therefore controls the yield of Fe_2O_3, you divide the

number of moles of Fe by 4 and the number of moles of O_2 by ___ before you compare them. What you have done is simply divide the number of moles for each reactant given in the problem by the coefficient that appears before that component in the balanced equation. Then you can compare and see which component is limiting. In any calculations, you will use only this "controlling" component, as in the following example.

3

How many moles of $BaSO_4$ are produced from a mixture of 3.5 mol H_2SO_4 and 2.5 mol $BaCl_2$, using the reaction

$$H_2SO_4 + BaCl_2 = BaSO_4 + 2HCl$$

First you must determine which is the limiting reactant, H_2SO_4 or $BaCl_2$. The coefficients in the balanced equation are both 1, so nothing more than a direct comparison of numbers of moles is necessary to see which is limiting.

Number moles $H_2SO_4 = 3.5$

Number moles $BaCl_2 = 2.5$

Since there are fewer moles of _____, *it* is the controlling reactant. You now solve the problem in the usual manner, using the $BaCl_2$ and forgetting about the amount of H_2SO_4, since it is in excess.

$BaCl_2$

? mol $BaSO_4$ = _____

2.5 mol $BaCl_2$

Multiplying by the bridge then gives you the answer.

$$? \text{ mol } BaSO_4 = 2.5 \text{ mol } BaCl_2 \times \frac{\text{mol } BaSO_4}{\text{mol } BaCl_2} = 2.5 \text{ mol } BaSO_4$$

Here is another example in which the coefficients are not all 1 and the amounts are given as weights and not moles. How many grams of $Al_2(SO_4)_3$ are produced by the reaction of 225 g $Al(OH)_3$ with 784 g H_2SO_4, using the reaction

$$2Al(OH)_3 + 3H_2SO_4 = Al_2(SO_4)_3 + 6H_2O$$

First you must determine the number of moles of each reactant.

$$? \text{ mol } Al(OH)_3 = 225 \text{ g } Al(OH)_3 \times \frac{\text{mol } Al(OH)_3}{78.0 \text{_____}}$$

g $Al(OH)_3$

$$= 2.88 \text{ mol } Al(OH)_3$$

$$? \text{ mol } H_2SO_4 = 784 \text{ g } H_2SO_4 \times \text{_____} = \text{\_\_\_\_\_} \text{ mol } H_2SO_4$$

$\dfrac{\text{mol } H_2SO_4}{98.1 \text{ g } H_2SO_4}$; 7.99

2	Since the balanced equation has a coefficient of 2 in front of the $Al(OH)_3$, in order to make a comparison between the two reactants you must divide the number of moles of $Al(OH)_3$ by ___. In the same way,
divide	there is a coefficient of 3 before the H_2SO_4, so you must _____
3	the number of moles of H_2SO_4 by ___.

$$\frac{2.88 \text{ mol } Al(OH)_3}{2} = 1.44, \qquad \frac{7.99 \text{ mol } H_2SO_4}{3} = 2.66$$

$Al(OH)_3$	You can see that the _____ is the limiting component, since its number is smaller. You can now solve the problem as you did in
$Al(OH)_3$	Section B, using only the data concerning the _____. You
excess	disregard the H_2SO_4 because it is in _____.

Do the problem yourself. Did you get an answer of 494 g $Al_2(SO_4)_3$? If you didn't, check your setup against the one below.

$$? \text{ g } Al_2(SO_4)_3 = 225 \text{ g } Al(OH)_3 \times \frac{\text{mol } Al(OH)_3}{78.0 \text{ g } Al(OH)_3}$$

$$\times \frac{1 \text{ mol } Al_2(SO_4)_3}{2 \text{ mol } Al(OH)_3} \times \frac{342.3 \text{ g } Al_2(SO_4)_3}{\text{mol } Al_2(SO_4)_3}$$

Try one more, but this time we will simply decide which of the reactants is limiting. Given 15.8 g $KMnO_4$, 6.80 g H_2O_2, and an excess of HCl reacting according to the following equation

$$2KMnO_4 + 5H_2O_2 + 6HCl = 2MnCl_2 + 5O_2 + 8H_2O + 2KCl$$

Which of the reactants is controlling (limiting)?

moles	The first thing you do is determine the number of _____ of each reactant.
15.8 g $KMnO_4$	$? \text{ mol } KMnO_4 = $ _____ $\times \dfrac{\text{mol } KMnO_4}{158 \text{ g } KMnO_4} = 0.100 \text{ mol } KMnO_4$
H_2O_2; g H_2O_2	$? \text{ mol }$ _____ $= 6.80 \text{ g } H_2O_2 \times \dfrac{\text{mol } H_2O_2}{34.0 \text{ _____}} = 0.200 \text{ mol } H_2O_2$
	In order to compare the two reactants to determine which is limiting,
divide	you _____ the number of moles of $KMnO_4$ by 2 and the
5	number of moles of H_2O_2 by ___. These numbers come from the
equation	balanced _____.
H_2O_2	$\dfrac{0.100 \text{ mol } KMnO_4}{2} = 0.0500, \qquad \dfrac{0.200 \text{ mol } \text{_____}}{5} = 0.0400$
H_2O_2	Consequently, the limiting reactant is _____.

SECTION D   Percent Yield

Unfortunately, not all chemical reactions produce as much product as
you would expect from the stoichiometry. Many reach a certain point
and go no farther. Chemists say that reactions reach *equilibrium*, and the
amounts of reactants and products remain constant. (This was mentioned
in Section B in Chapter 8.) Also, organic reactions can often produce
several different products, only one of which is of interest. Thus you will
not prepare as much of the compound as you would expect. And certain
laboratory procedures for separating and purifying the product result in
some loss. Consequently, we talk about this problem by expressing the
amount of product actually found as a percentage of the theoretically
(stoichiometrically) possible amount. This is called the *percent yield*:

$$\text{Percent yield} = \frac{\text{actual amount}}{\text{theoretical amount}} \times 100$$

Here is an example. Suppose that you react 1.27 g of Cu with an
excess of S and find that you have produced 1.80 g CuS. You want to
know the percent yield. First you calculate the theoretical amount of
CuS that could be formed. The reaction is $Cu + S = CuS$. So for the
theoretical amount, you start your calculation with

? g CuS = _____   1.27 g Cu

To get to moles, you use the _____ of ____: molecular weight; Cu

? g CuS = 1.27 g Cu $\times$ _____ $\dfrac{\text{mol Cu}}{63.5 \text{ g Cu}}$

Now you can "bridge" to CuS:

? g CuS = 1.27 ~~g Cu~~ $\times \dfrac{\text{mol Cu}}{63.5 \text{ g Cu}} \times$ _____ $\dfrac{\text{mol CuS}}{\text{mol Cu}}$

The _____ of CuS will get you to the desired grams of   molecular weight
CuS:

? g CuS = 1.27 ~~g Cu~~ $\times \dfrac{\text{mol Cu}}{63.5 \text{ g Cu}} \times \dfrac{\text{mol CuS}}{\text{mol Cu}} \times$ _____ $\dfrac{95.6 \text{ g CuS}}{\text{mol CuS}}$

$\qquad = 1.27 \times \dfrac{1}{63.5} \times 1 \times 95.6$ g CuS = _____ g CuS theoretical   1.91

Since you were told that only 1.80 g of CuS were actually produced, you
get the percent yield from

$$\text{Percent yield} = \frac{\text{_____ amount}}{\text{_____ amount}}$$ actual

theoretical

$$= \frac{\text{_____}}{\text{_____}} \times 100 = \text{_____} \%$$ 1.80 g CuS ; 94.2

1.91 g CuS

The units of *amount* don't *have* to be grams. As long as both the actual amount and the theoretical amount are in the same units, you'll still get the same percent yield. Percent is dimensionless.

For example, when N_2O_4 is heated, it decomposes to NO_2:

$$N_2O_4 = 2NO_2$$

But it was found that when 0.10 mol of N_2O_4 was heated, only 0.09 mol of NO_2 was formed. If you want to know the percent yield, you must first determine the theoretical yield. Since you have amounts in moles, you work in _____.

moles

0.10

$$? \text{ mol } NO_2 = \text{_____} \text{ mol } N_2O_4$$

You "bridge" from the equation

$\dfrac{2 \text{ mol } NO_2}{\text{mol } N_2O_4}$

$$? \text{ mol } NO_2 = 0.10 \text{ mol } N_2O_4 \times \text{_____}$$

0.20

$$= \text{_____} \text{ mol } NO_2 \text{ theoretical}$$

The percent yield will then be

moles

theoretical

$$\text{Percent yield} = \dfrac{\text{actual} \text{_____}}{\text{_____} \text{ moles}}$$

45

0.20 mol NO_2

$$= \dfrac{0.09 \text{ mol } NO_2}{\text{_____}} \times 100 = \text{_____} \% \text{ yield}$$

Incidentally, don't be disheartened and think that calculating the stoichiometry of reactions is a waste of time, because the reactions don't go all the way anyhow. All *metathesis* (double-replacement) reactions, and most ionic reactions in solution, *do* go to completion.

PROBLEM SET

As soon as you finish doing a problem, check the answer to it on the following pages immediately. The setups as well as the answers are shown, so that you can locate any error. Use the following balanced equations for the first 14 problems. The letter in the statement of the problem tells which equation the problem refers to. Problems 15 and 16 have the balanced equation in them, but the last six problems don't have any balanced equations. The reactants and products are given (by name mostly, not by formula—to make the problem a little harder). You will have to balance the equations yourself.

(a) $MgCO_3 = MgO + CO_2$
(b) $2Na + Cl_2 = 2NaCl$
(c) $Fe_2O_3 + 2Al = Al_2O_3 + 2Fe$
(d) $2Al(OH)_3 + 3H_2SO_4 = Al_2(SO_4)_3 + 6H_2O$
(e) $2KMnO_4 + 5H_2C_2O_4 + 6HCl = 2MnCl_2 + 10\,CO_2 + 2KCl + 8H_2O$

Section A

1. How many moles of CO_2 are produced according to equation (a) by the reaction of 6.0 mol of $MgCO_3$?

2. Suppose that 1.6 mol $Al_2(SO_4)_3$ are produced by reaction (d). How many moles of H_2O are also produced?

3. According to equation (e), 1.5 mol of $KMnO_4$ react completely with how many moles of $H_2C_2O_4$?

Section B

4. How many grams of Al_2O_3 are produced by the complete reaction of 0.20 mol of Al, according to equation (c)?

5. When 0.45 mol CO_2 is produced by equation (e), how many grams of H_2O are also produced?

6. The complete reaction of 4.6 g Na, according to equation (b), yields how many grams of NaCl?

7. How many grams of H_2SO_4 are required for the complete reaction of 65.0 g of $Al(OH)_3$ according to equation (d)?

8. According to equation (a), 4.0 kg $MgCO_3$ yield how many grams of CO_2?

9. How many grams of HCl are required for the complete reaction of 316 g $KMnO_4$ according to equation (e)?

Section C

10. How many moles of Al_2O_3 can be produced by the reaction of 7.0 mol Fe_2O_3 and 3.0 mol Al according to equation (c)?

11. How many moles of $MnCl_2$ can be produced by the reaction of 5.0 mol $KMnO_4$, 3.0 mol $H_2C_2O_4$, and 22 mol HCl, according to equation (e)?

12. How many grams of NaCl can be produced by reacting 100 g Na with 100 g Cl_2 according to equation (b)?

13. How many grams of Fe are produced by reacting 2.00 kg Al with 300 g Fe_2O_3 following the reaction of equation (c)?

14. How many grams of which reactant are left over in Problem 13?

Section D

15. What is the percent yield for the reaction $MgCO_3 = MgO + CO_2$, given that 6.0 mol of $MgCO_3$ produce 5.8 mol of CO_2?

16. Gaseous H_2S dissociates into H_2 and S gases at very high temperatures: $H_2S = H_2 + S$. When 0.620 g of H_2S was held at $2000°C$, it was found that 13 mg of H_2 were produced. What is the percent yield?

Miscellaneous Problems from Various Sections

17. Ammonia, NH_3, is produced by the reaction of nitrogen, N_2, with hydrogen, H_2. How many moles of hydrogen are needed to produce 3.0 kg of NH_3? Write the balanced equation.

18. The first step in the Ostwald process for manufacturing nitric acid is the reaction of ammonia, NH_3, with oxygen, O_2, to produce nitric oxide, NO, and water. The reaction consumes 595 g of ammonia. How many grams of water are produced? Write the balanced equation.

19. Suppose that 135.0 g of aluminum metal (Al) reacted with 2.00 mol $TiCl_4$. How many moles of aluminum chloride are produced? The other product is titanium metal. You need to have the balanced equation.

20. Sodium reacts violently with water to produce hydrogen and sodium hydroxide. How many grams of hydrogen are produced by the reaction of 400 mg of sodium with water?

21. Water gas, a mixture of carbon monoxide and hydrogen, is produced by the reaction of carbon with water. If 360 g of H_2O is reacted with 5 mol C, how much of which reactant will be left over? You need to write a balanced equation.

22. A standard analytical procedure used to determine the NaCl content in a sample is to react Cl^- with Ag^+ and precipitate silver chloride. Suppose that 0.350 g of silver chloride is produced. How many grams of sodium chloride were there in the sample?

PROBLEM SET ANSWERS

The first 14 problems are divided into sections corresponding to the sections in the chapter. You should be able to do Problems 1, 2, and 3. If you weren't able to, redo Section A. If you missed only one problem in Section B (Problems 4–9), you are doing well. If you missed more, you had better redo the section. Section C is such a nuisance that if you got only three correct, that will have to do. If you had trouble with more than three of Problems 10–14, you had better redo Section C. The last six problems are scattered types. But more than likely, any errors you made were caused either by writing incorrect formulas for the compounds, or by incorrectly balancing the equations.

1. 6.0 mol CO_2 $\left(\text{? mol } CO_2 = 6.0 \, \text{mol MgCO}_3 \times \dfrac{\text{mol } CO_2}{\text{mol MgCO}_3} \right)$

2. 9.6 mol H_2O $\left(\text{? mol } H_2O = 1.6 \, \text{mol Al}_2\text{(SO}_4\text{)}_3 \times \dfrac{6 \, \text{mol } H_2O}{\text{mol Al}_2\text{(SO}_4\text{)}_3} \right)$

3. 3.8 mol $H_2C_2O_4$ $\left(\text{? mol } H_2C_2O_4 = 1.5 \, \text{mol KMnO}_4 \times \dfrac{5 \, \text{mol } H_2C_2O_4}{2 \, \text{mol KMnO}_4} \right)$

4. 10 g Al_2O_3 $\left(\text{? g } Al_2O_3 = 0.20 \, \text{mol Al} \times \dfrac{\text{mol Al}_2O_3}{2 \, \text{mol Al}} \times \dfrac{102 \, \text{g } Al_2O_3}{\text{mol Al}_2O_3} \right)$

5. 6.5 g H_2O $\left(\text{? g } H_2O = 0.45 \text{ mol } CO_2 \times \dfrac{8 \text{ mol } H_2O}{10 \text{ mol } CO_2} \times \dfrac{18.0 \text{ g } H_2O}{\text{mol } H_2O}\right)$

6. 12 g NaCl $\left(\text{? g NaCl} = 4.6 \text{ g Na} \times \dfrac{\text{mol Na}}{23.0 \text{ g Na}} \times \dfrac{2 \text{ mol NaCl}}{2 \text{ mol Na}} \times \dfrac{58.5 \text{ g NaCl}}{\text{mol NaCl}}\right)$

7. 122 g H_2SO_4 $\left(\text{? g } H_2SO_4 = 65.0 \text{ g Al(OH)}_3 \times \dfrac{\text{mol Al(OH)}_3}{78.0 \text{ g Al(OH)}_3}\right.$

$\left. \times \dfrac{3 \text{ mol } H_2SO_4}{2 \text{ mol Al(OH)}_3} \times \dfrac{98.1 \text{ g } H_2SO_4}{\text{mol } H_2SO_4}\right)$

8. 2.1×10^3 g CO_2 $\left(\text{? g } CO_2 = 4.0 \text{ kg } MgCO_3 \times \dfrac{10^3 \text{ g } MgCO_3}{\text{kg } MgCO_3}\right.$

$\times \dfrac{\text{mol } MgCO_3}{84.3 \text{ g } MgCO_3} \times \dfrac{\text{mol } CO_2}{\text{mol } MgCO_3}$

$\left. \times \dfrac{44.0 \text{ g } CO_2}{\text{mol } CO_2}\right)$

9. 219 g HCl $\left(\text{? g HCl} = 316 \text{ g } KMnO_4 \times \dfrac{\text{mol } KMnO_4}{158 \text{ g } KMnO_4}\right.$

$\left. \times \dfrac{6 \text{ mol HCl}}{2 \text{ mol } KMnO_4} \times \dfrac{36.5 \text{ g HCl}}{\text{mol HCl}}\right)$

10. 1.5 mol Al_2O_3 $\left(\text{First determine which is limiting:}\right.$

$\dfrac{7.0 \text{ mol } Fe_2O_3}{1} = \dfrac{3.0 \text{ mol Al}}{2} = 1.5 \text{ (limiting)}$

Then solve $\quad \text{? mol } Al_2O_3 = 3 \text{ mol Al} \times \left.\dfrac{\text{mol } Al_2O_3}{2 \text{ mol Al}}\right)$

11. 1.2 mol $MnCl_2$ $\left(\text{First determine which is limiting:}\right.$

$\dfrac{5.0 \text{ mol } KMnO_4}{2} = 2.5, \qquad \dfrac{3.0 \text{ mol } H_2C_2O_4}{5} = 0.66 \text{ (limiting)}$

$\dfrac{22 \text{ mol HCl}}{6} = 3.7$

Then solve $\quad \text{? mol } MnCl_2 = 3.0 \text{ mol } H_2C_2O_4 \times \left.\dfrac{2 \text{ mol } MnCl_2}{5 \text{ mol } H_2C_2O_4}\right)$

12. 165 g NaCl $\left(\text{First determine which is limiting:}\right.$

$\text{mol Na} = 100 \text{ g Na} \times \dfrac{\text{mol Na}}{23.0 \text{ g Na}} = 4.35 \text{ mol Na}; \dfrac{4.35 \text{ mol Na}}{2} = 2.17$

$\text{mol } Cl_2 = 100 \text{ g } Cl_2 \times \dfrac{\text{mol } Cl_2}{71.0 \text{ g } Cl_2} = 1.41 \text{ mol } Cl_2 ; 1.41 \text{ mol } Cl_2 = 1.41 \text{ (limiting)}$

Then solve $\quad \text{? NaCl} = 100 \text{ g } Cl_2 \times \dfrac{\text{mol } Cl_2}{71.0 \text{ g } Cl_2} \times \dfrac{2 \text{ mol NaCl}}{\text{mol } Cl_2} \times \left.\dfrac{58.5 \text{ g NaCl}}{\text{mol NaCl}}\right)$

13. 210 g Fe $\left(\text{First determine which is limiting:}\right.$

$\text{mol Al} = 2.00 \text{ kg Al} \times \dfrac{10^3 \text{ g Al}}{\text{kg Al}} \times \dfrac{\text{mol Al}}{27.0 \text{ g Al}} = 74.0 \text{ mol Al}; \dfrac{74.0 \text{ mol Al}}{2} = 37.0$

$$\text{mol } Fe_2O_3 = 300 \text{ g } \cancel{Fe_2O_3} \times \frac{\text{mol } Fe_2O_3}{159.6 \text{ g } \cancel{Fe_2O_3}} = \frac{1.89 \text{ mol } Fe_2O_3}{1} \text{ (limiting)}$$

Then solve $\quad ? \text{ g Fe} = 300 \text{ g } \cancel{Fe_2O_3} \times \frac{\cancel{\text{mol } Fe_2O_3}}{159.6 \text{ g } \cancel{Fe_2O_3}}$

$$\times \frac{2 \cancel{\text{ mol Fe}}}{\cancel{\text{mol } Fe_2O_3}} \times \frac{55.8 \text{ g Fe}}{\cancel{\text{mol Fe}}} \Big)$$

14. 1.90 kg Al   (Since you found in Problem 13 that Fe_2O_3 is limiting and is therefore completely used up, all you need do is find out how much Al is used by the Fe_2O_3 and subtract this amount from the amount of Al you started with.

$$? \text{ g Al} = 300 \text{ g } \cancel{Fe_2O_3} \times \frac{\cancel{\text{mol } Fe_2O_3}}{159.6 \text{ g } \cancel{Fe_2O_3}} \times \frac{2 \cancel{\text{ mol Al}}}{\cancel{\text{mol } Fe_2O_3}} \times \frac{27.0 \text{ g Al}}{\cancel{\text{mol Al}}}$$

$$= 102 \text{ g Al are used.}$$

Since 2000 g Al were present, then 1898 g Al must remain. Rounded off to the correct number of significant figures, this is 1.90 kg.)

15. 97% $\Big($ $? \text{ mol } CO_2 = 6.0 \cancel{\text{ mol } MgCO_3} \times \frac{\text{mol } CO_2}{\cancel{\text{mol } MgCO_3}}$

$$= 6.0 \text{ mol } CO_2 \text{ theoretical}$$

$$\text{Percent yield} = \frac{5.8 \text{ mol } CO_2}{6.0 \text{ mol } CO_2} \times 100 = 97\% \text{ yield} \Big)$$

16. 35.4% $\Big($ $? \text{ g } H_2 = 0.620 \cancel{\text{ g } H_2S} \times \frac{\cancel{\text{mol } H_2S}}{34.1 \cancel{\text{ g } H_2S}} \times \frac{\cancel{\text{mol } H_2}}{\cancel{\text{mol } H_2S}} \times \frac{2.02 \text{ g } H_2}{\cancel{\text{mol } H_2}}$

$$= 0.0367 \text{ g } H_2 \text{ theoretical}$$

$$\text{Percent yield} = \frac{13 \cancel{\text{ mg } H_2} \times \frac{10^{-3} \text{ g } H_2}{\cancel{\text{mg } H_2}}}{0.0367 \text{ g } H_2} \times 100 = 35.4\% \text{ yield} \Big)$$

17. $2.6 \times 10^2 \text{ mol } H_2$ $\Big(3H_2 + N_2 = 2NH_3$

$$? \text{ mol } H_2 = 3.0 \cancel{\text{ kg } NH_3} \times \frac{10^3 \cancel{\text{ g } NH_3}}{\cancel{\text{kg } NH_3}} \times \frac{\cancel{\text{mol } NH_3}}{17.0 \cancel{\text{ g } NH_3}} \times \frac{3 \text{ mol } H_2}{2 \cancel{\text{ mol } NH_3}}$$

$$= 264.7 \text{ mol } H_2 \text{ or } 2.6 \times 10^2 \text{ mol } H_2 \Big)$$
(to the correct significant figures)

18. $945 \text{ g } H_2O$ $\Big(4NH_3 + 5O_2 = 4NO + 6H_2O$

$$\text{g } H_2O = 595 \cancel{\text{ g } NH_3} \times \frac{\cancel{\text{mol } NH_3}}{17.0 \cancel{\text{ g } NH_3}} \times \frac{6 \cancel{\text{ mol } H_2O}}{4 \cancel{\text{ mol } NH_3}} \times \frac{18.0 \text{ g } H_2O}{\cancel{\text{mol } H_2O}}$$

$$= 945 \text{ g } H_2O \Big)$$

19. $2.67 \text{ mol } AlCl_3$ $\Big(4Al + 3TiCl_4 = 4AlCl_3 + 3Ti$

First determine which is the limiting reactant.

$$? \text{ mol Al} = 135.0 \cancel{\text{ g Al}} \times \frac{\text{mol Al}}{27.0 \cancel{\text{ g Al}}} = 5.00 \text{ mol Al}$$

$$\frac{5.00 \text{ mol Al}}{4} = 1.24, \quad \frac{2.00 \text{ mol TiCl}_4}{3} = 0.667$$

Thus the $TiCl_4$ is the limiting reactant.

$$? \text{ mol AlCl}_3 = 2.00 \text{ mol TiCl}_4 \times \frac{4 \text{ mol AlCl}_3}{3 \text{ mol TiCl}_4} = 2.67 \text{ mol AlCl}_3 \Big)$$

20. 0.0176 g H_2 $\quad \Big(2Na + 2H_2O = 2NaOH + H_2$

$$? \text{ g } H_2 = 400 \text{ mg Na} \times \frac{10^{-3} \text{ g Na}}{\text{mg Na}} \times \frac{\text{mol Na}}{23.0 \text{ g Na}} \times \frac{\text{mol } H_2}{2 \text{ mol Na}} \times \frac{2.02 \text{ g } H_2}{\text{mol } H_2}$$

$$= 0.0176 \text{ g } H_2\Big)$$

21. 15 mol (or 270 g) H_2O $\quad \Big(C + H_2O = CO + H_2$

$$? \text{ mol } H_2O = 360 \text{ g } H_2O \times \frac{\text{mol } H_2O}{18.0 \text{ g } H_2O} = 20.0 \text{ mol } H_2O$$

Since there are only 5.0 mol C, only 5.0 mol H_2O are used. Thus 20.0 mol H_2O present − 5.0 mol H_2O used = 15.0 mol H_2O left

$$? \text{ g } H_2O = 15.0 \text{ mol } H_2O \times \frac{18.0 \text{ g } H_2O}{\text{mol } H_2O} = 270 \text{ g } H_2O\Big)$$

22. 0.143 g NaCl $\quad \Big(NaCl + AgNO_3 = AgCl + NaNO_3$

$$? \text{ g NaCl} = 0.350 \text{ g AgCl} \times \frac{\text{mol AgCl}}{143.4 \text{ g AgCl}} \times \frac{\text{mol NaCl}}{\text{mol AgCl}} \times \frac{58.5 \text{ g NaCl}}{\text{mol NaCl}}$$

$$= 0.143 \text{ g NaCl}\Big)$$

CHAPTER 11

Complicated Stoichiometry

PRETEST

These questions may take you as long as 30 minutes to do. As you will see, they are fairly complicated.

1. How many pounds of NaOH are required to produce 500 g Na_2SO_4 according to the reaction

   $H_2SO_4 + 2NaOH = Na_2SO_4 + 2H_2O$

2. How many tons of water (H_2O) are produced when 18.0 tons of Na_2SO_4 are produced according to the equation given in question 1?

3. It was found that a 35.0 g sample of iron ore contains 19.0 g Fe. How many grams of H_2 are produced by 25.0 g of ore according to the reaction

   $2Fe + 6HCl = 2FeCl_3 + 3H_2$

4. What is the percent purity of a sample of iron ore if a 50.0 g sample of the impure ore produces 2.00 g H_2 according to the reaction given in question 3?

5. What is the percent purity of a sample of iron ore if 2.00 lb of the ore require 5.15 L of an HCl solution (density = 0.990 g/mL), which contains 12.4 g HCl in 50.0 g of solution? The reaction is the same as that given in question 3.

6. How many avoirdupois ounces (16 oz/lb) of Ag can be prepared by the reaction of $20 worth of Ag_2S ore? The ore is 54.0 wt % Ag_2S and costs $5/kg. The reaction used is

   $Ag_2S + Zn + 2H_2O = 2Ag + H_2S + Zn(OH)_2$

7. How many liters of H_2 gas at 1.0 atm pressure and 0°C are needed to react with 6.40 g O_2 according to the equation shown below? The density of H_2 gas at this temperature and pressure is 0.0893 g/L.

   $2H_2 + O_2 = 2H_2O$

8. How many grams of $CaCO_3$ are produced if 2.90 g $Fe_2(CO_3)_3$ are reacted with an excess of HCl to produce $FeCl_3$ and CO_2, as shown in the first equation below, and then the CO_2 gas is bubbled into a solution of $Ca(OH)_2$ to produce $CaCO_3$, as shown in the second equation?

   $Fe_2(CO_3)_3 + 6HCl = 2FeCl_3 + 3CO_2 + 3H_2O$

   $CO_2 + Ca(OH)_2 = CaCO_3 + H_2O$

PRETEST ANSWERS

If you did all eight questions and got the correct answers you are really doing well. Each not only involves one new type of conversion factor, but also may have in it the conversion factors of the preceding questions. Consequently, you should start in the body of the chapter at the section where you first had trouble. If you were able to do questions 1 and 2 correctly, you can skip Section A. If you got questions 1 through 4, you can skip Sections A and B. If you got questions 1–5 and 7, you can skip Sections A, B, and C. If you got questions 1 through 7 correctly, you can skip Sections A, B, C, and E. And if you missed only question 8, all you need do is Section F.

Incidentally, all the questions, except 5, are used as examples in the text, so you will see how they are worked out.

1. 0.620 lb NaOH
2. 4.56 tons H_2O
3. 0.735 g H_2
4. 73.7% Fe
5. 71.0% Fe
6. 66.3 oz Ag
7. 9.05 L H_2
8. 2.99 g $CaCO_3$

The principle used to solve the problems in this chapter will be exactly the same as that used for the simpler problems in Chapter 10. The only difference is that you will be using more conversion factors in solving each problem. All stoichiometry problems are basically the same. You are asked a question about a molecule (or atom) and are given information about another. They are related to each other through a balanced equation from which you can set up a bridge conversion factor. The bridge contains the two substances expressed in moles.

To solve a problem, you first note the dimensions of the answer. Then write down the dimensions of the information given and, using conversion factors, convert this to moles of that substance so that you can multiply by the bridge. You now have moles of the substance in the answer which you can convert to the desired units of the question asked.

In the following sections, we will separate the types of conversions you must use to get to moles so that you can use the bridge.

SECTION A   Nonmetric Units

We can start with question 1 of the pretest. How many pounds of NaOH are required to produce 500 g of Na_2SO_4 according to the balanced equation

$$H_2SO_4 + 2NaOH = Na_2SO_4 + 2H_2O$$

How many pounds NaOH?   The question asked is "_____"

500 g Na_2SO_4 The information given is _____.

The bridge conversion factor from the balanced equation which relates the substance in the question asked to the substance in the information given is _____.

$\dfrac{2 \text{ mol NaOH}}{\text{mol Na}_2\text{SO}_4}$

So starting in the usual way, you write

   ? lb NaOH = 500 g Na$_2$SO$_4$

Now you want to convert grams of Na$_2$SO$_4$ to _____ of Na$_2$SO$_4$ so that you can use the bridge. The conversion factor that does

moles

this is _____, the reciprocal of the molecular weight.
   Now, if you multiply by the bridge conversion factor, the unwanted

$\dfrac{\text{mol Na}_2\text{SO}_4}{142.1 \text{ g Na}_2\text{SO}_4}$

dimension is _____. You must now convert this to the

moles NaOH

dimensions of the answer, _____.
   You can get from moles NaOH to grams NaOH using the conversion

pounds NaOH

factor _____. You can then get from grams NaOH to the dimensions of the answer by using the conversion factor

$\dfrac{40.0 \text{ g NaOH}}{\text{mol NaOH}}$

_____. The only dimension left that does not cancel is pounds NaOH, which is the dimension of the answer. So the answer is

$\dfrac{\text{lb NaOH}}{454 \text{ g NaOH}}$

$$? \text{ lb NaOH} = 500 \, \cancel{\text{g Na}_2\text{SO}_4} \times \frac{\cancel{\text{mol Na}_2\text{SO}_4}}{142.1 \, \cancel{\text{g Na}_2\text{SO}_4}} \times \frac{2 \, \cancel{\text{mol NaOH}}}{\cancel{\text{mol Na}_2\text{SO}_4}}$$

$$\times \frac{40.0 \, \cancel{\text{g NaOH}}}{\cancel{\text{mol NaOH}}} \times \frac{\text{lb NaOH}}{454 \, \cancel{\text{g NaOH}}} = 0.620 \text{ lb NaOH}$$

This was the first question on the pretest. It is really quite simple.
   Here is another example in which the question asked is in metric units but the information given is not. How many grams of H$_2$O are produced by the reaction of 11 oz H$_2$SO$_4$ according to the reaction

   2NaOH + H$_2$SO$_4$ = Na$_2$SO$_4$ + 2H$_2$O

Starting in the usual way, you have

   ? g H$_2$O = _____

11 oz H$_2$SO$_4$

You must get from ounces of H$_2$SO$_4$ to _____ of H$_2$SO$_4$ so that you can use the bridge. Since the molecular weight is given in grams per mole, you must first get to grams of H$_2$SO$_4$. This takes two

moles

conversion factors: _____ × _____.
   Now that the unwanted dimension is grams H$_2$SO$_4$, you can get to

$\dfrac{\text{lb H}_2\text{SO}_4}{16 \text{ oz H}_2\text{SO}_4}, \quad \dfrac{454 \text{ g H}_2\text{SO}_4}{\text{lb H}_2\text{SO}_4}$

moles H$_2$SO$_4$ by using the inverted molecular weight, _____.
Finish the problem off and check your answer with the following:

$\dfrac{\text{mol H}_2\text{SO}_4}{98.1 \text{ g H}_2\text{SO}_4}$

$$? \text{ g H}_2\text{O} = 11 \, \cancel{\text{oz H}_2\text{SO}_4} \times \frac{\cancel{\text{lb H}_2\text{SO}_4}}{16 \, \cancel{\text{oz H}_2\text{SO}_4}} \times \frac{454 \, \cancel{\text{g H}_2\text{SO}_4}}{\cancel{\text{lb H}_2\text{SO}_4}}$$

$$\times \frac{\cancel{\text{mol H}_2\text{SO}_4}}{98.1 \, \cancel{\text{g H}_2\text{SO}_4}} \times \frac{2 \, \cancel{\text{mol H}_2\text{O}}}{\cancel{\text{mol H}_2\text{SO}_4}} \times \frac{18.0 \text{ g H}_2\text{O}}{\cancel{\text{mol H}_2\text{O}}}$$

$$= 1.1 \times 10^2 \text{ g H}_2\text{O} \quad \text{(Note significant figures.)}$$

What if you have a problem in which neither the question asked nor the information given is in metric units? You can, of course, convert both into and out of metric units, as the following example shows. (This was question 2 on the pretest.)

How many tons of H_2O are produced when 18.0 tons of Na_2SO_4 are prepared according to the equation

$$2NaOH + H_2SO_4 = Na_2SO_4 + 2H_2O$$

Starting in the usual way, you have

tons H_2O; 18.0 tons Na_2SO_4

?_____ = _____

You can get from tons Na_2SO_4 to grams Na_2SO_4 if you multiply by the two conversion factors

$$\frac{2000\ lb\ Na_2SO_4}{ton\ Na_2SO_4} \cdot \frac{454\ g\ Na_2SO_4}{lb\ Na_2SO_4}$$

_____ and _____

Now you use the inverted form of the molecular weight of Na_2SO_4 to get to moles of Na_2SO_4. Then you use the bridge to get to moles of H_2O. Once at moles of H_2O, you can get to grams of H_2O by using the

$$\frac{18.0\ g\ H_2O}{mol\ H_2O}$$

conversion factor _____, the molecular weight.

tons H_2O

There remain only the conversions to get grams H_2O to _____, the dimensions of the question asked. You can do this in two steps by using the conversion factors

$$\frac{lb\ H_2O}{454\ g\ H_2O} \cdot \frac{ton\ H_2O}{2000\ lb\ H_2O}$$

_____ and _____

You then find the answer through the steps below.

$$? \text{ tons } H_2O = 18.0\ \cancel{tons\ Na_2SO_4} \times \frac{2000\ \cancel{lb\ Na_2SO_4}}{\cancel{ton\ Na_2SO_4}} \times \frac{454\ \cancel{g\ Na_2SO_4}}{\cancel{lb\ Na_2SO_4}}$$

$$\times \frac{\cancel{mol\ Na_2SO_4}}{142.1\ \cancel{g\ Na_2SO_4}} \times \frac{2\ \cancel{mol\ H_2O}}{\cancel{mol\ Na_2SO_4}} \times \frac{18.0\ \cancel{g\ H_2O}}{\cancel{mol\ H_2O}}$$

$$\times \frac{\cancel{lb\ H_2O}}{454\ \cancel{g\ H_2O}} \times \frac{ton\ H_2O}{2000\ \cancel{lb\ H_2O}} = 4.56 \text{ tons } H_2O$$

You may have noticed that when you wrote down the numbers you had

$$2000 \times 454 \quad \text{and} \quad \frac{1}{2000} \times \frac{1}{454}$$

These numbers cancel completely, so it is not worth the effort to put them in at all. There is a short cut that you can take when you have all your units nonmetric.

Recall that when we defined the mole, we said that it is really called the *gram-mole*, and is defined as the number of molecules that weigh the same as the molecular weight in grams. We could have defined a different mole as being the number of molecules that weigh the molecular weight in pounds, or as being the number of molecules that weigh the molecular weight in tons. The first would be called a pound-mole and the second a ton-mole. The terms pound-mole and ton-mole do not mean pound $\times$ mole or ton $\times$ mole, and you cannot cancel the individual parts of these names. They are to be treated as single units. The bridge conversion factor that you set up from the balanced equation could be in pound-moles or ton-moles, as you choose.

Thus, in the balanced equation $H_2 + Cl_2 = 2HCl$, the bridge relating ton-moles of H_2 to HCl would be _____ and the bridge connecting pound-moles of H_2 and NH_3 in the equation $N_2 + 3H_2 = 2NH_3$ would be _____.

Of course, the molecular weight conversion factors are now expressed in pounds per pound-mole and tons per ton-mole. So the molecular weight of HCl in ton-moles would be _____ and the molecular weight of NH_3 in pound-moles would be

_____.

Let's try the preceding problem again, but since both the question asked and the information given are in tons, we'll use ton-moles. Starting in the usual way, then, we have

? tons H_2O = _____

The reciprocal of the molecular weight expressed in ton-moles for Na_2SO_4 is _____. It is now possible to use a bridge in ton-moles relating Na_2SO_4 to H_2O. It is _____. Ton-moles of H_2O can be related to the dimensions of the answer by using the molecular weight in tons, _____. You now have the dimension of the question asked as the only dimension left. The setup for the whole problem is now

$$? \text{ tons } H_2O = 18.0 \ \cancel{\text{tons } Na_2SO_4} \times \frac{\cancel{\text{ton-mol } Na_2SO_4}}{142.1 \ \cancel{\text{tons } Na_2SO_4}}$$

$$\times \frac{2 \ \cancel{\text{ton-mol } H_2O}}{\cancel{\text{ton-mol } Na_2SO_4}} \times \frac{18.0 \text{ tons } H_2O}{\cancel{\text{ton-mol } H_2O}}$$

$$= 18.0 \times \frac{1}{142.1} \times 2 \times 18.0 \text{ tons } H_2O$$

$$= 4.56 \text{ tons } H_2O$$

As you can see, this procedure has shortened the setup quite a bit by eliminating four conversion factors. However, since no more multiplications or divisions would be needed in the long way (all the numbers in the four extra conversion factors simply cancel), you really don't have to use the pound-moles and ton-moles.

$\dfrac{\text{ton-mol } H_2}{2 \text{ ton-mol } HCl}$

$\dfrac{3 \text{ lb-mol } H_2}{2 \text{ lb-mol } NH_3}$

$\dfrac{36.5 \text{ tons } HCl}{\text{ton-mol } HCl}$

$\dfrac{17.0 \text{ lb } NH_3}{\text{lb-mol } NH_3}$

18.0 tons Na_2SO_4

$\dfrac{\text{ton-mol } Na_2SO_4}{142.1 \text{ tons } Na_2SO_4}$

$\dfrac{2 \text{ ton-mol } H_2O}{\text{ton-mol } Na_2SO_4}$

$\dfrac{18.0 \text{ tons } H_2O}{\text{ton-mol } H_2O}$

SECTION B Impure Substances

Frequently the substance in the information given in the problem is not pure. Sometimes the substance in the question asked will not be pure. It is even possible to have problems in which both the substance asked for and the given substance are impure. This doesn't present too much difficulty, since you will always have conversion factors given in the problem that relate pure substance to impure substance.

Consider question 3 on the pretest. You are told that 35.0 g iron ore contain 19.0 g Fe. How many grams of H_2, then, can be produced by the reaction of 25.0 g of ore according to the balanced equation

$$2Fe + 6HCl = 2FeCl_3 + 3H_2$$

The iron ore given in the problem is impure, but the conversion factor

$\dfrac{19.0 \text{ g Fe}}{35.0 \text{ g ore}}$

given in the problem relating pure Fe to iron ore is _____. Thus, starting in the usual way, you have

g H_2 ; 25.0 g ore

$? \underline{\hspace{1cm}} = \underline{\hspace{2cm}}$

grams ore

The unwanted dimension is _____, but this is contained in the conversion factor given in the problem. Now you multiply by this factor.

$$? \text{ g } H_2 = 25.0 \text{ g ore} \times \frac{19.0 \text{ g Fe}}{35.0 \text{ g ore}}$$

grams Fe

This operation leaves _____ as the new unwanted dimension. This is the point at which you would have started the problem if you'd had only pure substances. Finish the problem and then compare your answer with the following setup:

$$? \text{ g } H_2 = 25.0 \text{ g ore} \times \frac{19.0 \text{ g Fe}}{35.0 \text{ g ore}} \times \frac{\text{mol Fe}}{55.8 \text{ g Fe}} \times \frac{3 \text{ mol } H_2}{2 \text{ mol Fe}} \times \frac{2.02 \text{ g } H_2}{\text{mol } H_2}$$

$$= 0.735 \text{ g } H_2$$

That is really all there is to this type of problem. You go from weight of impure substance to grams of pure substance. The reciprocal molecular weight converts grams to moles. Then, using the bridge, you cross over to the substance in the question asked in moles. Finally, you convert this to the desired units, such as grams, pounds, or whatever.

One word of caution: You must be careful to label the dimensions completely so that you will not cancel grams impure with grams pure. Don't make this mistake!

It is very common to express the purity of a substance in percent. This is the number of grams of pure substance in 100 g of impure substance. (If your knowledge of this is shaky, refer back to Section C in Chapter 4.) In these cases, the conversion factor given in the problem will be

100 g

grams of pure substance per _____ of impure substance. Setting up the problem is then the same as in the type you have already done.

Here is a problem to show you how this works. How many grams of air, which is 22.5 wt % O_2, are needed for the complete oxidation of 7.20 g pentane according to the reaction

$$C_5H_{12} + 8O_2 = 5CO_2 + 6H_2O$$

The conversion factor given in the problem is the weight percent of O_2
in air. You can write this as the fraction _____.

You might start your problems by writing the conversion factors
that you think you will need. First you will need a bridge that relates

pentane, C_5H_{12}, to _____. You get this from the balanced equation. It

is _____. Since you are dealing with weights of the two
reactants, you will also need the conversion factors that relate weights to

moles. These are the _____ weights,

$$\frac{72.0 \text{ g } C_5H_{12}}{\underline{\hspace{1cm}}} , \quad \frac{32.0 \underline{\hspace{1cm}}}{\text{mol } O_2}$$

Now you can start in the usual way:

? _____ = _____

[You should have noticed by now that the hardest step in setting up
any problem is the first one—determining the exact dimensions of the
question asked and noting the information given in the problem. Once
you are over this hump, everything falls into place.]

The unwanted dimension in the information given is _____.

You must convert this to _____ so that you can use your

_____. You do this by multiplying by the inverted form of the

_____ .

$$? \text{ g air} = 7.20 \text{ g } C_5H_{12} \times \frac{\text{mol } C_5H_{12}}{72.0 \text{ g } C_5H_{12}}$$

Now that the unwanted dimension is moles C_5H_{12}, you can multiply by
your bridge:

$$? \text{ g air} = 7.20 \text{ g } C_5H_{12} \times \frac{\text{mol } C_5H_{12}}{72.0 \text{ g } C_5H_{12}} \times \underline{\hspace{2cm}}$$

The unwanted dimension now is _____, which you can convert

to _____ by multiplying by the molecular weight.

$$? \text{ g air} = 7.20 \text{ g } C_5H_{12} \times \frac{\text{mol } C_5H_{12}}{72.0 \text{ g } C_5H_{12}} \times \frac{8 \text{ mol } O_2}{\text{mol } C_5H_{12}} \times \frac{32.0 \text{ g } O_2}{\text{mol } O_2}$$

You now have as the unwanted dimension grams of O_2, which can be
converted to the dimensions of the question asked by using the conversion factor given in the problem, the weight _____ of O_2 in air.

Right margin answers:

$\dfrac{22.5 \text{ g } O_2}{100 \text{ g air}}$

O_2
$\dfrac{\text{mol } C_5H_{12}}{8 \text{ mol } O_2}$

molecular

g O_2

mol C_5H_{12}

g air; 7.20 g C_5H_{12}

grams C_5H_{12}

moles C_5H_{12}

bridge

molecular weight

$\dfrac{8 \text{ mol } O_2}{\text{mol } C_5H_{12}}$

moles O_2

grams O_2

percent

$$? \text{ g air} = 7.20 \text{ g } \cancel{C_5 H_{12}} \times \frac{\cancel{\text{mol } C_5 H_{12}}}{72.0 \text{ g } \cancel{C_5 H_{12}}} \times \frac{8 \cancel{\text{mol } O_2}}{\cancel{\text{mol } C_5 H_{12}}}$$

$$\times \frac{32.0 \text{ g } \cancel{O_2}}{\cancel{\text{mol } O_2}} \times \frac{100 \text{ g air}}{22.5 \text{ g } \cancel{O_2}}$$

Notice that you must use the weight-percent conversion factor in its

inverted; cancel | _____ form in order to _____ the unwanted dimension and have only the dimension asked for in the question. The numerical answer is then

$$? \text{ g air} = 7.20 \times \frac{1}{72.0} \times 8 \times 32.0 \times \frac{100 \text{ g air}}{22.5} = 114 \text{ g air}$$

Often you will be asked for the percent purity of a sample. In such cases you must reword the question to: How many grams of pure substance are there in 100 g impure sample? The question asked and the information given are all in this one sentence.

Question 4 on the pretest is an example of this type. You are asked to find the percent purity of a sample of Fe ore, if a 50.0 g sample of the impure ore produces 2.00 g H_2 according to the reaction

$$2Fe + 6HCl = 2FeCl_3 + 3H_2$$

grams of Fe | Rewording the question, you ask how many _____

100 g | are in _____ impure ore. (If you have forgotten this method, go back to Section C in Chapter 4.) There is a conversion factor given in the

2.00 g H_2
$\dfrac{}{}$
50.0 g imp
$\dfrac{}{}$
2.00 g H_2 | problem. It says that there are 50.0 g impure ore per _____.

Writing this as a fraction, you have _____. (It's a good idea to write down all conversion factors given in the problem before you even start the solution.) The bridge you use from the balanced equation must

H_2
$\dfrac{2 \text{ mol Fe}}{3 \text{ mol } H_2}$ | relate Fe to _____ (the only other substance mentioned in the problem). It is _____.
Now you are ready to set up the problem.

100 g imp | ? g Fe = _____

grams impure
$\dfrac{2.00 \text{ g } H_2}{50.0 \text{ g imp}}$ | The unwanted dimension in the information given is _____ , and the only conversion factor that has this in it is _____. The unwanted dimension now is grams of H_2. Multiplying by the inverted

$\dfrac{\text{mol } H_2}{2.0 \text{ g } H_2}$ | form of the molecular weight, _____, you now have moles of H_2 and can use the bridge to get to moles of Fe. Multiplying by the bridge gives you

$\dfrac{2 \text{ mol Fe}}{3 \text{ mol } H_2}$ | $$? \text{ g Fe} = 100 \text{ g imp} \times \frac{2.00 \text{ g } \cancel{H_2}}{50.0 \text{ g imp}} \times \frac{\text{mol } H_2}{2.02 \text{ g } \cancel{H_2}} \times \underline{\qquad\qquad}$$

weight | Finally, if you multiply by the atomic _____ of Fe, all dimensions cancel except the dimension of the answer.

$$? \text{ g Fe} = 100 \text{ g imp} \times \frac{2.00 \text{ g H}_2}{50.0 \text{ g imp}} \times \frac{\text{mol H}_2}{2.02 \text{ g H}_2} \times \frac{2 \text{ mol Fe}}{3 \text{ mol H}_2} \times \frac{55.8 \text{ g Fe}}{\text{mol Fe}}$$

$$= 73.7 \text{ g Fe}/100 \text{ g imp} = 73.7\% \text{ Fe}$$

We'll see more problems of this type in Section C, and you'll have more practice in working them.

SECTION C Solutions and Density

A solution of a substance in a nonreacting solvent is really just another form of an impure substance. Thus, if you have a 24 wt % solution of HCl, you can write the conversion factor _____. You can solve the problem in the same way as in Section B. Try an example.

$$\frac{24 \text{ g HCl}}{100 \text{ g soln}}$$

How many grams of a 24 wt % solution of HCl are required to react with 75 g Fe according to the balanced equation

$$2\text{Fe} + 6\text{HCl} = 2\text{FeCl}_3 + 3\text{H}_2$$

The question asked is: How many _____? (Be careful: You are *not* asked, "How many grams HCl?") The information given is

_____.

grams solution

$$75 \text{ g Fe}$$
$$\frac{\text{mol Fe}}{55.8 \text{ g Fe}}$$

The first conversion factor you use is _____. You use this because you always head toward _____ of a compound so that you can use the bridge. The next conversion factor will be the bridge, which relates the compound in the question with the compound in the information given. That factor is _____. The next conversion factor by which you multiply is _____, since this gets you to the conversion factor given in the problem. Finally, when you multiply by the conversion factor _____ given in the problem, all the dimensions cancel except _____. Therefore the answer must be

moles

$$\frac{6 \text{ mol HCl}}{2 \text{ mol Fe}}$$
$$\frac{36.5 \text{ g HCl}}{\text{mol HCl}}$$

$$\frac{100 \text{ g soln}}{24 \text{ g HCl}}$$
grams solution

$$? \text{ g soln} = 75 \text{ g Fe} \times \frac{\text{mol Fe}}{55.8 \text{ g Fe}} \times \frac{6 \text{ mol HCl}}{2 \text{ mol Fe}} \times \frac{36.5 \text{ g HCl}}{\text{mol HCl}} \times \frac{100 \text{ g soln}}{24 \text{ g HCl}}$$

$$= 6.1 \times 10^2 \text{ g soln}$$

There are also problems that have volume units instead of weight units for the information given or the question asked (or even both). In such cases you always have the density as a conversion factor given in the problem. This allows you to convert from volume to weight. The units of density, you recall, are _____ or _____ for gases.

$$\frac{\text{g}}{\text{mL}} ; \frac{\text{g}}{\text{L}}$$

Question 7 in the pretest is a good example of this type. You are asked how many liters of H_2 gas at 1.0 atm and $0°C$ are needed to react with 6.40 g O_2 according to the equation

$$2\text{H}_2 + \text{O}_2 = 2\text{H}_2\text{O}$$

The density of H_2 gas at the temperature and pressure shown is 0.0893 g/L.

$$\frac{0.0893 \text{ g } H_2}{L \ H_2}$$

$$L \ H_2 ; 6.40 \text{ g } O_2$$

The density is the conversion factor given in the problem. Written as a fraction, it is _____. Thus, starting in the usual way, you have

? _____ = _____

$$\frac{mol \ O_2}{32.0 \text{ g } O_2}$$

$$\frac{2 \ mol \ H_2}{mol \ O_2}$$

$$\frac{2.02 \text{ g } H_2}{mol \ H_2}$$

grams H_2

liters H_2

$$\frac{L \ H_2}{0.0893 \text{ g } H_2}$$

You want to move toward the *bridge*, so the first conversion factor you multiply by is _____. Next you multiply by the bridge, which is _____. From this you can get the weight of H_2 in grams if you multiply by the molecular weight, _____. The unwanted dimension is _____, and the dimensions that the answer will have are _____.

But the density gives you this conversion, so finally you multiply by _____, and the only dimension left is the dimension of the answer.

$$? \ L \ H_2 = 6.40 \ \cancel{g \ O_2} \times \frac{\cancel{mol \ O_2}}{32.0 \ \cancel{g \ O_2}} \times \frac{2 \ \cancel{mol \ H_2}}{1 \ \cancel{mol \ O_2}} \times \frac{2.02 \ \cancel{g \ H_2}}{\cancel{mol \ H_2}} \times \frac{L \ H_2}{0.0893 \ \cancel{g \ H_2}}$$

$$= 9.05 \ L \ H_2 \quad \text{to the correct number of significant figures}$$

(If you don't remember why 9.048 was rounded off to 9.05, refer back to Chapter 1, Section C.)

It is common in problems that deal with solutions to have volume dimensions rather than weight units given. You always have a density of the solution to use as a conversion factor, but, once again, be careful to write complete dimensions. For example, if you have a solution with a density of 1.10 g/mL, you must write the conversion factor as 1.10 g soln/mL soln. Simply writing 1.10 g/mL will not do. You will become confused and cancel grams solution with grams something else. The following example shows this.

How many grams of Mg can be consumed by 25.0 mL of a 7.10 wt % solution of HCl whose density is 0.990 g/mL according to the reaction

$$Mg + 2HCl = MgCl_2 + H_2$$

When a problem starts to get complicated, remember to write down all the conversion factors given in the problem before you start the solution.

$$\frac{7.10 \text{ g } HCl}{100 \text{ g soln}}$$

$$\frac{0.990 \text{ g soln}}{mL \ soln}$$

There are two such factors in this problem: _____ and _____. Note how the density conversion factor is exactly written. Writing down the bridge conversion factor gives you something to head for. In this problem, the bridge connecting Mg and HCl is

$$\frac{1 \ mol \ Mg}{2 \ mol \ HCl}$$

$$\frac{24.3 \text{ g } Mg}{mol \ Mg} ; \frac{36.5 \text{ g } HCl}{mol \ HCl}$$

_____. You'll also probably need the molecular (or atomic) weights of the substances mentioned in the problem. These would be _____ and _____.

Now you are ready to start.

? g Mg = _____

The dimension you want to cancel out is _____. The only conversion factor you have that has this dimension in it is

_____. Now the unwanted dimension is _____.

The conversion factor that contains this dimension is _____.

Now the unwanted dimension is _____, but you can convert this to moles HCl (so that you can use the bridge) by multiplying by

_____.

Next you multiply by the bridge, which is the only conversion factor left containing the unwanted dimension, moles HCl. You now have a conversion factor dealing with Mg. And if you multiply by

_____, the only dimension left will be the dimension of the question asked. Thus the final answer is

$$? \text{ g Mg} = 25.0 \, \cancel{\text{mL soln}} \times \frac{0.990 \, \cancel{\text{g soln}}}{\cancel{\text{mL soln}}} \times \frac{7.10 \, \cancel{\text{g HCl}}}{100 \, \cancel{\text{g soln}}} \times \frac{\cancel{\text{mol HCl}}}{36.5 \, \cancel{\text{g HCl}}}$$

$$\times \frac{\cancel{\text{mol Mg}}}{2 \, \cancel{\text{mol HCl}}} \times \frac{24.3 \text{ g Mg}}{\cancel{\text{mol Mg}}} = 0.585 \text{ g Mg}$$

SECTION D Solution Concentrations in Molarity

Chemists have devised a method of expressing the concentration of a substance in a solution which—rather than introducing a complication—simplifies things when you are doing solution stoichiometry. They express the amount of dissolved material in terms of the number of moles present in one liter of solution. This is called the *molarity* (the symbol is *M*) of the solution. We will say a great deal more about molarity in Chapter 14. All you need know at this point is that the molarity is the *moles per liter*. As with all "per" expressions, it can be

written as a _____ factor. It is exactly moles of *solute* (the dissolved substance) per liter of solution, and has the dimensions

_____ solute

liter _____

Thus if you have a 6.0 *M* solution of sugar, you can write the conversion factor

_____

_____

Side column (answers):

25.0 mL soln

milliliters solution

$\dfrac{0.990 \text{ g soln}}{\text{mL soln}}$; g soln

$\dfrac{7.10 \text{ g HCl}}{100 \text{ g soln}}$

grams HCl

$\dfrac{\text{mol HCl}}{36.5 \text{ g HCl}}$

$\dfrac{24.3 \text{ g Mg}}{\text{mol Mg}}$

conversion

moles

solution

6.0 mol sugar

L soln

Let's see how you would use this in a common-type problem. How many liters of a 0.100 M solution of H_2SO_4 react with 0.250 g of NaOH according to the equation

$$H_2SO_4 + 2NaOH = Na_2SO_4 + 2H_2O$$

The question asked is "How many liters of solution?" and the information given is _____. The conversion factor given in

0.250 g NaOH

$$\frac{0.100\ mol\ H_2SO_4}{L\ soln}$$

the problem is the molarity of the H_2SO_4 solution, _____.
The bridge you will use relates moles of H_2SO_4 to moles of NaOH. It is

$$\frac{mol\ H_2SO_4}{2\ mol\ NaOH}$$

So, starting in the usual way, you have

0.250 g NaOH

? L soln = _____

moles

In order to use the bridge, you must convert the grams NaOH to _____

molecular weight

$$\frac{mol\ NaOH}{40.0\ g\ NaOH}$$

NaOH. You multiply by the inverted form of the _____.

? L soln = 0.250 g NaOH $\times$ _____

Now you can use the bridge:

$$\frac{mol\ H_2SO_4}{2\ mol\ NaOH}$$

? L soln = 0.250 ~~g NaOH~~ $\times \dfrac{mol\ NaOH}{40.0\ \text{~~g NaOH~~}} \times$ _____

moles H_2SO_4

Now the unwanted dimension is _____, but the conversion factor given in the problem, the molarity of the H_2SO_4, leads directly to the dimensions of the answer:

$$\frac{L\ soln}{0.100\ mol\ H_2SO_4}$$

? L soln = 0.250 ~~g NaOH~~ $\times \dfrac{\text{~~mol NaOH~~}}{40.0\ \text{~~g NaOH~~}} \times \dfrac{mol\ H_2SO_4}{2\ \text{~~mol NaOH~~}} \times$ _____

So the answer is

$$? \ L\ soln = 0.250 \times \frac{1}{40.0} \times \frac{1}{2} \times \frac{L\ soln}{0.100} = 0.0312\ L\ soln$$

The reason molarity is such a convenient conversion factor is that it contains moles of substance as one of its dimensions. It therefore allows us to use the moles of that substance in the bridge without any further conversion factors.

Let's try another problem in which both the reactants are in solution and the molarity of both solutions is given. How many milliliters of a 0.300 M solution of HCl react with 15.0 mL of a 0.425 M solution of $KMnO_4$ according to the equation

$$2KMnO_4 + 16HCl = 2MnCl_2 + 5Cl_2 + 8H_2O + 2KCl$$

The two molarities are the two conversion factors given in the problem, but you must be careful not to confuse liters of $KMnO_4$ solution with

HCl

liters of _____ solution. So every time you write the volume of a solution, you must follow it with the substance that the solution contains. Thus you can write liters of the $KMnO_4$ solution as $\underline{\text{L } KMnO_4}$ (you need not write soln; it is understood that volumes can only be of solution). And milliliters of HCl solution are _____. So you can write the two molarities as conversion factors:

$$\frac{0.300 \text{ mol HCl}}{\rule{3cm}{0.4pt}} \quad \text{and} \quad \frac{\rule{3cm}{0.4pt}}{\text{L } KMnO_4}$$

mL HCl

0.425 mol $KMnO_4$

L HCl

The bridge that you will need relates HCl to $KMnO_4$, and is

_____

16 mol HCl
2 mol $KMnO_4$

Starting in the usual way, you have

$? \underline{\hspace{3cm}} = \underline{\hspace{4cm}}$

mL HCl; 15.0 mL $KMnO_4$

In order to use the molarity conversion factor, you convert the milliliters $KMnO_4$ to _____.

$? \text{ mL HCl} = 15.0 \text{ mL } KMnO_4 \times \underline{\hspace{4cm}}$

Now you can use the molarity conversion factor, which will get you directly to _____.

liters $KMnO_4$

$\dfrac{10^{-3} \text{ L } KMnO_4}{\text{mL } KMnO_4}$

moles $KMnO_4$

$$? \text{ mL HCl} = 15.0 \;\cancel{\text{mL } KMnO_4} \times \frac{10^{-3} \;\cancel{\text{L } KMnO_4}}{\cancel{\text{mL } KMnO_4}} \times \frac{0.425 \text{ mol } KMnO_4}{\cancel{\text{L } KMnO_4}}$$

Once you are at moles $KMnO_4$, you can use the _____ conversion factor that gets you to _____:

bridge

moles HCl

$$? \text{ mL HCl} = 15.0 \;\cancel{\text{mL } KMnO_4} \times \frac{10^{-3} \;\cancel{\text{L } KMnO_4}}{\cancel{\text{mL } KMnO_4}} \times \frac{0.425 \;\cancel{\text{mol } KMnO_4}}{\cancel{\text{L } KMnO_4}}$$
$$\times \frac{16 \text{ mol HCl}}{2 \;\cancel{\text{mol } KMnO_4}}$$

The remaining unwanted dimension is moles HCl, but the _____ of the HCl solution gets you straight to volume of HCl solution:

molarity

$$? \text{ mL HCl} = 15.0 \;\cancel{\text{mL } KMnO_4} \times \frac{10^{-3} \;\cancel{\text{L } KMnO_4}}{\cancel{\text{mL } KMnO_4}} \times \frac{0.425 \;\cancel{\text{mol } KMnO_4}}{\cancel{\text{L } KMnO_4}}$$
$$\times \frac{16 \;\cancel{\text{mol HCl}}}{2 \;\cancel{\text{mol } KMnO_4}} \times \frac{\text{L HCl}}{0.300 \;\cancel{\text{mol HCl}}}$$

But the remaining volume of HCl solution is in liters, and you want mL HCl for your answer. So

$$? \text{ mL HCl} = 15.0 \;\cancel{\text{mL } KMnO_4} \times \frac{10^{-3} \;\cancel{\text{L } KMnO_4}}{\cancel{\text{mL } KMnO_4}} \times \frac{0.425 \;\cancel{\text{mol } KMnO_4}}{\cancel{\text{L } KMnO_4}}$$
$$\times \frac{16 \;\cancel{\text{mol HCl}}}{2 \;\cancel{\text{mol } KMnO_4}} \times \frac{\cancel{\text{L HCl}}}{0.300 \;\cancel{\text{mol HCl}}} \times \frac{\text{mL HCl}}{10^{-3} \;\cancel{\text{L HCl}}}$$

Now all dimensions cancel except the desired mL HCl:

$$? \text{ mL HCl} = 15.0 \times 10^{-3} \times 0.425 \times \frac{16}{2} \times \frac{1}{0.300} \times \frac{\text{mL HCl}}{10^{-3}} = 170 \text{ mL HCl}$$

We aren't going to say more about solution stoichiometry at this point because it's such an important topic that there will be an entire chapter (Chapter 15) devoted only to problems involving solutions.

SECTION E   Miscellaneous Complications

There are a number of complications that may appear in stoichiometry problems, such as cost in dollars, time required to consume or produce something, amount of heat required or released, or even such things as distances you can travel burning a certain amount of fuel, or number of insects that can be killed by the products of a reaction. Any time you have a chemical reaction and can write a balanced equation, you can have a stoichiometry problem concerning any aspect of the substances in the equation. However, the problem always gives you some conversion factor or factors that will allow you to get to or from this complication to moles of the substance, so that you can use your bridge.

Once again, write down all the conversion factors given in the problem before you even set it up. Then you can simply pick and choose from these in order to cancel the unwanted dimensions. Here is a simple example. The oxidation of S to SO_3 consumes 6.00 mol of O_2 per minute. How many minutes does it take to produce 14.0 mol of SO_3 according to the equation

$$2S = 3O_2 = 2SO_3$$

First you write down the conversion factor given in the problem:

$\frac{6.00 \text{ mol } O_2}{\text{min}}$

_____. Now you start in the usual way.

minutes; 14.0 mol SO_3

$\dfrac{\text{moles } SO_3}{}$

? _____ = _____

$\dfrac{\text{moles } O_2}{}$

The unwanted dimension is already in _____, so you can use

$\dfrac{3 \text{ mol } O_2}{2 \text{ mol } SO_3}$

the bridge conversion factor, which relates this to _____. You

$\text{moles } O_2$

multiply by _____. The unwanted dimension is now

_____, but the conversion factor given in the problem relates this directly to the dimensions that the answer must have. If you

$\dfrac{\text{min}}{6.00 \text{ mol } O_2}$

multiply by _____, the dimensions all cancel, except the dimension in the question asked. The answer must be

$$? \text{ min} = 14.0 \; \cancel{\text{mol } SO_3} \times \frac{3 \; \cancel{\text{mol } O_2}}{2 \; \cancel{\text{mol } SO_3}} \times \frac{\text{min}}{6.00 \; \cancel{\text{mol } O_2}} = 3.50 \text{ min}$$

You may find it hard to pick the correct bridge, since the substances involved may be slightly hidden. In the above problem, it is obvious that the bridge must contain moles SO_3, but you have to look around a little to realize that the dimensions of the answer, minutes, can be related only to O_2, since this is in the conversion factor given in the problem. If by chance you choose an incorrect bridge, you will quickly reach a dead end and realize that you have to go back and find another one.

The preceding example is simple to work because both the information given and the conversion factor in the problem are already in moles. In a more usual case, they will be stated in terms of weight and you must make the usual conversions to get to moles so that you can use the bridge. Here is an example.

The rate of O_2 consumption in the oxidation of S to SO_3 is 6.00 g O_2 per minute. How many seconds does it take to make 1.75 lb of SO_3 according to the reaction

$$2S + 3O_2 = 2SO_3$$

The conversion factor given in the problem is _____. Starting in the usual way, then, you have:

? _____ = _____

The unwanted dimension is _____, and you must, in order to

use a bridge, get to _____ SO_3. It takes two conversion factors

to do this. First you convert pounds SO_3 to _____ SO_3. You

can do this by multiplying by the conversion factor _____.
Once you have grams of SO_3, you can get to moles of SO_3 by multiply-

ing by the conversion factor, which is the _____ in its

inverted form. The conversion factor is _____.

The unwanted dimension is now moles of SO_3, and you can use the

bridge, _____, to get you to moles of O_2. The reason

that you choose this particular bridge is that the _____
given in the problem contains the substance O_2 in it. The unwanted

dimension is now _____. But the conversion factor given in the

problem has _____ O_2. To get to grams of O_2, you multiply by

the molecular weight of O_2, _____. The unwanted

dimension is now _____, and you can multiply by the conver-

sion factor given in the problem, _____.

The last unwanted dimension is _____, but the dimensions of

the answer to the question are _____. It is easy to get from

minutes to seconds by multiplying by _____. Finally, the only
dimension left is that of the answer, which must be

$\dfrac{6.00 \text{ g } O_2}{\text{min}}$

sec; 1.75 lb SO_3

pounds SO_3

moles

grams
$\dfrac{454 \text{ g } SO_3}{\text{lb } SO_3}$

molecular weight
$\dfrac{\text{mol } SO_3}{80.1 \text{ g } SO_3}$

$\dfrac{3 \text{ mol } O_2}{2 \text{ mol } SO_3}$
conversion factor

moles O_2

grams
$\dfrac{32.0 \text{ g } O_2}{\text{mol } O_2}$
grams O_2
$\dfrac{\text{min}}{6.00 \text{ g } O_2}$
minutes

seconds
$\dfrac{60 \text{ sec}}{\text{min}}$

$$? \text{ sec} = 1.75 \text{ lb SO}_3 \times \frac{454 \text{ g SO}_3}{\text{lb SO}_3} \times \frac{\text{mol SO}_3}{80.1 \text{ g SO}_3} \times \frac{3 \text{ mol O}_2}{2 \text{ mol SO}_3} \times \frac{32.0 \text{ g O}_2}{\text{mol O}_2}$$

$$\times \frac{\text{min}}{6.00 \text{ g O}_2} \times \frac{60 \text{ sec}}{\text{min}} = 4.76 \times 10^3 \text{ sec}$$

Industrial chemists need to know the cost of chemical reactions. In such specific problems there is always some conversion factor having either dollars or cents as one of its units. You approach these problems in the same way as you did the time problems. Here is an example.

Ammonia, NH_3, can be produced by the direct reaction of N_2 with H_2 according to the equation

$$N_2 + 3H_2 = 2NH_3$$

Suppose that H_2 costs \$0.01/kg. How much does the H_2 cost when you produce 6000 g NH_3?

$\dfrac{\$0.01}{\text{kg } H_2}$

The conversion factor given in the problem is _____. Starting in the usual way, then, you have

6000 g NH_3

$? \$ = $ _____

In order to use the bridge conversion factor, you convert grams NH_3 to

moles; $\dfrac{\text{mol } NH_3}{17.0 \text{ g } NH_3}$

_____ NH_3. You can do this by multiplying by _____, the molecular weight of NH_3 in its inverted form. The bridge you use must

moles NH_3; moles H_2

have _____ in the denominator and should have _____ in the numerator (on top) because the conversion factor given in the problem has to do with H_2. The bridge is therefore _____.

$\dfrac{3 \text{ mol } H_2}{2 \text{ mol } NH_3}$

The units for H_2 in the conversion factor given in the problem are

kilograms; kilograms

_____ H_2. Therefore you convert moles H_2 to _____ H_2, which can be done in two conversions. First you multiply by

molecular

2.02 g H_2/mol H_2, the _____ weight of H_2. Then you multiply

$\dfrac{\text{kg } H_2}{10^3 \text{ g } H_2}$

by _____. The unwanted dimension is now kilograms H_2. This appears in the conversion factor given in the problem. Multiplying by

dollars

this conversion factor leaves _____ as the only dimension. The answer, then, is

$$? \$ = 6000 \text{ g } NH_3 \times \frac{\text{mol } NH_3}{17.0 \text{ g } NH_3} \times \frac{3 \text{ mol } H_2}{2 \text{ mol } NH_3}$$

$$\times \frac{2.02 \text{ g } H_2}{\text{mol } H_2} \times \frac{\text{kg } H_2}{10^3 \text{ g } H_2} \times \frac{\$0.01}{\text{kg } H_2} = \$0.0107$$

(Notice that there are two significant figures in the answer. The price of the H_2 is not a measured quantity and has no uncertainty.) The 2.02 g H_2/mol H_2 controls the significant figures.

Let us look now at question 6 in the pretest, a complex one. If you start out by writing all the conversion factors in the problem, however, it won't be very difficult.

You are asked how many avoirdupois ounces (16 oz/lb) of Ag can be prepared by the reaction of $20 worth of Ag_2S ore. The ore is 54.0 wt % Ag_2S and costs $5/kg. The reaction used is

$$Ag_2S + Zn + 2H_2O = 2Ag + H_2S + Zn(OH)_2$$

First write down all the conversion factors in the problem.

$\dfrac{16 \text{ oz Ag}}{\rule{2cm}{0.4pt}}$	$\dfrac{54.0 \text{ g Ag}_2\text{S}}{\rule{2cm}{0.4pt}}$	$\dfrac{\$5.00}{\rule{2cm}{0.4pt}}$

lb Ag; 100 g ore; kg ore

The two substances mentioned in the problem are Ag and _____, so the bridge conversion factor from the balanced equation is

Ag_2S

_____. Since the problem deals with weights of both substances, you'll probably also need their molecular weight conversion factors.

$\dfrac{2 \text{ mol Ag}}{\text{mol Ag}_2\text{S}}$

$$\dfrac{107.9 \text{ g Ag}}{\rule{1.5cm}{0.4pt}} \qquad \dfrac{\rule{1cm}{0.4pt} \text{ g Ag}_2\text{S}}{\text{mol Ag}_2\text{S}}$$

247.9

$\underline{\rule{1.5cm}{0.4pt}}$

mol Ag

Now you are ready to start. The question asked is "_____

How many ounces Ag

_____?" The information given is _____.

$20

$?\underline{\rule{2cm}{0.4pt}} = \underline{\rule{1cm}{0.4pt}}$

oz Ag; $20

The unwanted dimension that you wish to cancel is _____. In your collection of conversion factors, there is only one that contains

dollars

this dimension, so you multiply by _____. The unwanted

$\dfrac{\text{kg ore}}{\$5}$

dimension is now _____, and the only conversion factor

kilograms ore

left that has this substance is _____. As you see, one weight of ore is in kilograms and the other is in grams. You must convert from one

$\dfrac{54.0 \text{ g Ag}_2\text{S}}{100 \text{ g ore}}$

to the other by multiplying by _____.
Now you can multiply by the conversion factor given in the problem,

$\dfrac{10^3 \text{ g ore}}{\text{kg ore}}$

_____. The new unwanted dimension is _____. In order to be able to use the bridge conversion factor, you must have

$\dfrac{54.0 \text{ g Ag}_2\text{S}}{100 \text{ g ore}}$; grams Ag$_2$S

_____ Ag_2S, which you can get if you multiply by the molecular weight conversion factor, _____(in its inverted form).

moles
$\dfrac{\text{mol Ag}_2\text{S}}{247.9 \text{ g Ag}_2\text{S}}$

Now multiply by the bridge conversion factor, _____. The unwanted dimension is now in the substance of the question asked. All that you must do is get to the weight dimension, ounces. First you get to

$\dfrac{2 \text{ mol Ag}}{\text{mol Ag}_2\text{S}}$

grams Ag by multiplying by _____, the atomic weight of Ag. Then you get from grams Ag to pounds Ag by multiplying by

$\dfrac{107.9 \text{ g Ag}}{\text{mol Ag}}$

_____. Finally, you get from pounds Ag to ounces Ag by

$\dfrac{\text{lb Ag}}{454 \text{ g Ag}}$

multiplying by the conversion factor _____. The only

$\dfrac{16 \text{ oz Ag}}{\text{lb Ag}}$

ounces Ag

dimension left now is _____, and that is the dimension that the answer must have. Therefore the answer must be

$$? \text{ oz Ag} = \$20 \times \frac{\text{kg ore}}{\$5} \times \frac{10^3 \text{ g ore}}{\text{kg ore}} \times \frac{54.0 \text{ g Ag}_2\text{S}}{100 \text{ g ore}} \times \frac{\text{mol Ag}_2\text{S}}{247.9 \text{ g Ag}_2\text{S}}$$

$$\times \frac{2 \text{ mol Ag}}{\text{mol Ag}_2\text{S}} \times \frac{107.9 \text{ g Ag}}{\text{mol Ag}} \times \frac{\text{lb Ag}}{454 \text{ g Ag}} \times \frac{16 \text{ oz Ag}}{\text{lb Ag}}$$

$$= 66.3 \text{ oz Ag}$$

Occasionally you may have a problem that asks for the total cost of the reactants in a chemical equation. What this amounts to is more than one problem. You simply determine the cost of each reactant separately and then add them all up. An example of this type appears in the problem set at the end of this chapter.

SECTION F Consecutive Reactions

Sometimes you have a problem in which the chemical reaction occurs in a series of steps, one after the other. This is called a series of *consecutive reactions.* You may be asked about a product in the last reaction and be given information about a reactant in the first. In such a case, you write balanced equations for all the steps and use more than a single bridge. An example will make this clear.

How many moles of $CaCO_3$ are produced when 0.10 mol $Fe_2(CO_3)_3$ are reacted with an excess of HCl according to equation I to produce $FeCl_3$, H_2O, and CO_2, and then the CO_2 is reacted with $Ca(OH)_2$ according to equation II to produce $CaCO_3$?

(I) $6HCl + Fe_2(CO_3)_3 = 3CO_2 + 2FeCl_3 + 3H_2O$

(II) $CO_2 + Ca(OH)_2 = CaCO_3 + H_2O$

How many moles $CaCO_3$

0.10 mol $Fe_2(CO_3)_3$

0.10 mol $Fe_2(CO_3)_3$

The question asked is "_____?" The information given is _____.

? mol $CaCO_3$ = _____

You are already at moles, so you are ready for a bridge conversion factor. You therefore look for the substance that is common to both

CO_2

equations. It is _____. Your bridge must contain both moles $Fe_2(CO_3)_3$ and moles CO_2. You can get such a bridge from equation I; it is

$\dfrac{3 \text{ mol } CO_2}{\text{mol } Fe_2(CO_3)_3}$

moles CO_2

_____.

Your unwanted dimension is now _____, and your answer

moles $CaCO_3$

must have the dimensions _____. You look for a bridge that relates these two. There is one that you can get from equation II: it is

$\dfrac{\text{mol } CaCO_3}{\text{mol } CO_2}$

moles $CaCO_3$

_____. If you multiply by this second bridge, the only dimension left is _____, which is the dimension of the answer.

$$? \text{ mol CaCO}_3 = 0.10 \, \cancel{\text{mol Fe}_2\text{CO}_3} \times \frac{3 \, \cancel{\text{mol CO}_2}}{\cancel{\text{mol Fe}_2\text{CO}_3}} \times \frac{\text{mol CaCO}_3}{\cancel{\text{mol CO}_2}}$$

$$= 0.30 \text{ mol CaCO}_3$$

This problem is essentially question 8 on the pretest, except that the amounts are changed and are all expressed in moles. In question 8 you convert from grams to moles by using the _____ weight conversion factors. Try to do it and see if you come up with the correct answer.

    molecular

Notice that if you combine the two bridge conversion factors,

$$\frac{3 \, \cancel{\text{mol CO}_2}}{\text{mol Fe}_2(\text{CO}_3)_3} \times \frac{\text{mol CaCO}_3}{\cancel{\text{mol CO}_2}}$$

moles CO_2 cancels. This leaves a new conversion factor,

$$\frac{3 \text{ mol CaCO}_3}{\text{mol Fe}_2(\text{CO}_3)_3}$$

You could have gotten this combined bridge by simply adding the two equations. However, equation II has to be multiplied through by 3 so that the CO_2's cancel.

$$6\text{HCl} + \text{Fe}_2(\text{CO}_3)_3 = 3\text{CO}_2 + 2\text{FeCl}_3 + 3\text{H}_2\text{O}$$
$$\underline{3\text{CO}_2 + 3\text{Ca(OH)}_2 = 3\text{CaCO}_3 + 3\text{H}_2\text{O}}$$
$$6\text{HCl} + \text{Fe}_2(\text{CO}_3)_3 + 3\text{Ca(OH)}_2 = 2\text{FeCl}_3 + 3\text{CaCO}_3 + 6\text{H}_2\text{O}$$

This procedure of adding the equations is preferable if the substance in which you are interested appears on the same side of two equations. An example of this is the production of H_2 by the burning of C. For example, find how many grams of H_2 are formed by burning 100 g C following the two consecutive reactions:

$$\text{C} + \text{H}_2\text{O} = \text{H}_2 + \text{CO}, \qquad \text{CO} + \text{H}_2\text{O} = \text{H}_2 + \text{CO}_2$$

Notice that H_2 appears on the right-hand side of both equations. Therefore you add them. (No adjustment is necessary, since the CO cancels as the equations stand.) Adding them gives you the combined equation

$$\text{C} + \underline{\hspace{1.5cm}} = \underline{\hspace{1.5cm}} + \text{CO}_2$$

The bridge relating C to H_2 from this combined equation is _____. You can now solve the problem in the normal way using this bridge.

$$? \text{ g H}_2 = \underline{\hspace{3cm}}$$

Next you multiply by _____. Next you multiply by _____.

Next you multiply by _____. The only dimension left is

_____. So the answer must be _____.

$2\text{H}_2\text{O}; 2\text{H}_2$

$\dfrac{2 \text{ mol H}_2}{\text{mol C}}$

100 g C

$\dfrac{\text{mol C}}{12.0 \text{ g C}}$, $\dfrac{2 \text{ mol H}_2}{\text{mol C}}$

$\dfrac{2.02 \text{ g H}_2}{\text{mol H}_2}$

grams H_2; 33.7 g H_2

You should be rather good at doing these now so try the problem set that follows. Remember, writing down all conversion factors given in the problem before you start is a big help. Then you just pick and choose from these.

PROBLEM SET

This is a very long problem set. To save time, don't bother to solve the problems for numerical answers. Simply set them up. On the following pages you will find the setups and also the setups repeated without the dimensions that have canceled.

Some of the problems are rather complicated. If you think that you are an expert at the setups, you can try them. If not, skip them. They are marked with an asterisk before the number. You need not bother to work out all the molecular weights. The following table gives you the ones you need.

	g/mol		g/mol		g/mol		g/mol
Ag	107.9	CO	28.0	HNO_3	63.0	NaOH	40.0
Al	27.0	C_2H_5OH	46.0	H_2S	34.1	NO	30.0
$AlCl_3$	133.5	$C_6H_{12}O_6$	180.0	H_2SO_4	98.1	P	31.0
$BiCl_3$	315.5	Cl_2	71.0	KCl	74.6	S	32.1
Bi_2S_3	514.3	$H_2C_2O_4$	90.0	$KMnO_4$	158.0	SiO_2	60.1
$Ca_3(PO_4)_2$	310.3	HCl	36.5	Na	23.0	SO_2	64.1
C	12.0	H_2	2.02	$NaAlO_2$	82.0	SO_3	80.1

The first 14 problems use the balanced equations shown below. The problem tells you which equation is referred to.

(a) $2AlCl_3 + 3H_2SO_4 = Al_2(SO_4)_3 + 6HCl$
(b) $2Na + 2H_2O = 2NaOH + H_2$
(c) $2KMnO_4 + 5H_2C_2O_4 + 6HCl = 2MnCl_2 + 10CO_2 + 8H_2O + 2KCl$
(d) $11H_2O + 12CO_2 = C_{12}H_{22}O_{11} + 12O_2$ (photosynthesis)
(e) $2Al + 2NaOH + 2H_2O = 2NaAlO_2 + 3H_2$
(f) $2BiCl_3 + 3H_2S = Bi_2S_3 + 6HCl$

1. How many grams of H_2 can be obtained from the reaction of 4.0 lb Na according to reaction (b)?

2. If 16.0 oz of Al are reacted with an excess of NaOH as in reaction (e), how many kilograms of H_2 are formed?

3. What is the maximum number of grams of Bi_2S_3 that can be obtained by the reaction of 1.10 lb $BiCl_3$? [Use (f).]

*4. If you want to completely react 5.0 tons of Al, how many kilograms of NaOH do you need? [Use (e).]

5. How many tons of H_2SO_4 are required to react completely with 16.0 tons $AlCl_3$, using (a)?

6. You have a sample of $KMnO_4$ weighing 7.20 oz. How many ounces of $H_2C_2O_4$ does it consume, following reaction (c)?

7. A 75 g sample of bauxite, an aluminum ore, contains 8.0 g Al. What weight, in grams, of H_2 can be prepared by reacting 425 g bauxite? (e)

8. How many grams of H_2SO_4 are there in 25.0 g of an H_2SO_4 solution, if 90.0 g of the solution are required to react with 52.0 g $AlCl_3$ (a)?

*9. A 6.0 oz sample of impure $KMnO_4$ contains only 90.0 g $KMnO_4$. How many grams of an 85% pure sample of $H_2C_2O_4$ are needed to react with 14.0 oz of the impure $KMnO_4$ (c)?

10. How many pounds of an 85.0% pure sample of H_2S are required to completely react with 4.0 lb $BiCl_3$ (f)?

11. What is the percent purity of an $H_2C_2O_4$ solution if 35.0 g of the solution produces 0.30 g KCl (c)?

*12. What is the percent purity of an Al ore if 4.0 kg of the ore produces 0.80 lb H_2 (e)?

13. How many grams of $NaAlO_2$ can be produced by the reaction of 1.50 L of a 0.250 *M* solution of NaOH with an excess of Al (e)?

14. How many milliliters of a 0.250 *M* $H_2C_2O_4$ solution react with 35.0 mL of 2.00 *M* $KMnO_4$ solution (c)?

The following 16 problems use the balanced equations shown below.

(g) $C_6H_{12}O_6 = 2C_2H_5OH + 2CO_2$
(h) $Ca_3(PO_4)_2 + 3SiO_2 + 5C = 3CaSiO_3 + 5CO + 2P$
(i) $3Ag + 4HNO_3 = 3AgNO_3 + NO + 2H_2O$
(j) $2KMnO_4 + 16HCl = 2MnCl_2 + 5Cl_2 + 8H_2O + 2KCl$
(k) (1) $S + O_2 = SO_2$ (2) $2SO_2 + O_2 = 2SO_3$
     (3) $SO_3 + H_2O = H_2SO_4$ (4) $H_2SO_4 + 2NaOH = Na_2SO_4 + 2H_2O$

15. Glucose, $C_6H_{12}O_6$, can be enzymatically reduced to alcohol, C_2H_5OH, according to reaction (g). It takes 3.0 hr to produce 5.0 kg alcohol. How many days does it take to consume 2.0 tons sugar?

16. How many grams of Cl_2 can be produced by oxidizing 300 mL of an HCl solution that is 25.0 wt % HCl and has a density of 1.13 g/mL (j)?

17. What is the density of Cl_2 gas in grams per liter if the Cl_2 gas produced when 31.6 g $KMnO_4$ is reacted with an excess of HCl [according to equation (j)] occupies 4.40 L?

18. What is the percent purity of a sample of silver (Ag) ore if a 5.00 g sample of the ore reacts with 11.0 mL of an HNO_3 solution that is 17.00 wt % HNO_3 and whose density is 1.10 g/mL (i)?

19. Suppose that 1.00 mol of SO_2 gas occupies 22.4 L at a temperature of 0°C and a pressure of 1.0 atm. What volume of SO_2 gas at this temperature and pressure is produced by the oxidation of 1.00 m³ of S, whose density is 1.8 mL (k-1)?

*20. How many milliliters of an H_2SO_4 solution whose density is 1.40 g/mL and that is 50 wt % H_2SO_4 react with 35.0 mL of a 20.0 wt % solution of NaOH whose density is 0.90 g/mL (k-4)?

21. What is the density (in grams per liter) of the CO produced if a 550 g sample of $Ca_3(PO_4)_2$ that is 60.0% pure gives 110 L of CO gas (h)?

22. Suppose that the price of glucose, $C_6H_{12}O_6$, is $0.15/lb. What would it cost to prepare 150 kg of alcohol, C_2H_5OH (g)?

23. What is the cost of the coke, pure C, needed for the preparation of 4.0 tons of phosphorus, P, if the cost of coke is $15/ton (h)?

24. What is the total cost of the raw materials needed to produce 10.0 lb of P, if the cost of $Ca_3(PO_4)_2$ is $0.12/lb, the cost of SiO_2 is $0.05/kg, and the cost of C is $15/ton (h)?

*25. How many dollars' worth of Ag can be obtained from 3.0 yd^3 of silver ore whose density is 8.0 g/mL if 2.0 lb of the ore can produce 2.2 L of NO? The density of the NO is 1.34 g/L, and the price of Ag is $8.00/oz (16 oz/lb) (i)?

26. Sulfur, S, is oxidized to SO_3 according to equations (k-1) and (k-2). How many grams of SO_3 can be produced by the oxidization of 3.0 lb S?

*27. Suppose that 128 g of S are oxidized to SO_3 according to equations (k-1) and (k-2) and the SO_3 is reacted with water to form 3.0 L of solution according to equation (k-3). What is the weight percent H_2SO_4 in the solution you have made? The density of the solution is 1.08 g/mL.

28. How many milliliters of a 30.0 wt % solution of NaOH are required to react (k-4) with the H_2SO_4 produced from reactions (k-1), (k-2), and (k-3) if you start with 16.05 g S? Density of the NaOH solution is 1.33 g/mL.

*29. Suppose that 35.0 g of a sample of impure S is oxidized to SO_3 [(k-1) and (k-2)] and the SO_3 is dissolved in excess H_2O. This solution requires 500 mL of a solution of NaOH that contains 4.0 mol of NaOH per liter of NaOH solution (k-4) for complete reaction. What is the percent purity of the sample of S?

*30. How many hours would it take to produce $400 worth of alcohol, C_2H_5OH, given that you can produce 5.0 kg alcohol in 3.0 hr and the price of glucose is $0.15/lb? The selling price of alcohol is twice the cost of its raw material, glucose ($C_6H_{12}O_6$) (g).

PROBLEM SET ANSWERS

The setups for the problems are shown below, and the numbers which you end up multiplying and dividing are given. Your answer is correct even if you do not have the numbers in exactly the same order as shown. Simply check and see that all the numbers are present in both the numerators and denominators of your solution. If you can do any 15 of these complicated problems, you are doing well and can proceed to the next chapter.

1. $? \text{ g H}_2 = 4.0 \text{ lb Na} \times \dfrac{454 \text{ g Na}}{\text{lb Na}} \times \dfrac{\text{mol Na}}{23.0 \text{ g Na}} \times \dfrac{\text{mol H}_2}{2 \text{ mol Na}} \times \dfrac{2.02 \text{ g H}_2}{\text{mol H}_2}$

$= 4.0 \times 454 \times \dfrac{1}{23.0} \times \dfrac{1}{2} \times 2.02 \text{ g H}_2$

2. $? \text{ kg H}_2 = 16.0 \text{ oz Al} \times \dfrac{\text{lb Al}}{16 \text{ oz Al}} \times \dfrac{454 \text{ g Al}}{\text{lb Al}} \times \dfrac{\text{mol Al}}{27.0 \text{ g Al}} \times \dfrac{3 \text{ mol H}_2}{2 \text{ mol Al}} \times \dfrac{2.02 \text{ g H}_2}{\text{mol H}_2}$

$\times \dfrac{\text{kg H}_2}{10^3 \text{ g H}_2} = 16.0 \times \dfrac{1}{16} \times 454 \times \dfrac{1}{27.0} \times \dfrac{3}{2} \times 2.0 \times \dfrac{1}{10^3} \text{ kg H}_2$

3. $? \text{ g Bi}_2\text{S}_3 = 1.10 \text{ lb BiCl}_3 \times \dfrac{454 \text{ g BiCl}_3}{\text{lb BiCl}_3} \times \dfrac{\text{mol BiCl}_3}{315.5 \text{ g BiCl}_3} \times \dfrac{\text{mol Bi}_2\text{S}_3}{2 \text{ mol BiCl}_3}$

$\times \dfrac{514.3 \text{ g Bi}_2\text{S}_3}{\text{mol Bi}_2\text{S}_3} = 1.10 \times 454 \times \dfrac{1}{315.5} \times \dfrac{1}{2} \times 514.3 \text{ g Bi}_2\text{S}_3$

4. $? \text{ kg NaOH} = 5.0 \text{ tons Al} \times \dfrac{2000 \text{ lb Al}}{\text{ton Al}} \times \dfrac{454 \text{ g Al}}{\text{lb Al}} \times \dfrac{\text{mol Al}}{27.0 \text{ g Al}}$

$\times \dfrac{2 \text{ mol NaOH}}{2 \text{ mol Al}} \times \dfrac{40.0 \text{ g NaOH}}{\text{mol NaOH}} \times \dfrac{\text{kg NaOH}}{10^3 \text{ g NaOH}}$

$= 5.0 \times 2000 \times 454 \times \dfrac{1}{27.0} \times \dfrac{2}{2} \times 40.0 \times \dfrac{1}{10^3} \text{ kg NaOH}$

5. $? \text{ tons H}_2\text{SO}_4 = 16.0 \text{ tons AlCl}_3 \times \dfrac{\text{ton-mol AlCl}_3}{133.5 \text{ tons AlCl}_3} \times \dfrac{3 \text{ ton-mol H}_2\text{SO}_4}{2 \text{ ton-mol AlCl}_3}$

$\times \dfrac{98.1 \text{ tons H}_2\text{SO}_4}{\text{ton-mol H}_2\text{SO}_4} = 16.0 \times \dfrac{1}{133.5} \times \dfrac{3}{2} \times 98.1 \text{ tons H}_2\text{SO}_4$

(If you didn't use ton-moles, you will have both 2000 and 454 appearing in the numerator and denominator and canceling.)

6. $? \text{ oz H}_2\text{C}_2\text{O}_4 = 7.20 \text{ oz KMnO}_4 \times \dfrac{\text{oz-mol KMnO}_4}{158.0 \text{ oz KMnO}_4} \times \dfrac{5 \text{ oz-mol H}_2\text{C}_2\text{O}_4}{2 \text{ oz-mol KMnO}_4}$

$\times \dfrac{90.0 \text{ oz H}_2\text{C}_2\text{O}_4}{\text{oz-mol H}_2\text{C}_2\text{O}_4} = 7.20 \times \dfrac{1}{158.0} \times \dfrac{5}{2} \times 90.0 \text{ oz H}_2\text{C}_2\text{O}_4$

(If you didn't use ounce-moles, you will have 16 and 454 in both numerator and denominator and canceling.)

7. $? \text{ g H}_2 = 425 \text{ g bauxite} \times \dfrac{8.0 \text{ g Al}}{75 \text{ g bauxite}} \times \dfrac{\text{mol Al}}{27.0 \text{ g Al}} \times \dfrac{3 \text{ mol H}_2}{2 \text{ mol Al}} \times \dfrac{2.02 \text{ g H}_2}{\text{mol H}_2}$

$= 425 \times \dfrac{8.0}{75} \times \dfrac{1}{27.0} \times \dfrac{3}{2} \times 2.02 \text{ g H}_2$

8. $? \text{ g H}_2\text{SO}_4 = 25.0 \text{ g soln} \times \dfrac{52.0 \text{ g AlCl}_3}{90.0 \text{ g soln}} \times \dfrac{\text{mol AlCl}_3}{133.5 \text{ g AlCl}_3} \times \dfrac{3 \text{ mol H}_2\text{SO}_4}{2 \text{ mol AlCl}_3}$

$\times \dfrac{98.1 \text{ g H}_2\text{SO}_4}{\text{mol H}_2\text{SO}_4} = 25.0 \times \dfrac{52.0}{90.0} \times \dfrac{1}{133.5} \times \dfrac{3}{2} \times 98.1 \text{ g H}_2\text{SO}_4$

9. $? \text{ g imp H}_2\text{C}_2\text{O}_4 = 14.0 \text{ oz imp KMnO}_4 \times \dfrac{90.0 \text{ g KMnO}_4}{6.0 \text{ oz imp KMnO}_4}$

$\times \dfrac{\text{mol KMnO}_4}{158 \text{ g KMnO}_4} \times \dfrac{5 \text{ mol H}_2\text{C}_2\text{O}_4}{2 \text{ mol KMnO}_4}$

$\times \dfrac{90.0 \text{ g H}_2\text{C}_2\text{O}_4}{\text{mol H}_2\text{C}_2\text{O}_4} \times \dfrac{100 \text{ g imp H}_2\text{C}_2\text{O}_4}{85 \text{ g H}_2\text{C}_2\text{O}_4}$

$= 14.0 \times \dfrac{90.0}{6.0} \times \dfrac{1}{158} \times \dfrac{5}{2} \times 90.0 \times \dfrac{100}{85} \text{ g imp H}_2\text{C}_2\text{O}_4$

10. $? \text{ lb imp H}_2\text{S} = 4.0 \text{ lb BiCl}_3 \times \dfrac{\text{lb-mol BiCl}_3}{315.5 \text{ lb BiCl}_3} \times \dfrac{3 \text{ lb-mol H}_2\text{S}}{2 \text{ lb-mol BiCl}_3}$

$\times \dfrac{34.1 \text{ lb H}_2\text{S}}{\text{lb-mol H}_2\text{S}} \times \dfrac{100 \text{ lb imp H}_2\text{S}}{85.0 \text{ lb H}_2\text{S}}$

$= 4.0 \times \dfrac{1}{315.5} \times \dfrac{3}{2} \times 34.1 \times \dfrac{100}{85.0} \text{ lb imp H}_2\text{S}$

11. $? \text{ g H}_2\text{C}_2\text{O}_4 = 100 \text{ g soln} \times \dfrac{0.30 \text{ g KCl}}{35.0 \text{ g soln}} \times \dfrac{\text{mol KCl}}{74.6 \text{ g KCl}} \times \dfrac{5 \text{ mol H}_2\text{C}_2\text{O}_4}{2 \text{ mol KCl}}$

$\times \dfrac{90.0 \text{ g H}_2\text{C}_2\text{O}_4}{\text{mol H}_2\text{C}_2\text{O}_4} = 100 \times \dfrac{0.30}{35.0} \times \dfrac{1}{74.6} \times \dfrac{5}{2} \times 90.0 \text{ g H}_2\text{C}_2\text{O}_4$

12. $? \text{ g Al} = 100 \text{ g ore} \times \dfrac{\text{kg ore}}{10^3 \text{ g ore}} \times \dfrac{0.80 \text{ lb H}_2}{4.0 \text{ kg ore}} \times \dfrac{454 \text{ g H}_2}{\text{lb H}_2}$

$\times \dfrac{\text{mol H}_2}{2.02 \text{ g H}_2} \times \dfrac{2 \text{ mol Al}}{3 \text{ mol H}_2} \times \dfrac{27.0 \text{ g Al}}{\text{mol Al}}$

$= 100 \times \dfrac{1}{10^3} \times \dfrac{0.80}{4.0} \times 454 \times \dfrac{1}{2.02} \times \dfrac{2}{3} \times 27.0 \text{ g Al per 100 g ore}$

13. $? \text{ g NaAlO}_2 = 1.50 \text{ L soln} \times \dfrac{0.250 \text{ mol NaOH}}{\text{L soln}} \times \dfrac{2 \text{ mol NaAlO}_2}{2 \text{ mol NaOH}}$

$\times \dfrac{82.0 \text{ g NaAlO}_2}{\text{mol NaAlO}_2} = 1.50 \times 0.250 \times \dfrac{2}{2} \times 82.0 \text{ g NaAlO}_2$

14. $? \text{ mL H}_2\text{C}_2\text{O}_4 = 35.0 \text{ mL KMnO}_4 \times \dfrac{10^{-3} \text{ L KMnO}_4}{\text{mL KMnO}_4} \times \dfrac{2.00 \text{ mol KMnO}_4}{\text{L KMnO}_4}$

$\times \dfrac{5 \text{ mol H}_2\text{C}_2\text{O}_4}{2 \text{ mol KMnO}_4} \times \dfrac{\text{L H}_2\text{C}_2\text{O}_4}{0.250 \text{ mol H}_2\text{C}_2\text{O}_4} \times \dfrac{\text{mL H}_2\text{C}_2\text{O}_4}{10^{-3} \text{ L H}_2\text{C}_2\text{O}_4}$

$= 35.0 \times 10^{-3} \times 2.00 \times \dfrac{5}{2} \times \dfrac{1}{0.250} \times \dfrac{\text{mL H}_2\text{C}_2\text{O}_4}{10^{-3}}$

15. $? \text{ days} = 2.0 \text{ tons } C_6H_{12}O_6 \times \dfrac{2000 \text{ lb } C_6H_{12}O_6}{\text{ton } C_6H_{12}O_6} \times \dfrac{454 \text{ g } C_6H_{12}O_6}{\text{lb } C_6H_{12}O_6}$

$\times \dfrac{\text{mol } C_6H_{12}O_6}{180.0 \text{ g } C_6H_{12}O_6} \times \dfrac{2 \text{ mol } C_2H_5OH}{\text{mol } C_6H_{12}O_6} \times \dfrac{46.0 \text{ g } C_2H_5OH}{\text{mol } C_2H_5OH}$

$\times \dfrac{\text{kg } C_2H_5OH}{10^3 \text{ g } C_2H_5OH} \times \dfrac{3.0 \text{ hr}}{5.0 \text{ kg } C_2H_5OH} \times \dfrac{\text{day}}{24 \text{ hr}}$

$= 2.0 \times 2000 \times 454 \times \dfrac{1}{180.0} \times 2 \times 46.0 \times \dfrac{1}{10^3} \times \dfrac{3.0}{5.0} \times \dfrac{1}{24} \text{ days}$

16. $? \text{ g } Cl_2 = 300 \text{ mL soln} \times \dfrac{1.13 \text{ g soln}}{\text{mL soln}} \times \dfrac{25.0 \text{ g HCl}}{100 \text{ g soln}} \times \dfrac{\text{mol HCl}}{36.5 \text{ g HCl}}$

$\times \dfrac{5 \text{ mol } Cl_2}{16 \text{ mol HCl}} \times \dfrac{71.0 \text{ g } Cl_2}{\text{mol } Cl_2}$

$= 300 \times 1.13 \times \dfrac{25.0}{100} \times \dfrac{1}{36.5} \times \dfrac{5}{16} \times 71.0 \text{ g } Cl_2$

17. $\dfrac{? \text{ g } Cl_2}{\text{L } Cl_2} = \dfrac{31.6 \text{ g KMnO}_4}{4.40 \text{ L } Cl_2} \times \dfrac{\text{mol KMnO}_4}{158 \text{ g KMnO}_4} \times \dfrac{5 \text{ mol } Cl_2}{2 \text{ mol KMnO}_4} \times \dfrac{71.0 \text{ g } Cl_2}{\text{mol } Cl_2}$

$= \dfrac{31.6}{4.40} \times \dfrac{1}{158} \times \dfrac{5}{2} \times 71.0 \dfrac{\text{g } Cl_2}{\text{L } Cl_2}$

18. $? \text{ g Ag} = 100 \text{ g imp} \times \dfrac{11.0 \text{ mL soln}}{5.00 \text{ g imp}} \times \dfrac{1.10 \text{ g soln}}{\text{mL soln}} \times \dfrac{17.00 \text{ g HNO}_3}{100 \text{ g soln}}$

$\times \dfrac{\text{mol HNO}_3}{63.0 \text{ g HNO}_3} \times \dfrac{3 \text{ mol Ag}}{4 \text{ mol HNO}_3} \times \dfrac{107.9 \text{ g Ag}}{\text{mol Ag}}$

$= 100 \times \dfrac{11.0}{5.00} \times 1.10 \times \dfrac{17.00}{100} \times \dfrac{1}{63.0} \times \dfrac{3}{4} \times 107.9 \text{ g Ag}/100 \text{ g imp}$

19. $? \text{ L } SO_2 = 1.00 \text{ m}^3 \text{ S} \times \dfrac{\text{cm}^3 \text{ S}}{(10^{-2})^3 \text{ m}^3 \text{ S}} \times \dfrac{\text{mL S}}{\text{cm}^3 \text{ S}} \times \dfrac{1.8 \text{ g S}}{\text{mL S}}$

$\times \dfrac{\text{mol S}}{32.1 \text{ g S}} \times \dfrac{\text{mol } SO_2}{\text{mol S}} \times \dfrac{22.4 \text{ L } SO_2}{1.00 \text{ mol } SO_2}$

$= 1.00 \times \dfrac{1}{10^{-6}} \times 1 \times 1.8 \times \dfrac{1}{32.1} \times 1 \times \dfrac{22.4}{1.00} \text{ L } SO_2$

20. $? \text{ mL } H_2SO_4 \text{ soln} = 35.0 \text{ mL NaOH soln} \times \dfrac{0.90 \text{ g NaOH soln}}{\text{mL NaOH soln}}$

$\times \dfrac{20.0 \text{ g NaOH}}{100 \text{ g NaOH soln}} \times \dfrac{\text{mol NaOH}}{40.0 \text{ g NaOH}} \times \dfrac{\text{mol } H_2SO_4}{2 \text{ mol NaOH}}$

$\times \dfrac{98.1 \text{ g } H_2SO_4}{\text{mol } H_2SO_4} \times \dfrac{100 \text{ g } H_2SO_4 \text{ soln}}{50 \text{ g } H_2SO_4}$

$\times \dfrac{\text{mL } H_2SO_4 \text{ soln}}{1.40 \text{ g } H_2SO_4 \text{ soln}}$

$= 35.0 \times 0.90 \times \dfrac{20.0}{100} \times \dfrac{1}{40.0} \times \dfrac{1}{2} \times 98.1 \times \dfrac{100}{50} \times \dfrac{1 \text{ mL } H_2SO_4 \text{ soln}}{1.40}$

(In this type of problem, be especially careful not to confuse the two different solutions. Write the complete dimension to indicate which solution it is.)

21. $$\frac{?\ g\ CO}{L\ CO} = \frac{550\ g\ imp}{110\ L\ CO} \times \frac{60.0\ g\ Ca_3(PO_4)_2}{100\ g\ imp} \times \frac{mol\ Ca_3(PO_4)_2}{310.3\ g\ Ca_3(PO_4)_2}$$

$$\times \frac{5\ mol\ CO}{mol\ Ca_3(PO_4)_2} \times \frac{28.0\ g\ CO}{mol\ CO}$$

$$= \frac{550}{110} \times \frac{60.0}{100} \times \frac{1}{310.3} \times 5 \times 28.0\ \frac{g\ CO}{L\ CO}$$

22. $$?\ \$ = 150\ kg\ C_2H_5OH \times \frac{10^3\ g\ C_2H_5OH}{kg\ C_2H_5OH} \times \frac{mol\ C_2H_5OH}{46.0\ g\ C_2H_5OH}$$

$$\times \frac{mol\ C_6H_{12}O_6}{2\ mol\ C_2H_5OH} \times \frac{180.0\ g\ C_6H_{12}O_6}{mol\ C_6H_{12}O_6} \times \frac{lb\ C_6H_{12}O_6}{454\ g\ C_6H_{12}O_6}$$

$$\times \frac{\$0.15}{lb\ C_6H_{12}O_6}$$

$$= 150 \times 10^3 \times \frac{1}{46.0} \times \frac{1}{2} \times 180.0 \times \frac{1}{454} \times \$0.15$$

23. $$?\ \$ = 4.0\ tons\ P \times \frac{ton\text{-}mol\ P}{31.0\ tons\ P} \times \frac{5\ ton\text{-}mol\ C}{2\ ton\text{-}mol\ P} \times \frac{12.0\ tons\ C}{ton\text{-}mol\ C} \times \frac{\$15}{ton\ C}$$

$$= 4.0 \times \frac{1}{31.0} \times \frac{5}{2} \times 12.0 \times \$15$$

(This problem is much easier to solve when you use ton-moles. If you didn't, you will have 2000 and 454 canceling.)

24. This is really three problems. You calculate the cost of $Ca_3(PO_4)_2$, then the cost of SiO_2, and finally the cost of C. When you have done that, you add all the costs. Only the separate cost calculations are shown below. Since we are not solving to a final numerical answer, it isn't possible to add them.

$$?\ \$ = 10.0\ lb\ P \times \frac{lb\text{-}mol\ P}{31.0\ lb\ P} \times \frac{lb\text{-}mol\ Ca_3(PO_4)_2}{2\ lb\text{-}mol\ P}$$

$$\times \frac{310.3\ lb\ Ca_3(PO_4)_2}{lb\text{-}mol\ Ca_3(PO_4)_2} \times \frac{\$0.12}{lb\ Ca_3(PO_4)_2}$$

$$= 10.0 \times \frac{1}{31.0} \times \frac{1}{2} \times 310.3 \times \$0.12$$

$$?\ \$ = 10.0\ lb\ P \times \frac{kg\ P}{2.2\ lb\ P} \times \frac{kg\text{-}mol\ P}{31.0\ kg\ P} \times \frac{3\ kg\text{-}mol\ SiO_2}{2\ kg\text{-}mol\ P}$$

$$\times \frac{60.1\ kg\ SiO_2}{kg\text{-}mol\ SiO_2} \times \frac{\$0.05}{kg\ SiO_2}$$

$$= 10.0 \times \frac{1}{2.2} \times \frac{1}{31.0} \times \frac{3}{2} \times 60.1 \times \$0.05$$

$$?\ \$ = 10.0\ lb\ P \times \frac{ton\ P}{2000\ lb\ P} \times \frac{ton\text{-}mol\ P}{31.0\ tons\ P} \times \frac{5\ ton\text{-}mol\ C}{2\ ton\text{-}mol\ P}$$

$$\times \frac{12.0\ tons\ C}{ton\text{-}mol\ C} \times \frac{\$15}{ton\ C} = 10.0 \times \frac{1}{2000} \times \frac{1}{31.0} \times \frac{5}{2} \times 12.0 \times \$15$$

[Notice that, to simplify this type of problem, you convert immediately from pounds P to whatever weight unit the cost was expressed in: pounds, kilograms, or tons. Then you express the bridge in pound-moles, kilogram-moles, and ton-moles.]

25. This is the type of complex problem which you should always start by writing all the conversion factors given in the problem. Here they are.

$$\frac{8.0 \text{ g ore}}{\text{mL ore}}, \quad \frac{2.0 \text{ lb ore}}{2.2 \text{ L NO}}, \quad \frac{1.34 \text{ g NO}}{\text{L NO}}, \quad \frac{\$2.05}{\text{oz Ag}}, \quad \frac{16 \text{ oz Ag}}{\text{lb Ag}}$$

$$? \$ = 3.0 \text{ yd}^3 \text{ ore} \times \frac{(36)^3 \text{ in}^3 \text{ ore}}{\text{yd}^3 \text{ ore}} \times \frac{(2.54)^3 \text{ cm}^3 \text{ ore}}{\text{in}^3 \text{ ore}} \times \frac{\text{mL ore}}{\text{cm}^3 \text{ ore}}$$

$$\times \frac{8.0 \text{ g ore}}{\text{mL ore}} \times \frac{\text{lb ore}}{454 \text{ g ore}} \times \frac{2.2 \text{ L NO}}{2.0 \text{ lb ore}} \times \frac{1.34 \text{ g NO}}{\text{L NO}} \times \frac{\text{mol NO}}{30.0 \text{ g NO}}$$

$$\times \frac{3 \text{ mol Ag}}{\text{mol NO}} \times \frac{107.9 \text{ g Ag}}{\text{mol Ag}} \times \frac{\text{lb Ag}}{454 \text{ g Ag}} \times \frac{16 \text{ oz Ag}}{\text{lb Ag}} \times \frac{\$8.00}{\text{oz Ag}}$$

$$= 3.0 \times (36)^3 \times (2.54)^3 \times 1 \times 8.0 \times \frac{1}{454} \times \frac{2.2}{2.0} \times 1.34 \times \frac{1}{30.0}$$

$$\times 3 \times 107.9 \times \frac{1}{454} \times 16 \times \$8.00$$

26. Since the substances in the problem do not appear on the same side of the equations, we can use two bridges.

$$? \text{ g SO}_3 = 3.0 \text{ lb S} \times \frac{454 \text{ g S}}{\text{lb S}} \times \frac{\text{mol S}}{32.1 \text{ g S}} \times \frac{\text{mol SO}_2}{\text{mol S}} \times \frac{2 \text{ mol SO}_3}{2 \text{ mol SO}_2} \times \frac{80.1 \text{ g SO}_3}{\text{mol SO}_3}$$

$$= 3.0 \times 454 \times \frac{1}{32.1} \times 1 \times \frac{2}{2} \times 80.1 \text{ g SO}_3$$

27. $? \text{ g H}_2\text{SO}_4 = 100 \text{ g soln} \times \frac{\text{mL soln}}{1.08 \text{ g soln}} \times \frac{10^{-3} \text{ L soln}}{\text{mL soln}}$

$$\times \frac{128 \text{ g S}}{3.0 \text{ L soln}} \times \frac{\text{mol S}}{32.1 \text{ g S}} \times \frac{\text{mol SO}_2}{\text{mol S}}$$

$$\times \frac{2 \text{ mol SO}_3}{2 \text{ mol SO}_2} \times \frac{\text{mol H}_2\text{SO}_4}{\text{mol SO}_3} \times \frac{98.1 \text{ g H}_2\text{SO}_4}{\text{mol H}_2\text{SO}_4}$$

$$= 100 \times \frac{1}{1.08} \times 10^{-3} \times \frac{128}{3.0} \times \frac{1}{32.1} \times 1 \times \frac{2}{2} \times 1 \times 98.1 \text{ g H}_2\text{SO}_4 / 100 \text{ g soln}$$

28. $? \text{ mL NaOH soln} = 16.05 \text{ g S} \times \frac{\text{mol S}}{32.1 \text{ g S}} \times \frac{\text{mol SO}_2}{\text{mol S}} \times \frac{2 \text{ mol SO}_3}{2 \text{ mol SO}_2}$

$$\times \frac{\text{mol H}_2\text{SO}_4}{\text{mol SO}_3} \times \frac{2 \text{ mol NaOH}}{\text{mol H}_2\text{SO}_4} \times \frac{40.0 \text{ g NaOH}}{\text{mol NaOH}}$$

$$\times \frac{100 \text{ g NaOH soln}}{30.0 \text{ g NaOH}} \times \frac{\text{mL NaOH soln}}{1.33 \text{ g NaOH soln}}$$

$$= 16.05 \times \frac{1}{32.1} \times 1 \times \frac{2}{2} \times 1 \times 2 \times 40.0 \times \frac{100}{30.0}$$

$$\times \frac{\text{mL NaOH}}{1.33}$$

29. $? \text{ g S} = 100 \text{ g imp} \times \dfrac{500 \text{ mL NaOH soln}}{35.0 \text{ g imp}} \times \dfrac{10^{-3} \text{ L NaOH soln}}{\text{mL NaOH soln}}$

$\times \dfrac{4.0 \text{ mol NaOH}}{\text{L NaOH soln}} \times \dfrac{\text{mol H}_2\text{SO}_4}{2 \text{ mol NaOH}} \times \dfrac{\text{mol SO}_3}{\text{mol H}_2\text{SO}_4}$

$\times \dfrac{2 \text{ mol SO}_2}{2 \text{ mol SO}_3} \times \dfrac{\text{mol S}}{\text{mol SO}_2} \times \dfrac{32.1 \text{ g S}}{\text{mol S}}$

$= 100 \times \dfrac{500}{35.0} \times 10^{-3} \times 4.0 \times \dfrac{1}{2} \times 1 \times \dfrac{2}{2} \times 1 \times 32.1 \text{ g S}/100 \text{ g imp}$

30. You start by saying, "If the alcohol is worth $400, and it is twice as expensive as the sugar, then the sugar must have cost $200." Then you can set up the problem.

$? \text{ hr} = \$200 \times \dfrac{\text{lb C}_6\text{H}_{12}\text{O}_6}{\$0.15} \times \dfrac{454 \text{ g C}_6\text{H}_{12}\text{O}_6}{\text{lb C}_6\text{H}_{12}\text{O}_6} \times \dfrac{\text{mol C}_6\text{H}_{12}\text{O}_6}{180 \text{ g C}_6\text{H}_{12}\text{O}_6}$

$\times \dfrac{2 \text{ mol C}_2\text{H}_5\text{OH}}{\text{mol C}_6\text{H}_{12}\text{O}_6} \times \dfrac{46.0 \text{ g C}_2\text{H}_5\text{OH}}{\text{mol C}_2\text{H}_5\text{OH}} \times \dfrac{\text{kg C}_2\text{H}_5\text{OH}}{10^3 \text{ g C}_2\text{H}_5\text{OH}}$

$\times \dfrac{3 \text{ hr}}{5.0 \text{ kg C}_2\text{H}_5\text{OH}}$

$= 200 \times \dfrac{1}{0.15} \times 454 \times \dfrac{1}{180} \times 2 \times 46.0 \times \dfrac{1}{10^3} \times \dfrac{3}{5.0} \text{ hr}$

CHAPTER 12

Gas Laws

PRETEST

If you have had some previous experience with gas-law calculations, try this pretest. You should be able to complete it in 40 minutes.

1. How many atmospheres pressure is 380 mmHg pressure?

2. What is the temperature in kelvins when the temperature is $30°C$?

3. You have 2.0 L of hydrogen gas at $27°C$ and a pressure of 6.0 atm. You increase its pressure to 9.0 atm and lower its temperature to $-23°C$. What will its new volume be?

4. Suppose that 300 mL of oxygen gas at a temperature of 350 K and a pressure of 780 torr has its volume reduced to 0.150 L and its pressure raised to 3.0 atm. What is its new temperature in degrees Celsius?

5. What must the pressure be in millimeters of mercury (mmHg) to have a gas at $18°C$ occupy the same volume as it would at STP?

6. How many moles of an ideal gas occupy 2.0 L at a temperature of $27°C$ and a pressure of 0.5 atm?

7. When 16.0 g of He gas are contained in a 2.0 L vessel at a temperature of 100 K, what is its pressure in atmospheres?

8. What is the density of O_2 gas in grams per liter at $27°C$ and 152 cmHg pressure? (Note: cmHg, not mmHg.)

9. You are given 3.0 L of an unknown gas at a pressure of 1.0 atm and a temperature of $27°C$. It weighs 2.4 g. What is its molecular weight?

10. What is the molecular weight of a gas whose density is 4.50 g/L at $-73°C$ and a pressure of 380 mmHg?

11. What is the pressure in a 4.0 L container that is at $27°C$ and contains 2.0 mol of H_2 and 5.0 mol of He gas?

12. What is the molecular weight of a gas if 70.0 mL of it collected over water at a temperature of $27°C$ and a total pressure of 756.5 mmHg weighs 0.080 g when dry? The vapor pressure of water at this temperature is 26.5 mmHg.

13. A sample of N_2 gas diffuses through a small hole at a rate of 2.6 mL/min. At what rate would Cl_2 gas diffuse through the same hole under the same conditions?

14. Suppose that 20.0 mL of an unknown gas diffuse through a pinhole in the same time that it takes 40.0 mL of CH_4 to diffuse through the same pinhole under the same conditions. What is the molecular weight of the unknown gas?

PRETEST ANSWERS

All the questions are worked out in the body of the chapter. If you were able to do questions 1 and 2, you can skip Section A. Questions 3, 4, and 5 are covered in Section B. If you got them right, you may skip that section. If you got numbers 6, 7, and 8, you may skip Section C. Section D contains the solutions to 9 and 10. If you were able to do numbers 11 and 12, you may skip Section E. Section F contains the subject of questions 13 and 14.

1. 0.500 atm
2. 303 K
3. 1.1 L
4. 238°C
5. 810 mmHg
6. 0.04 mol
7. 16 atm
8. 2.60 g/L
9. 20 g/mol
10. 148 g/mol
11. 4.3×10^1 atm
12. 29 g/mol
13. 1.6 mL/min
14. 64 g/mol

The temperature, pressure, volume, and number of moles of most gases, particularly at high temperature and low pressure, have the same relationship. This relationship can be expressed mathematically as

$$\boxed{PV = nRT}$$

This expression is called an *equation of state* or the *ideal-gas equation*. It says that the product of the pressure P times the volume V equals the number of moles n times a constant R times the absolute temperature T. (We are going to use the SI notation, K, for absolute temperature, rather than the metric °K.)

For example, if you have 1 mol of a gas at 1 atm pressure and 273 K (0°C), it will occupy 22.4 L. Putting these values in the ideal-gas equation, you have

(1.00 atm)(22.4 L) = (1.00 mol)(R)(273 K)

Solving for R by cross-multiplying, you get

$$\frac{(1.00 \text{ atm})(22.4 \text{ L})}{(1.00 \text{ mol})(273 \text{ K})} = \frac{0.0820 \text{ (L) (atm)}}{\text{(mol)(K)}} = R$$

The R is, of course, a conversion factor, but it has an odd set of dimen-

four

$\frac{g}{mL}$

sions. There are _____ different units. However, density has two different units, _____; and molecular weight has two different units

$\dfrac{g}{mol}$

_____. Problems in Chapter 16 use the molal boiling and freezing point constants, which are conversion factors with three different units.

The numerical value for a conversion factor depends on which units you use. For example, there are 5280 feet per mile or 63,360 inches per mile. We'll try to be consistent and use only the units liters, atmospheres, moles, and kelvins for *R*. Then you have to remember only one number: 0.0820.

SECTION A   Gas Dimensions

The ideal-gas equation holds true only if the temperature is expressed in kelvins. This is also called the *absolute temperature.* It is simple to change from degrees Celsius to kelvins. You add 273.16 to the temperature in degrees Celsius and you have the temperature in kelvins.

$$K = 273.16 + °C$$

We will use just 273 (three significant figures). A temperature of 100°C is thus equal to _____ K. A temperature of 533 K is equal to _____°C; and a temperature of –10°C is equal to _____ K.

373; 260

263

There are many ways of expressing pressure. When you talk about the air pressure in the tires on your car, you use _____ per square inch (psi). In countries that use metric units, people talk about kilograms per square meter. In most scientific work, there also are two other units used. One is millimeters Hg, or *torrs*, which represents the height of a column of mercury that the pressure can support. The other unit used is *atmospheres* (atm) which is the average barometric pressure at sea level.

pounds

The conversion factors needed to get from one unit to another are

$$\frac{760 \text{ mmHg}}{atm} \text{ , } \frac{76 \text{ cmHg}}{atm} \text{ , } \frac{760 \text{ torr}}{atm} \text{ , } \frac{14.7 \text{ psi}}{atm}$$

Using the numerical value of *R* as 0.0820, you always express pressure in _____. Thus a pressure of 1520 torr equals _____ atm, and a pressure of 380 mmHg equals _____ atm.

atmospheres; 2.000

0.500

[We will not use the SI unit *pascal* (Pa) for pressure. It uses only SI base units (1 Pa = 1 kg/m sec^2) and there are 101.325 Pa/atm. This is an inconveniently large number of digits.]

SECTION B   Change-of-Conditions Problems

There are actually only two types of gas-law problems. We will examine the simplest first. You have a sample of gas at certain conditions of temperature, pressure, and volume (called its *variables* or *parameters*).

You change one or two of the variables and want to know what the value of the third variable will be after the change. These problems are so easy that we won't use dimensional analysis.

The reason that these are so simple is that you are talking about the same sample of gas, so that the number of moles (n) is the same both before and after the change. That is, n is a constant. R is always a constant, and so you can write the ideal-gas equation as

$$\frac{PV}{T} = nR = \text{constant}$$

Now consider the gas under its initial set of conditions. The pressure is P' (read "P prime"), the volume is V', and the temperature is T'. You can then write

nR

$$\frac{P'V'}{T'} = \underline{\qquad}$$

If you change the variables so that the pressure is P'' (read "P double prime"), the volume is V'', and the temperature is T'', you can write

nR

$$\frac{P''V''}{T''} = \underline{\qquad}$$

Since both these expressions equal the same nR, they must equal each other.

$$\frac{P'V'}{T'} = \frac{P''V''}{T''}$$

In a change-of-conditions problem, you have the initial conditions P', V', and T' all given. You also have two of the final variables given and have to calculate only the third. It is very simple. However, so that you don't get confused, take the time to write a small table.

$P' = $ $P'' = $

$V' = $ $V'' = $

$T' = $ $T'' = $

Fill in all the values given in the problem, using a question mark for the unknown parameter. Be careful that the temperature is expressed in kelvins (K). You cannot work in any other units, although pressure and volume can be in any units so long as they are the same for the initial as for the final condition. (All conversion factors cancel that way.) Once you have the table set up, you can put the values in the expression

$\dfrac{P''V''}{T''}$

$$\frac{P'V'}{T'} = \underline{\qquad}$$

and solve for the unknown final parameter.

Question 3 in the pretest is a simple example of this type of problem. You have 2.0 L of hydrogen gas at 27°C and a pressure of 6.0 atm. You increase its pressure to 9.0 atm and lower its temperature to −23°C. What will its new volume be?

First you set up your table:

$P' = 6.0$ atm $P'' = 9.0$ atm
$V' = 2.0$ L $V'' = ?$ L
$T' = 300$ K $T'' = $ _____   250 K

Then you have

$$\frac{P'V'}{T'} = \frac{P''V''}{\underline{\quad}}$$ T''

Putting in the values from the table, you have

$$\frac{(6.0\ \text{atm})(2.0\ \text{L})}{300\ \text{K}} = \frac{(9.0\ \text{atm})(?\ \text{L})}{\underline{\quad}}$$ 250 K

Cross-multiplying gives you

$$\frac{(6.0\ \text{atm})(2.0\ \text{L})(\underline{\quad})}{(9.0\ \text{atm})(\underline{\quad})} = ?\ \text{L}$$ 250 K
300 K

$$\left(\frac{6.0}{9.0}\right)\left(\frac{250}{300}\right)(2.0\ \text{L}) = 1.1\ \text{L} = ?\ \text{L}$$

Perhaps now is a good time to learn an algebraic trick called *cross multiplication.* If you know this already, you can skip the next short section.

Cross Multiplication

When you have any equality (equation), you can multiply or divide both sides of the equality by the same factor and still have an equality. An example will prove this. Consider

$$\frac{6}{8} = \frac{3}{4} \quad \text{or} \quad 0.75 = 0.75$$

If you multiply both sides by 2, you have

$$\frac{12}{8} = \frac{6}{4} \quad \text{or} \quad 1.5 = 1.5 \quad \text{(Both sides are still equal.)}$$

Or, if you divide both sides by 3, you have

$$\frac{6}{3 \times 8} = \frac{3}{3 \times 4} \quad \text{or} \quad \frac{6}{24} = \frac{3}{12} \quad \text{or} \quad 0.25 = 0.25$$

(Both sides are still equal.)

Let us do this now with letters rather than numbers. Consider

$$ax = by$$

We will divide both sides by a:

$$\frac{ax}{a} = \frac{by}{a}$$

The *a*'s on the left cancel, leaving

$$x = \frac{by}{a}$$

In effect, what you have done is move the *a* from the numerator on the left to the denominator on the right. You have cross-multiplied it.

$$\textcircled{a}\,x = \frac{by}{\textcircled{a}}$$

Consider another equality:

$$\frac{x}{a} = by$$

If you multiply both sides by *a*, then

$$\frac{ax}{a} = aby$$

The *a*'s on the left cancel, leaving

$$x = aby$$

In effect, what you have done is move the *a* from the denominator on the left to the numerator on the right. You have cross-multiplied.

$$\frac{x}{\textcircled{a}} = \textcircled{a}\ by$$

Consequently, if you ever want to solve an equation for just one of its parts, all you have to do is cross-multiply in such a way as to have the component you wish to solve for alone in the numerator on one side of the equation. To solve the preceding example, we cross-multiplied as follows.

$$\frac{(6.0\ \text{atm})(2.0\ \text{L})}{300\text{K}} = \frac{(9.0\ \text{atm})\ (?\ \text{L})}{250\ \text{K}}$$

or

$$\frac{(6.0\ \text{atm})(250\ \text{K})(2.0\ \text{L})}{(9.0\ \text{atm})(300\ \text{K})} = ?\ \text{L}$$

As you can see from the foregoing, any units for pressure and volume are fine as long as they are the same for the final state of the gas as for the initial. The temperature, however, must be in kelvins. Question 4 on the pretest illustrates this point.

Suppose that 300 mL of oxygen gas at a temperature of 350 K and a pressure of 780 torr has its volume reduced to 0.150 L and its pressure raised to 3.0 atm. What will its temperature be in degrees Celsius?

First you set up your table with everything in the same units.

$P' = 780$ torr $\quad P'' = 3.0 \text{ atm} \times \dfrac{760 \text{ torr}}{\text{atm}} = $ _____ torr

$V' = 300$ mL $\quad V'' = 0.150 \text{ L} \times \dfrac{\text{mL}}{10^{-3} \text{ L}} = $ _____ mL

$T' = 350$ K $\quad\quad T'' = ?$ __

2280

150

K

Then you have the relationship

$$\dfrac{P'V'}{T'} = \underline{}$$

$\dfrac{P''V''}{T''}$

Putting in the values from the table (and since the dimensions all match, you needn't bother to put them in),

$$\dfrac{(780)(300)}{350} = \dfrac{\rule{3cm}{0.4pt}}{? \text{ K}}$$

$(2280)(150)$

Cross-multiplying gives you

$$? \text{ K} = \dfrac{(350)\rule{2.5cm}{0.4pt}}{(780)(300)} = \underline{} \text{ K}$$

$(2280)(150)$; 511

To get this temperature to degrees celsius, you must _____ 273.

subtract

$? \,^{\circ}C = $ _____ $^{\circ}C$

238

For convenience, chemists have selected a temperature and a pressure that they call either *standard conditions* (SC) or *standard temperature and pressure* (STP) for gases. This is 1.00 atm and 273 K (_ $^{\circ}C$). You may recall that this is the temperature and pressure we used to get a value for R. At STP, then, 1.00 mol of a gas occupies _____ L.

0

22.4

Question 5 on the pretest refers to these conditions. What must the pressure be in millimeters of mercury (mmHg) to have a gas at 18$^{\circ}C$ occupy the same volume as it would at STP? This problem is a little strange because there is no volume given. You simply know that it is the same in the initial state as in the final state. You can assume that they are both 1 L, since they cancel in any event.

$\quad$ (STP)

$P' = 1.00$ atm $\quad\quad\quad P'' = ?$ _____

atm

$V' = 1$ L $\quad\quad\quad\quad V'' = 1$ L

$T' = $ _____ $\quad\quad\quad T'' = 291$ K

273 K

Putting these values into

$$\dfrac{P'V'}{T'} = \underline{} \quad \text{you have} \quad \dfrac{(1.0)(1)}{273} = \dfrac{(?\underline{})(1)}{291}$$

$\dfrac{P''V''}{T''}$; atm

Cross-multiplying gives you

$$\dfrac{(1.0)(1)(291)}{(1)(273)} = \rule{3cm}{0.4pt}$$

1.07 atm

To get the pressure in millimeters of mercury, you solve the following equation.

$\dfrac{760 \text{ mmHg}}{\text{atm}}$; 810

? mmHg = 1.07 atm × _____ = _____ mmHg

This is all there is to changing a sample of gas from one set of parameters to another.

SECTION C   Determining an Unknown Parameter

The second type of gas-law problem does not involve changing the conditions of a sample of gas. Instead, it asks you to determine one of the four parameters that define a gas (pressure, volume, temperature, or moles), the other three being given. This means that you have three separate pieces of information given. It is sometimes hard to choose which one of the three to start with. One easy way to decide is to look at the ideal-gas equation

$$PV = nRT$$

and see which two pieces of information given are both on the same side of the equals sign. You can start with both of these. You always then use the conversion factor R, the ideal-gas constant, which is

L; atm

mol; K

$$\dfrac{0.0820\ (\underline{})(\underline{})}{(\underline{})(\underline{})}$$

Let's try question 6 on the pretest to see how this goes. How many moles of an ideal gas occupy 2.0 L at a temperature of 27°C and a pressure of 0.5 atm? The question asked is: "How many _____ ?"

moles

0.5 atm

2.0 L; kelvins

300

v

The information given is that pressure equals _____ and volume equals _____. The temperature, which must be converted to _____, is _____ K. The two pieces of given information that are both on the same side of the equation of state are P and ___. You may therefore start with both of these.

2.0 L

? mol = (0.5 atm)(_____)

Next you can use the conversion factor for R to cancel these two

$\dfrac{(\text{mol})(\text{K})}{0.0820\ (\text{L})(\text{atm})}$

K

$\dfrac{1}{300\ \text{K}}$

unwanted dimensions. You multiply by _____, its inverted form, leaving the unwanted dimension ___. You can cancel this by multiplying by _____, which is the reciprocal of the information given. Now the only dimension left is the dimension of the answer, which must be

$$? \text{ mol} = (0.5\ \cancel{\text{atm}})(2.0\ \cancel{\text{L}}) \times \dfrac{(\text{mol})(\cancel{\text{K}})}{0.0820\ (\cancel{\text{L}})(\cancel{\text{atm}})} \times \dfrac{1}{300\ \cancel{\text{K}}} = 0.04 \text{ mol}$$

This example is particularly easy since all the parameters in the information given are in the same dimensions as in the gas-law conversion factor, R. The next example is not so simple. Suppose that 3.0 mol of an ideal gas are at a temperature of $-23°C$ and a pressure of 380 torr. What volume in milliliters does the gas occupy?

The question asked is: "How many _____?" The information   milliliters

given is the temperature (which must be in kelvins) _____; the pressure,   250 K

_____; and the number of moles, _____. The two   380 torr; 3.0 mol

given parameters that are on the same side of the ideal-gas equation are

__ and __. So you start with these two. n; T

   $? \text{ mL} = (3.0 \text{ mol})(\underline{\qquad})$ 250 K

Now you can cancel the two unwanted dimensions by multiplying by $\dfrac{0.0820 \text{ (L)(atm)}}{\text{(mol)(K)}}$

the ideal-gas constant R, which is _____. The unwanted

dimensions are now _____ and _____. However, the   liters; atmospheres

units of pressure in the information given are _____. In order to cancel   torrs
the unwanted dimension, you must convert atmospheres to torrs by

multiplying by the conversion factor _____. Now you can use $\dfrac{760 \text{ torr}}{\text{atm}}$
your third piece of information given, by multiplying by its reciprocal,

_____. The unwanted dimension now remaining is _____. $\dfrac{1}{380 \text{ torr}}$; liters

Since you want your answer to be in _____, you multiply by   milliliters

_____. The answer is thus $\dfrac{\text{mL}}{10^{-3} \text{ L}}$

$$? \text{ mL} = (3.0 \text{ mol})(250 \text{ K}) \times \frac{0.0820 \text{ (L)(atm)}}{\text{(mol)(K)}} \times \frac{760 \text{ torr}}{\text{atm}}$$

$$\times \frac{1}{380 \text{ torr}} \times \frac{\text{mL}}{10^{-3} \text{ L}} = 1.2 \times 10^5 \text{ mL}$$

Sometimes you have n in the information given as weight of gas rather than moles of gas. However, the conversion is very simple, as question 7 in the pretest demonstrates.

When 16.0 g of He gas is contained in a 2.0 L vessel at a temperature of 100 K, what is its pressure in atmospheres? The question asked is:

"How many _____?" The information given is the temperature,   atmospheres

_____; the volume, _____; and the amount of He, _____. Of the   100 K; 2.0 L; 16.0 g

four variables—$P, V, T,$ and n—the pieces of information given are __, __ T; V

and, indirectly, __. The two that are on the same side of the ideal-gas n

equation are __ and __. So you start with these. T; n

   $? \text{ atm} = (100 \text{ K})(\underline{\qquad})$ 16.0 g He

It is obvious that if you are going to cancel the amount of He by using the gas-law conversion factor R, the amount of He has to have the dimensions _____. It is simple to get from grams to moles. All you do is multiply by _____, the inverted form of the molecular weight.

moles

$\dfrac{\text{mol He}}{4.00 \text{ g He}}$

$\dfrac{0.0820 \text{ (L)(atm)}}{\text{(mol)(K)}}$

Now you can multiply by the gas-law conversion factor R, _____, to cancel the two unwanted dimensions.

liters

$\dfrac{1}{2.0 \text{ L}}$

The unwanted dimension remaining is _____. But since you have this in the third piece of the information given, you multiply by _____. Now the only dimension left is the dimension of the answer, which must be

$$? \text{ atm} = (100\,\cancel{K})(16.0\,\cancel{\text{g He}}) \times \frac{\cancel{\text{mol He}}}{4.00\,\cancel{\text{g He}}} \times \frac{0.0820\,\cancel{(L)}(atm)}{\cancel{(mol He)}\cancel{(K)}} \times \frac{1}{2.0\,\cancel{L}}$$

16

$$= \underline{} \text{ atm}$$

Question 8 on the pretest shows you another way that the information can be given to you. What is the density of O_2 gas in grams per liter at $27°C$ and 152 cmHg pressure?

Probably the best way to do this problem is to reword it to ask "How many grams of O_2 are there in _____ at $27°C$ and a pressure of 152 cmHg?" The question now asked is "How many _____?" The information given is the volume, _____; the temperature, _____ (it must be in kelvins); and the pressure, _____. The two parameters that are on the same side of the equation of state are __ and __. So you write

1.0 L

grams of O_2

1.0 L; 300 K

152 cmHg

P; V

$$? \text{ g } O_2 = (152 \text{ cmHg})(\underline{})$$

1.0 L

Before you can cancel the pressure term, you must convert it to the dimension _____ by multiplying by _____. Now you can use your gas-law conversion factor. You multiply by its inverted form, _____. You can cancel the unwanted dimension kelvins by using the third piece of information given. You multiply by the reciprocal of the temperature, _____. The unwanted dimension now is _____. You can convert this to the dimensions of your answer by multiplying by _____, the molecular weight. The answer is therefore

atmospheres; $\dfrac{\text{atm}}{76 \text{ cmHg}}$

$\dfrac{\text{(mol)(K)}}{0.0820 \text{ (L)(atm)}}$

$\dfrac{1}{300 \text{ K}}$

moles O_2

$\dfrac{32.0 \text{ g } O_2}{\text{mol } O_2}$

$$? \text{ g } O_2 = (152)(1.0) \left(\frac{1}{76}\right)\left(\frac{1}{0.0820}\right)\left(\frac{1}{300}\right)(32.0) \text{ g } O_2$$

$$= 2.6 \text{ g in } 1.0 \text{ L} \quad so \quad d = 2.6 \text{ g/L}$$

Note that you could have solved the problem directly for density, which is a conversion factor, by starting with the gas-law conversion factor and working from there.

$$? d_{O_2} = \frac{? \text{ g } O_2}{\text{L } O_2} = \frac{\cancel{(\text{mol})}\cancel{(\text{K})}}{0.0820 \text{ (L)}\cancel{(\text{atm})}} \times \frac{32.0 \text{ g } O_2}{\cancel{\text{mol } O_2}}$$

$$\times \frac{152 \cancel{\text{ cmHg}}}{300 \cancel{\text{ K}}} \times \frac{\cancel{\text{atm}}}{76 \cancel{\text{ cmHg}}}$$

You set the gas-law conversion factor so that it gives liters in the denominator, and cancel with the information given.

SECTION D   Determining Molecular Weight

By reversing the process of the last two examples, you can determine the molecular weight of an ideal gas if you know its pressure, volume, temperature, and weight. Question 9 on the pretest is a straightforward example of this calculation.

You are given 3.0 L of an unknown gas at a pressure of 1.0 atm and a temperature of 27°C. It weighs 2.4 g. What is its molecular weight? You can do this problem by directly solving for the conversion factor for molecular weight, grams per mole. However, to make sure you start out right, it is better to reword the question as: "How many grams equal one mole?" Then you have the question asked and the information given all in one place.

In these problems, there is always one conversion factor given: the weight of unknown per volume of unknown. This might be given as the density, or, as in our problem, directly as

$$\frac{\text{_____ g } X}{\text{_____ L } X}$$

2.4
3.0

In our problem you also have two pieces of information given: the pressure, _____, and the temperature, _____. Of course you know the gas-law conversion factor, _____. So, starting in the usual way, with the question asked and the information given, you have

1.0 atm; 300 K
$$\frac{0.0820 \text{ (L)}(\text{atm})}{(\text{mol})(\text{K})}$$

$$? \text{_____} = \text{_____}$$

g X; mol X

The gas-law conversion factor cancels the unwanted mol X:

$$? \text{ g } X = \cancel{\text{mol }} X \times \frac{0.0820(\text{L})(\text{atm})}{\cancel{(\text{mol})}(\text{K})}$$

And the conversion factor given in the problem cancels the unwanted dimension liter:

$$? \text{ g } X = \cancel{\text{mol }} X \times \frac{0.0820 \text{ (L)}(\text{atm})}{\cancel{(\text{mol})}(\text{K})} \times \text{_____}$$

$$\frac{2.4 \text{ g } X}{3.0 \text{ L}}$$

This gives you g X in the numerator, which is what you want for your answer. But two unwanted dimensions remain: _____ and

atm

K

$$300 \text{ K}; \frac{1}{1.0 \text{ atm}}$$

__. These can be canceled using the information given. You multiply by

_____ and _____. (Notice that you must use the reciprocal of the pressure so that the dimensions cancel.)

$$? \text{ g } X = \cancel{\text{mol}} X \times \frac{0.0820 \ \cancel{(L)} \cancel{(\text{atm})}}{\cancel{(\text{mol})} \cancel{(K)}} \times \frac{2.4 \text{ g } X}{3.0 \cancel{L}} \times 300 \cancel{K} \times \frac{1}{1.0 \ \cancel{\text{atm}}}$$

$$= 20 \text{ g } X$$

You now know that one mole of X weighs 20 g. That is, its molecular weight is 20 g/mol.

If you like to remember formulas, you can see that the above is really

$$\frac{? \text{ g } X}{\text{mol } X} = \frac{(\text{weight sample})}{V} \times 0.0820 \times \frac{T}{P} = \frac{(g)RT}{PV}$$

But make sure that you express everything in kelvins, liters, and atmospheres!

We can do question 10 on the pretest by using this formula. What is the molecular weight of a gas whose density is 4.50 g/L at $-73°$C and a pressure of 380 mmHg?

g (the weight of gas) $= 4.50$ g

1.00

V (the volume of gas) $=$ _____ L

200

T (the temperature) $=$ _____ K

0.500

P (the pressure) $=$ _____ atm

0.0820; 148

0.500

$$\frac{? \text{ g } X}{\text{mol } X} = \frac{(g)RT}{PV} = \frac{4.50 \times \text{\_\_\_\_\_} \times 200}{\text{\_\_\_\_\_} \times 1.00} = \frac{\text{\_\_\_\_\_} \text{ g } X}{\text{mol } X}$$

SECTION E Mixtures of Gases; Dalton's Law

When you mix several gases in a container, each gas acts as if the others were not present. Each gas exerts its own pressure (called the *partial pressure p*). Consequently, the total pressure in the container equals the sum of all the partial pressures. This is called *Dalton's law.* Mathematically, it is

$$P = p_1 + p_2 + p_3 + \cdots$$

The p_1, p_2, p_3 refer to the partial pressures (individual pressures) of gas 1, gas 2, gas 3, and so forth.

Question 11 illustrates this phenomenon. What is the pressure in a 4.0 L container that is at 27°C and which contains 2.0 mol of H_2 gas and 5.0 mol of He gas? You must determine the pressure of each gas separately. Then you simply add all the pressures to get the total pressure.

$? \text{ atm } H_2 = (2.0 \text{ mol } H_2)(300 \text{ K}) \times 0.0820(\underline{\hspace{2cm}})$

$$\frac{(L)(atm)}{(mol\ H_2)(K)}$$

$\times \underline{\hspace{2cm}} = \underline{\hspace{1cm}} \text{ atm } H_2$

$\dfrac{1}{4.0\ L}; \ 12.3$

$? \text{ atm He} = \underline{\hspace{1cm}} \text{ atm } \underline{\hspace{1cm}}$

$30.8; \ He$

(You calculate this, then check your answer.) To get the total pressure in the container, you must _____ the partial pressures. When you express this to the correct number of significant figures, you get

add

$P_{total} = \underline{\hspace{2.5cm}} \text{ atm}$

$43, \text{ or } 4.3 \times 10^1$

This comes up experimentally when you collect a gas over some liquid. Every liquid has a certain amount of its own gas phase associated with it. The amount (pressure) depends on the temperature. You can look up this vapor pressure of liquids in a table, either in your text or in the *Handbook of Chemistry and Physics* (Boca Raton, Florida: The Chemical Rubber Co.). Experimentally, you will measure the total pressure. However, this total pressure is due to the gas plus the vapor of the liquid. In order to get the pressure of the gas, you must _____ the liquid's vapor pressure from the total pressure that was determined experimentally.

subtract

Question 12 in the pretest is a good example of this type of problem. What is the molecular weight of a gas if 70.0 mL of it collected over water at a temperature of 27°C has a total pressure of 756.5 mmHg? The vapor pressure of water at this temperature is 26.5 mmHg. The weight of dry gas is 0.080 g. First you calculate the partial pressure of the unknown gas by _____ the partial pressure of water from the total pressure.

subtracting

$p_x = 756.5 \text{ mmHg} \underline{\hspace{2cm}} = \underline{\hspace{1cm}} \text{ mmHg}$

$- 26.5 \text{ mmHg}; \ 730.0$

If you want to use the memorized formula for molecular weight,

$$\frac{? g X}{mol\ X} = \frac{(g)\ RT}{PV}$$

you must convert the pressure to _____.

atmospheres

$p_x \text{ atm} = 730.0 \text{ mmHg} \times \underline{\hspace{2cm}} = \underline{\hspace{2cm}}$

$\dfrac{atm}{760.0\ mmHg}; \ 0.9605 \text{ atm}$

To save one step, you can express the volume immediately in liters. To get the molecular weight, you have

$$\frac{? g X}{mol\ X} = \frac{(0.080)(0.0820)(\underline{\hspace{1cm}})}{(0.9605)(\underline{\hspace{1cm}})} = \underline{\hspace{1cm}} \frac{g\ X}{mol\ X}$$

$300; \ 29$

0.0700

(Notice the number of significant figures.)

A somewhat more complicated problem is the following. A 6.0 L container at 127°C has a total pressure of 12.0 atm. It contains 0.50 g of He. How many moles of O_2 does it contain if that is the only other gas present?

The first thing you must determine is the partial pressure of He.

4.0 g He

$$p_{He} \text{ atm} = (0.50 \text{ g He})(400 \text{ K}) \left(\dfrac{\text{mol He}}{\underline{\hspace{1cm}}}\right)$$

$$\dfrac{(L)(atm)}{(mol\ He)(K)}$$

$$\times\ 0.0820\ (\underline{\hspace{2cm}}) \times \left(\dfrac{1}{6.0\ L}\right)$$

0.68

$$= \underline{\hspace{1cm}} \text{ atm}$$

subtracting

Now you can determine the partial pressure of O_2 by _____ the

total

partial pressure of He from the _____ pressure:

0.68 atm

$$p_{O_2} = 12.0 \text{ atm} - \underline{\hspace{2cm}} = 11.3 \text{ atm}$$

Now it is simple to calculate the number of moles of O_2:

$$\dfrac{(mol\ O_2)(K)}{0.0820\ (L)(atm)}$$

$$? \text{ mol } O_2 = (11.3 \text{ atm})(6.0 \text{ L})(\underline{\hspace{2cm}})$$

$$\dfrac{1}{400\ K};\ 2.1$$

$$\times (\underline{\hspace{1cm}}) = \underline{\hspace{1cm}} \text{ mol } O_2$$

SECTION F Diffusion of Gases

The rate at which a gas diffuses is inversely proportional to the square root of its molecular weight (or as it was first noted, to its density). This is called *Graham's law.* The diffusion of two different gases (at the same pressure and temperature) can be mathematically stated as

$$\dfrac{r_1}{r_2} = \dfrac{\sqrt{M_2}}{\sqrt{M_1}} = \sqrt{\dfrac{M_2}{M_1}}$$

where r_1 and r_2 are the rates of diffusion of gases 1 and 2 and M_1 and M_2 are their respective molecular weights.

Question 13 on the pretest is typical of the type of problem you will come across. A sample of N_2 gas diffuses through a small hole at a rate of 2.6 mL/min. At what rate would Cl_2 gas diffuse through the same hole under the same conditions?

28.0

The molecular weight of N_2 is _____ g/mol. The molecular weight of

71.0; 2.6

Cl_2 is _____ g/mol. And the rate of diffusion of N_2 is _____ mL/min. Using Graham's law, you have

$$\dfrac{r_{Cl_2}}{r_{N_2}} = \sqrt{\dfrac{M_{N_2}}{M_{Cl_2}}}$$

Putting in the specified values, you have

$$\dfrac{r_{Cl_2}}{2.6 \text{ mL/min}} = \sqrt{\dfrac{28.0}{71.0}} = \sqrt{0.394} = 0.628$$

Cross-multiplying gives you

$$r_{Cl_2} = 2.6 \times 0.628 \text{ mL/min} = 1.6 \text{ mL/min}$$

(If you cannot determine square roots on your calculator, Section B in Chapter 1 shows how to do it with logarithms.)

Question 14 on the pretest is another way of presenting the problem. Here the rates of diffusion are known and you want to determine the molecular weight.

"Suppose that 20.0 mL of an unknown gas diffuses through a pinhole in the same time that it takes 40.0 mL of CH_4 to diffuse through the same pinhole under the same conditions. What is the molecular weight of the unknown gas?" The rate is amount per time. (A word of caution: The rate can be expressed in moles per unit time or volume per unit time, but *not* in grams per unit time.) Since both gases diffuse in the same time, we can say that the rates are expressed simply as the amount.

Rate CH_4 = _____ Rate X = _____	40.0 mL; 20.0 mL
Molecular weight CH_4 = _____ g/mol	16.0

Therefore we can write Graham's law as

$$\frac{r_{CH_4}}{\underline{\quad}} = \frac{\sqrt{M\underline{\quad}}}{\sqrt{M\underline{\quad}}}$$

X

r_X; CH_4

Putting in the specified values, we have

$$\frac{\underline{\qquad}}{\underline{\qquad}} = \frac{\sqrt{M_X}}{\underline{\quad}} \qquad \text{so} \qquad \underline{\qquad} = \sqrt{M_X}$$

$\dfrac{40.0\ \text{mL}}{20.0\ \text{mL}}$; $\sqrt{16}$; 8.00

$$\underline{\qquad} = M_X \qquad \text{(in grams per mole)}$$

64.0

PROBLEM SET

You may want to omit problems marked with an asterisk, since they involve extremely complicated calculations.

1. If 333 mL of an ideal gas at a temperature of 25°C and a pressure of 750 torr has its temperature lowered to −11°C and its pressure lowered to 730 mmHg, what will its new volume be in milliliters?

2. A cylinder fitted with a movable piston contains 2.0 L of an ideal gas. The pressure of the gas is 750 mmHg and its temperature is 0.0°C. The piston is lifted so that the volume in the cylinder becomes 2.5 L and the pressure drops to 740 mmHg. What will the temperature be in °C?

3. A sample of an ideal gas occupies 0.50 L at 0.99 atm pressure and a temperature of 273 K. It is expanded until its volume is 755 mL and its temperature is 0°C. What will its pressure be in torrs?

4. The first step in the Ostwald process for the production of nitric acid produces NO gas from the oxidation of NH_3. If 333 mL of NO at a temperature of 298 K and a pressure of 755 mmHg are prepared, what volume NO will this be equal to at STP?

*5. How many cubic feet will a sample of gas that occupies 2.5 ft^3 at 68°F occupy at 300°F if the pressure is held constant?

6. What pressure in atmospheres will be exerted by 0.312 mol of an ideal gas at 42°C in a volume of 58.4 mL?

7. How many moles of a gas occupy 5.0 L at STP?

8. At what temperature in kelvins must you hold 12.8 g of O_2 gas to have a pressure of 0.20 atm and a volume of 0.50 L?

9. What is the density in grams per liter of Cl_2 gas at STP?

10. A chemist produces N_2 gas by heating $(NH_4)_2Cr_2O_7$ and traps the gas in a 6.0 L container held at −23°C. The pressure of the gas is 2.0 atm. How many grams of N_2 were prepared?

11. What is the molecular weight of a gas, given that 268 mL of it, at 69°C and 17.9 torr pressure, weigh 0.0156 g?

12. The density of an unknown gas is found to be 2.66 g/L at 289.3 K and a pressure of 0.980 atm. What is the molecular weight of the gas?

13. It is found that at 25°C the density of N_2O gas is 1.73 g/L. At what pressure is it being kept (in atmospheres)?

*14. What volume in milliliters does 1.00 g of C_8H_{18} occupy at 735 torr and 210.2°F?

15. How many grams of acetylene, C_2H_2, occupy 1.00 L at 24°C and a pressure of 742 mmHg?

16. What is the total pressure if a 3.0 L container at −123°C contains 2.8 g N_2, 2.2 g CO_2, and 0.020 g H_2?

17. How many moles of a gas are present if, when the gas is collected over water at 26°C, it has a total pressure of 755 torr and a volume of 3.5 L? The vapor pressure of water at this temperature is 25 mmHg.

*18. A sample of H_2 gas collected over water at 25°C and 1.0 atm total pressure occupies 0.10 L. What volume would the dry gas occupy at this same pressure and temperature? The partial pressure of water at 25°C is 23.8 mmHg.

19. A sample of O_2 diffuses through a small opening at the rate of 3.00 mL/min. At what rate would H_2 diffuse through the same hole under the same conditions?

20. An unknown gas is found to diffuse through a pinhole at the rate of 3.0 mL/min. Through the same pinhole, He is found to diffuse at the rate of 12.0 mL in 2.0 min. What is the molecular weight of the unknown gas?

21. A storage tank for holding nitrogen is equipped with a relief valve that will open and allow gas to escape if the pressure reaches 2.00 atm. The tank registers a pressure of 1.74 atm at 35°C. At what temperature will the valve first open?

22. A chemist determines the molecular weight of an unknown volatile liquid by placing the liquid in a flask with a small opening so that vapor may escape. The flask containing the liquid is then placed in a bath that brings its temperature to 99°C and the liquid is all converted to the gas, the excess gas escaping from the flask. The flask is cooled and the remaining gas condenses back to 0.389 g of liquid. The volume of the flask is 265 mL and the barometric pressure is 740 torr. What is the molecular weight of the unknown liquid? [Note: Since the flask is open to the atmosphere, the pressure inside the flask will be the same as barometric pressure.]

23. A 2.0 L container at 27°C contains 1.5 mol of He gas. Enough O_2 gas is added to bring the total pressure of the container to 23.40 atm. How many grams of O_2 were added?

24. If we used water rather than mercury in a barometer and the pressure was 740 mmHg, how high a column, in feet of water, would be in the water-filled barometer? The density of water is 0.996 g/mL and the density of mercury is 13.5 g/mL.

25. An empty 1.0 L flask is filled by adding to it the contents of a 2.0 L flask of O_2 at 2.0 atm pressure and the contents of a 0.5 L flask of He, also at 2.0 atm. All flasks are at room temperature, 22°C. What will be the pressure of the mixture of gases in the 1.0 L flask?

PROBLEM SET ANSWERS

You should be able to do Problems 1–4. If you cannot, redo Section B. If you do not get four correct answers from Problems 6–10, check Section C. Problems 11 and 12 are simple and you should get them both. They are covered in Section D. You should be able to do at least four problems in the group 13–18 and both Problems 19 and 20, which are covered in Section F. Problems 20–25 cover all areas of the chapter. You will have to decide which section talks about them. They are more difficult. If you can manage three of the five, you're doing fine.

1. $P' = 750$ torr $P'' = 730$ torr $\dfrac{(750)(333)}{298} = \dfrac{(730)(? \text{ mL})}{262}$

   $V' = 333$ mL $V'' = ?$ mL

   $T' = 298$ K $T'' = 262$ K $?$ mL $= 301$ mL

2. $P' = 750$ mmHg $P'' = 740$ mmHg $\dfrac{(750)(2.0)}{273} = \dfrac{(740)(2.5)}{? \text{ K}}$

   $V' = 2.0$ L $V'' = 2.5$ L

   $T' = 273$ K $T'' = ?$ K $?$ K $= 337$ K

   $? \,°C = 64°C$

3. $P' = 0.99$ atm $P'' = ?$ atm $\dfrac{(0.99)(0.50)}{273} = \dfrac{(? \text{ atm})(0.755)}{273}$

   $V' = 0.50$ L $V'' = 0.755$ L

   $T' = 273$ K $T'' = 273$ K $?$ atm $= 0.66$ atm

   $?$ torr $= 5.0 \times 10^2$ torr

4. $P' = 755$ mmHg $P'' = 760$ mmHg $\dfrac{(755)(333)}{298} = \dfrac{(760)(?\text{ mL})}{273}$

   $V' = 333$ mL $V'' = ?$ mL

   $T' = 298$ K $T'' = 273$ K $?$ mL $= 303$ mL

5. $P' = 1$ $P'' = 1$ $\dfrac{(1)(2.5)}{293} = \dfrac{(1)(?\text{ ft}^3)}{422}$

   $V' = 2.5$ ft^3 $V'' = ?$ ft^3

   $T' = 293$ K $T'' = 422$ K $?$ ft^3 $= 3.6$ ft^3

6. $?$ atm $= (0.312 \cancel{\text{ mol}})(315 \cancel{\text{ K}}) \times \dfrac{0.0820 \cancel{(L)}(\text{atm})}{\cancel{(\text{mol})}\cancel{(K)}} \times \dfrac{1}{0.0584 \cancel{L}} = 138$ atm

7. $?$ mol $= (5.0 \cancel{L})(1.0 \cancel{\text{ atm}}) \times \dfrac{(\text{mol})\cancel{(K)}}{0.0820 \cancel{(L)}\cancel{(\text{atm})}} \times \dfrac{1}{273 \cancel{K}} = 0.22$ mol

8. $?$ K $= (0.50 \cancel{L})(0.20 \cancel{\text{ atm}}) \times \dfrac{(\text{mol})(K)}{0.0820 \cancel{(L)}\cancel{(\text{atm})}} \times \dfrac{32.0 \cancel{\text{ g }O_2}}{\cancel{\text{mol }O_2}} \times \dfrac{1}{12.8 \cancel{\text{ g }O_2}} = 3.0$ K

9. $?$ g Cl$_2$ $= (1.00 \cancel{L})(1.00 \cancel{\text{ atm}}) \times \dfrac{\cancel{(\text{mol})}\cancel{(K)}}{0.0820 \cancel{(L)}\cancel{(\text{atm})}} \times \dfrac{71.0 \text{ g Cl}_2}{\cancel{\text{mol Cl}_2}} \times \dfrac{1}{273 \cancel{K}}$

   $= 3.17$ g Cl$_2$ in 1.00 L

   $\left(\text{You might also remember that at STP there are 22.4 L/mol.}\right.$

   $\left.\dfrac{?\text{ g}}{L} = \dfrac{\cancel{\text{mol}}}{22.4 \text{ L}} \times \dfrac{71.0 \text{ g}}{\cancel{\text{mol}}} = \dfrac{3.17 \text{ g}}{L}\right)$

10. $?$ g N$_2$ $= (6.0 \cancel{L})(2.0 \cancel{\text{ atm}}) \times \dfrac{\cancel{(\text{mol})}\cancel{(K)}}{0.0820 \cancel{(L)}\cancel{(\text{atm})}} \times \dfrac{1}{250 \cancel{K}} \times \dfrac{28.0 \text{ g N}_2}{\cancel{\text{mol N}_2}} = 16$ g N$_2$

11. $?$ g $X = \cancel{\text{mol }X} \times \dfrac{0.0820 \cancel{(L)}\cancel{(\text{atm})}}{\cancel{(\text{mol})}\cancel{(K)}} \times \dfrac{0.0156 \text{ g }X}{0.268 \cancel{L}\cancel{X}} \times \dfrac{1}{17.9 \cancel{\text{ torr}}} \times \dfrac{760 \cancel{\text{ torr}}}{\cancel{\text{ atm}}} \times 342 \cancel{K}$

   $= 69.3$ g X in 1 mol X

If you prefer to remember the formula,

$$\text{Molecular weight} = \frac{g\,RT}{PV}$$

then

$$P(\text{atm}) = 17.9 \cancel{\text{ torr}} \times \frac{\text{atm}}{760 \cancel{\text{ torr}}} = 0.0236 \text{ atm}$$

$$\frac{?\text{ g }X}{\text{mol }X} = \frac{(0.0156)(0.0820)(342)}{(0.0236)(0.268)} = 69.2 \text{ g/mol}$$

12. $?$ g $X = \cancel{\text{mol }X} \times \dfrac{0.0820 \cancel{(L)}\cancel{(\text{atm})}}{\cancel{(\text{mol})}\cancel{(K)}} \times \dfrac{2.66 \text{ g }X}{\cancel{L}\cancel{X}} \times \dfrac{1}{0.980 \cancel{\text{ atm}}} \times 289.3 \cancel{K}$

    $= 64.4$ g X per mol X

Or from the formula

$$\text{Molecular weight} = \frac{g\,RT}{PV}$$

the weight is 2.66 g and the volume is 1.00 L.

Molecular weight $= \dfrac{(2.66)(0.0820)(289.3)}{(0.980)(1.00)} = 64.4$ g/mol

13. $? \text{ atm} = \dfrac{1.73 \text{ g } N_2O}{L} \times \dfrac{0.0820 \text{ (L)(atm)}}{\text{(mol)(K)}} \times \dfrac{\text{mol } N_2O}{44.0 \text{ g } N_2O} \times 298 \text{ K} = 0.961$ atm

14. First get the temperature in °C.

$$°C = \frac{5}{9}(°F - 32) = \frac{5}{9}(178.2) = 99.0°C$$

Then convert it to K: $K = 273 + °C = 372$ K.

$? \text{ mL} = (1.00 \text{ g } C_8H_{18})(372 \text{ K}) \times \dfrac{0.0820 \text{(L)(atm)}}{\text{(mol)(K)}} \times \dfrac{\text{mol } C_8H_{18}}{114.0 \text{ g } C_8H_{18}}$

$\times \dfrac{760 \text{ torr}}{\text{atm}} \times \dfrac{1}{735 \text{ torr}} \times \dfrac{\text{mL}}{10^{-3} \text{ L}} = 277$ mL

15. $? \text{ g } C_2H_2 = (1.00 \text{ L})(742 \text{ mmHg}) \times \dfrac{\text{atm}}{760 \text{ mmHg}} \times \dfrac{\text{(mol)(K)}}{0.0820 \text{ (L)(atm)}}$

$\times \dfrac{26.0 \text{ g } C_2H_2}{\text{mol } C_2H_2} \times \dfrac{1}{297 \text{ K}} = 1.04 \text{ g } C_2H_2$

16. $? \text{ atm } N_2 = (150 \text{ K})(2.8 \text{ g } N_2) \times \dfrac{\text{mol } N_2}{28.0 \text{ g } N_2} \times \dfrac{0.0820 \text{ (L)(atm)}}{\text{(mol)(K)}} \times \dfrac{1}{3.0 \text{ L}}$

$= 0.41$ atm

$? \text{ atm } CO_2 = (150 \text{ K})(2.2 \text{ g } CO_2) \times \dfrac{\text{mol } CO_2}{44.0 \text{ g } CO_2} \times \dfrac{0.0820 \text{ (L)(atm)}}{\text{(mol)(K)}} \times \dfrac{1}{3.0 \text{ L}}$

$= 0.205$ atm

$? \text{ atm } H_2 = (150 \text{ K})(0.020 \text{ g } H_2) \times \dfrac{\text{mol } H_2}{2.02 \text{ g } H_2} \times \dfrac{0.0820 \text{ (L)(atm)}}{\text{(mol)(K)}} \times \dfrac{1}{3.0 \text{ L}}$

$= 0.041$ atm

$P_{\text{total}} = p_{N_2} + p_{CO_2} + p_{H_2} = 0.41 + 0.205 + 0.041 = 0.66$ atm

17. $p_{\text{gas}} = 755 \text{ torr} - 25 \text{ torr} = 730$ torr

$? \text{ mol} = (3.5 \text{ L})(730 \text{ torr}) \times \dfrac{\text{atm}}{760 \text{ torr}} \times \dfrac{\text{(mol)(K)}}{0.0820 \text{ (L)(atm)}} \times \dfrac{1}{299 \text{ K}}$

$= 0.14$ mol

18. First determine the number of moles of gas, as in Problem 17.

$? \text{ mol} = (0.10 \text{ L}) \times (760 - 23.8) \text{ mmHg}$

$\times \dfrac{\text{atm}}{760 \text{ mmHg}} \times \dfrac{\text{(mol)(K)}}{0.0820 \text{ (L)(atm)}} \times \dfrac{1}{298 \text{ K}}$

$= 0.0040$ mol

Then determine the volume that this many moles will occupy.

$? \text{ L} = (0.0040 \text{ mol})(298 \text{ K}) \times \dfrac{0.0820 \text{ (L)(atm)}}{\text{(mol)(K)}} \times \dfrac{1}{1.0 \text{ atm}} = 0.098$ L

19. $\dfrac{r_{H_2}}{r_{O_2}} = \sqrt{\dfrac{M_{O_2}}{M_{H_2}}}$ $\dfrac{r_{H_2}}{3.00 \text{ mL/min}} = \sqrt{\dfrac{32}{2}} = \sqrt{16} = 4.0$

   $r_{H_2} = 3.00 \text{ mL/min} \times 4.0 = 12 \text{ mL/min}$

20. $\dfrac{r_{He}}{r_X} = \dfrac{\sqrt{M_X}}{\sqrt{M_{He}}}$ $\dfrac{12.0 \text{ mL/2.0 min}}{3.0 \text{ mL/min}} = \dfrac{\sqrt{M_X}}{\sqrt{4.0}}$

    $\dfrac{6.0}{3.0} = \dfrac{\sqrt{M_X}}{2.0}$ $4.0 = \sqrt{M_X}$ $16 \text{ g/mol} = M_X$

21. $P' = 1.74$ atm $P'' = 2.00$ atm $\dfrac{(1.74)(1)}{308} = \dfrac{(2.00)(1)}{? \text{ K}}$

    $V' = 1$ $V'' = 1$ (volume

    constant) $? \text{ K} = 354 \text{ K}$

    $T' = 308 \text{ K}$ $T'' = ? \text{ K}$ $?°C = 81°C$

22. The question resolves itself to determining the molecular weight if you know that the volume is 265 mL, the pressure is 740 torr, the temperature is 99°C, and the weight is 0.389 g.

    $? \text{ g } X = \cancel{\text{mol } X} \times \dfrac{0.0820 \, \cancel{(L)}\cancel{(atm)}}{\cancel{(mol \, X)}\cancel{(K)}} \times \dfrac{0.389 \text{ g } X}{0.265 \, \cancel{L}} \times \dfrac{372 \, \cancel{K}}{740 \, \cancel{torr}} \times \dfrac{760 \, \cancel{torr}}{\cancel{atm}}$

    $= 46.0 \text{ g } X \text{ per mol } X$

    If you prefer to remember the formula,

    $$\text{Molecular weight} = \dfrac{g \, X}{\text{mol } X} = \dfrac{g \, RT}{PV} = \dfrac{(0.389)(0.0820)(372)}{\left(\dfrac{740}{760}\right)(0.265)}$$

    $$= 46.0 \, \dfrac{g \, X}{\text{mol } X}$$

23. First you determine the pressure of He.

    $? \text{ atm He} = 1.5 \, \cancel{\text{mol He}} \times \dfrac{(0.0820)\cancel{(L)}(\text{atm He})}{\cancel{(mol \, He)}\cancel{(K)}} \times \dfrac{300 \, \cancel{K}}{2.0 \, \cancel{L}} = 18.45 \text{ atm He}$

    The total pressure equals the sum of the partial pressures:

    $P_{total} = p_{He} + p_{O_2}$

    So the pressure of O_2 is

    $p_{O_2} = P_{total} - p_{He} = 23.40 \text{ atm total} - 18.45 \text{ atm He} = 4.95 \text{ atm } O_2$

    You then calculate the weight of O_2 knowing that the volume and temperature are the same as before:

    $? \text{ g } O_2 = 2.0 \, \cancel{L} \times \dfrac{\cancel{(mol \, O_2)}\cancel{(K)}}{0.0820 \, \cancel{(L)}\cancel{(atm \, O_2)}} \times \dfrac{4.95 \, \cancel{\text{atm } O_2}}{300 \, \cancel{K}} \times \dfrac{32.0 \text{ g } O_2}{\cancel{\text{mol } O_2}} = 12.9 \text{ g } O_2$

24. The pressure is equal to the height of a column of liquid times the density of the liquid that the pressure can support. Therefore you can write (since the pressure is the same for the water and the mercury)

Height mercury $\times$ density mercury = height water $\times$ density water

Or, by cross-multiplying, you get

Height water = height mercury $\times$ $\dfrac{\text{density mercury}}{\text{density water}}$

$$= 740 \text{ mm} \times \frac{13.5}{0.996} = 1.00 \times 10^4 \text{ mm}$$

Converting this to feet, you have

$$? \text{ ft} = 1.00 \times 10^4 \cancel{\text{ mm}} \times \frac{10^{-3} \cancel{\text{ m}}}{\cancel{\text{ mm}}} \times \frac{\cancel{\text{ cm}}}{10^{-2} \cancel{\text{ m}}} \times \frac{\cancel{\text{ inch}}}{2.54 \cancel{\text{ cm}}} \times \frac{\text{ft}}{12 \cancel{\text{ inch}}} = 32.8 \text{ ft}$$

25. The easiest method is to calculate the number of moles of He and O_2 and determine the pressure that the total number of moles will produce in the 1.0 L flask.

$$? \text{ mol } O_2 = 2.0 \cancel{\text{ L } O_2} \times \frac{(\text{mol } O_2)\cancel{(K)}}{0.0820 \cancel{(L\ O_2)}\cancel{(\text{atm})}} \times \frac{2.0 \cancel{\text{ atm } O_2}}{295 \cancel{\text{ K}}} = 0.165 \text{ mol } O_2$$

$$? \text{ mol He} = 0.50 \cancel{\text{ L He}} \times \frac{(\text{mol He})\cancel{(K)}}{0.0820 \cancel{(L\ He)}\cancel{(\text{atm})}} \times \frac{2.0 \cancel{\text{ atm}} \text{ He}}{295 \cancel{\text{ K}}} = 0.041 \text{ mol He}$$

Total moles = 0.165 + 0.041 = 0.21 mol, to the correct significant figures.

Then you can get the pressure in the 1.0 L flask.

$$? \text{ atm} = 0.21 \cancel{\text{ mol}} \times \frac{0.0820 \cancel{(L)}(\text{atm})}{\cancel{(\text{mol})}\cancel{(K)}} \times \frac{295 \cancel{\text{ K}}}{1.0 \cancel{\text{ L}}} = 5.1 \text{ atm}$$

CHAPTER 13

Stoichiometry Involving Gases

PRETEST

These questions should be fairly easy if you remember that in all stoichiometry problems you head toward a bridge conversion factor in moles. You should be able to do these in 20 minutes.

1. How many liters of O_2 gas at 2.0 atm pressure and $27°C$ are consumed by the complete reaction of 8.0 g H_2 according to the reaction

    $2H_2 + O_2 = 2H_2O$

2. How many grams of water can be produced by the complete reaction of 5.0 L of O_2 at 6.0 atm pressure and $27°C$ according to the same reaction as in question 1?

3. How many liters of H_2 gas at $123°C$ and 740 mmHg pressure are required to react with 5.0 L of O_2 at 2.0 atm pressure and $27°C$ according to the same reaction as in question 1?

4. How many liters of CO are formed by the reaction of 5.0 L of O_2 if both gases are at 2.0 atm and 300 K? The reaction is

    $2C + O_2 = 2CO$

5. How many liters of HCl gas at 2.0 atm pressure and 300 K are needed to produce 5.0 L of CO_2 gas at 1520 torr and $27°C$ according to the equation

    $CaCO_3 + 2HCl = CaCl_2 + CO_2 + H_2O$

6. What is the percent purity of a sample of iron ore if 1.25 g of sample, when reacted with an excess of HCl, produces 220 mL of H_2 gas collected over water at a total pressure of 1.00 atm and $26°C$? The vapor pressure of water at this temperature is 25 mmHg. The reaction is

    $2Fe + 6HCl = 2FeCl_3 + 3H_2$

7. How many liters of Cl_2 gas at STP are produced by the reaction of 30.0 mL of a 12.0 M solution of HCl? (That is, there are 12.0 mol HCl per liter of solution.) There is an excess of $KMnO_4$. The reaction is

    $2KMnO_4 + 16HCl = 2KCl + 2MnCl_2 + 5Cl_2 + 8H_2O$

PRETEST ANSWERS

If you didn't answer all the questions correctly, locate them in the body of the chapter and see where you went wrong. If you are totally lost with respect to how to approach these problems, you had better go through the entire chapter. It is quite short. If you got them all correct, go straight to the problem set at the end of the chapter.

1. 25 L O_2
2. 43 g H_2O
3. 27 L H_2
4. 10 L CO
5. 10 L HCl
6. 25.8% pure
7. 2.52 L Cl_2

When you have chemical reactions involving gases, it is possible to have amounts of reactants or products not expressed in weights or moles, but rather as the volume (or pressure or temperature) of a gas. Chapter 12 showed how to convert from the variables that define a gas to get moles or grams.

Actually, there are two main types of problems. In the first, all three parameters that define the gas (P, V, and T) are given. We will examine this type in Section A.

SECTION A Problems in which P, V, and T Are Given

It is easy to recognize this type of problem, since you find pressure, volume, and temperature of the gas in the information given. You must solve the problem in two steps. First, from the P, V, and T you can

moles calculate the _____ of the gas. Once you have this, you use it as your information given in the second step, which is to solve for the substance of interest. Question 2 on the pretest is an example of this type.

How many grams of H_2O can be produced by the complete reaction of 5.0 L of O_2 gas at 6.0 atm pressure and 27°C according to the reaction

$$2H_2 + O_2 = 2H_2O$$

5.0 L As you see, all three variables are given. The volume is _____, the

temperature; 300 pressure is 6.0 atm, and the _____ is _____ K. You therefore

moles solve for the number of _____ of O_2.

(mol)(K); $\dfrac{1}{300\ K}$

$$? \text{ mol } O_2 = (5.0 \text{ L})(6.0 \text{ atm}) \times \frac{\rule{2cm}{0.4pt}}{0.0820\ (L)(atm)} \times \rule{3cm}{0.4pt}$$

1.2

$$= \rule{1.5cm}{0.4pt} \text{ mol } O_2$$

The 1.2 mol O_2 is now given information, so you can do step 2:

$$? \text{ g } H_2O = 1.2 \text{ mol } O_2 \times \frac{\rule{3cm}{0.4pt}}{\text{mol } O_2} \times \frac{18.0 \text{ g } H_2O}{\text{mol } H_2O}$$

$$= 43 \text{ g } H_2O \quad \text{(to two significant figures)}$$

2 mol H_2O

You can see how simple step 2 is. Your information given is in moles, which brings you immediately to your bridge conversion factor from the balanced equation.

If you are good at dimensional analysis, you should see that it is possible to do this problem in only one step. But note this: You must label eacn dimension to show to what substance it refers. (This is even more important when you are doing problems with two different gases.) Let's see how the problem would go. The question asked is: "How many grams of water?" and the information given is 5.0 L O_2. You start in the normal way:

$$? \rule{2cm}{0.4pt} = \rule{2cm}{0.4pt}$$

g H_2O; 5.0 L O_2

Then you use the gas-law conversion factor to cancel the unwanted dimension of the information given:

$$? \text{ g } H_2O = 5.0 \text{ L } O_2 \times \rule{3cm}{0.4pt}$$

$$\frac{\text{mol } O_2 \text{ K}_{O_2}}{0.0820 \text{ L } O_2 \text{ atm}_{O_2}}$$

(Notice that, since the gas-law conversion factor is referring to O_2 gas, all

dimensions are for O_2.) What you really want is _____ so

mol O_2

that you can use the bridge in the balanced equation. Therefore _____

K_{O_2}

and _____ are unwanted dimensions. But this is information given in the problem; there are 6.0 atm$_{O_2}$ and 300 K$_{O_2}$. You just put these in in their proper place to cancel the unwanted dimensions.

atm$_{O_2}$

$$? \text{ g } H_2O = 5.0 \, \cancel{\text{L } O_2} \times \frac{\text{mol } O_2 \text{ K}_{O_2}}{0.0820 \, \cancel{\text{L } O_2} \text{ atm}_{O_2}} \times \frac{\rule{2cm}{0.4pt}}{\rule{2cm}{0.4pt}}$$

6.0 atm$_{O_2}$

300 K$_{O_2}$

Now the only unwanted dimension is _____, and so you can use your bridge to get to the desired H_2O:

moles O_2

$$? \text{ g } H_2O = 5.0 \, \cancel{\text{L } O_2} \times \frac{\text{mol } O_2 \, \cancel{\text{K}_{O_2}}}{0.0820 \, \cancel{\text{L } O_2} \, \cancel{\text{atm}_{O_2}}} \times \frac{6.0 \, \cancel{\text{atm}_{O_2}}}{300 \, \cancel{\text{K}_{O_2}}} \times \rule{2cm}{0.4pt}$$

$$\frac{2 \text{ mol } H_2O}{\text{mol } O_2}$$

Now all that remains to be done is to get from moles H_2O to _____, the dimensions of the answer. But this is easy. You just use the

grams H_2O

_____ of H_2O:

molecular weight

$$? \text{ g } H_2O = 5.0 \, \cancel{\text{L } O_2} \times \frac{\cancel{\text{mol } O_2} \, \cancel{\text{K}_{O_2}}}{0.0820 \, \cancel{\text{L } O_2} \, \cancel{\text{atm}_{O_2}}} \times \frac{6.0 \, \cancel{\text{atm}_{O_2}}}{300 \, \cancel{\text{K}_{O_2}}}$$

$$\times \frac{2 \text{ mol } H_2O}{\cancel{\text{mol } O_2}} \times \rule{2cm}{0.4pt}$$

$$\frac{18.0 \text{ g } H_2O}{\text{mol } H_2O}$$

All dimensions now cancel except for the desired grams H_2O, and so the answer will be

$$? \text{ g } H_2O = 5.0 \times \frac{1}{0.0820} \times \frac{6.0}{300} \times 2 \times 18.0 \text{ g } H_2O$$

$$= 43.9 \text{ g } H_2O, \text{ or } 44 \text{ g } H_2O$$

(to the correct number of significant figures). [This points up something very interesting. When we used the two-step method, the answer was 43 g H_2O to the correct number of significant figures. This happened because we rounded off after the first step (mol O_2). We said that there were 1.2 mol O_2 (to two significant figures). Actually there were 1.2195 mol O_2. Had we used this value and not rounded off until after step 2, we would have gotten the same 44 g H_2O as in the second method. So it is best not to round off until the very end. Of course, since the last signifi- cant figure is a guess anyhow, whether it is 43 g H_2O or 44 g H_2O is a moot point.]

Question 6 on the pretest is simply a more complex example of this type. What is the percent purity of an iron-containing sample if 1.25 g of sample, when reacted with an excess of HCl, produces 220 mL of H_2 gas collected over water at a total pressure of 1.00 atm and 26°C? The vapor pressure of water at this temperature is 25 mmHg. The reaction is

$$2Fe + 6HCl = 2FeCl_3 + 3H_2$$

When you look at the problem, you will note that all three variables,

P; V; T	__, __ and __, are given for the H_2 gas. Your first step, then, is to
moles	calculate the number of _____ of H_2. Write down these values for the variables.
299; 0.220	$T =$ _____ K $V =$ _____ L
735	$p_{H_2} =$ _____ mmHg (You get this by subtracting the partial
water; total	pressure of _____ from the _____ pressure expressed in mmHg.)
0.220 L	$? \text{ mol } H_2 = ($_____$)(735 \text{ mmHg})$
$\dfrac{\text{atm}}{760 \text{ mmHg}}; \dfrac{1}{299 \text{ K}}$	$\times$ _____ $\times \dfrac{(\text{mol})(K)}{0.0820(L)(\text{atm})} \times$ _____
8.68×10^{-3}	$=$ _____ $\text{mol } H_2$
8.68×10^{-3}	You now know that 1.25 g of impure sample produces _____ mol of H_2. This is the conversion factor given in the problem. Rewording the problem to ask how many grams of Fe there are in 100 g impure sample, you start with
100 g imp	$? \text{ g Fe} =$ _____
	To remove the unwanted dimension grams impure, you can multiply by
$\dfrac{8.68 \times 10^{-3} \text{ mol } H_2}{1.25 \text{ g imp}}$	_____ , the conversion factor given in the problem. Now
bridge	you can use your _____. So you multiply by the conversion
$\dfrac{2 \text{ mol Fe}}{3 \text{ mol } H_2}$	factor, _____. Finally, to get the desired dimension grams Fe,
$\dfrac{55.8 \text{ g Fe}}{\text{mol Fe}}$	you multiply by _____. The answer, then, is

$$? \text{ g Fe} = 100 \times \frac{8.68 \times 10^{-3}}{1.25} \times \frac{2}{3} \times 55.8 \text{ g Fe} = 25.8 \text{ g Fe} = 25.8\%$$

This problem is more difficult because it involves a percent purity, but in the first step you must still calculate the number of moles of the gas. It then becomes part of the conversion factor given in the problem.

Once again the problem can be done in a single step. The question of the purity of the sample can be reworded to ask "How many grams of

Fe in _____?" This provides both the question asked and the infor- | 100 g imp
mation given. The conversion factor and additional information in the problem is that 220 mL of H_2 gas were produced by 1.25 g impure sample and that the H_2 gas was at 299 K and 735 mmHg (gotten by subtracting the partial pressure of water from the total pressure).

Starting in the usual way, you write

$$? \text{ g Fe} = \text{_____}$$

100 g imp

The conversion factor relating grams impure to milliliters H_2 cancels the unwanted dimension:

$$? \text{ g Fe} = 100 \text{ g imp} \times \text{_____}$$

$\dfrac{220 \text{ mL } H_2}{1.25 \text{ g imp}}$

You will want to use the gas-law conversion factor to cancel the volume of H_2, but the volume must be expressed in _____. So

liters H_2

$$? \text{ g Fe} = 100 \cancel{\text{ g imp}} \times \frac{220 \cancel{\text{ mL } H_2}}{1.25 \cancel{\text{ g imp}}} \times \frac{10^{-3} \text{ L } H_2}{\cancel{\text{mL } H_2}}$$

Now you can multiply by the gas-law conversion factor to cancel L H_2

$$? \text{ g Fe} = 100 \cancel{\text{ g imp}} \times \frac{220 \cancel{\text{ mL } H_2}}{1.25 \cancel{\text{ g imp}}} \times \frac{10^{-3} \text{ L } H_2}{\cancel{\text{mL } H_2}} \times \text{\_\_\_\_}$$

$\dfrac{\text{mol } H_2 \text{ K}_{H_2}}{0.0820 \text{ L } H_2 \text{ atm}_{H_2}}$

Once again you use the pressure and temperature of the H_2 given in the problem to cancel these unwanted dimensions.

$$? \text{ g Fe} = 100 \cancel{\text{ g imp}} \times \frac{220 \cancel{\text{ mL } H_2}}{1.25 \cancel{\text{ g imp}}} \times \frac{10^{-3} \cancel{\text{ L } H_2}}{\cancel{\text{mL } H_2}}$$
$$\times \frac{\text{mol } H_2 \text{ K}_{H_2}}{0.0820 \cancel{\text{ L } H_2} \text{ atm}_{H_2}} \times \text{\_\_\_\_}$$

$\dfrac{735 \text{ mmHg}_{H_2}}{299 \text{ K}_{H_2}}$

There is a small problem. The pressure given in the problem is expressed in millimeters of mercury, but the pressure in the gas-law conversion

factor is in _____. We must convert, using the relationship

atmospheres

$$\frac{\text{\_\_\_\_ mmHg}}{\text{atm}}$$

760

Using this conversion factor removes both the atmospheres and the millimeters of mercury. Watch the way this goes:

$$? \text{ g Fe} = 100 \, \text{g imp} \times \frac{220 \, \text{mL H}_2}{1.25 \, \text{g imp}} \times \frac{10^{-3} \, \text{L H}_2}{\text{mL H}_2}$$

$$\times \frac{\text{mol H}_2 \, \text{K}_{\text{H}_2}}{0.0820 \, \text{L H}_2 \, \text{atm}_{\text{H}_2}} \times \frac{735 \, \text{mmHg}_{\text{H}_2}}{299 \, \text{K}_{\text{H}_2}} \times \underline{\hspace{2cm}}$$

$$\frac{\text{atm}_{\text{H}_2}}{760 \, \text{mmHg}_{\text{H}_2}}$$

Now the only remaining dimension is moles H_2. So you can use the

bridge

_____ from the balanced equation.

$$? \text{ g Fe} = 100 \, \text{g imp} \times \frac{220 \, \text{mL H}_2}{1.25 \, \text{g imp}} \times \frac{10^{-3} \, \text{L H}_2}{\text{mL H}_2} \times \frac{\text{mol H}_2 \, \text{K}_{\text{H}_2}}{0.0820 \, \text{L H}_2 \, \text{atm}_{\text{H}_2}}$$

$$\frac{2 \, \text{mol Fe}}{3 \, \text{mol H}_2}$$

$$\times \frac{735 \, \text{mmHg}_{\text{H}_2}}{299 \, \text{K}_{\text{H}_2}} \times \frac{\text{atm}_{\text{H}_2}}{760 \, \text{mmHg}_{\text{H}_2}} \times \underline{\hspace{2cm}}$$

moles Fe

You now have as the only unwanted dimension _____ and it is easy to get to the desired grams Fe. You just multiply by the atomic

$$\frac{55.8 \, \text{g Fe}}{\text{mol Fe}}$$

weight, _____.

So, after canceling all dimensions, you have

$$? \text{ g Fe} = 100 \times \frac{220}{1.25} \times 10^{-3} \times \frac{1}{0.0820} \times \frac{735}{299} \times \frac{1}{760} \times \frac{2}{3} \times 55.8 \text{ g Fe}$$

$$= 25.8 \text{ g Fe in 100 g imp} \quad \text{or} \quad 25.8\% \text{ pure}$$

Whether you prefer to do these problems in two separate steps or in one long series of conversions is up to you. There are some short cuts that you can take, such as immediately expressing the volume in liters and the pressure in atmospheres.

SECTION B Problems in which One of the Variables, *P*, *V*, or *T*, Is Missing

The second type of problem you will see when you are dealing with reactions of gases is one in which only two of the three variables of a gas are given. You are asked to find the third. Once again it is possible either to do the problems in one step or to break them up into two separate steps. However, unlike the problems in the preceding section, where all variables are given, if you try to solve for the unknown temperature you will find that no one-step procedure will produce K in the numerator.

You can solve only for $\frac{1}{K}$. Problem 13, at the end of this chapter, is an example of this special case. None of the examples that follow have this complication. We'll see the one-step solution as well as the two-step method.

When you work with the two-step method, the first step uses the balanced equation to determine the number of moles of the gas. Then in the second step, you use the gas-law calculation to determine the missing variable.

Question 1 on the pretest is a simple example of this type of problem. How many liters of O_2 gas at 2.0 atm pressure and 27°C are consumed by the complete reaction of 8.0 g H_2 according to the reaction

$$2H_2 + O_2 = 2H_2O$$

As you can see, the problem gives only two of the variables for the O_2 and asks you to determine the third. So, your first step is to determine the number of moles of O_2 from the equation.

$$? \text{ mol } O_2 = 8.0 \text{ g } H_2 \times \underline{\hspace{2cm}} \times \frac{\text{mol } O_2}{2 \text{ mol } H_2} = \underline{\hspace{1cm}} \text{ mol } O_2$$

$\dfrac{\text{mol } H_2}{2.02 \text{ g } H_2}$; 1.98

Now you can determine the volume of O_2, since you know the number of moles, the pressure, and the temperature.

$$? \text{ L } O_2 = (1.98 \text{ mol})(\underline{\hspace{1cm}}) \times \frac{0.0820 \text{ (L)(atm)}}{\text{(mol)}(K)}$$

300 K

$$\times \underline{\hspace{2cm}} = 24 \text{ L } O_2 \text{ to correct significant figures}$$

$\dfrac{1}{2.0 \text{ atm}}$

The one-step solution would have as the question asked, "How many

$\underline{\hspace{4cm}}$?" and 8.0 g H_2 as the information given:

liters of O_2

$$? \text{ L } O_2 = \underline{\hspace{2cm}}$$

8.0 g H_2

Then, as in any stoichiometry problem, so that you can use the bridge,

you want $\underline{\hspace{1cm}}$ of the substance. It is easy to get from grams H_2 to

moles

moles H_2. You use the $\underline{\hspace{3cm}}$.

molecular weight

$$? \text{ L } O_2 = 8.0 \text{ g } H_2 \times \underline{\hspace{2cm}}$$

$\dfrac{\text{mol } H_2}{2.02 \text{ g } H_2}$

Now you can use the bridge from the balanced equation

$$? \text{ L } O_2 = 8.0 \text{ g } H_2 \times \frac{\text{mol } H_2}{2.02 \text{ g } H_2} \times \underline{\hspace{2cm}}$$

$\dfrac{\text{mol } O_2}{2 \text{ mol } H_2}$

Once you have moles of the desired substance, you can use the gas-law conversion factor to get to the desired variable of the gas.

$$? \text{ L } O_2 = 8.0 \text{ g } H_2 \times \frac{\text{mol } H_2}{2.02 \text{ g } H_2} \times \frac{\text{mol } O_2}{2 \text{ mol } H_2} \times \underline{\hspace{2cm}}$$

$\dfrac{0.0820 \text{ L } O_2 \text{ atm}_{O_2}}{\text{mol } O_2 \text{ K}_{O_2}}$

You now have the dimension you want for the answer, liters O_2, but the

unwanted dimensions $\underline{\hspace{1cm}}$ and $\underline{\hspace{1cm}}$ remain. But these two are

atm_{O_2}; K_{O_2}

given in the problem. The pressure of the O_2 is $\underline{\hspace{2cm}}$ and its

2.0 atm

temperature is $\underline{\hspace{1cm}}$ K. You need only multiply by them so that they cancel the unwanted dimensions.

300

$$? \text{ L } O_2 = 8.0 \text{ g } H_2 \times \frac{\text{mol } H_2}{2.02 \text{ g } H_2} \times \frac{\text{mol } O_2}{2 \text{ mol } H_2}$$

$$\times \frac{0.0820 \text{ L } O_2 \text{ atm}_{O_2}}{\text{mol } O_2 \text{ K}_{O_2}} \times \underline{\hspace{2cm}}$$

$\dfrac{300 \text{ K}_{O_2}}{2.0 \text{ atm}_{H_2}}$

The only dimension left is liters O_2, which is the dimension you want for your answer. So the answer is (canceling all dimensions)

$$? \text{ L } O_2 = 8.0 \times \frac{1}{2.02} \times \frac{1}{2} \times 0.0820 \text{ L } O_2 \times \frac{300}{2.0}$$

$$= 24.3 \text{ L } O_2 \quad \text{or, to the correct significant figures, 24 L } O_2$$

Question 7 on the pretest is the same type, but it has some added complications. How many liters of Cl_2 gas at STP are produced by the reaction of 30.0 mL of a 12.0 M solution of HCl? (That is, there are 12.0 mol HCl per liter of solution.) There is an excess of $KMnO_4$. The reaction is

$$2KMnO_4 + 16HCl = 2KCl + 2MnCl_2 + 5Cl_2 + 8H_2O$$

$T; P$

Only two of the three variables for the Cl_2 gas are given: __ and __. You see this by noting that the gas is at STP. If you do this by the two-step method, you start by solving for the number of moles of Cl_2 from the balanced equation. The information given is the amount of solution.

30.0 mL soln

$$? \text{ mol } Cl_2 = \underline{\hspace{3cm}}$$

The conversion factor given in the problem is the number of moles of

$\dfrac{12.0 \text{ mol HCl}}{\text{L soln}}$

HCl per liter of solution, _____. Therefore your first step is to

liters

convert milliliters solution to _____ solution so that you can use this

$\dfrac{10^{-3} \text{ L soln}}{\text{mL soln}}$

conversion factor. You multiply by _____. Now you can use the conversion factor given in the problem. You multiply by

$\dfrac{12.0 \text{ mol HCl}}{\text{L soln}}$

_____. Finally, you can use the bridge to get to the dimensions

$\dfrac{5 \text{ mol } Cl_2}{16 \text{ mol HCl}}$

of the answer. You multiply by _____. The answer, then, is

$$? \text{ mol } Cl_2 = 30.0 \times 10^{-3} \times 12.0 \times \frac{5 \text{ mol } Cl_2}{16} = 0.1125 \text{ mol } Cl_2$$

(We'll round off at the end.) Now you can determine the volume of

273; 1.00

Cl_2 gas, knowing that STP means that T is _____ K and P is _____ atm.

273 K

$$? \text{ L } Cl_2 = (0.112 \text{ mol})(\underline{\hspace{1cm}}) \times \frac{0.0820(L)(atm)}{(mol)(K)}$$

$\dfrac{1}{1.00 \text{ atm}}$; 2.52

$$\times \underline{\hspace{2cm}} = \underline{\hspace{1cm}} \text{ L } Cl_2$$

If you prefer to do the problem in one step, try it, and check the numbers you have after canceling the dimensions. You should have

$$\text{L } Cl_2 = 30.0 \times 10^{-3} \times 12.0 \times \frac{5}{16} \times 0.0820 \text{ L } Cl_2$$

$$\times \frac{273}{1.00} = 2.52 \text{ L } Cl_2$$

There is an even easier way if you are at STP and remember the conversion factor (good *only* at STP) 22.4 L/mol. Using the two-step method, after you calculate the moles of Cl_2 you get to liters directly.

$$? \text{ L } Cl_2 = 0.1125 \; \cancel{\text{mol } Cl_2} \times \frac{22.4 \text{ L } Cl_2}{\cancel{\text{mol } Cl_2}}$$

Using the one-step method, once you have only moles Cl_2 as your unwanted dimension, you can go to liters Cl_2 using the same conversion factor.

$$? \text{ L } Cl_2 = 30.0 \; \cancel{\text{mL soln}} \times \frac{10^{-3} \; \cancel{\text{L soln}}}{\cancel{\text{mL soln}}}$$

$$\times \frac{12.0 \; \cancel{\text{mol HCl}}}{\cancel{\text{L soln}}} \times \frac{5 \; \cancel{\text{mol } Cl_2}}{16 \; \cancel{\text{mol HCl}}} \times \frac{22.4 \text{ L } Cl_2}{\cancel{\text{mol } Cl_2}}$$

This 22.4 L/mol is a handy conversion factor; but remember, it is true only at STP.

SECTION C   Problems Concerning Two Gases, One with P, V, and T Given, and the Other with Only Two Variables Given

Sometimes you have a problem that involves two gases, one with all three variables given and the other with only two. You are asked to calculate the third variable. This type of problem can be done in three steps. First you calculate the number of moles of the gas whose three variables are known. Then you use this number of moles to determine the number of moles of the other gas. Finally, you use the number of moles of the other gas and the two variables given to calculate the missing variable.

You can also do the problems in one step, but be careful to write the units of the pressure, volume, and temperature so that you know *which* gas has these particular values for the variables. It is easy to get mixed up.

If this discussion seems complicated, an example should clear it up. Question 3 on the pretest asks how many liters of H_2 gas at $123°C$ and 740 mmHg pressure are required to react with 5.0 L of O_2 at 2.0 atm pressure and $27°C$ according to the equation

$$2H_2 + O_2 = 2H_2O$$

First we will do it by the three-step method. Looking at the problem, you see that all three variables are given for _____ gas. Thus you start by calculating the number of _____ of O_2.

$$? \text{ mol } O_2 = (5.0 \text{ L})(\text{_____}) \times \frac{(\text{mol})(K)}{0.0820(\text{L})(\text{atm})}$$

$$\times \text{_____} = 0.41 \text{ mol } O_2$$

In the second step you calculate the number of moles of the other gas, H_2, from the balanced equation:

$$? \text{ mol } H_2 = 0.41 \text{ mol } O_2 \times \text{_____} = \text{_____} \text{ mol } H_2$$

O_2

moles

2.0 atm

$\dfrac{1}{300 \text{ K}}$

$\dfrac{2 \text{ mol } H_2}{\text{mol } O_2}$; 0.82

The third step is to calculate the number of liters of H_2 at the temperature and pressure given in the problem:

396 K

$$? L\ H_2 = (0.82\ \text{mol})(\underline{\quad}) \times \frac{0.0820\ (L)\,(atm)}{(mol)\,(K)}$$

$\dfrac{1}{740\ \text{mmHg}}$; 27

$$X\ \underline{\quad} \times \frac{760\ \text{mmHg}}{atm} = \underline{\quad} L\ H_2$$

Now we'll do it in just one step. But watch how carefully the temperatures, pressures, and volumes are labeled. The question asked is "How many liters of H_2?" and the information given is 5.0 L O_2:

5.0 L O_2

$$? L\ H_2 = \underline{\qquad}$$

moles

As in any stoichiometry problem, you want to get to _____ so that you can use the bridge. To get from the volume of a gas to moles of the

gas-law

gas, you use the _____ conversion factor. But, in this case, the conversion factor will refer only to the O_2 gas. So

$$? L\ H_2 = 5.0\ L\ O_2 \times \frac{\text{mol } O_2\ K_{O_2}}{0.0820\ \underline{\qquad}}$$

L O_2 atm$_{O_2}$

You have canceled the unwanted liters O_2 and have the desired moles

K_{O_2}; atm$_{O_2}$

O_2, but the unwanted dimensions _____ and _____ remain. However, the temperature and pressure of the O_2 are given in the problem:

300; 2.0

_____ K_{O_2} and _____ atm$_{O_2}$. (You must be sure to pick the right ones so that you can cancel the unwanted dimensions.)

$\dfrac{2.0\ \text{atm}_{O_2}}{300\ K_{O_2}}$

$$? L\ H_2 = 5.0\ L\ O_2 \times \frac{\text{mol } O_2\ K_{O_2}}{0.0820\ L\ O_2\ \text{atm}_{O_2}} \times \underline{\qquad}$$

The only dimension left now is moles O_2, so you can bridge to H_2:

$\dfrac{2\ \text{mol } H_2}{\text{mol } O_2}$

$$? L\ H_2 = 5.0\ L\ O_2 \times \frac{\text{mol } O_2\ K_{O_2}}{0.0820\ L\ O_2\ \text{atm}_{O_2}} \times \frac{2.0\ \text{atm}_{O_2}}{300\ K_{O_2}} \times \underline{\qquad}$$

Finally you have moles of H_2. You can get to liters of the gas by using

gas-law

the _____ conversion factor for H_2

$$? L\ H_2 = 5.0\ L\ O_2 \times \frac{\text{mol } O_2\ K_{O_2}}{0.0820\ L\ O_2\ \text{atm}_{O_2}} \times \frac{2.0\ \text{atm}_{O_2}}{300\ K_{O_2}}$$

L H_2 atm$_{H_2}$

$$\times \frac{2\ \text{mol } H_2}{\text{mol } O_2} \times \frac{0.0820\ \underline{\qquad}}{\underline{\qquad}}$$

mol H_2 K_{H_2}

You now have the dimensions you want for your answer: liters H_2.

atm$_{H_2}$; K_{H_2}

But you also have the unwanted dimensions _____ and _____. However, values for these are given in the problem. So to cancel, you continue:

$$? \text{ L H}_2 = 5.0 \,\cancel{\text{L O}_2} \times \frac{\cancel{\text{mol O}_2}\,\cancel{\text{K}_{O_2}}}{0.0820 \,\cancel{\text{L O}_2}\,\cancel{\text{atm}_{O_2}}} \times \frac{2.0 \,\cancel{\text{atm}_{O_2}}}{300 \,\cancel{\text{K}_{O_2}}}$$

$$\times \frac{2 \,\cancel{\text{mol H}_2}}{\cancel{\text{mol O}_2}} \times \frac{0.0820 \text{ L H}_2 \text{ atm}_{H_2}}{\cancel{\text{mol H}_2}\,\text{K}_{H_2}} \times \underline{\qquad\qquad}$$

$$\frac{396 \text{ K}_{H_2}}{740 \text{ mmHg}_{H_2}}$$

The pressure given in the problem for the H_2 wasn't in atm_{H_2}, which we needed for the cancelation, but in mmHg_{H_2}. There is still one more conversion factor needed that will cancel the unwanted atm_{H_2}, as well as the mmHg_{H_2}:

$$? \text{ L H}_2 = 5.0 \,\cancel{\text{L O}_2} \times \frac{\cancel{\text{mol O}_2}\,\cancel{\text{K}_{O_2}}}{0.0820 \,\cancel{\text{L O}_2}\,\cancel{\text{atm}_{O_2}}} \times \frac{2.0 \,\cancel{\text{atm}_{O_2}}}{300 \,\cancel{\text{K}_{O_2}}} \times \frac{2 \,\cancel{\text{mol H}_2}}{\cancel{\text{mol O}_2}}$$

$$\times \frac{0.0820 \text{ L H}_2 \text{ atm}_{H_2}}{\cancel{\text{mol H}_2}\,\cancel{\text{K}_{H_2}}} \times \frac{396 \,\cancel{\text{K}_{H_2}}}{740 \text{ mmHg}_{H_2}} \times \underline{\qquad\qquad}$$

$$\frac{760 \text{ mmHg}_{H_2}}{\text{atm}_{H_2}}$$

Canceling all the dimensions leaves

$$? \text{ L H}_2 = 5.0 \times \frac{1}{0.0820} \times \frac{2.0}{300} \times 2 \times 0.0820 \text{ L H}_2 \times \frac{396}{740} \times 760$$

$$= 27 \text{ L H}_2$$

Incidentally, notice that the 0.0820 appears in both the numerator and the denominator, so it cancels out. This will always be so in this type of problem. In the next section we're going to look at a special case in which not only does the 0.0820 cancel but also the temperatures and pressures.

SECTION D   Problems in which Both Gases Are at the Same Temperature and Pressure

In special cases you may have two gases, one with P, V, and T given, the other without V, and both at the same temperature and pressure. There is a big simplification you can make. We will look at question 4 in the pretest, since both gases are at the same pressure and temperature.

How many liters of CO are formed by the reaction of 5.0 L of O_2 if both gases are at 2.0 atm and 300 K? The reaction is

$$2C + O_2 = 2CO$$

We will try this immediately in the one-step method. First,

$$? \text{ L CO} = \underline{\qquad\qquad}$$

5.0 L O_2

Then you put in the _____ conversion factor:

gas-law

$$? \text{ L CO} = 5.0 \text{ L O}_2 \times \frac{\text{mol O}_2 \text{ K}_{O_2}}{0.0820 \,\underline{\qquad\qquad}}$$

L O_2 atm$_{O_2}$

K_{O_2}; atm_{O_2}

2.0 atm_{O_2}

300 K_{O_2}

bridge

$\dfrac{2 \text{ mol CO}}{\text{mol } O_2}$

gas-law

L CO atm_{CO}

K_{CO}

atm_{CO}; K_{CO}

$\dfrac{300\ K_{CO}}{2.0\ atm_{CO}}$

To cancel the unwanted _____ and _____, you insert the values for these variables given in the problem:

$$? \text{L CO} = 5.0\ \cancel{L\ O_2} \times \frac{\text{mol } O_2\ K_{O_2}}{0.0820\ \cancel{L\ O_2}\ atm_{O_2}} \times \frac{\rule{3cm}{0.4pt}}{\rule{3cm}{0.4pt}}$$

Now you use the _____ from the balanced equation.

$$? \text{L CO} = 5.0\ \cancel{L\ O_2} \times \frac{\text{mol } O_2\ \cancel{K_{O_2}}}{0.0820\ \cancel{L\ O_2}\ \cancel{atm_{O_2}}} \times \frac{2.0\ \cancel{atm_{O_2}}}{300\ \cancel{K_{O_2}}} \times \rule{2cm}{0.4pt}$$

To get the desired dimension of the answer, liters CO, you once again use the _____ conversion factor:

$$? \text{L CO} = 5.0\ \cancel{L\ O_2} \times \frac{\cancel{\text{mol } O_2}\ \cancel{K_{O_2}}}{0.0820\ \cancel{L\ O_2}\ \cancel{atm_{O_2}}}$$

$$\times \frac{2.0\ \cancel{atm_{O_2}}}{300\ \cancel{K_{O_2}}} \times \frac{2\ \cancel{\text{mol CO}}}{\cancel{\text{mol } O_2}} \times \frac{0.0820\ \rule{2cm}{0.4pt}}{\cancel{\text{mol CO}}\ \rule{1.5cm}{0.4pt}}$$

You now have the dimension desired for the answer, but you still have the unwanted dimensions _____ and _____. But these can be canceled with the information given in the problem for CO gas.

$$? \text{L CO} = 5.0\ \cancel{L\ O_2} \times \frac{\cancel{\text{mol } O_2}\ \cancel{K_{O_2}}}{0.0820\ \cancel{L\ O_2}\ \cancel{atm_{O_2}}} \times \frac{2.0\ \cancel{atm_{O_2}}}{300\ \cancel{K_{O_2}}}$$

$$\times \frac{2\ \cancel{\text{mol CO}}}{\cancel{\text{mol } O_2}} \times \frac{0.0820\ \text{L CO } atm_{CO}}{\cancel{\text{mol CO}}\ K_{CO}} \times \rule{2cm}{0.4pt}$$

Now all the unwanted dimensions cancel, leaving

$$? \text{L CO} = 5.0 \times \frac{1}{0.0820} \times \frac{2.0}{300} \times \frac{2}{1} \times 0.0820 \text{ L CO} \times \frac{300}{2.0}$$

Now, not only does the 0.0820 appear in both the numerator and denominator, but the 300 K and 2.0 atm for the temperatures and pressures also appear in both, and cancel. So, if the temperature and pressure of both gases are the same, it doesn't matter what their values are. They cancel anyway. The only thing that affects the volume of gas produced is the volume of the other gas that reacted and the bridge conversion factor. As a matter of fact, when you have two gases both at the same temperature and pressure, you can write the bridge in terms of liters rather than moles. So, in this problem, you could have written as a bridge:

$$\frac{2 \text{ L CO}}{\text{L } O_2} \qquad \text{instead of} \qquad \frac{2 \text{ mol CO}}{\text{mol } O_2}$$

The problem would then be set up, starting in the usual way:

? L CO = _____ | 5.0 L O_2

The unwanted dimension in the information given, liters O_2, is in the new bridge. So you multiply to cancel the liters O_2.

? L CO = 5.0 L O_2 × _____ | $\dfrac{2\ L\ CO}{L\ O_2}$

The only dimension left is the dimension of the answer, which must be

? L CO = 5.0 × 2 L CO = 10 L CO

This relationship between volumes of gases reacting at the same temperature and pressure leads to *Gay-Lussac's law*, which says that when gases combine they do so in simple, whole-number multiples of the volumes.

Question 5 on the pretest is another example in which the two gases are at the same temperature and pressure. You must look carefully to see that this is so.

How many liters of HCl gas at 2.0 atm pressure and 300 K are needed to produce 5.0 L of CO_2 at 1520 torr and 27°C according to the equation

$$CaCO_3 + 2HCl = CaCl_2 + CO_2 + H_2O$$

If you convert 1520 torr to atmospheres,

? atm = 1520 torr × _____ = _____ atm | $\dfrac{atm}{760\ torr}$; 2.00

and you convert 27°C to K,

? K = 27°C _____ = _____ K | + 273; 300

you see that both the HCl and the CO_2 are at the same _____ | pressure

and _____. You can use the short-cut method: | temperature

? L HCl = _____ | 5.0 L CO_2

Next you use your bridge conversion factor expressed in liters:

? L HCl = 5.0 L CO_2 × _____ | $\dfrac{2\ L\ HCl}{L\ CO_2}$

and the only dimension left is _____. Your answer must therefore be | liters HCl

? L HCl = 5.0 × 2 L HCl = _____ | 10 L HCl

[A word about using stepwise solutions rather than the one-step method: For the one-step method, you must be very careful about labeling the variables. Also the one-step method doesn't really give you a feeling for what is going on. However, you should always use it when the gas is at STP (using the conversion factor 22.4 L/mol) or when you have two gases both at the same temperature and pressure (substituting liters for moles in the bridge conversion factor). You cannot use the one-step method in problems involving two gases when you wish to determine the temperature of one of the gases. Try to do Problem 13 and you will see the difficulty.]

The worked-out answers to all the problems that follow [except 2 (at STP) and 11 (both gases at the same *T* and *P*)] are done in steps. If you choose to do them by a one-step method, your answer, of course, should be the same with a possible variation of 1 in the last significant figure.

PROBLEM SET

1. How many liters of Cl_2 gas at 40°C and 755 mmHg pressure are produced by the complete reaction of 20.0 g of MnO_2 according to the reaction

$$MnO_2 + 4HCl = MnCl_2 + Cl_2 + 2H_2O$$

2. Calcium carbonate, if it is heated strongly, decomposes into calcium oxide and carbon dioxide:

$$CaCO_3 = CaO + CO_2$$

When 2270 g of calcium carbonate are heated, how many liters of carbon dioxide at STP are produced?

3. How many liters of O_2 gas at 1.00 atm pressure and 27°C are consumed by the oxidation of 100 g NH_3 according to the reaction

$$4NH_3 + 5O_2 = 4NO + 6H_2O$$

4. If you ran the same reaction as in Problem 3, and collected the NO produced by the 100 g of NH_3 over water at 29°C and 1.0 atm total pressure, how many liters of NO would you have? The vapor pressure of water at 29°C is 30 mmHg.

5. How many milliliters of O_2 gas at 25°C and 243.1 torr pressure are produced by the decomposition of 1.25 g of 97.5% pure $KClO_3$ according to the reaction

$$2KClO_3 = 2KCl + 3O_2$$

6. How many grams of C react with 500 mL of O_2 gas at 30°C and a pressure of 740 mmHg according to the reaction

$$2C + O_2 = 2CO$$

7. The Solvay process for the production of $NaHCO_3$ uses the following reaction:

$$NaCl + NH_3 + H_2O + CO_2 = NaHCO_3 + NH_4Cl$$

How many pounds of $NaHCO_3$ are produced by the complete reaction of 20.0 L of NH_3 gas at 750 torr pressure and 27°C?

8. You want to prepare 3.0 L of Cl_2 gas at 0°C and a pressure of 0.50 atm according to the reaction

$$2KMnO_4 + 16HCl = 2MnCl_2 + 5Cl_2 + 2KCl + 8H_2O$$

How many grams of $KMnO_4$ would you need?

9. How many milliliters of 10.0 M solution (that is, it contains 10.0 mol of HCl per liter of solution) are needed for the production of 15.0 L of CO_2 at 2.0 atm pressure and -23°C following the reaction

$$CaCO_3 + 2HCl = CaCl_2 + CO_2 + H_2O$$

10. What is the molarity of an HCl solution, given that 50 mL of the solution are required to produce 50 mL of Cl_2 gas at 27°C and 1520 torr according to the reaction

$$MnO_2 + 4HCl = MnCl_2 + Cl_2 + 2H_2O$$

11. The first step in the Ostwald process to produce nitric acid is the oxidation of ammonia:

    $$4NH_3 + 5O_2 = 4NO + 6H_2O$$

    How many liters of NO gas are produced when 5.0 L of O_2 react? Both gases are at 27°C and 2.0 atm pressure.

*12. How many gallons of water (density = 1.00 g/mL) are needed to yield 1.0×10^6 ft³ of H_2 gas at 20.0 atm pressure and 77°F according to the reaction

    $$2H_2O + C = 2H_2 + CO_2$$

13. What will be the temperature in degrees Celsius of the NO_2 gas formed from the reaction of 2.0 L of O_2 gas at 0°C and 700 mmHg according to the reaction

    $$2NO + O_2 = 2NO_2$$

    The NO_2 has a volume of 10.0 L and its pressure is 0.50 atm.

14. What is the volume of NH_3 collected over water at a total pressure of 743 torr and a temperature of 25°C from the reaction of 0.15 mol of H_2 according to the equation

    $$N_2 + 3H_2 = 2NH_3$$

    The vapor pressure of water at 25°C is 24 mmHg.

*15. What volume of air (4.0 vol % CO_2) at 1.05 atm pressure and 27°C is required to yield 4.0 lb of $C_{12}H_{22}O_{11}$ according to the reaction

    $$11H_2O + 12CO_2 = C_{12}H_{22}O_{11} + 12O_2$$

16. The combustion of butane, C_4H_{10}, in oxygen produces carbon dioxide and water. Suppose that you have only 35.0 L of oxygen at 1140 torr pressure and 127°C. How many grams of butane can you consume? You will have to write the balanced equation yourself.

17. An 18.0 g sample of impure MnO_2 is reacted with an excess of HCl solution and Cl_2 gas is produced according to the reaction

    $$MnO_2 + 4HCl(aq) = MnCl_2(aq) + Cl_2(g) + 2H_2O$$

    What is the percent purity of the MnO_2 sample if 1.50 L Cl_2 at 770 torr and 10°C was collected?

18. What is the molarity (moles per liter) for the solution prepared by dissolving the NH_3 produced by the reaction of 5.0 L N_2 gas at −73°C and 1520 mmHg pressure with an excess of H_2 gas in 2500.0 mL of water? (Assume that the volume of water and volume of solution are the same.) The reaction is

    $$N_2 + 3H_2 = 2NH_3$$

19. How many liters of NO_2 at STP can be prepared by the reaction of 6.0 L NO and 2.0 L O_2 (also both at STP) according to the reaction

    $$2NO + O_2 = 2NO_2$$

20. How much air (20 vol % O_2) is required to consume the H_2 produced by the reaction of 0.46 g Na? The air is at 20°C and 0.98 atm pressure. The reactions are

$$2Na + 2H_2O = 2NaOH + H_2 \quad \text{and} \quad 2H_2 + O_2 = 2H_2O$$

21. Before dry-cell batteries were invented, coal miners used *carbide* lamps to see in the mines. Water was allowed to drip on chunks of calcium carbide (CaC_2) and the acetylene (C_2H_2) gas that was formed exited from a small opening, where it was lit. As it burned, it produced a small flame. The acetylene (which was at barometric pressure, 745 torr, and 5°C, the temperature in the mine) burned at the rate of 1.50 L/hr. What weight (in grams) of CaC_2 had to be put in the lamp so that it would continue burning through an 8.0 hr work shift? The reaction is

$$CaC_2 + 2H_2O = C_2H_2 + Ca(OH)_2$$

*22. Dinitrogen tetroxide gas dissociates partially into two molecules of nitrogen dioxide gas:

$$N_2O_4 = 2NO_2$$

When 0.10 mol of N_2O_4 is placed in a container, it is found that the resulting mixture of the two gases occupies 3.6 L at a pressure of 1.0 atm and a temperature of 30°C. How many moles of N_2O_4 remain. [*Hint*: Calculate the total number of moles of both gases present and then assume that x mol of the N_2O_4 dissociate into $2x$ mol of NO_2. For the next few steps, you are on your own.]

*23. Gaseous hydrocarbons (compounds that contain only C and H), when burned with oxygen, produce carbon dioxide and water. With all the gases at STP, it is found that a 1.4 L sample of the gaseous hydrocarbon consumes 9.1 L O_2 and produces 5.6 L CO_2 gas and 5.63 g of liquid water. What is the molecular formula of the hydrocarbon? [Hint: Determine the number of moles of each component. Then find their equivalent ratio in whole numbers. These will be the coefficients for the balanced equation __C_xH_y + __O_2 = __CO_2 + __H_2O. You can then find the values of x and y.]

24. Both Mg and Al react with HCl(*aq*) to produce H_2 gas. Suppose that you take a 1.25 g sample of a Mg-Al alloy that is 45% Mg and react it with an excess of HCl. How many liters of H_2 gas at 27°C and 745 mmHg pressure are formed? Write the balanced equations for both reactions.

25. What volume of air (20 vol % O_2) at 740 torr and 22°C is needed to burn 21 L propane gas (C_3H_8) at 37°C and 2.3 atm pressure according to the equation

$$C_3H_8 + 5O_2 = 3CO_2 + 4H_2O$$

PROBLEM SET ANSWERS

Since you were able to handle Chapter 12 on gas laws and Chapter 10 on stoichiometry, this chapter should be easy, unless you get lost in the mass of numbers. You should be able to do at least 13 of the first 15 problems. Problems 16 through 25 are expressed in a more general way (the way that you will probably find them in your textbook). You will have to read them carefully to understand exactly what is asked for and what is given. Two of the problems, marked with asterisks, are hard, but even so you should be able to do at least seven of the last ten problems.

1. $? \text{ mol Cl}_2 = 20.0 \text{ g MnO}_2 \times \dfrac{\text{mol MnO}_2}{86.9 \text{ g MnO}_2} \times \dfrac{\text{mol Cl}_2}{\text{mol MnO}_2} = 0.230 \text{ mol Cl}_2$

$? \text{ L Cl}_2 = (0.230 \text{ mol})(313 \text{ K}) \times \dfrac{0.0820(\text{L})(\text{atm})}{(\text{mol})(\text{K})} \times \dfrac{1}{755 \text{ mmHg}}$

$\times \dfrac{760 \text{ mmHg}}{\text{atm}} = 5.94 \text{ L Cl}_2$

2. $? \text{ L CO}_2 = 2270 \text{ g CaCO}_3 \times \dfrac{\text{mol CaCO}_3}{100.1 \text{ g CaCO}_3} \times \dfrac{\text{mol CO}_2}{\text{mol CaCO}_3}$

$\times \dfrac{22.4 \text{ L CO}_2}{\text{mol CO}_2} \text{ (at STP)}$

$= 508 \text{ L CO}_2$

3. $? \text{ mol O}_2 = 100 \text{ g NH}_3 \times \dfrac{\text{mol NH}_3}{17.0 \text{ g NH}_3} \times \dfrac{5 \text{ mol O}_2}{4 \text{ mol NH}_3} = 7.35 \text{ mol O}_2$

$? \text{ L O}_2 = (7.35 \text{ mol})(300 \text{ K}) \times \dfrac{0.0820(\text{L})(\text{atm})}{(\text{mol})(\text{K})} \times \dfrac{1}{1.00 \text{ atm}} = 181 \text{ L O}_2$

4. $? \text{ mol NO} = 100 \text{ g NH}_3 \times \dfrac{\text{mol NH}_3}{17.0 \text{ g NH}_3} \times \dfrac{4 \text{ mol NO}}{4 \text{ mol NH}_3} = 5.88 \text{ mol NO}$

$p_{NO} = 760 \text{ mmHg} - 30 \text{ mmHg} = 730 \text{ mmHg}$

$? \text{ L NO} = (5.88 \text{ mol})(302 \text{ K}) \times \dfrac{0.0820(\text{L})(\text{atm})}{(\text{mol})(\text{K})} \times \dfrac{1}{730 \text{ mmHg}} \times \dfrac{760 \text{ mmHg}}{\text{atm}}$

$= 152 \text{ L NO} = 1.52 \times 10^2 \text{ L NO}$

5. $? \text{ mol O}_2 = 1.25 \text{ g imp} \times \dfrac{97.5 \text{ g KClO}_3}{100 \text{ g imp}} \times \dfrac{\text{mol KClO}_3}{122.6 \text{ g KClO}_3} \times \dfrac{3 \text{ mol O}_2}{2 \text{ mol KClO}_3}$

$= 1.49 \times 10^{-2} \text{ mol O}_2$

$? \text{ mL O}_2 = (1.49 \times 10^{-2} \text{ mol})(298 \text{ K}) \times \dfrac{0.0820(\text{L})(\text{atm})}{(\text{mol})(\text{K})} \times \dfrac{1}{243.1 \text{ torr}}$

$\times \dfrac{760 \text{ torr}}{\text{atm}} \times \dfrac{10^3 \text{ mL}}{\text{L}} = 1.14 \times 10^3 \text{ mL O}_2$

6. $? \text{ mol O}_2 = (0.500 \text{ L})(740 \text{ mmHg}) \times \dfrac{\text{atm}}{760 \text{ mmHg}} \times \dfrac{(\text{mol})(\text{K})}{0.0820(\text{L})(\text{atm})} \times \dfrac{1}{303 \text{ K}}$

$\qquad\qquad = 0.0196 \text{ mol O}_2$

$? \text{ g C} = 0.0196 \text{ mol O}_2 \times \dfrac{2 \text{ mol C}}{\text{mol O}_2} \times \dfrac{12.0 \text{ g C}}{\text{mol C}} = 4.70 \times 10^{-1} \text{ g C}$

7. $? \text{ mol NH}_3 = (20.0 \text{ L})(750 \text{ torr}) \times \dfrac{\text{atm}}{760 \text{ torr}} \times \dfrac{(\text{mol})(\text{K})}{0.0820(\text{L})(\text{atm})} \times \dfrac{1}{300 \text{ K}}$

$\qquad\qquad = 0.802 \text{ mol NH}_3$

$? \text{ lb NaHCO}_3 = 0.802 \text{ mol NH}_3 \times \dfrac{\text{mol NaHCO}_3}{\text{mol NH}_3} \times \dfrac{84.0 \text{ g NaHCO}_3}{\text{mol NaHCO}_3}$

$\qquad\qquad \times \dfrac{\text{lb NaHCO}_3}{454 \text{ g NaHCO}_3} = 0.148 \text{ lb NaHCO}_3$

8. $? \text{ mol Cl}_2 = (3.0 \text{ L})(0.50 \text{ atm}) \times \dfrac{(\text{mol})(\text{K})}{0.0820(\text{L})(\text{atm})} \times \dfrac{1}{273 \text{ K}}$

$\qquad\qquad = 0.067 \text{ mol Cl}_2$

$? \text{ g KMnO}_4 = 0.067 \text{ mol Cl}_2 \times \dfrac{2 \text{ mol KMnO}_4}{5 \text{ mol Cl}_2} \times \dfrac{158.0 \text{ g KMnO}_4}{\text{mol KMnO}_4}$

$\qquad\qquad = 4.2 \text{ g KMnO}_4$

9. $? \text{ mol CO}_2 = (15.0 \text{ L})(2.0 \text{ atm}) \times \dfrac{(\text{mol})(\text{K})}{0.0820(\text{L})(\text{atm})} \times \dfrac{1}{250 \text{ K}}$

$\qquad\qquad = 1.46 \text{ mol CO}_2$

$? \text{ mL soln} = 1.46 \text{ mol CO}_2 \times \dfrac{2 \text{ mol HCl}}{\text{mol CO}_2} \times \dfrac{\text{L soln}}{10.0 \text{ mol HCl}} \times \dfrac{\text{mL soln}}{10^{-3} \text{ L soln}}$

$\qquad\qquad = 292 \text{ mL soln} = 2.9 \times 10^2 \text{ mL soln}$ (one nonsignificant figure)

10. $? \text{ mol Cl}_2 = (0.050 \text{ L})(1520 \text{ torr}) \times \dfrac{\text{atm}}{760 \text{ torr}} \times \dfrac{(\text{mol})(\text{K})}{0.0820(\text{L})(\text{atm})} \times \dfrac{1}{300 \text{ K}}$

$\qquad\qquad = 4.06 \times 10^{-3} \text{ mol Cl}_2$ (one nonsignificant figure)

$? \text{ mol HCl} = 1 \text{ L soln} \times \dfrac{4.06 \times 10^{-3} \text{ mol Cl}_2}{0.050 \text{ L soln}} \times \dfrac{4 \text{ mol HCl}}{\text{mol Cl}_2}$

$\qquad\qquad = 0.33 \text{ mol HCl in 1 L soln, or } 0.33 \, M$

11. Since both gases are at the same *P* and *T*, you may substitute liters for moles.

$? \text{ L NO} = 5.0 \text{ L O}_2 \times \dfrac{4 \text{ L NO}}{5 \text{ L O}_2} = 4.0 \text{ L NO}$

12. First determine the volume in liters.

$? \text{ L} = 1.0 \times 10^6 \text{ ft}^3 \times \dfrac{(12)^3 \text{ in}^3}{\text{ft}^3} \times \dfrac{(2.54)^3 \text{ cm}^3}{\text{in}^3} \times \dfrac{\text{mL}}{\text{cm}^3} \times \dfrac{10^{-3} \text{ L}}{\text{mL}}$

$\qquad\qquad = 2.8 \times 10^7 \text{ L}$

Then find the temperature in kelvins.

$$? \,^{\circ}C = \frac{5}{9}(^{\circ}F - 32) = \frac{5}{9}(77 - 32) = 25^{\circ}C$$

$$? \, K = (25 + 273) = 298 \, K$$

$$? \text{ mol } H_2 = (2.8 \times 10^7 \,\cancel{L})(20.0\,\cancel{atm}) \times \frac{(mol)(\cancel{K})}{0.0820(\cancel{L})(\cancel{atm})} \times \frac{1}{298\,\cancel{K}}$$

$$= 2.3 \times 10^7 \text{ mol } H_2$$

$$? \text{ gal } H_2O = 2.3 \times 10^7 \,\cancel{mol\,H_2} \times \frac{2\,\cancel{mol\,H_2O}}{2\,\cancel{mol\,H_2}} \times \frac{18.0\,\cancel{g\,H_2O}}{\cancel{mol\,H_2O}}$$

$$\times \frac{\cancel{mL\,H_2O}}{1.00\,\cancel{g\,H_2O}} \times \frac{\cancel{qt\,H_2O}}{946\,\cancel{mL\,H_2O}} \times \frac{gal\,H_2O}{4\,\cancel{qt\,H_2O}}$$

$$= 1.1 \times 10^5 \text{ gal } H_2O$$

13. $$? \text{ mol } O_2 = (2.0\,\cancel{L})(700\,\cancel{mmHg}) \times \frac{\cancel{atm}}{760\,\cancel{mmHg}} \times \frac{(mol)(\cancel{K})}{0.0820(\cancel{L})(\cancel{atm})} \times \frac{1}{273\,\cancel{K}}$$

$$= 0.0823 \text{ mol } O_2$$

$$? \text{ mol } NO_2 = 0.0823 \,\cancel{mol\,O_2} \times \frac{2 \text{ mol } NO_2}{\cancel{mol\,O_2}} = 0.165 \text{ mol } NO_2$$

$$? \, K = (10.0\,\cancel{L})(0.50\,\cancel{atm}) \times \frac{(\cancel{mol})(K)}{0.0820(\cancel{L})(\cancel{atm})} \times \frac{1}{0.164\,\cancel{mol}} = 372 \, K$$

$$? \,^{\circ}C = 372 - 273 = 99^{\circ}C$$

14. $$? \text{ mol } NH_3 = 0.15 \,\cancel{mol\,H_2} \times \frac{2 \text{ mol } NH_3}{3 \,\cancel{mol\,H_2}} = 0.10 \text{ mol } NH_3$$

To get the partial pressure of NH_3:

$$p_{NH_3} = 743 \text{ torr} - \left(24 \,\cancel{mmHg} \times \frac{1 \text{ torr}}{\cancel{mmHg}}\right) = 719 \text{ torr}$$

$$? \text{ L } NH_3 = (0.10\,\cancel{mol})(298\,\cancel{K}) \times \frac{0.0820(L)(\cancel{atm})}{(\cancel{mol})(\cancel{K})} \times \frac{760\,\cancel{torr}}{\cancel{atm}} \times \frac{1}{719\,\cancel{torr}}$$

$$= 2.6 \text{ L } NH_3$$

15. $$? \text{ mol } CO_2 = 4.0 \,\cancel{lb\,C_{12}H_{22}O_{11}} \times \frac{454\,\cancel{g\,C_{12}H_{22}O_{11}}}{\cancel{lb\,C_{12}H_{22}O_{11}}} \times \frac{\cancel{mol\,C_{12}H_{22}O_{11}}}{342.0\,\cancel{g\,C_{12}H_{22}O_{11}}}$$

$$\times \frac{12 \text{ mol } CO_2}{\cancel{mol\,C_{12}H_{22}O_{11}}} = 64 \text{ mol } CO_2$$

Here you must be very careful. You will need to know the pressure of the CO_2 and you are given only the pressure of air. However, if you know the volume percent of one gas in a mixture, Dalton's law tells you that the partial pressure of the gas is directly related to its volume fraction. Thus the CO_2 pressure is 4.0% of the total pressure (or 4.0/100 volume fraction).

$$\frac{4.0}{100} \times 1.05 \text{ atm} = 0.042 \text{ atm } CO_2$$

$$? \text{ L air} = (64 \text{ mol CO}_2)(300 \text{ K}) \times \frac{0.0820(\text{L CO}_2)(\text{atm CO}_2)}{(\text{mol CO}_2)(\text{K})}$$

$$\times \frac{1}{0.042 \text{ atm CO}_2} \times \frac{100 \text{ L air}}{4.0 \text{ L CO}_2}$$

$$= 9.4 \times 10^5 \text{ L air}$$

16. $2C_4H_{10} + 13O_2 = 8CO_2 + 10H_2O$

$$? \text{ g C}_4\text{H}_{10} = 35.0 \text{ L O}_2 \times \frac{(\text{mol O}_2)(\text{K})}{0.0820(\text{L O}_2)(\text{atm})} \times \frac{\text{atm}}{760 \text{ torr}}$$

$$\times \frac{1140 \text{ torr}}{400 \text{ K}} \times \frac{2 \text{ mol C}_4\text{H}_{10}}{13 \text{ mol O}_2} \times \frac{58.0 \text{ g C}_4\text{H}_{10}}{\text{mol C}_4\text{H}_{10}}$$

$$= 14.3 \text{ g C}_4\text{H}_{10}$$

17. $? \text{ g MnO}_2 = 100 \text{ g imp} \times \frac{1.50 \text{ L Cl}_2}{18.0 \text{ g imp}} \times \frac{(\text{mol Cl}_2)(\text{K})}{0.0820(\text{L Cl}_2)(\text{atm})}$

$$\times \frac{\text{atm}}{760 \text{ torr}} \times \frac{770 \text{ torr}}{283 \text{ K}} \times \frac{\text{mol MnO}_2}{\text{mol Cl}_2} \times \frac{86.9 \text{ g MnO}_2}{\text{mol MnO}_2}$$

$$= 31.6 \text{ g MnO}_2 \text{ in 100 g imp, or 31.6\% pure}$$

18. This can be solved in one step by asking how many moles of NH_3 are in one liter of solution. Then you use the conversion factor given in the problem, which says that there are 5.0 L N_2 per 2500.0 mL of solution. But you might be more comfortable solving for the number of moles of NH_3 that are prepared in the first step, and then, in the second step, dividing this by the volume of solution expressed in liters. We will show the second procedure.

$$? \text{ mol NH}_3 = 5.0 \text{ L N}_2 \times \frac{(\text{mol N}_2)(\text{K})}{0.0820(\text{L N}_2)(\text{atm})} \times \frac{\text{atm}}{760 \text{ mmHg}}$$

$$\times \frac{1520 \text{ mmHg}}{200 \text{ K}} \times \frac{2 \text{ mol NH}_3}{\text{mol N}_2} = 1.2 \text{ mol NH}_3$$

$$? \text{ mol NH}_3 = 1 \text{ L soln} \times \frac{\text{mL soln}}{10^{-3} \text{ L soln}} \times \frac{1.2 \text{ mol NH}_3}{2500.0 \text{ mL soln}}$$

$$= 0.48 \text{ mol NH}_3 \text{ in 1 L, or 0.48 } M$$

19. This is a problem of determining the reactant that is controlling (Chapter 10, Section D). However, life is made simple by the fact that all gases are at the same temperature and pressure and you can treat the volumes in the same way as the moles. Thus, to determine which of the reactants is controlling, you divide its volume by its coefficient in the balanced equation:

$$\frac{6.0 \text{ L NO}}{2} = 3.0$$

$$\frac{2.0 \text{ L O}_2}{1} = 2.0 \quad \text{(This is smaller and therefore controlling.)}$$

Then

$$? \text{ L NO}_2 = 2.0 \text{ L O}_2 \times \frac{2 \text{ L NO}_2}{\text{L O}_2} = 4.0 \text{ L NO}_2$$

20. The same thing happens in this problem as in Problem 15. The partial pressure of one gas in a mixture of gases equals its *volume fraction* (volume percent divided by 100) times the total pressure. Also the reaction occurs in two consecutive steps, so that there will be two bridges. If your answer doesn't agree with the answer given here, see if you can follow the dimensional-analysis setup as it is shown.

$$? \text{ L air} = 0.46 \text{ g Na} \times \frac{\text{mol Na}}{23.0 \text{ g Na}} \times \frac{\text{mol } H_2}{2 \text{ mol Na}} \times \frac{\text{mol } O_2}{2 \text{ mol } H_2}$$

$$\times \frac{0.0820(\text{L } O_2)(\text{atm } O_2)}{(\text{mol } O_2)(\text{K})} \times \frac{100 \text{ L air}}{20 \text{ L } O_2} \times \frac{293 \text{ K}}{\left(\dfrac{20}{100}\right)(0.98 \text{ atm } O_2)}$$

$$= 3.1 \text{ L air}$$

21. This problem is easily done in one continuous dimensional-analysis solution.

$$? \text{ g CaC}_2 = 8.0 \text{ hr} \times \frac{1.50 \text{ L } C_2H_2}{\text{hr}} \times \frac{(\text{mol } C_2H_2)(\text{K})}{0.0820(\text{L } C_2H_2)(\text{atm})} \times \frac{\text{atm}}{760 \text{ torr}}$$

$$\times \frac{745 \text{ torr}}{278 \text{ K}} \times \frac{\text{mol CaC}_2}{\text{mol } C_2H_2} \times \frac{64.1 \text{ g CaC}_2}{\text{mol CaC}_2} = 33 \text{ g CaC}_2$$

   If you prefer to break the problem up into steps, you can solve for the number of liters of C_2H_2 used in 8.0 hr and then from the ideal-gas equation calculate the number of moles of C_2H_2. This equals the number of moles of CaC_2, and from this you can get the weight.

22. First you calculate the total number of moles of gases present:

$$\text{Total moles} = 3.6 \text{ L} \times \frac{(\text{mol})(\text{K})}{0.0820(\text{L})(\text{atm})} \times \frac{1.0 \text{ atm}}{303 \text{ K}} = 0.145 \text{ mol}$$

(We will carry an extra significant figure for the moment.) Assume that x number of moles of N_2O_4 dissociate. Then count the total number of moles in terms of x. The N_2O_4 that remains will equal what you started with initially minus what has dissociated:

mol $N_2O_4 = 0.10 - x$

In the dissociation, you form $2NO_2$ for each N_2O_4 dissociating:

mol $NO_2 = 2x$

So the total number of moles is

Total moles $= 0.10 - x + 2x = (0.10 + x)$ mol

Equating the two expressions for total moles, you get

0.145 mol $= (0.10 + x)$ mol

Subtracting 0.10 from both sides gives

0.045 mol $= x$

and the amount of N_2O_4 remaining equals the initial amount minus x:

mol N_2O_4 (remaining) $= (0.10 - 0.045)$ mol $= 0.055$ mol

23. The best approach is to calculate the number of moles of each of the substances. Since the gases are at STP, you can use the conversion factor

$\dfrac{22.4 \text{ L}}{\text{mol}}$ (We will keep one nonsignificant figure.)

$? \text{ mol } C_x H_y = 1.4 \text{ L} \times \dfrac{\text{mol}}{22.4 \text{ L}} = 0.0625 \text{ mol } C_x H_y$

$? \text{ mol } O_2 = 9.1 \text{ L} \times \dfrac{\text{mol}}{22.4 \text{ L}} = 0.406 \text{ mol } O_2$

$? \text{ mol } CO_2 = 5.6 \text{ L} \times \dfrac{\text{mol}}{22.4 \text{ L}} = 0.250 \text{ mol } CO_2$

You can find the moles of water using its molecular weight:

$? \text{ mol } H_2 O = 5.63 \text{ g } H_2 O \times \dfrac{\text{mol}}{18.0 \text{ g } H_2 O} = 0.312 \text{ mol } H_2 O$

The numbers of moles represent the coefficients in the balanced equation:

$0.0625 \, C_x H_y + 0.406 \, O_2 = 0.250 \, CO_2 + 0.312 \, H_2 O$

However, balanced equations always have whole numbers. So we can use the same technique we used in getting the empirical formulas: We divide each of the numbers by the smallest (0.0625 in this case):

$1 \, C_x H_y + 6.5 \, O_2 = 4 CO_2 + 5 H_2 O$

However there is still one nonwhole number, 6.5. You get rid of this by by multiplying all the coefficients by 2.

$2 C_x H_y + 13 O_2 = 8 CO_2 + 10 H_2 O$

Since all the C atoms from $2 C_x H_y$ appear in $8 CO_2$, the value of x must be 4. Since all the H atoms from $2 C_x H_y$ appear in $10 H_2 O$, the value for y must be 10. Therefore the formula is $C_4 H_{10}$.

24. You must first have the two balanced equations for the reactions:

$Mg + 2HCl = MgCl_2 + H_2$ $\qquad$ $2Al + 6HCl = 2AlCl_3 + 3H_2$

The best approach is to calculate the number of moles of H_2 that are produced by each reaction, add them, and then calculate the volume needed by this total number of moles.

From Mg: $? \text{ mol } H_2 = 1.25 \, \cancel{\text{g alloy}} \times \dfrac{45 \, \cancel{\text{g Mg}}}{100 \, \cancel{\text{g alloy}}} \times \dfrac{\cancel{\text{mol Mg}}}{24.3 \, \cancel{\text{g Mg}}} \times \dfrac{\text{mol } H_2}{\cancel{\text{mol Mg}}}$

$= 0.23 \text{ mol } H_2$

From Al: $? \text{ mol } H_2 = 1.25 \, \cancel{\text{g alloy}} \times \dfrac{55 \, \cancel{\text{g Al}}}{100 \, \cancel{\text{g alloy}}} \times \dfrac{\cancel{\text{mol Al}}}{27.0 \, \cancel{\text{g Al}}} \times \dfrac{3 \text{ mol } H_2}{2 \, \cancel{\text{mol Al}}}$

$= 0.038 \text{ mol } H_2$

(Notice that if the alloy is 45% Mg, then it is 55% Al.)

The total number of moles of H_2 is 0.23 mol + 0.038 mol = 0.27 mol, to the correct significant figures, so

$$? \text{ L} = 0.27 \, \cancel{\text{mol}} \times \frac{0.0820 \, (\text{L})\cancel{(\text{atm})}}{\cancel{(\text{mol})}\cancel{(\text{K})}} \times \frac{760 \, \cancel{\text{mmHg}}}{\cancel{\text{atm}}} \times \frac{300 \, \cancel{\text{K}}}{745 \, \cancel{\text{mmHg}}} = 6.8 \text{ L}$$

25. This problem is similar to Problem 20. The partial pressure of O_2 in air equals its volume fraction times the total pressure. This is the type of problem that concerns volumes of two different gases at different temperatures and pressures. You can solve it either in one step with careful labeling of units or in three separate steps. We'll do it both ways.

One step:

$$? \text{ L air} = 21 \, \cancel{\text{L C}_3\text{H}_8} \times \frac{(\cancel{\text{mol C}_3\text{H}_8})(\cancel{\text{K}_{\text{C}_3\text{H}_8}})}{0.0820 \, (\cancel{\text{L C}_3\text{H}_8})(\cancel{\text{atm}_{\text{C}_3\text{H}_8}})}$$

$$\times \frac{2.3 \, \cancel{\text{atm}_{\text{C}_3\text{H}_8}}}{310 \, \cancel{\text{K}_{\text{C}_3\text{H}_8}}} \times \frac{5 \, \text{mol O}_2}{\cancel{\text{mol C}_3\text{H}_8}} \times \frac{0.0820 \, (\cancel{\text{L O}_2})(\cancel{\text{atm}_{\text{O}_2}})}{(\cancel{\text{mol O}_2})(\cancel{\text{K}_{\text{O}_2}})}$$

$$\times \frac{100 \, \text{L air}}{20 \, \cancel{\text{L O}_2}} \times \frac{760 \, \cancel{\text{torr}_{\text{O}_2}}}{\cancel{\text{atm}_{\text{O}_2}}} \times \frac{295 \, \cancel{\text{K}_{\text{O}_2}}}{\left(\dfrac{20}{100}\right)(740 \, \cancel{\text{torr}_{\text{O}_2}})}$$

$$= 5.9 \times 10^3 \text{ L air}$$

Three steps:

$$? \text{ mol C}_3\text{H}_8 = 21 \, \cancel{\text{L C}_3\text{H}_8} \times \frac{(\text{mol C}_3\text{H}_8)(\cancel{\text{K}_{\text{C}_3\text{H}_8}})}{0.0820 \, (\cancel{\text{L C}_3\text{H}_8})(\cancel{\text{atm}_{\text{C}_3\text{H}_8}})} \times \frac{2.3 \, \cancel{\text{atm}_{\text{C}_3\text{H}_8}}}{310 \, \cancel{\text{K}_{\text{C}_3\text{H}_8}}}$$

$$= 1.9 \text{ mol C}_3\text{H}_8$$

$$? \text{ mol O}_2 = 1.9 \, \cancel{\text{mol C}_3\text{H}_8} \times \frac{5 \, \text{mol O}_2}{\cancel{\text{mol C}_3\text{H}_8}} = 9.5 \text{ mol O}_2$$

$$? \text{ L air} = 9.5 \, \cancel{\text{mol O}_2} \times \frac{0.0820 \, (\cancel{\text{L O}_2})(\cancel{\text{atm}_{\text{O}_2}})}{(\cancel{\text{mol O}_2})(\cancel{\text{K}_{\text{O}_2}})} \times \frac{760 \, \cancel{\text{torr}_{\text{O}_2}}}{\cancel{\text{atm}_{\text{O}_2}}}$$

$$\times \frac{293 \, \cancel{\text{K}_{\text{O}_2}}}{\left(\dfrac{20}{100}\right)(740 \, \cancel{\text{torr}_{\text{O}_2}})} \times \frac{100 \, \text{L air}}{20 \, \cancel{\text{L O}_2}} = 5.9 \times 10^3 \text{ L air}$$

CHAPTER 14

Solution Concentrations: Molarity, Formality, Molality

PRETEST

Before you try these questions, you will need three new conversion factors.

$$M = \text{molarity} = \frac{\text{moles solute}}{\text{liter solution}}$$

$$F = \text{formality} = \frac{\text{formula weights solute}}{\text{liter solution}}$$

$$m = \text{molality} = \frac{\text{moles solute}}{\text{kilogram solvent}}$$

The following problems are quite simple if you know the conversions. You should be able to finish them in 20 minutes.

1. How many grams of HCl are in 3.00 L of 1.5 M HCl?

2. To what volume in liters must you dilute 11.7 g NaCl to prepare a 0.10 F solution of NaCl?

3. What is the molarity of an NaCl solution that was prepared by adding 23.4 g NaCl to 100.0 g water? The density of the resulting solution is 1.20 g/mL.

4. How many liters of a 0.70 M solution of HCl would you have to take in order to have 85.0 g HCl?

5. How many milliliters of a 6.0 M solution must you dilute to 3.0 L to have a 0.10 M HCl solution?

6. What is the formality of a solution prepared by adding enough water to 250 mL of a 6.0 F solution of KCl to bring the volume up to 5.0 L?

7. What will be the molarity of a solution prepared by mixing 3.0 L of 0.10 M HCl with 1.0 L of 6.0 M HCl?

8. What is the molality of an alcohol solution that was prepared by dissolving 23.0 g of alcohol (C_2H_5OH) in 200.0 g water?

9. What is the molality of a 6.0 M solution of HCl whose density is 1.1 g/mL?

PRETEST ANSWERS

If you answered all nine questions correctly, skip Chapter 14 and go directly to the problem set at the end of the chapter. If you missed questions 1, 3, or 4, you had better go through Section A. If you missed question 2, check Section B for the definition of formality. If you missed questions 5 and 6, work through Section D in the chapter. And if you missed question 7, do Section E. Questions 8 and 9 come from Section C. Incidentally, you will find all these questions worked out as examples in the chapter.

1. 164 g HCl (or 1.6×10^2 to the correct number of significant figures)
2. 2.0 L
3. 3.89 *M* (If you got 4.80 *M*, you did the problem incorrectly; you did not add the weight of solvent to the weight of solute.)

4. 3.3 L   5. 50 mL   6. 0.30 *F*
7. 1.6 *M*   8. 2.50 *m*   9. 6.8 *m*

You have already been introduced to the concept of *molarity* in Section D, Chapter 11. You were told that it is a very convenient way of expressing concentrations of substances in solution. You can describe a solution as so many weight percent of one material in a total amount of solution (either weight or volume) or as so many grams of one substance and so many of the other. However, this may not be the most convenient way. When you are doing stoichiometry problems, the bridge is

moles   always in _____. Therefore it would really be much easier if, instead of expressing a concentration in weight percent, you had expressed the concentration in moles to start with.

What we are going to do in this chapter is to examine three different ways of expressing concentrations which are commonly used because they simplify your calculations. These are *molarity*, *formality* and *molality*.

[There is a fourth method for expressing concentrations, called *normality*. However, determining it can pose more difficulties than is warranted by the convenience its use provides in doing stoichiometry problems. Consequently, your instructor may feel that you need not consider normality. As a matter of fact, one of the units in its concentration term, *equivalent*, does not exist in SI. However, if your instructor is interested in this concentration unit, he or she will explain how it works.]

We will skip it and start with molarity.

SECTION A Molarity

When you work in a laboratory, it is much faster and easier to measure a volume of a solution than to weigh it on a balance. Consequently, a

concentration expression was set up that would have the amount of the
substance of interest expressed in _____ because moles are convenient, | moles
and the amount of solution expressed in the metric unit of volume,
which is liters.

When concentrations are expressed this way, they are called *molarity*
and abbreviated *M*. Thus a 6.0 *M* solution of sugar has _____ mol of | 6.0
sugar in __ L of solution. Writing this as a fractional conversion factor: | 1

$$\frac{6.0 \text{ mol sugar}}{\text{liter solution}}$$

If you have 3 mol of acetic acid dissolved in a liter of solution, you call
this a _____ or _____ solution. | 3 *M*; 3 molar

There are several new terms that you will be using when you work with
solutions. First there is the *solute*, which is the substance that you
dissolve. You dissolve the solute in a *solvent*, and the resulting mixture is
called the *solution*. Generally, the solute is present in smaller amounts
than the solvent.

Now we can give molarity its exact definition:

$$\text{Molarity} = M = \frac{\text{moles solute}}{\text{liter solution}}$$

As you can see, it is a conversion factor.

How about trying a few problems? They are really quite easy.

How many moles of HNO_3 are in exactly 300. mL of a 6.0 *M* solution
of HNO_3? Starting in the usual way, you write

? _____ = _____ | mol HNO_3; 300. mL soln

If you write 6.0 *M* HNO_3 as a conversion factor given in the problem,
you have _____. In order to be able to use this | $\frac{6.0 \text{ mol } HNO_3}{\text{L soln}}$
conversion factor, you must convert the unwanted dimension, milliliters
solution, to _____. You can do this by multiplying by | liters solution
_____. Now you can multiply by the molarity, the | $\frac{10^{-3} \text{ L solution}}{\text{mL solution}}$
conversion factor given in the problem, and the only dimension left that
does not cancel is _____. Therefore the answer must be | mol HNO_3

$$? \text{ mol } HNO_3 = 300. \text{ mL soln} \times \frac{10^{-3} \text{ L soln}}{\text{mL soln}} \times \frac{6.0 \text{ mol } HNO_3}{\text{L soln}}$$

$$= 1.8 \text{ mol } HNO_3$$

Try problem 1 on the pretest, which asks: How many grams of HCl are
in 3.00 L of 1.5 *M* HCl? Starting in the usual way, you have

? _____ = _____ | g HCl; 3.00 L soln

$\dfrac{1.5 \text{ mol HCl}}{\text{L soln}}$

moles HCl

grams HCl

$\dfrac{36.5 \text{ g HCl}}{\text{mol HCl}}$

The conversion factor given in the problem is the molarity of the HCl solution, _____. When you multiply by this conversion factor, you cancel the unwanted dimension, liters solution, but you leave a new unwanted dimension, _____. However, the dimension of the answer is _____, so you must multiply by the conversion factor _____, which is the molecular weight of HCl. The answer is

$$? \text{ g HCl} = 3.00 \, \cancel{\text{L soln}} \times \dfrac{1.5 \, \cancel{\text{mol HCl}}}{\cancel{\text{L soln}}} \times \dfrac{36.5 \text{ g HCl}}{\cancel{\text{mol HCl}}} = 164 \text{ g HCl}$$

Question 4 on the pretest is the same sort of problem turned the other way around. How many liters of a 0.70 *M* solution of HCl would you have to take in order to have 85.0 g HCl? (Remember to write your own setup as you fill in the answers to the blanks.)
Starting in the usual way, you have

L soln; 85.0 g HCl

$? _____ = _____$

grams HCl

$\dfrac{0.70 \text{ mol HCl}}{\text{L soln}}$

$\dfrac{\text{mol HCl}}{36.5 \text{ g HCl}}$

moles HCl

$\dfrac{\text{L soln}}{0.70 \text{ mol HCl}}$

liters solution

The unwanted dimension is _____. The conversion factor given in the problem is the molarity of the solution, _____. You can get to this from the unwanted dimension of the information given by multiplying by _____, the inverted form of the molecular weight. The unwanted dimension left is _____, which appears in the conversion factor, the molarity, given in the problem.

Next you multiply by _____, and the only remaining dimension is _____, which is the dimension that the answer must have. Therefore the answer is

$$? \text{ L soln} = 85.0 \, \cancel{\text{g HCl}} \times \dfrac{\cancel{\text{mol HCl}}}{36.5 \, \cancel{\text{g HCl}}} \times \dfrac{\text{L soln}}{0.70 \, \cancel{\text{mol HCl}}} = 3.3 \text{ L soln}$$

How can you determine the molarity of a solution? You can do this most easily by using the same type of short cut we used in Chapter 4. There we used it for density, which is also a conversion factor. You say that your answer must have the dimensions of molarity, which are

$\dfrac{\text{mol solute}}{\text{L soln}}$

_____. Then you set up the problem so that all dimensions except these two cancel. A simple example will make this clear.
What is the molarity of an acetic acid solution if there are 3.0 mol of acetic acid in 4.0 L of solution? Since the answer will be in terms of

$\dfrac{\text{mol acetic acid}}{\text{L soln}}$

molarity, its dimensions must be _____. So you write

$$? M = \dfrac{? \text{ mol acetic acid}}{\text{L soln}} =$$

$\dfrac{3.0 \text{ mol acetic acid}}{4.0 \text{ L soln}}$

The conversion factor given in the problem is _____.
If you write this in, you see that you are left with exactly the dimensions that the answer must have:

$$? M = \frac{? \text{ mol acetic acid}}{\text{L soln}} = \frac{3.0 \text{ mol acetic acid}}{4.0 \text{ L soln}} = 0.75 \ M$$

Of course, you won't always get all the information given to you in such an easy form. However, simply keep multiplying by the appropriate conversions that will get you to the two dimensions of the answer.

Here is another problem. What is the molarity of a solution that contains 8.0 g NaOH in 300.0 mL of solution? Since the question asked

is the molarity, then the answer will have the dimensions _____.

$\dfrac{\text{mol NaOH}}{\text{L soln}}$

The conversion factor given in the problem is _____. You start by writing

$\dfrac{8.0 \text{ g NaOH}}{300 \text{ mL soln}}$

$$? M = \frac{? \text{ mol NaOH}}{\text{liters soln}} = \frac{8.0 \text{ g NaOH}}{300 \text{ mL soln}}$$

Now you convert grams NaOH to _____ NaOH and milliliters solution

to _____ solution. You can do this by multiplying by the two conver-

sion factors _____ and _____. The answer, then, is

moles

liters

$\dfrac{\text{mol NaOH}}{40.0 \text{ NaOH}} , \dfrac{\text{mL soln}}{10^{-3} \text{ L soln}}$

$$? M = \frac{? \text{ mol NaOH}}{\text{L soln}} = \frac{8.0 \text{ g NaOH}}{300 \text{ mL soln}} \times \frac{\text{mol NaOH}}{40.0 \text{ g NaOH}} \times \frac{\text{mL soln}}{10^{-3} \text{ L soln}}$$

$$= \frac{0.67 \text{ mol NaOH}}{\text{L soln}} = 0.67 \ M$$

Sometimes difficulties are put in your way. Instead of having information about the volume of solution, you will have information about the amounts of solvent. In these cases you must add the amount of solute to the amount of solvent to get the amount of solution. Question 3 on the pretest is a typical example. It asks you to find the molarity of an NaCl solution that was prepared by adding 23.4 g NaCl to 100.0 g water. The density of the resulting solution was 1.20 g/mL.

First you must determine the weight of the solution. It is equal to the

weight of _____ plus the weight of water. So you have _____ g solution. Now you can get the conversion factor given in the problem as

NaCl; 123.4

$$\frac{23.4 \text{ g NaCl}}{\text{_____ g soln}}$$

123.4

You are also given another conversion factor in the problem,

_____, the density of the solution. Now the problem is straightforward.

$\dfrac{1.20 \text{ g soln}}{\text{mL soln}}$

$$? M = \frac{? \text{_____}}{\text{L soln}} = \frac{23.4 \text{ g NaCl}}{123.4 \text{ g soln}}$$

mol NaCl

Next you convert grams NaCl to _____ NaCl, which you can do by

multiplying by the conversion factor _____. Then you convert

grams solution to _____ solution. You can do this in two steps. First

moles

$\dfrac{\text{mol NaCl}}{58.5 \text{ g NaCl}}$

liters

$$\frac{1.20 \text{ g soln}}{\text{mL soln}}$$

milliliters solution

$$\frac{\text{mL soln}}{10^{-3} \text{ L soln}}$$

you can multiply by _____ to remove the unwanted dimension grams solution. The new unwanted dimension in the denominator is

_____. You can get to the desired liters solution by multiplying by _____.

The answer, then, is

$$? M = \frac{? \text{ mol NaCl}}{\text{L soln}} = \frac{23.4 \text{ g NaCl}}{123.4 \text{ g soln}} \times \frac{\text{mol NaCl}}{58.5 \text{ g NaCl}}$$

$$\times \frac{1.20 \text{ g soln}}{\text{mL soln}} \times \frac{\text{mL soln}}{10^{-3} \text{ L soln}}$$

$$= \frac{3.89 \text{ mol NaCl}}{\text{L soln}} = 3.89 \ M$$

If you didn't add the weight of the NaCl to the weight of water to get the weight of solution, you would have found the incorrect answer, 4.80 M, mentioned in the pretest answers. This is another example of why you must label your dimensions completely.

This problem is about as complicated as they can become. Incidentally, the density of the solution is not the same as the density of solvent. However, for very dilute solutions in which the amount of solute is very small, you can assume that the addition of a little solute won't change the density of the solvent significantly.

Be very careful if you are given a problem of this type with the amounts of solute and solvent specified in volumes. In many cases, the total volume of a solution is *not* simply the sum of the volumes of the components. If this sort of problem arises, you must have the densities of the solute and solvent available so that you can determine their weights, which are additive. This type of problem is so rare that we won't even do an example.

However, we *will* use a slightly different method of determining molarity. You've seen the method before. It is the technique of rewording the question when you are asked to find the value of a conversion factor. You make the conversion factor both the question asked and the information given. Thus if you want the molarity—that is, the moles per liter of solution—the question asked becomes "How many

L soln

1.00 L soln

moles?" The information given is 1.00 _____. You really want to know how many moles equals _____. (Incidentally, the 1.00 is a defined number and doesn't affect the significant figures. As a matter of fact, you could write just *liter*, with an understood 1.00 in front of it.)

Let's try the second example we just had. What is the molarity of a solution that contains 8.0 g NaOH in exactly 300 mL of solution? You start from the reworded question:

mol NaOH; 1.00 L soln

? _____ = _____

liters solution

The unwanted dimension is _____, but the conversion factor given in the problem has milliliters solution. You must convert:

$$\frac{\text{mL soln}}{10^{-3} \text{ L soln}}$$

? mol NaOH = 1.00 L soln × _____

Now you can use the conversion factor given in the problem:

$$? \text{ mol NaOH} = 1.00 \text{ } \cancel{\text{L soln}} \times \frac{\text{mL soln}}{10^{-3} \text{ } \cancel{\text{L soln}}} \times \underline{\hspace{3cm}}$$

8.0 g NaOH
300 mL soln

It is easy to convert from the remaining unwanted dimension, grams NaOH, to the units of the answer, moles NaOH. You use the _____ _____ of NaOH in its inverted form:

molecular weight

$$? \text{ mol NaOH} = 1.00 \text{ } \cancel{\text{L soln}} \times \frac{\cancel{\text{mL soln}}}{10^{-3} \text{ } \cancel{\text{L soln}}}$$

$$\times \frac{8.0 \text{ g NaOH}}{300 \text{ } \cancel{\text{mL soln}}} \times \underline{\hspace{3cm}}$$

mol NaOH
40.0 g NaOH

The only dimension left is moles NaOH, which is what you want. So canceling dimensions leaves

$$? \text{ mol NaOH} = 1.00 \times \frac{1}{10^{-3}} \times \frac{8.0}{300} \times \frac{\text{mol NaOH}}{40.0}$$

$$= 0.67 \text{ mol NaOH in 1.00 L soln.}$$

This means that there are 0.67 mol per liter, so the solution is 0.67 *M*.

Here is Question 3 on the pretest again, worked by rewording the question. You were asked to find the molarity of an NaCl solution prepared by adding 23.4 g NaCl to 100.0 g water. The density of the solution was 1.20 g/mL. As before, you must determine the total weight of the solution. It is _____ g. Once again this gives you a conversion factor:

123.4

$$\frac{23.4 \text{ g} \underline{\hspace{1.5cm}}}{123.4 \text{ g} \underline{\hspace{1.5cm}}}$$

NaCl

soln

The density of the solution gives you another conversion factor,

$$\frac{1.20 \text{ g} \underline{\hspace{1.5cm}}}{\text{mL} \underline{\hspace{1.5cm}}}$$

soln

soln

Since you have a weight of NaCl in the conversion factor, you will undoubtedly have to use the _____ of NaCl. This is another conversion factor,

molecular weight

$$\frac{58.5 \text{ g} \underline{\hspace{1.5cm}}}{\underline{\hspace{2cm}}}$$

NaCl

mol NaCl

(It's a very good idea to write the conversion factors given in the problem as well as any others you think you might need *before starting the problem.* Then, as you solve it, you can just pick and choose from these.)

mol

1.00 L

1.00 L soln

Now you reword the question to ask "How many _____ NaCl are in _____ soln? Then you start the problem:

? mol NaCl = _____

The unwanted dimension is a volume of the solution. The only conver-

density

sion factor available with a volume is the _____ of the solution, but it is in milliliters. So you have to convert liters to milliliters before you use the density. Do both steps next:

g soln

$$? \text{ mol NaCl} = 1.00 \text{ L soln} \times \frac{\text{mL soln}}{\text{_____ L soln}} \times \frac{1.20 \text{ _____}}{\text{_____}}$$

10^{-3}; mL soln

grams solution

The unwanted dimension is now _____. You still have a conversion factor from the problem with this dimension. Multiply by it.

$\dfrac{23.4 \text{ g NaCl}}{123.4 \text{ g soln}}$

$$? \text{ mol NaCl} = 1.00 \text{ L soln} \times \frac{\text{mL soln}}{10^{-3} \text{ L soln}} \times \frac{1.20 \text{ g soln}}{\text{mL soln}} \times \text{_____}$$

Now that the remaining unwanted dimension is grams NaCl, it is easy enough to get to the moles NaCl you want for the answer:

$$? \text{ mol NaCl} = 1.00 \text{ L soln} \times \frac{\text{mL soln}}{10^{-3} \text{ L soln}}$$

$\dfrac{\text{mol NaCl}}{58.5 \text{ g NaCl}}$

$$\times \frac{1.20 \text{ g soln}}{\text{mL soln}} \times \frac{23.4 \text{ g NaCl}}{123.4 \text{ g soln}} \times \text{_____}$$

After you cancel dimensions, this gives you, for your answer:

$$? \text{ mol NaCl} = 1.00 \times \frac{1}{10^{-3}} \times 1.20 \times \frac{23.4}{123.4} \times \frac{\text{mol NaCl}}{58.5}$$

$$= 3.89 \text{ mol NaCl in 1.00 L soln.}$$

Therefore your answer is 3.89 *M*.

SECTION B   Formality

By this point in the chapter some chemists and chemistry teachers might question our choice of words, saying that we have been using "molarity" when we should have been using "formality." This is the same sort of problem that arose in Section A of Chapter 7. Some compounds do not exist as molecules, but rather as ions held together by electrostatic bonding. If you define moles as being 6.02×10^{23} *molecules*, you cannot speak of moles of ionic compounds. You can use only *gram-formula weight*, which is the weight of 6.02×10^{23} of the ions that make up the formula of the compound. (The SI system neatly avoids this distinction. It defines a mole as being the mass of the same number of units of whatever is mentioned as there are C atoms in 12.000 g of carbon-12 isotope.) Those who want to be super-precise do not use the concentration molarity for substances that are dissociated in water solution. An example would be HCl dissolved in water. If you had 1.00 mol of HCl gas (molecules) dissolved in 1.00 L of water, you would call the solution 1.00 F (that is, 1.00 formal) because the HCl molecules have dissociated into H^+ and Cl^- ions. Formality is defined as

$$\text{Formality} = F = \frac{\text{formula weights solute}}{\text{liter solution}}$$

Question 2 on the pretest asks to what volume in liters you must dilute 11.7 g NaCl to prepare a 0.10 F solution of NaCl. Starting in the usual way, you have

? _____ = 11.7 g NaCl   L soln

The conversion factor given in the problem is that you have a 0.10 F solution. This can be written as

0.10 _____ NaCl   formula wt
         L soln

The formula weight of NaCl (which is exactly equal to what we have been calling molecular weight) is

         58.5 g NaCl
_____

_____ NaCl   formula wt

Now you can solve the problem. In order to cancel the unwanted

dimension, you multiply by _____. This leaves the $\dfrac{\text{formula wt NaCl}}{\text{58.5 g NaCl}}$

unwanted dimension _____. You can cancel this   formula wt NaCl
dimension and get to the dimensions of your answer by multiplying by

_____. The answer is therefore $\dfrac{\text{L soln}}{\text{0.10 formula wt NaCl}}$

$$? \text{ L soln} = 11.7 \, \cancel{\text{g NaCl}} \times \frac{\cancel{\text{formula wt NaCl}}}{58.5 \, \cancel{\text{g NaCl}}} \times \frac{\text{L soln}}{0.10 \, \cancel{\text{formula wt NaCl}}}$$

$$= 2.0 \text{ L soln}$$

Since you calculate molarity and formality problems in exactly the same way, you might simply interchange M (molarity) and F (formality), and moles with formula weights, when you are faced with problems containing F (formality).

SECTION C Molality

There is one more concentration unit that deals with moles of solute. This is *molality*, which is defined as

$$\text{Molarity} = m = \frac{\text{moles solute}}{\text{1000 g solvent}} = \frac{\text{moles solute}}{\text{kilogram solvent}}$$

Notice that the moles of solute are not related to the volume of solution, but rather to the weight of solvent. This type of relationship finds use in only a special class of problems. These problems are related to what are called the *colligative properties* of solutions, which will be covered in Chapter 16. For the moment, it is necessary that you know only how to handle simple problems expressing the concentration in molality.

We can start right out with question 8 on the pretest, which asks for the molality of an alcohol solution prepared by dissolving 23.0 g alcohol (C_2H_5OH) in 200.0 g water. The specific question asked is "_____

_____" Therefore the dimensions that the answer will have

What is the molality?

kilogram

are moles alcohol/_____ water. You can write

$$? m = \frac{? \text{ mol alcohol}}{\text{kg water}} =$$

$\dfrac{23.0 \text{ g alcohol}}{200 \text{ g water}}$

moles

kilograms

$\dfrac{\text{mol alcohol}}{46.0 \text{ g alcohol}} , \dfrac{10^3 \text{ g water}}{\text{kg water}}$

The conversion factor given in the problem is _____.

You can convert grams of alcohol to _____ of alcohol and convert

grams of water to _____ of water by multiplying by two

conversion factors: _____ and _____. And so the answer is

$$? m = \frac{? \text{ mol alcohol}}{\text{kg water}} = \frac{23.0 \text{ g alcohol}}{200 \text{ g water}} \times \frac{\text{mol alcohol}}{46.0 \text{ g alcohol}} \times \frac{10^3 \text{ g water}}{\text{kg water}}$$

$$= 2.5 \, m$$

If you prefer to reword the question when the desired information is itself a conversion factor, you can ask for molality as "How many moles

kilogram

of solute are dissolved in a _____ of solvent?" In the example we just had, you would start the solution of the problem this way:

mol alcohol; kg water

$$? \underline{\hspace{3cm}} = \underline{\hspace{2cm}}$$

Since the conversion factor in the problem has grams of water, you must first convert the unwanted dimension, kilograms of water,

$$? \text{ mol alcohol} = \text{kg water} \times \frac{10^3 \text{ g water}}{\underline{\hspace{1.5cm}}}$$

kg water

Then you can use the conversion factor from the problem,

$\dfrac{23.0 \text{ g alcohol}}{200 \text{ g water}}$

$$? \text{ mol alcohol} = \text{kg water} \times \frac{10^3 \text{ g water}}{\text{kg water}} \times \underline{\hspace{2cm}}$$

grams alcohol

The remaining unwanted dimension is _____, which you convert to the moles alcohol you want for the answer:

$$? \text{ mol alcohol} = \text{kg water} \times \frac{10^3 \text{ g water}}{\text{kg water}} \times \frac{23.0 \text{ g alcohol}}{200 \text{ g water}}$$

$\dfrac{\text{mol alcohol}}{46.0 \text{ g alcohol}}$

$$\times \underline{\hspace{3cm}}$$

When you cancel all the dimensions appearing in the numerator and denominator, you are left with

$$? \text{ mol alcohol} = 1 \times 10^3 \times \frac{23.0}{200} \times \frac{\text{mol alcohol}}{46.0}$$

$$= 2.50 \text{ mol alcohol in 1 kg water}$$

So there are 2.50 mol alcohol per kilogram water, and the solution is 2.50 *m*.

There are many ways to present the conversion factor given in the problem. Take, for example, the problem of finding the molality of a 30.0 wt % solution of ethylene glycol in water. (The formula for ethylene glycol is $C_2H_6O_2$.) Here you must use the fact that the sum of

the percentages always adds up to _____, so that if the solution is | 100

30.0 wt % ethylene glycol, it must be _____ wt % water. In fact, you can | 70.0

make a new conversion factor from this: There are _____ g ethylene | 30.0

glycol per _____ g water. Or, writing this as a fraction, _____. | 70.0; $\dfrac{30.0 \text{ g ethylene glycol}}{70.0 \text{ g water}}$

Now the solution to the problem is the same as in the first example:

$$? \, m = \frac{? \underline{\quad\quad} \text{ ethylene glycol}}{\underline{\quad\quad\quad}}$$

| mol

| kg water

If you now put in the conversion factor that you developed from the information given in the problem, you have

$$? \, m = \frac{? \text{ mol ethylene glycol}}{\text{kg water}} = \frac{30.0 \text{ g ethylene glycol}}{70.0 \text{ g water}}$$

You must then convert grams ethylene glycol to _____ ethylene glycol | moles

and grams water to _____ water. You do this by using the two | kilograms

conversion factors, _____ and _____: | $\dfrac{\text{mol ethylene glycol}}{62.0 \text{ g ethylene glycol}}$; $\dfrac{10^3 \text{ g water}}{\text{kg water}}$

$$? \, m = \frac{30.0 \text{ g ethylene glycol}}{70.0 \text{ g water}} \times \frac{\text{mol ethylene glycol}}{62.0 \text{ g ethylene glycol}} \times \frac{10^3 \text{ g water}}{\text{kg water}}$$

$$= 6.91 \, m$$

You can go from a concentration given in molarity to a concentration in molality, but you must know the density of the solution. To find the molality, you must know the weights of solute and solvent separately. If you know the density of the solution, you can calculate how much a liter of it weighs. If you know the number of moles of solute in a liter (the molarity), you can calculate the weight of solute in a liter. Then, if you subtract the weight of solute in a liter from the total weight of a liter of solution, you will have the weight of solvent in a liter of solution. This seems complicated, but only because it must be done in several steps. Working question 9 on the pretest will clarify this type of problem.

What is the molality of a 6.0 *M* solution of HCl whose density is 1.1 g/mL? First find the weight of a liter of the solution:

$$\frac{1.1 \text{ g soln}}{\text{mL soln}}$$

milliliters

$$\frac{\text{mL soln}}{10^{-3} \text{ L soln}} , \frac{1.1 \text{ g soln}}{\text{mL soln}}$$

1100

$$\frac{6.0 \text{ mol HCl}}{\text{L soln}}$$

moles HCl

$$\frac{36.5 \text{ g HCl}}{\text{mol HCl}}$$

6.0

881

mol HCl

kg water

$$\frac{6.0 \text{ mol HCl}}{881 \text{ g water}}$$

grams; kilograms

$$\frac{10^3 \text{ g water}}{\text{kg water}}$$

? g soln = 1.00 L soln

You can do this using the density of the solution, _____, after first converting liters of solution to _____ of solution. The answer is

? g soln = 1.00 L soln X _____ X _____

= _____ g soln

Next you must calculate the weight of HCl in 1.00 L solution:

? g HCl = 1.00 L soln

Since you know that the solution is 6.0 M HCl, you have a conversion factor given. It is _____. If you multiply by this, you cancel liter solution but have the new unwanted dimension, _____. However, it is simple to get from this unwanted dimension to the dimension of the answer by multiplying by _____. And so the answer is

$$? \text{ g HCl} = 1.0 \text{ L soln} \times \frac{6.0 \text{ mol HCl}}{\text{L soln}} \times \frac{36.5 \text{ g HCl}}{\text{mol HCl}} = 219 \text{ g HCl}$$

You now know the weight of one liter of solution and the weight of HCl that it contains. You can find the weight of water in the solution by subtraction:

? g water = 1100 g soln − 219 g HCl = 881 g water

You now know that 1.0 L of the solution that contains _____ mol HCl also contains _____ g water. This is the conversion factor you need to solve the problem. Now you start the usual way:

$$? \, m = \frac{? \, \rule{3cm}{0.4pt}}{\rule{3cm}{0.4pt}}$$

You put in the conversion factor, which you have calculated:

$$? \, m = \frac{? \text{ mol HCl}}{\text{kg water}} = \rule{3cm}{0.4pt} .$$

All you must do now is convert _____ water to _____ water. Therefore you multiply by _____. And so the answer is

$$? \, m = \frac{6.0 \text{ mol HCl}}{881 \text{ g water}} \times \frac{10^3 \text{ g water}}{\text{kg water}} = 6.8 \, m$$

The same type of problem could be asked the other way around. You could be given a solution of known molality and asked to calculate the molarity. In this case, you determine the weight of solute per kilogram solvent and add this weight to the 1000 g solvent. Then, using the density of solution, calculate what volume of solution this composite weight of solution equals. You then know the number of moles of solute per this calculated volume of solution and can easily calculate the molarity.

These complicated problems occur infrequently. We won't spend any more time on them, but will go on to more usual calculations.

SECTION D   Preparing Dilute Solutions from Concentrated Ones

Very often in a laboratory you have a solution of known molarity or molality and want to use it to prepare a certain volume of a more dilute solution. Calculating how much to use is simple, except that you are working with *two* solutions: one the initial solution that you have and the other the final solution that you wish to prepare. You must be very careful to label your dimensions so that you don't confuse the two. We will work through question 5 on the pretest, which should make this clear.

How many milliliters of a 6.0 *M* solution would you dilute to 3.0 L in order to have a 0.10 *M* HCl solution? It is easiest to call the 6.0 *M* HCl solution the "initial solution" and the 0.10 *M* HCl solution the "final solution." There are two molarities given in the problem, and therefore two conversion factors:

$$\frac{6.0 \text{ mol HCl}}{\text{L init soln}} \quad \text{and} \quad \frac{0.10 \text{ mol HCl}}{\underline{\hspace{2cm}}}$$

final soln

The information given in the problem is that the volume of the final

solution will be _____. Thus, starting in the usual way, you write

3.0 L

$$? \text{ mL} \underline{\hspace{2cm}} = 3.0 \text{ L} \underline{\hspace{2cm}}$$

init soln; final soln

Notice how careful you must be to distinguish between the two solutions.

To cancel the unwanted dimension _____, you multiply by the

L final soln

conversion factor _____. The unwanted dimension is

$\dfrac{0.10 \text{ mol HCl}}{\text{L final soln}}$

now _____. But this appears in the other conversion factor, so

mol HCl

you can multiply by _____. The unwanted dimension

$\dfrac{\text{L init soln}}{6.0 \text{ mol HCl}}$

is now _____. But you want your answer to have the

liters init solution

dimensions _____. You can make this conversion by

milliliters init solution

multiplying by _____. Your setup should therefore appear as

$\dfrac{\text{mL init soln}}{10^{-3} \text{ L init soln}}$

$$? \text{ mL init soln} = 3.0 \; \cancel{\text{L final soln}} \times \frac{0.10 \; \cancel{\text{mol HCl}}}{\cancel{\text{L final soln}}}$$

$$\times \frac{\cancel{\text{L init soln}}}{6.0 \; \cancel{\text{mol HCl}}} \times \frac{\text{mL init soln}}{10^{-3} \; \cancel{\text{L init soln}}}$$

$$= 50 \text{ mL init soln}$$

These problems are easy if you *always* write down the complete dimension of each solution to which you are referring.

Now let's do question 6 in the pretest. What is the formality of a solution prepared by adding enough water to 250 mL of a 6.0 *F* solution of KCl to bring the volume up to 5.0 L? Although this problem concerns formality, we're going to solve it as if it were molarity, since the two are numerically identical.

$\dfrac{6.0 \text{ mol KCl}}{\text{L init soln}}$

The problem contains two conversion factors: One is _____, the molarity (formality) of the initial solution. The other is that there

init; final

$\dfrac{250 \text{ mL init soln}}{5.0 \text{ L final soln}}$

are 250 mL of _____ solution in 5.0 L _____ solution. This can be written fractionally as _____.

Since you want the molarity of the final solution, you write

$$? \, M \text{ final solution} = \frac{? \text{ mol KCl}}{\text{L final soln}}$$

Now you can pick either of the conversion factors given in the problem to start. It must have one of the dimensions of the answer in the correct place. Suppose that you start with the first,

$$\frac{6.0 \text{ mol KCl}}{\text{L init soln}}$$

This gives you moles KCl in the numerator, but has an unwanted

liters init soln

dimension in the denominator, which is _____.
However, your second conversion factor allows you to cancel this. You

$\dfrac{0.250 \text{ L init soln}}{5.0 \text{ L final soln}}$

therefore multiply by _____. (Notice how 250 mL was changed to 0.250 L. You will start doing this by yourself soon. It saves writing one more conversion factor.)

The dimensions that remain are the dimensions of the answer:

$$? \, F = ? \, M = \frac{? \text{ mol KCl}}{\text{L final soln}} = \frac{6.0 \text{ mol KCl}}{\cancel{\text{L init soln}}} \times \frac{0.250 \; \cancel{\text{L init soln}}}{5.0 \text{ L final soln}}$$

$$= 0.30 \, F$$

Working the problem by rewording the question produces almost the same setup:

L final soln

$$? \text{ Formula weight KCl} = ? \text{ mol KCl} = \underline{\hspace{4cm}}$$

Then there is no question which of the two conversion factors to use first. You choose the one containing the unwanted liters final soln:

$\dfrac{250 \text{ mL init soln}}{5.0 \text{ L final soln}}$

$$? \text{ mol KCl} = \text{L final soln} \times \underline{\hspace{3cm}}$$

The second conversion factor has liters initial soln as one of its dimensions, so before you use the first conversion factor, save yourself a step by writing 250 mL init soln and 0.250 L init soln. So you write

$$? \text{ mol KCl} = \text{L final soln} \times \frac{0.250 \text{ L init soln}}{5.0 \text{ L final soln}}$$

Now you can get directly to the dimensions of the answer. You multiply by _____.

$\dfrac{6.0 \text{ mol KCl}}{\text{L init soln}}$

After canceling all dimensions that appear in both the numerator and denominator, you have

$$? \text{ mol KCl} = \frac{0.250}{5.0} \times 6.0 \text{ mol KCl} = 0.30 \text{ mol KCl present per L soln}$$

Therefore the solution is 0.30 *M*, which is the same as 0.30 *F*.

Sometimes the concentration of the initial solution isn't given to you in molarity, molality, or formality, but this doesn't make the problem much more difficult. For example, how many grams of a 35 wt % solution of NaOH must be diluted to 25.0 mL to have a 0.10 *M* solution of NaOH?

In this case the initial NaOH solution is shown as a weight percent. The conversion factor that you have from this is _____.

$\dfrac{35 \text{ g NaOH}}{100 \text{ g init soln}}$

The concentration of the final solution, however, is given in molarity. The conversion factor you have from this is _____. The information given in the problem is that you want to prepare 25.0 mL _____.

$\dfrac{0.10 \text{ mol NaOH}}{\text{L final soln}}$

final soln

Starting in the usual way, then, you write

$? \text{ g init soln} = 25.0 \text{ mL final soln}$

One of the conversion factors has a volume of final solution in it, but it is expressed in _____ of final solution. However, you can save yourself one conversion factor by expressing 1 liter as _____ mL. Then you can multiply by

liters

10^3

$\dfrac{0.10 \text{ mol NaOH}}{10^3 \text{ mL final soln}}$

The unwanted dimension is now _____, but the other

moles NaOH

conversion factor given in the problem uses _____ NaOH.

grams

Consequently, you convert moles NaOH to _____ NaOH by multiplying by the conversion factor _____. Now you can multiply by the conversion factor given in the problem, _____.

grams

$\dfrac{40.0 \text{ g NaOH}}{\text{mol NaOH}}$

$\dfrac{100 \text{ g init soln}}{35 \text{ g NaOH}}$

The only dimension remaining is _____, and so the answer must be

grams initial solution

$$? \text{ g init soln} = 25.0 \text{ mL final soln} \times \frac{0.10 \text{ mol NaOH}}{10^3 \text{ mL final soln}}$$

$$\times \frac{40.0 \text{ g NaOH}}{\text{mol NaOH}} \times \frac{100 \text{ g init soln}}{35 \text{ g NaOH}}$$

$$= 0.29 \text{ g init soln}$$

For practice, we'll try another problem, which is a little more complicated. Instead of finding the weight of initial solution, you will find the volume.

How many milliliters of a 10 wt % solution of NaOH whose density is 1.1 g/mL will you need to prepare 300 mL of a 0.20 *M* solution of NaOH? There are three conversion factors given in the problem, and you start by writing these down:

init soln	$\dfrac{10 \text{ g NaOH}}{\rule{3cm}{0.4pt}}$ $\dfrac{1.1 \text{ g} \rule{2cm}{0.4pt}}{}$
init soln; init soln	$\dfrac{100 \text{ g} \rule{2cm}{0.4pt}}{}$ $\dfrac{\text{mL} \rule{2cm}{0.4pt}}{}$
	$\dfrac{0.20 \text{ mol NaOH}}{}$
final soln	$\text{L} \rule{3cm}{0.4pt}$

Since the solute is expressed in grams in one of the conversion factors, you will also need a conversion factor that converts grams NaOH to

moles; molecular   _____ NaOH. This conversion factor is the _____ weight. It can be written as

$$\frac{40.0 \text{ g NaOH}}{}$$

mol   _____ NaOH

Now you are ready to start. The question asked is "How many

initial   milliliters _____ solution?" The information given is 0.300 L

final   _____ soln. (Notice how 300 mL was immediately written as 0.300 L; this saves one conversion factor.) Starting in the usual way, you have

$$? \text{ mL init soln} = 0.300 \text{ L final soln}$$

From the conversion factors that you have written, you pick one with

liters final solution   _____ in the denominator. So you multiply by

$\dfrac{0.20 \text{ mol NaOH}}{\text{L final soln}}$; moles NaOH   _____. The unwanted dimension now is _____.

grams   Since the other conversion factor given in the problem is in _____ NaOH, you multiply by the molecular-weight conversion factor,

$\dfrac{40.0 \text{ g NaOH}}{\text{mol NaOH}}$; $\dfrac{100 \text{ g init soln}}{10 \text{ g NaOH}}$ _____. Now you can multiply by _____. The

grams initial solution   unwanted dimension is now _____. But the last conversion factor, the density, will get you to the dimensions of the answer.

$\dfrac{\text{mL init soln}}{1.1 \text{ g init soln}}$ You multiply by _____. The only dimension remaining is the dimension of the answer, which must be

? mL init soln = 0.300 L~~final soln~~ $\times \dfrac{0.20 \text{~~mol NaOH~~}}{\text{L~~final soln~~}}$

$$\times \dfrac{40.0 \text{~~g NaOH~~}}{\text{~~mol NaOH~~}} \times \dfrac{100 \text{~~g init soln~~}}{10 \text{~~g NaOH~~}} \times \dfrac{\text{mL init soln}}{1.1 \text{~~g init soln~~}}$$

= 22 mL init soln

SECTION E   Mixtures of Solutions of Different Molarity

In a laboratory you sometimes mix two or more solutions of the same substance that are of different concentrations. In order to calculate the concentration of the resulting mixture, you must know the total amount of solute present and the total volume of the mixture.

You may prefer to do this calculation in a series of steps. First you calculate the amount of solute in each sample taken. Next you add these together. Then you add the volumes of the samples. Using these two composite amounts, you can calculate the concentration of the mixture. We will do question 7 on the pretest as an example of this type.

What will be the molarity of a solution prepared by mixing 3.0 L of 0.10 *M* HCl with 1.0 L of 6.0 *M* HCl?

First you calculate the number of moles of HCl in each solution.

? mol HCl = 3.0 L~~0.1 M~~ $\times \dfrac{0.1 \text{ mol HCl}}{\text{L~~0.1 M~~}}$ = 0.30 mol HCl

? mol HCl = 1.0 L~~6.0 M~~ $\times \dfrac{6.0 \text{ mol HCl}}{\text{L~~6.0 M~~}}$ = 6.0 mol HCl

(Notice the way the solutions are labeled to distinguish between them.)

Total mol HCl = 0.30 mol HCl + 6.0 mol HCl = 6.3 mol HCl

Then you calculate the total volume of the final solution, which is equal to the sum of the volume of the two solutions:

Total volume = 3.0 L + 1.0 L = 4.0 L

Now that you have the number of moles of HCl and the volume of solution, you can easily calculate the molarity:

? *M* = $\dfrac{? \text{ mol HCl}}{\text{L soln}}$ = $\dfrac{6.3 \text{ mol HCl}}{4.0 \text{ L soln}}$ = $\dfrac{1.6 \text{ mol HCl}}{\text{L soln}}$ = 1.6 *M*

PROBLEM SET

These problems are relatively simple. Some have nonmetric units to refresh your memory on these conversions. Check the answer to each problem as you finish it, to see that you are on the right track.

1. How many moles of ethyl alcohol, C_2H_5OH, are present in 65 mL of a 1.5 M solution?

2. How many liters of a 6.0 M solution of acetic acid, CH_3COOH, contain 0.0030 mol acetic acid?

3. Suppose that you want to have 14.6 g HCl. How many milliliters of a 0.10 F solution of HCl will you need?

4. How many grams of a 0.50 F solution of HCl would contain 7.3 g HCl? The density of the solution is 1.2 g/mL.

5. If you dissolve 1.96 g H_2SO_4 in enough water to bring the volume to 250 mL, what is the molarity of the solution?

6. What is the weight percent formic acid, HCOOH, in a 0.40 M solution of formic acid? The solution's density is 1.15 g/mL.

7. What is the formality of a 25 wt % solution of NaOH? The solution's density is 1.3 g/mL.

8. If you want to prepare 600 mL of a 1.50 M solution of acetic acid, CH_3COOH, how many grams of the acid must you take?

9. How many grams of sodium formate, HCOONa, are in 5.0 mL of a 1.4 F solution of sodium formate?

10. How many ounces of NaCl are required to prepare 4.0 L of 0.10 F NaCl solution?

11. If you have 5.0 gal of 0.10 F KOH solution, how many pounds of KOH are present?

12. How many grams of ethyl alcohol, C_2H_5OH, must you add to 30.0 g water to prepare a 0.10 M solution?

13. What is the molality of a 40 wt % solution of NaOH?

14. You want 85 g of KOH. How many grams of a 3.0 M solution of KOH will provide it?

15. What is the wt % $AlCl_3$ in a 0.20 M solution of $AlCl_3$?

16. To what volume must you bring 25.0 mL of a 6.00 M solution of sugar in order to prepare a 0.150 M solution of sugar?

17. If you want to prepare 450 mL of 0.35 M formic acid, how many milliliters of 5.0 M formic acid, HCOOH, must you use?

18. What is the molarity of a solution prepared by adding 3.0 L of water (solvent) to 1.7 L of 0.50 M sugar solution?

19. What is the molality of a 0.10 M solution of ethylene glycol, $C_2H_6O_2$? The solution's density is 0.90 g/mL.

20. What is the molarity of the solution that results from mixing 50 mL of 10.0 M HCl with 2.0 L of 0.10 M HCl?

21. When you evaporate 250.0 mL of a 0.585 F solution of NaCl to dryness, how many grams of NaCl should you find?

22. If you dissolve 0.70 mol of HCl in enough water to prepare 250 mL of solution, what is the molarity of the solution you've prepared?

23. What is the percent purity of an impure sample of $Na_2C_2O_4$, if you dissolve 30.0 g of the impure material in enough water to make 100.0 mL of solution, and get a solution that is 2.00 M $Na_2C_2O_4$?

24. A solution is prepared by adding 2.0 L of 6.0 M HCl to 500 mL of a 9.0 M HCl solution. What is the molarity of the resulting solution? Assume that the volumes are additive.

25. You want to prepare 3.00 L of 6.00 M CH_3COOH (acetic acid). How many grams of pure CH_3COOH must you use?

26. What is the formality of a solution that is 11.1% (by weight) $CaCl_2$? The density of the solution is 1.15 g/mL.

27. What is the molality of the solution in Problem 26?

28. How many grams of a 90.0% pure sample of H_3PO_4 must you add to enough water to have 6.00 L of solution so that the solution is 1.50 M H_3PO_4?

29. When you have 250 mL of a 2.8 M solution of HCl, how many moles of HCl do you have?

PROBLEM SET ANSWERS

You should be able to do at least nine out of the first eleven problems correctly. If you cannot, redo Sections A and B in the chapter. Problems 12, 13, 14, 15, and 19 are covered in Section C in the chapter. If you miss more than two of these, redo the section. You should be able to do problems 16 and 17. If you cannot, check back to Section D. Problems 18 and 20 are covered in Section E, and you should be able to do at least one of them. Problems 21 to 30 come from all over the chapter.

1. ? mol alcohol = 65 mL soln $\times \dfrac{10^{-3} \text{ L soln}}{\text{mL soln}} \times \dfrac{1.5 \text{ mol alcohol}}{\text{L soln}}$

   = 0.098 mol alcohol

2. ? L soln = 0.0030 mol acetic acid $\times \dfrac{\text{L soln}}{6.0 \text{ mol acetic acid}}$

           = 5.0×10^{-4} L soln

3. ? mL soln = 14.6 g HCl $\times \dfrac{\text{formula wt HCl}}{36.5 \text{ g HCl}} \times \dfrac{10^3 \text{ mL soln}}{0.10 \text{ formula wt HCl}}$

           = 4.0×10^3 mL soln

4. ? g soln = 7.3 g HCl $\times \dfrac{\text{formula wt HCl}}{36.5 \text{ g HCl}} \times \dfrac{10^3 \text{ mL soln}}{0.50 \text{ formula wt HCl}} \times \dfrac{1.2 \text{ g soln}}{\text{mL soln}}$

           = 4.8×10^2 g soln

5. $? M = \dfrac{? \text{ mol } H_2SO_4}{\text{L soln}} = \dfrac{1.96 \text{ g } H_2SO_4}{250 \text{ mL soln}} \times \dfrac{\text{mL soln}}{10^{-3} \text{ L soln}} \times \dfrac{\text{mol } H_2SO_4}{98.1 \text{ g } H_2SO_4}$

           = 0.0799 *M*

    or

     $? \text{ mol } H_2SO_4 = \text{L soln} \times \dfrac{1.96 \text{ g } H_2SO_4}{0.250 \text{ L soln}} \times \dfrac{\text{mol } H_2SO_4}{98.1 \text{ g } H_2SO_4}$

           = 0.0799 mol H_2SO_4 in one liter of solution

    Therefore it is 0.0799 mol/L = 0.0799 *M*

6. ? g HCOOH = 100 g soln $\times \dfrac{\text{mL soln}}{1.15 \text{ g soln}} \times \dfrac{10^{-3} \text{ L soln}}{\text{mL soln}} \times \dfrac{0.40 \text{ mol HCOOH}}{\text{L soln}}$

           $\times \dfrac{46.0 \text{ g HCOOH}}{\text{mol HCOOH}}$

           = 1.6 g HCOOH in 100 g soln or 1.6%

7. $? F = \dfrac{? \text{ formula wt NaOH}}{\text{L soln}} = \dfrac{25 \text{ g NaOH}}{100 \text{ g soln}} \times \dfrac{\text{formula wt NaOH}}{40.0 \text{ g NaOH}} \times \dfrac{1.3 \text{ g soln}}{\text{mL soln}}$

           $\times \dfrac{\text{mL soln}}{10^{-3} \text{ L soln}}$

     = 8.1 *F*

    or

     ? formula wt NaOH = 1.00 L soln $\times \dfrac{\text{mL soln}}{10^{-3} \text{ L soln}}$

           $\times \dfrac{1.3 \text{ g soln}}{\text{mL soln}} \times \dfrac{25 \text{ g NaOH}}{100 \text{ g soln}} \times \dfrac{\text{formula wt NaOH}}{40.0 \text{ g NaOH}}$

           = 8.1 formula wt NaOH in one liter of solution

    Therefore it is 8.1 *F*.

8. ? g acetic acid = 0.600 L soln $\times \dfrac{1.50 \text{ mol acetic acid}}{\text{L soln}} \times \dfrac{60.0 \text{ g acetic acid}}{\text{mol acetic acid}}$

           = 54.0 g acetic acid

9. ? g HCOONa = 5.0 m̶L̶ ̶s̶o̶l̶n̶ $\times \dfrac{10^{-3} \text{ L̶ ̶s̶o̶l̶n̶}}{\text{m̶L̶ ̶s̶o̶l̶n̶}}$

$\times \dfrac{1.4 \text{ f̶o̶r̶m̶u̶l̶a̶ ̶w̶t̶ ̶H̶C̶O̶O̶N̶a̶}}{\text{L̶ ̶s̶o̶l̶n̶}} \times \dfrac{68.0 \text{ g HCOONa}}{\text{f̶o̶r̶m̶u̶l̶a̶ ̶w̶t̶ ̶H̶C̶O̶O̶N̶a̶}}$

= 0.48 g HCOONa

10. ? oz NaCl = 4.0 L̶ ̶s̶o̶l̶n̶ $\times \dfrac{0.10 \text{ f̶o̶r̶m̶u̶l̶a̶ ̶w̶t̶ ̶N̶a̶C̶l̶}}{\text{L̶ ̶s̶o̶l̶n̶}}$

$\times \dfrac{58.5 \text{ g̶ ̶N̶a̶C̶l̶}}{\text{f̶o̶r̶m̶u̶l̶a̶ ̶w̶t̶ ̶N̶a̶C̶l̶}} \times \dfrac{\text{l̶b̶ ̶N̶a̶C̶l̶}}{454 \text{ g̶ ̶N̶a̶C̶l̶}} \times \dfrac{16 \text{ oz NaCl}}{\text{l̶b̶ ̶N̶a̶C̶l̶}}$

= 0.82 oz NaCl

11. ? lb KOH = 5.0 g̶a̶l̶ ̶s̶o̶l̶n̶ $\times \dfrac{4 \text{ q̶t̶ ̶s̶o̶l̶n̶}}{\text{g̶a̶l̶ ̶s̶o̶l̶n̶}} \times \dfrac{0.946 \text{ L̶ ̶s̶o̶l̶n̶}}{\text{q̶t̶ ̶s̶o̶l̶n̶}}$

$\times \dfrac{0.10 \text{ f̶o̶r̶m̶u̶l̶a̶ ̶w̶t̶ ̶K̶O̶H̶}}{\text{L̶ ̶s̶o̶l̶n̶}} \times \dfrac{56.1 \text{ g̶ ̶K̶O̶H̶}}{\text{f̶o̶r̶m̶u̶l̶a̶ ̶w̶t̶ ̶K̶O̶H̶}} \times \dfrac{\text{lb KOH}}{454 \text{ g̶ ̶K̶O̶H̶}}$

= 0.23 lb KOH

12. ? g ethyl alcohol = 30.0 g̶ ̶w̶a̶t̶e̶r̶ $\times \dfrac{\text{k̶g̶ ̶w̶a̶t̶e̶r̶}}{10^3 \text{ g̶ ̶w̶a̶t̶e̶r̶}} \times \dfrac{0.10 \text{ m̶o̶l̶ ̶e̶t̶h̶y̶l̶ ̶a̶l̶c̶o̶h̶o̶l̶}}{\text{k̶g̶ ̶w̶a̶t̶e̶r̶}}$

$\times \dfrac{46.0 \text{ g ethyl alcohol}}{\text{m̶o̶l̶ ̶e̶t̶h̶y̶l̶ ̶a̶l̶c̶o̶h̶o̶l̶}}$

= 0.14 g ethyl alcohol

13. $? m = \dfrac{? \text{ mol NaOH}}{\text{kg solvent}} = \dfrac{40 \text{ g̶ ̶N̶a̶O̶H̶}}{60 \text{ g̶ ̶s̶o̶l̶v̶e̶n̶t̶}} \times \dfrac{\text{mol NaOH}}{40.0 \text{ g̶ ̶N̶a̶O̶H̶}} \times \dfrac{10^3 \text{ g̶ ̶s̶o̶l̶v̶e̶n̶t̶}}{\text{kg solvent}}$

$= 17 \, m$

(Notice that a 40 wt % soln of NaOH is 40 g NaOH per 60 g solvent.)

or

? mol NaOH = 10^3 g̶ ̶s̶o̶l̶v̶e̶n̶t̶ $\times \dfrac{40 \text{ g̶ ̶N̶a̶O̶H̶}}{60 \text{ g̶ ̶s̶o̶l̶v̶e̶n̶t̶}} \times \dfrac{\text{mol NaOH}}{40.0 \text{ g̶ ̶N̶a̶O̶H̶}}$

= 17 mol NaOH in one kilogram solvent

Therefore it is 17 *m*.

14. This is a hard one! First you must determine the weight of KOH associated with 1 kg (1000 g) of solvent:

? g KOH = 1 k̶g̶ ̶s̶o̶l̶v̶e̶n̶t̶ $\times \dfrac{3.0 \text{ m̶o̶l̶ ̶K̶O̶H̶}}{\text{k̶g̶ ̶s̶o̶l̶v̶e̶n̶t̶}} \times \dfrac{56.1 \text{ g KOH}}{\text{m̶o̶l̶ ̶K̶O̶H̶}}$ = 168 g KOH

Then you can set up a new conversion factor, which says that there are 168 g KOH per (1000 + 168) g solution. Now you can solve the problem.

? g soln = 85 g̶ ̶K̶O̶H̶ $\times \dfrac{1168 \text{ g soln}}{168 \text{ g̶ ̶K̶O̶H̶}}$ = 5.9 $\times 10^2$ g soln

15. This problem is similar to Problem 14. First you must determine the weight of $AlCl_3$ associated with a kilogram of solvent.

$$? \text{ g } AlCl_3 = \text{kg solvent} \times \frac{0.20 \text{ mol } AlCl_3}{\text{kg solvent}} \times \frac{133.5 \text{ g } AlCl_3}{\text{mol } AlCl_3} = 26.7 \text{ g } AlCl_3$$

Then you set up a new conversion factor which says that there are 26.7 g $AlCl_3$ per (1000 + 26.7) g solution. Now you can solve the problem.

$$? \text{ g } AlCl_3 = 100 \text{ g soln} \times \frac{26.7 \text{ g } AlCl_3}{1026.7 \text{ g soln}} = 2.6 \text{ g } AlCl_3 \text{ in } 100 \text{ g soln or } 2.6\%$$

16. $? \text{ L final soln} = 0.0250 \text{ L init soln} \times \dfrac{6.00 \text{ mol sugar}}{\text{L init soln}} \times \dfrac{\text{L final soln}}{0.150 \text{ mol sugar}}$

$$= 1.00 \text{ L final soln}$$

17. $? \text{ mL init soln} = 450 \text{ mL final soln} \times \dfrac{0.35 \text{ mol HCOOH}}{10^3 \text{ mL final soln}}$

$$\times \frac{\text{L init soln}}{5.0 \text{ mol HCOOH}} \times \frac{\text{mL init soln}}{10^{-3} \text{ L init soln}}$$

$$= 32 \text{ mL init soln}$$

18. $? M = \dfrac{? \text{ mol sugar}}{\text{L final soln}} = \dfrac{0.50 \text{ mol sugar}}{\text{L init soln}} \times \dfrac{1.7 \text{ L init soln}}{4.7 \text{ L final soln}} = 0.18 \dfrac{\text{mol sugar}}{\text{L final soln}}$

$$= 0.18 \ M$$

(Notice that the volume of the final solution is the volume of the initial solution + volume solvent added.)

19. This is one of those problems in which you must calculate the weight of solute in a liter of solution, then the weight of a liter of solution, then the weight of solute by subtraction.

$$? \text{ g ethylene glycol} = 1.00 \text{ L soln} \times \frac{0.10 \text{ mol ethylene glycol}}{\text{L soln}}$$

$$\times \frac{62.0 \text{ g ethylene glycol}}{\text{mol ethylene glycol}} = 6.2 \text{ g ethylene glycol}$$

$$? \text{ g soln} = 1.0 \text{ L soln} \times \frac{\text{mL soln}}{10^{-3} \text{ L soln}} \times \frac{0.90 \text{ g soln}}{\text{mL soln}} = 900 \text{ g soln}$$

$? \text{ g solvent} = 900 \text{ g soln} - 6.2 \text{ g ethylene glycol} = 893.8 \text{ g solvent}$

The new conversion factor that you have thus calculated is

$$\frac{0.10 \text{ mol ethylene glycol}}{893.8 \text{ g solvent}}$$

$$? \ m = \frac{? \text{ mol ethylene glycol}}{\text{kg solvent}}$$

$$= \frac{0.10 \text{ mol ethylene glycol}}{893.8 \text{ g solvent}} \times \frac{10^3 \text{ g solvent}}{\text{kg solvent}}$$

$$= 0.11 \ m$$

Or you can do this last step by rewording the question:

$$? \text{ mol ethylene glycol} = 10^3 \, \cancel{\text{g solvent}} \times \frac{0.10 \text{ mol ethylene glycol}}{893.8 \, \cancel{\text{g solvent}}}$$

$$= 0.11 \text{ mol ethylene glycol with one kg of solvent}$$

Therefore the solution is 0.11 *m*.

20. Here you calculate the number of moles of HCl in each sample separately.

$$? \text{ mol HCl} = 0.050 \, \cancel{\text{L}_{10M}} \times \frac{10 \text{ mol HCl}}{\cancel{\text{L}_{10M}}} \qquad = \qquad 0.50 \text{ mol HCl}$$

$$? \text{ mol HCl} = 2.0 \, \cancel{\text{L}_{0.10M}} \times \frac{0.10 \text{ mol HCl}}{\cancel{\text{L}_{0.10M}}} \qquad = \qquad \underline{0.20 \text{ mol HCl}}$$

$$\text{Total mol HCl} = 0.70 \text{ mol HCl}$$

$$\text{Total vol} \quad = 2.05 \text{ L}$$

$$? \, M = \frac{? \text{ mol HCl}}{\text{L total}} = \frac{0.70 \text{ mol HCl}}{2.05 \text{ L total}} = 0.34 \, M$$

21. $? \text{ g NaCl} = 0.250 \, \cancel{\text{L soln}} \times \dfrac{0.585 \, \cancel{\text{formula wt NaCl}}}{\cancel{\text{L soln}}} \times \dfrac{58.5 \text{ g NaCl}}{\cancel{\text{formula wt NaCl}}}$

$$= 8.56 \text{ g NaCl}$$

22. $\dfrac{? \text{ mol HCl}}{\text{L soln}} = \dfrac{0.70 \text{ mol HCl}}{0.250 \text{ L soln}} = 2.8 \, M$

23. $? \text{ g Na}_2\text{C}_2\text{O}_4 = 100 \, \cancel{\text{g imp}} \times \dfrac{0.1000 \, \cancel{\text{L soln}}}{30.0 \, \cancel{\text{g imp}}} \times \dfrac{2.00 \, \cancel{\text{mol Na}_2\text{C}_2\text{O}_4}}{\cancel{\text{L soln}}}$

$$\times \frac{134 \text{ g Na}_2\text{C}_2\text{O}_4}{\cancel{\text{mol Na}_2\text{C}_2\text{O}_4}} = 89.3 \text{ g Na}_2\text{C}_2\text{O}_4 \text{ in } 100 \text{ g imp}$$

Therefore the material is 89.3% pure.

24. We can do this in steps. First find the number of moles of HCl in each addition:

$$? \text{ mol HCl} = 2.0 \, \cancel{\text{L}_{6.0M}} \times \frac{6.0 \text{ mol HCl}}{\cancel{\text{L}_{6.0M}}} \qquad = \qquad 12 \text{ mol HCl}$$

$$? \text{ mol HCl} = 0.500 \, \cancel{\text{L}_{9.0M}} \times \frac{9.0 \text{ mol HCl}}{\cancel{\text{L}_{9.0M}}} \qquad = \qquad \underline{4.5 \text{ mol HCl}}$$

$$\text{Total HCl} = 16.5 \text{ mol HCl}$$

The total volume is 2.0 L + 0.500 L = 2.5 L.
   Thus the molarity of the final solution is

$$\frac{? \text{ mol HCl}}{\text{L total}} = \frac{16.5 \text{ mol HCl}}{2.5 \text{ L total}} = 6.6 \, M$$

25. ? g CH_3COOH = 3.00 L soln $\times \dfrac{6.00 \text{ mol } CH_3COOH}{\text{L soln}} \times \dfrac{60.0 \text{ g } CH_3COOH}{\text{mol } CH_3COOH}$

$\qquad$ = 1.08×10^3 g CH_3COOH

(Note the way that the number was written to have the correct number of significant figures.)

26. ? $F = \dfrac{? \text{ formula wt}}{\text{L soln}} = \dfrac{11.1 \text{ g } CaCl_2}{100 \text{ g soln}} \times \dfrac{\text{formula wt } CaCl_2}{111.1 \text{ g } CaCl_2}$

$\qquad \times \dfrac{1.15 \text{ g soln}}{\text{mL soln}} \times \dfrac{\text{mL soln}}{10^{-3} \text{ L soln}} = 1.15\ F$

If you prefer to reword the question, the solution is:

? formula wt = 1.00 L soln $\times \dfrac{\text{mL soln}}{10^{-3} \text{ L soln}} \times \dfrac{1.15 \text{ g soln}}{\text{mL soln}}$

$\qquad \times \dfrac{11.1 \text{ g } CaCl_2}{100 \text{ g soln}} \times \dfrac{\text{formula wt } CaCl_2}{111.1 \text{ g } CaCl_2}$

$\qquad$ = 1.15 formula wt $CaCl_2$ in 1.00 L solution

The solution is therefore 1.15 *F.*

27. For this you need a conversion factor relating grams $CaCl_2$ to grams H_2O. It is

$\dfrac{11.1 \text{ g } CaCl_2}{88.9 \text{ g } H_2O}$

So you write:

? $m = \dfrac{? \text{ mol } CaCl_2}{\text{kg } H_2O} = \dfrac{11.1 \text{ g } CaCl_2}{88.9 \text{ g } H_2O} \times \dfrac{\text{mol } CaCl_2}{111.1 \text{ g } CaCl_2} \times \dfrac{10^3 \text{ g } H_2O}{\text{kg } H_2O}$

$\qquad = \dfrac{1.12 \text{ mol } CaCl_2}{\text{kg } H_2O} = 1.12\ m$

If you prefer to reword the problem to ask "How many moles of $CaCl_2$ are there with 1.00 kg H_2O?" you have

? mol $CaCl_2$ = 1.00 kg $H_2O \times \dfrac{10^3 \text{ g } H_2O}{\text{kg } H_2O} \times \dfrac{11.1 \text{ g } CaCl_2}{88.9 \text{ g } H_2O} \times \dfrac{\text{mol } CaCl_2}{111.1 \text{ g } CaCl_2}$

$\qquad$ = 1.12 mol $CaCl_2$ per kg H_2O

This solution is then 1.12 *m.*

28. ? g imp = 6.00 L soln $\times \dfrac{1.50 \text{ mol } H_3PO_4}{\text{L soln}} \times \dfrac{98.0 \text{ g } H_3PO_4}{\text{mol } H_3PO_4} \times \dfrac{100 \text{ g imp}}{90.0 \text{ g } H_3PO_4}$

$\qquad$ = 980 g imp

29. ? mol HCl = 0.250 L soln $\times \dfrac{2.8 \text{ mol HCl}}{\text{L soln}}$ = 0.70 mol HCl

pH: The Way Chemists Express the Strength of Acidic and Basic Solutions

PRETEST

If you can handle this material, you should be able to complete this test in 10 minutes, if your calculator can take logarithms and get antilogarithms. It might take 15 minutes if you have to use the table of logarithms in Appendix I. Brackets, [ ], are the conventional way to show concentration in mole/liter.

1. What is the $[H^+]$ in water if the $[OH^-]$ is 1.0×10^{-8} M?
2. What is the $[OH^-]$ in water if the $[H^+]$ is 3.33×10^{-6} M?
3. Is a solution in which the $[H^+]$ is 0.000035 M acidic or basic?
4. What is the pH when the $[H^+]$ is 1.0×10^{-3} M?
5. What is the pH when the $[H^+]$ is 0.000730 M?
6. What is the pH when the $[OH^-]$ is 0.000534 M?
7. What is the $[H^+]$ when the pH is 3.32?
8. What is the $[OH^-]$ when the pH is 11.68?
9. What is the pH when the pOH is 8.50?

PRETEST ANSWERS

If you understand the material, you should have answered all these correctly. If you did, go directly to the problem set at the end of the chapter. Otherwise, you had better work through the chapter. Many of these are used as examples.

1. 1.0×10^{-6} M
2. 3.00×10^{-9} M
3. Acidic
4. 3.00
5. 3.137
6. 10.73
7. 4.8×10^{-4} M
8. 4.8×10^{-3}
9. 5.50

SECTION A   What pH Means

When chemists discuss the strengths of aqueous (in water) solutions of acids and bases, what they are really concerned with is the concentration of H^+ ions. This may vary from as low a concentration as 1.0×10^{-14} M up to as much as 1.0 M. To make it easier to work with numbers that can vary over so many magnitudes, chemists have worked out a system that uses only numbers between 1 and 14. In this system, they express the concentration of H^+ ions as the exponent to which 10 is raised when the concentration is expressed in exponential notation. Recall from Section B in Chapter 1 that this is exactly what the logarithm to the base 10 is. If you have forgotten what logarithms mean and how to work with them, you had better check back to Chapter 1.

To simplify things still further, you give the log of the H^+ ion concentration a negative sign. When you take the logarithm of something and give it a negative sign, you call this a p function. The p means "–log." Thus, if you are interested in the p function of the H^+ ion concentration, you can write

$$pH = -\log [H^+] \qquad \text{or} \qquad [H^+] = 10^{-pH}$$

In the same way, you can write

-log; $^{-pOH}$

$$pOH = \underline{\hspace{1cm}} [OH^-] \qquad \text{or} \qquad [OH^-] = 10^{\overline{\hspace{2cm}}}$$

We are going to use the conventional system of showing the molarity, M, of a substance by putting the substance in [ ]. Thus if you are talking about the molarity of H^+ ions in a solution, you write $[H^+]$. If you are talking about the molarity of $KMnO_4$ in a solution, you write $[KMnO_4]$.

Not only concentrations have p functions. Any number can have a "–log." Thus you can express the dissociation constant for water, K_w, as a p function.

$$pK_w = -\log K_w$$

Similarly, you can express the dissociation constant of a weak acid, K_a, as a p function.

-log K_a

$$pK_a = \underline{\hspace{3cm}}$$

In general, you only use these p functions for very small numbers. For example, $K_w = 1 \times 10^{-14}$ and K_a for acetic acid is 1.8×10^{-5}.

SECTION B   Calculating the pH from the H⁺ Ion Concentration

Let's see how this system works with pH. Suppose that you have a solution in which the H^+ ion concentration is 1×10^{-3} M. To get the pH, you write

1×10^{-3}; –3.0; 3.0

$$pH = -\log[H^+] = -\log [\underline{\hspace{2cm}}] = -(\underline{\hspace{1cm}}) = \underline{\hspace{1cm}}$$

(If you weren't able to get from log 1 × 10^{-3} to -3.0 go back to Section B, Chapter 1. You should also remember the sign convention. Two negative numbers multiplied or divided give a positive number.)

If the H⁺ ion concentration were 0.00001 *M*, to get the pH you would write

pH = _____ = $-$ log [_____] = $-$(_____) = _____ $-\log [\text{H}^+]$; 1 × 10^{-5} ; -5.0; 5.0

This is pretty easy. You can check yourself by doing the following on a separate sheet of paper and comparing your answers with those in the margin. Determine the pH for the following [H⁺].

[H⁺]	pH	
1 × 10^{-5} *M*	_____	5.0
0.000001 *M*	_____	6.0
1 *M*	_____	0.0
1 × 10^{-13} *M*	_____	13.0
0.1 *M*	_____	1.0

Now that you are able to find the pH when you know the concentration of H⁺ ions in a solution, let's look at what these pH values mean. If

the [H⁺] is 0.1 *M*, the pH is _____.   1.0

   If the [H⁺] is one-tenth as much, 0.01 *M*, the pH is _____.   2.0

   If the pH is one-hundredth as much, 0.001 *M*, the pH is _____.   3.0

   And if the pH is one-thousandth as much, 0.0001 *M*, the pH is _____.   4.0

The lower the [H⁺], the _____ the pH. Since the strength of   greater
an acid is directly related to the concentration of H⁺ ions, the weaker the

acid solution, the _____ the pH. An acid with a pH of 1 is 10   greater

times _____ than an acid with a pH of 2. An acid solution with   stronger

a pH of 6 is _____ times weaker than one with a pH of 4.   100

So far we have considered only acid solutions whose H⁺ ion concentrations can be expressed with the digits 0 and 1. We are now going to consider other concentrations.

Suppose that you want to know the pH of a solution whose [H⁺] is 0.050 *M*.

   pH = _____ [H⁺] $-\log$

and you put in the value for [H⁺]

   pH = $-$log [_____]   0.050

If your calculator has a log *x* key, you can go directly to the answer. You press 0.050 and then log *x*. The value -1.30103 will appear. Just change the sign. If your calculator does not get logarithms, you will have to use the table of four-place logarithms in Appendix I, and the method shown below. First you write the number whose log you want, expressed in scientific notation:

5.0×10^{-2}

$$pH = -\log [\underline{\hspace{3cm}}]$$

As you did in Section B, Chapter 1, you break this up into the sum of two logarithms.

$\log 10^{-2}$

$$pH = -(\log 5.0 + \underline{\hspace{3cm}})$$

You can find the log of 5.0 in the table of logarithms. You know that

-2

$\log 10^{-2}$ equals _____.

0.70

$$pH = -[\underline{\hspace{2cm}} + (-2)]$$

(Notice that you may write only two significant figures on the log because the 5.0 has just two significant figures. This is true whether you use the log table or a calculator.)

Making the algebraic addition gives you

-1.30

$$pH = -(\underline{\hspace{2cm}})$$

and the two minus signs cancel to give you

1.30

$$pH = \underline{\hspace{2cm}}$$

(You have to round off the 1.30103 value from the calculator to 1.30 as well.)

It should be obvious that the pH is between 1 and 2, since the $[H^+]$ is

1.0

less than 0.1 M, which has a pH of _____, but more than 0.01 M, which

2.0

has a pH of _____. So the pH has to be between 1.0 and 2.0.

Here's another problem, to make sure that you have this. You want the pH of an acid solution whose $[H^+]$ is 0.0000356 M. If your calculator gives logs, you key in 0.0000356, then the log x key, and the value -4.44855 appears directly. If the number whose log you want exceeds the number of digits your calculator will accept, you must express the number in scientific notation. For example, an eight-digit calculator will not accept 0.000000356 and you must get the log of 3.56×10^{-7}. If you must use the log tables, you start by writing

$-\log[H^+]$

$$pH = \underline{\hspace{3cm}}$$

Writing the $[H^+]$ in scientific notation, you have

3.56×10^{-5}

$$pH = -\log [\underline{\hspace{3cm}}]$$

You then break this up into the sum of two logarithms:

$3.56; \ 10^{-5}$

$$pH = -(\log \underline{\hspace{2cm}} + \log \underline{\hspace{2cm}})$$

When you look up log 3.56 in the table of logarithms and put in the log 10^{-5}, you have

$0.551; \ -5$

$$pH = -[\underline{\hspace{2cm}} + (\underline{\hspace{2cm}})]$$

(Notice the number of significant figures on log 3.56.) Making the algebraic addition, you have

pH = -(_____)   which gives you the pH   pH = _____

-4.449; 4.449

The -4.44855 value that a calculator gives also has to be rounded off to 4.449, and its sign has to be changed.

To see that you are doing this correctly, determine the pH for the following H⁺ ion concentrations. (Watch your significant figures. Remember, the pH value can only have as many digits *after* the decimal as there are significant figures in the [H⁺].)

$[H^+]$	pH
0.153 *M*	_____
0.000730 *M*	_____
0.00265 *M*	_____

0.815

3.137

2.577

SECTION C   Determining the H⁺ Ion Concentration from the pH

Let's see how you can reverse this process. Instead of finding the pH from [H⁺], find the H⁺ ion concentration from the pH. Suppose that the pH of a solution is 3.0. You know that the relationship between pH and [H⁺] is

pH = _____   Therefore you can write   3.0 = _____

$-\log [H^+]$; $-\log [H^+]$

There is no way of dealing with negative logarithms, so what you do is multiply both sides of the equation by -1. The + becomes a -, and the - becomes a +. So

-3.0 = log [H⁺]

If you want to determine the number whose logarithm you know, you must find the antilogarithm. (Once again, if you are rusty on any of this, refer to Chapter 1, Section B. There is an entire part that deals with antilogarithms.) You write

antilog (-3.0) = [H⁺]

The antilogarithm of -3.0 is simply _____. So you can write

1×10^{-3}

1 X 10⁻³ *M* = [H⁺]

Here is another example. You want to know the H⁺ ion concentration in a solution that has a pH of 11.0. You start by writing the relationship between pH and [H⁺].

pH = _____

$-\log [H^+]$

And you then put in the value for the pH from the problem:

_____ = -log [H⁺]

11.0

Next you change the signs on both sides:

-11.0

_____ = log [H$^+$]

antilogarithm

Now all you have to do is to find the _____ of -11.0.

1×10^{-11}

It is _____. So you know that

1×10^{-11} M

_____ = [H$^+$]

This is easy as long as the pH is a whole number greater than 1. If it has something written to the right of the decimal, getting the anti-logarithm is harder.

Your calculator may be able to get you antilogarithms. This is a big help. However, different models have different ways of doing this. Some models have a 10^x key. Since the logarithm to the base 10 is the exponent to which 10 is raised, you enter the number whose antilogarithm you want and then press the 10^x key. Other models have a 2nd-function key. Pressing this and then the log key gets you the antilogarithm. However, the number whose antilog you want is a negative number. Some calculators allow you to enter a negative number, but others require that you determine the antilog of the positive number and then divide 1 by this value. (That is, you take the reciprocal, for which there may also be a key labeled $\frac{1}{x}$.) When everything else fails, consult the operating manual for your calculator. But remember, you want the antilog for the negative of the pH.

If your calculator cannot get antilogarithms, use the table of four-place logarithms in Appendix I, and proceed as follows.

Suppose that you want to know the concentration of H$^+$ ions in a solution that has a pH of 3.40. You start in the usual way, by writing

$-$log [H$^+$]

pH = _____

and you put in the value of the pH in the problem.

$-$log [H$^+$]

3.40 = _____

Changing the signs on both sides gives you

-3.40

_____ = log [H$^+$]

Since you cannot get the antilogarithm of a nonwhole negative number, you must change the -3.40 to $(-4 + 0.60)$. Now you can write

antilog $(-4 + 0.60)$ = [H$^+$]

Since the antilogarithm of two terms added is the same as the antilogarithm of one times the antilogarithm of the other, you write

[H$^+$]

antilog $-4 \times$ antilog 0.60 = _____

10^{-4}

You know that antilog -4 is _____. You can look up the antilog of

0.60 in a table of logarithms. It is _____. So you can write

$$\text{_____} = [H^+]$$

3.98

3.98×10^{-4} *M*

It is better to express this as 4.0×10^{-4} *M* to have the correct number of significant figures. If you used a calculator, the answer of 3.98089×10^{-4} *M* must be rounded off and then expressed in scientific notation: 4.0×10^{-4} *M*. You can show only as many significant figures in the [H⁺] as there are digits *after* the decimal in the pH.

Here's another example. You want to know what concentration of H⁺ ions has a pH of 11.68. You always start by writing the relationship between pH and [H⁺] :

$$pH = \text{_____}$$

$-\log [H^+]$

Inserting the value from the question, you have

$$\text{_____} = -\log [H^+]$$

11.68

Then, changing signs on both sides of the equation, you have

$$\text{_____} = \log [H^+]$$

−11.68

At this point you can use your calculator to get the antilog of −11.68. It will read 2.0893×10^{-12} *M* or 2.1×10^{-12} *M* (to the correct significant figures). If you are using log tables, you proceed as follows. Writing the −11.68 as a negative whole number and a positive number less than 1 gives you

$$\text{_____} = \log [H^+]$$

$-12 + 0.32$

You must determine the _____ of (−12 + 0.32) in order to get the value of [H⁺]. Since the antilogarithm of the sum of two numbers equals the antilogarithm of one _____ the antilogarithm of the other, you can write

antilogarithm

times

$$\text{antilog} -12 \text{\_\_} \text{antilog} .32 = [H^+]$$

$\times$

(We've omitted the 0 before the decimal since that's how it appears in the table.) The antilog −12 equals _____. When you look up antilog .32 in the table of logarithms, you find that it is _____. So you can write

10^{-12}

2.09

$$\text{_____} = [H^+]$$

2.09×10^{-12} *M*

You can express this only as 2.1×10^{-12} *M*, considering the significant figures.

So that you can check to see that you are able to do these, determine the [H⁺] that is equivalent to the following pH values. The answers are given in the margin. (Watch the significant figures, and remember, [ ] means *M*.)

	pH	[H⁺]
5.9×10^{-10} M	9.23	
4.2×10^{-7} M	6.37	
7.9×10^{-3} M	2.10	
2.9×10^{-2} M	1.54	

You can write these answers to only two significant figures, since there were only two figures after the decimal in the pH values.

SECTION D   Calculating the pOH from the OH⁻ Ion Concentration

When a base, such as NaOH, is dissolved in water, it breaks up into ions— a positive ion and an OH⁻ ion. The strength of the base depends on the concentration of the OH⁻ ions, just as the strength of acids depends on the concentration of H⁺ ions. And just as you used a p function with the [H⁺], you can use a p function to express the concentration of the OH⁻ ion, [OH⁻]:

$$pOH = -\log [OH^-]$$

The calculations are identical. Thus, if you know that the [OH⁻] is 1×10^{-3} M, then you can calculate the pOH:

OH⁻; 1×10^{-3}; –3.0; 3.0

$$pOH = -\log [\underline{\hspace{1cm}}] = -\log [\underline{\hspace{2cm}}] = -(\underline{\hspace{1cm}}) = \underline{\hspace{0.5cm}}$$

If the concentration of OH⁻ ions cannot be expressed simply as 10 to a whole-number power, you have to use either a table of logarithms or a calculator to get the pOH value. For example, if you want the pOH of a solution that has an OH⁻ ion concentration of 0.00065 M, and you are using a calculator, you punch in the number and then log x.

If you are using the table of four-place logarithms, you proceed as follows. [We're going to start to separate these solutions using log tables to find logarithms and antilogarithms by putting them in a box so they won't distract those of you who are using a calculator. The correct answer will always appear at the end and outside the box.]

$-\log [OH^-]$	pOH = _____
$-\log [6.5 \times 10^{-4}]$	= _____ (in scientific notation)
0.8129	= $-[\underline{\hspace{2cm}} + (-4)]$

3.19   = _____ (to the correct significant figures)

If you can determine the pH from [H⁺], then you can just as easily determine pOH from [OH⁻].

SECTION E Determining the OH⁻ Ion Concentration from the pOH

You use exactly the same method to get the concentration of OH⁻ ions from the pOH that you used to get the concentration of H⁺ ions from the pH. For example, suppose you want to know the concentration of OH⁻ ions that gives you a pOH of 4.0. You start by writing the relationship between pOH and $[OH^-]$:

pOH = _____ $-\log [OH^-]$

Then you put in the value for the pOH:

4.0 = _____ $-\log [OH^-]$

To get the logarithm, you have to _____ the signs on both sides of the equation: change

_____ = $\log [OH^-]$ -4.0

And now to get the value for $[OH^-]$, you must find the _____ of -4.0:   antilogarithm

_____ $(-4.0) = [OH^-]$ antilog

This gives you _____ $M = [OH^-]$. 1×10^{-4}

If your pOH is not a simple whole number, then you have to use a calculator or the table of logarithms to get the antilogarithm of the pOH value, as you did with pH. We will not go through this procedure, as you have had plenty of practice with finding the $[H^+]$ from pH.

SECTION F The Water Equilibrium

Pure water itself breaks up very slightly into ions. One molecule of water produces one H⁺ ion and one OH⁻ ion. The equation for this is

$H_2O \rightarrow H^+ + OH^-$

This type of reaction is called a *dynamic equilibrium*, since as fast as the water molecules break up, H⁺ and OH⁻ ions are recombining to produce more water molecules. You can show this by using two arrows.

$H_2O \rightleftarrows H^+ + OH^-$

Since the ions are being used as fast as they are being formed, their concentration remains constant. In pure water the concentration of H⁺ ions is 10^{-7} M. The concentration of OH⁻ ions is also _____, 10^{-7} M
since as many H⁺ ions are produced as OH⁻ ions. Thus the pH of pure water is ____, and the pOH of pure water is also ____. The chemist says   7; 7
that pure water is *neutral*, that is, neither acidic nor basic. A neutral solution therefore has a pH of ____.   7

Any solution that has a H⁺ ion concentration greater than that of pure water, 10^{-7} M, is said to be *acidic*. Any solution that has a H⁺ ion concentration less than that of pure water is called *basic*. Thus any

less

more

solution with a pH _____ than 7 is acidic and any solution with a

pH _____ than 7 is basic.

Because of the nature of dynamic equilibria, if the concentration of any of the substances in equilibrium is changed, the entire system responds to this change to minimize its effect. This is called the *Le Chatelier principle* or the *Law of Mass Action.* It is expressed mathematically by stating that the product of the concentrations remains constant. In the case of water, you write

$$K_w = 1.0 \times 10^{-14} = [\text{H}^+][\text{OH}^-]$$

K_w is called the dissociation constant for water, and its value is 1.0×10^{-14}. You can see how this works for pure water whose [H⁺]

1.0×10^{-7}

1.0×10^{-7} and [OH⁻] is also _____.

$$K_w = 1.0 \times 10^{-14} = [\text{H}^+][\text{OH}^-]$$

1.0×10^{-7}, 1.0×10^{-7}

$$= [\underline{\hspace{2cm}}][\underline{\hspace{2cm}}]$$

If you have an acid solution whose [H⁺] ion concentration is increased to 1.0×10^{-5} (remember, 10^{-5} is larger than 10^{-7}), you have

$$K_w = 1.0 \times 10^{-14} = [\text{H}^+][\text{OH}^-]$$
$$= [10^{-5}][\text{OH}^-]$$

or

$$\frac{1.0 \times 10^{-14}}{1.0 \times 10^{-5}} = [\text{OH}^-]$$

1.0×10^{-9}

$$\underline{\hspace{2cm}} = [\text{OH}^-]$$

To have the product of the two concentrations still equal to 1.0×10^{-14}, the [OH⁻] must decrease to 1.0×10^{-9} M. Thus, increasing the [H⁺] decreases the [OH⁻]. (Remember, 10^{-9} is smaller than 10^{-7}. As a matter of fact, if you have forgotten how to manipulate these powers of 10, check back to Section A in Chapter 1.)

Here is another example. You have a basic solution where the OH⁻ ion concentration is 10^{-2} M. (We will omit the 1.0 in this example since it only tells us our significant figures.) You can calculate the H⁺ ion concentration in the following way.

-14

10^{-2}

$$K_w = 10^{\underline{\hspace{1cm}}} = [\text{H}^+][\text{OH}^-]$$
$$= [\text{H}^+][\underline{\hspace{1cm}}]$$

or

10^{-14}

$$\frac{\underline{\hspace{1cm}}}{10^{-2}} = [\text{H}^+]$$

10^{-12}

$$\underline{\hspace{2cm}} = [\text{H}^+]$$

SECTION G   The Relationship Between pH and pOH

Consider our solution in the preceding section with a $[H^+] = 1.0 \times 10^{-5}\ M$. Its pH is

$$pH = -\log [H^+] = -\log (\underline{\hspace{2cm}}) = \underline{\hspace{1cm}}$$

1.0×10^{-5}; 5.00

If the $[H^+] = 1.0 \times 10^{-5}\ M$, then the $[OH^-]$ is

$$K_w = 1.0 \times 10^{-14} = [H^+][OH^-] = (\underline{\hspace{2cm}})[OH^-]$$

$1.0 \times 10^{-5}\ M$

$$\underline{\hspace{2cm}} = [OH^-]$$

$1.0 \times 10^{-9}\ M$

Its pOH is

$$pOH = -\log [OH^-] = -\log (\underline{\hspace{2cm}}) = \underline{\hspace{1cm}}$$

1.0×10^{-9}; 9.00

The sum of the pH and pOH is \underline{\hspace{1cm}}. This is *always* true:

14.00

$$pH + pOH = 14.00$$

We can see how this comes about. Suppose that you use the water equilibrium expression,

$$K_w = 1.0 \times 10^{-14} = [H^+][OH^-]$$

and you take the p values for everything. Since the p value means "$-\log$," you have

$$-\log K_w = -\log 1.0 \times 10^{-14} = -\log [H^+][OH^-]$$

You can break up the $\log [H^+][OH^-]$ into the sum of two logarithms:

$$\log [H^+][OH^-] = \log [H^+] + \log [OH^-]$$

so that you can write

$$-\log K_w = -\log 1.0 \times 10^{-14} = -(\log [H^+] + \log [OH^-])$$

Moving the minus sign inside the parentheses gives you

$$-\log K_w = -\log 1.0 \times 10^{-14} = -\log [H^+] - \log [OH^-]$$

But you know that $\log 1.0 \times 10^{-14}$ equals \underline{\hspace{1cm}}, that $-\log [H^+]$ is the

-14.00

pH, and $-\log [OH^-]$ is the \underline{\hspace{1cm}}. You also know that $-\log K_w$ equals pK_w. So we can write

pOH

$$pK_w = -(-14.00) = pH + pOH \quad \text{or} \quad pK_w = \underline{\hspace{1cm}} = pH + pOH$$

14.00

This is the reason that

$$pH + pOH = 14.00$$

This is a very useful relationship. It allows you to go from pH to pOH for a given solution by a simple subtraction. For example, if a solution has a pOH of 3.00 you can find its pH by

$$pH + pOH = 14.00, \quad pH + \underline{\hspace{1cm}} = 14.00$$

3.00

Subtracting 3.00 from both sides of the equation, you have

$$pH + 3.00 - 3.00 = 14.00 - \underline{\hspace{1cm}} \quad \text{and finally} \quad pH = \underline{\hspace{1cm}}$$

3.00; 11.00

What this amounts to (if your algebra is in good order, you know this already) is

pH	$pH = 14.00 - pOH$ and in the same way $pOH = 14.00 - \underline{}$

Try this one for practice. What is the pH for a solution whose pOH is 2.5? All you do is write

pOH; 2.5; 11.5	$pH = 14.00 - \underline{} = 14.00 - \underline{} = \underline{}$

Because it is so easy to get from pOH to pH, chemists rarely speak of pOH values. Instead, they give all solutions, both acidic and basic, a pH value. However, when they are dealing with a basic solution, it is easiest to calculate the pOH. Then they can easily change the pOH to pH.

Here is how this is done. Suppose that you have a solution with an OH^- ion concentration of 1×10^{-5} M and you want the pH. The first thing you do is determine the pOH:

$-\log [OH^-]$; $-\log [1 \times 10^{-5}]$; 5.0	$pOH = \underline{} = \underline{} = \underline{}$
14.00	Now to determine the pH, all you do is subtract 5.0 from $\underline{}$.
14.00; 9.0	$pH = \underline{} - 5.0 = \underline{}$

Here's another problem for practice. You want the pH for a solution that has a concentration of OH^- ions of 0.000731 M. First you

pOH	determine the $\underline{}$.
$[OH^-]$	$pOH = -\log \underline{}$
7.31×10^{-4}	$= -\log \underline{}$ (in scientific notation)
3.136	$= \underline{}$
subtract	Now to find the pH, you $\underline{}$ the pOH from 14.00:
3.136; 10.86	$pH = 14.00 - \underline{} = \underline{}$

You can also use the changing of pH to pOH when you have a pH value and you want to know the OH^- ion concentration. For example, let's say that you want to know $[OH^-]$ when the pH is 9.77. Since you are interested in $[OH^-]$, you change pH to pOH.

14.00; 14.00; 4.23	$pOH = \underline{} - pH = \underline{} - 9.77 = \underline{}$

Now you can calculate the OH^- ion concentration from the pOH.

$[OH^-]$	$pOH = -\log \underline{}$
4.23	$\underline{} = -\log [OH^-]$
-4.23	$\underline{} = \log [OH^-]$
-5	$(\underline{} + 0.77) = \log [OH^-]$
antilog	$\underline{} (-5 + 0.77) = [OH^-]$

_____ = [OH⁻] | 5.8885 X 10⁻⁵ *M*

But using our rule for significant figures that you can only have as many digits in the molarity as you have after the decimal in the pH (pOH), then

_____ = [OH⁻] | 5.9 X 10⁻⁵ *M*

PROBLEM SET

Section B–Section E

1. What is the pH of a system with $[H^+] = 1.0 \times 10^{-3}$ *M*?

2. What is the pH of a system with $[H^+] = 0.0000000013$ *M*?

3. What is the pH of a system with H^+ ion concentration of 3.72×10^{-11} *M*?

4. What is the $[H^+]$ of a system whose pH is 4.62?

5. What is the $[H^+]$ of a system whose pH is 11.65?

6. What is the pOH of a system in which $[OH^-]$ is 3.82×10^{-9} *M*?

7. What is the $[OH^-]$ of a system with a pOH = 9.21?

Section F

8. What is the $[OH^-]$ in a water system if the $[H^+]$ is 2.5×10^{-6} *M*?

9. Is the system in Question 8 acidic or basic?

10. What is the $[H^+]$ of a system in water if the $[OH^-]$ is 0.000005 *M*?

11. Is the system in Question 10 acidic or basic?

12. What will $[H^+] \times [OH^-]$ always equal at 25°C in a water system? How is this number symbolized?

Section G

13. The pH + pOH at 25°C always equals what in a water system?

14. What is the pOH of a system with a pH = 6.54?

15. What is the pH of a system with an $[OH^-] = 0.00032$ *M*?

Miscellaneous Problems

16. Assume that you can dissolve 4.0 g of $Ba(OH)_2$ in enough water to produce exactly 100 mL of aqueous solution. Each molecule of $Ba(OH)_2$ dissociates into two OH^- ions. What is the pH of the solution?

17. You prepare 2.00 L of HCl solution by dissolving HCl gas in water. Its pH is 2.65. How many liters of the gas at STP were dissolved in the solution? [If you have forgotten your gas-law calculations (Chapter 12), at STP a mole of an ideal gas occupies 22.4 L.]

18. How many times greater is the concentration of KOH in a solution with a pH of 9.9 than it is in a solution with a pH of 8.6?

19. Your stomach starts its digestive process by the action of *gastric juices*, a mixture of HCl with pepsin, rennin, and mucin. The pH of gastric juice is about 1.40. Assuming that all of the H^+ comes from the HCl, what is the HCl concentration in moles/liter? HCl is totally dissociated.

20. Not all acids dissociate completely. Many are *weak electrolytes*, and only a small percentage of their molecules yield H^+ ions. Determine the percentage of HF molecules dissociated if it was found that a 0.10 *M* solution has a pH of 2.09.

PROBLEM SET ANSWERS

Answers are shown right after problem numbers. [*Note*: Those parts of the solutions in boxes are not needed if you use a calculator.]

Section B–Section E

1. 3.00 $pH = -\log [H^+] = -\log (1.0 \times 10^{-3}) \boxed{= -[0 + (-3)]}$
   $= +3.00$

2. 8.89 $[H^+] = 0.0000000013 = 1.3 \times 10^{-9}$
   $pH = -\log [H^+] = \boxed{-[\log (1.3) + (-9)] = -(0.114 - 9)}$
   $= -(-8.886) = 8.89$ (to the correct significant figures)

3. 10.429 $pH = -\log [H^+] = \boxed{-[\log (3.72) + (-11)] = -(0.571 - 11)}$
   $= -(-10.429) = 10.429$

4. 2.4×10^{-5} *M* $pH = -\log [H^+];$ $4.62 = -\log [H^+];$ $-4.62 = \log [H^+]$
   $\boxed{-5 + 0.38 = \log [H^+]};$ $2.4 \times 10^{-5} = [H^+]$

5. 2.2×10^{-12} *M* $pH = -\log [H^+];$ $11.65 = -\log [H^+];$ $-11.65 = \log [H^+];$
   $\boxed{-12 + 0.35 = \log [H^+]};$ 2.24×10^{-12} *M* $= [H^+]$
   or 2.2×10^{-12} *M* (to the correct significant figures)

6. 8.418 $pOH = -\log [OH^-] = -\log 3.82 \times 10^{-9}$
   $\boxed{= -[\log 3.82 + (-9)] = -(0.582 - 9) = -(-8.418)} = 8.418$

7. 6.2×10^{-10} *M* $pOH = -\log [OH^-];$ $9.21 = -\log [OH^-];$ $-9.21 = \log [OH^-]$
   $\boxed{-10 + 0.79 = \log [OH^-];$ $6.17 \times 10^{-10} = [OH^-]}$
   or 6.2×10^{-10} *M* (to the correct significant figures)

Section F

8. 4.0×10^{-9} *M* $[OH^-] = \dfrac{1.0 \times 10^{-14}}{[H^+]} = \dfrac{1.0 \times 10^{-14}}{2.5 \times 10^{-6}} = 4.0 \times 10^{-9}$

9. The system is acidic, since the $[OH^-]$ is less than 1.0×10^{-7}.

10. 2×10^{-9} *M* $[OH^-] = 0.000005\ M = 5 \times 10^{-6}\ M$

    $[H^+] = \dfrac{1.0 \times 10^{-14}}{[OH^-]} = \dfrac{1.0 \times 10^{-14}}{5 \times 10^{-6}} = 2 \times 10^{-9}$

11. The system is basic, since the $[H^+]$ is less than 1.0×10^{-7}.

12. $1.00 \times 10^{-14} = K_w$ $[H^+][OH^-] = 1.00 \times 10^{-14}$

Section G

13. 14.00   14.00 is pK_w at 25°C.

14. 7.46 $pOH + pH = 14.00$; $pOH = 14.00 - pH = 14.00 - 6.54 = 7.46$

15. 10.51   First find the pOH: $pOH = -\log[OH^-] = -\log 3.2 \times 10^{-4}$

    $\boxed{= -[\log 3.2 + (-4)] = -(0.5051 - 4) = -(-3.4949)}$

    $= 3.49485$ (calculator)

    Then get the pH from $14.00 - pOH = pH$.

    $14.00 - 3.4949 = pH = 10.5051 = 10.51$ (to the correct significant figures)

Miscellaneous Problems

16. 13.67   First you find the moles OH^- per liter:

    $? \text{ mole } OH^- = 1.00\ \text{L} \times \dfrac{\text{mL}}{10^{-3}\ \text{L}} \times \dfrac{4.0\ \text{g Ba(OH)}_2}{100\ \text{mL}} \times \dfrac{\text{mol Ba(OH)}_2}{171.3\ \text{g Ba(OH)}_2}$

    $\times \dfrac{2\ \text{mol } OH^-}{\text{mol Ba(OH)}_2} = 0.47\ \text{mol } OH^- = 0.47\ M\ OH^-$

    Then you find the pOH:

    $pOH = -\log[OH^-] = -\log 4.7 \times 10^{-1}$

    $\boxed{= -(\log 4.7 + \log 10^{-1}) = -(0.67 - 1)}$

    $= -(-0.33) = 0.33$

    The pH is then $14.00 - pOH = 14.00 - 0.33 = 13.67$.

17. 0.099 L HCl(*g*)

    First you determine the moles of H^+ per liter from the pH:

    $pH = -\log[H^+]$ $2.65 = -\log[H^+]$ $\boxed{-3 + 0.35 = \log[H^+]}$

    $2.2 \times 10^{-3} = [H^+]$

The moles of HCl per liter is the same as the moles of H^+ per liter. Therefore the solution is

2.2×10^{-3} mol HCl/L soln

The problem is now solved by dimensional analysis:

$$? \text{ L HCl}(g) = 2.00 \text{ L soln} \times \frac{2.2 \times 10^{-3} \text{ mol HCl}}{\text{L soln}} \times \frac{22.4 \text{ L HCl}(g)_{STP}}{\text{mol HCl}}$$

$$= 0.099 \text{ L HCl}(g)_{STP}$$

18. 20   The simplest method is to use the fact that $[OH^-] = 10^{-pOH}$ and the pOH = 14.00 − pH.

Thus for pH = 9.9,   pOH = 14.00 − 9.9 = 4.1 so $[OH^-] = 10^{-4.1}$

For pH = 8.6, pOH = 14.00 − 8.6 = 5.4 and $[OH^-] = 10^{-5.4}$

The times greater the pH 9.9 is than the pH 8.6 is the ratio of the two:

$$\frac{10^{-4.1}}{10^{-5.4}} = 10^{-4.1} \times 10^{+5.4} = 10^{+1.3}$$

To get a normal number for $10^{1.3}$, you simply take the antilog of 1.3:

$10^{1.3} = \text{antilog } 1.3 \boxed{= \text{antilog } (1 + 0.3)} = 2 \times 10^1$

So the concentration of KOH in the 9.9 pH solution is 20 times greater.

19. 4.0×10^{-2} *M*   The concentration of HCl is the same as the concentration of H^+, that is, $[H^+]$. You get this from the pH.

pH = −log $[H^+]$ 1.40 = −log $[H^+]$ −1.40 = log $[H^+]$

$\boxed{-2 + 0.60 = \log [H^+]}$

$4.0 \times 10^{-2} = [H^+] = [HCl] = 4.0 \times 10^{-2}$ *M*

20. 8.1%   This problem is similar to the percentage-yield problems you did in Chapter 10. You must calculate the number of moles that react (dissociate), divide this by the total number of moles that could have dissociated, and multiply by 100. For each mole of HF that dissociates, you get one mole of H^+. Therefore you calculate the concentration of H^+ from the pH:

pH = 2.09 = −log $[H^+]$ −2.09 = log $[H^+]$

$\boxed{-3 + 0.91 = \log [H^+]}$

$8.1 \times 10^{-3} = [H^+]$

This says that 8.1×10^{-3} mol HF per liter dissociate. But the solution had 0.10 mol HF per liter. Therefore the percentage that dissociates is:

$$\text{Percent} = \frac{8.1 \times 10^{-3}}{0.10} \times 100 = 8.1\%$$

CHAPTER 16

Stoichiometry Involving Solutions

PRETEST

These are stoichiometry problems in which one or more of the reactants is in solution. If you were able to handle Chapters 10 and 14, you should be able to complete this in 30 minutes.

1. How many liters of a 0.30 M solution of HCl do you need for the complete reaction of 500.0 g of $CaCO_3$ according to the reaction
   $2HCl + CaCO_3 = CaCl_2 + H_2O + CO_2$?

2. How many liters of H_2 gas at STP can be produced by the reaction of 10.0 mL of 6.00 M HCl with an excess of Mg metal? You will recall that there are 22.4 L/mol for an ideal gas at STP.

3. What is the percent purity of an impure sample of Na_2CO_3, given that a 6.00 g sample of the impure material titrates to an equivalence point with 52.0 mL of 1.50 M HCl solution? The reaction produces NaCl, CO_2, and H_2O.

4. It takes 22.31 mL of an NaOH solution to react with 2.114 g of potassium hydrogen phthalate, KHP. What is the molarity of the NaOH solution? The reaction is KHP + NaOH = KNaP + H_2O. (The molecular weight of KHP is 204.2 g/mol.)

5. How many liters of a 0.30 M solution of HCl are required for complete reaction with 5.0 L of a 0.10 M solution of Na_2CO_3? The reaction is
   $Na_2CO_3 + 2HCl = 2NaCl + CO_2 + H_2O$.

6. It is found that 10.00 mL of an NaCl solution titrates to an equivalence point with 18.35 mL of 0.103 M $Hg_2(NO_3)_2$. What is the molarity of the NaCl solution? The reaction considered is
   $Hg_2(NO_3)_2 + 2NaCl = Hg_2Cl_2(s) + 2NaNO_3$.

7. An impure sample of $CaCO_3$ is reacted with an excess of 1.50 M HCl solution. The excess HCl solution is then back-titrated with an NaOH solution to its equivalence point. The data are:

   Weight impure $CaCO_3$ = 2.20 g
   Volume 1.50 M HCl   = 45.0 mL
   Volume NaOH for back-titration = 14.0 mL

If 30.0 mL HCl titrates to its equivalence point with 22.0 mL NaOH, what is the percent purity of the $CaCO_3$ sample? The balanced equation for the reaction is

$$CaCO_3 + 2HCl = CaCl_2 + CO_2 + H_2O$$

PRETEST ANSWERS

You should have done at least 6 out of the 7 questions correctly in order to go directly to the Problem Set at the end of the chapter. If you missed only question 7, go over Section D. Missing more than one question means that you had better go through the whole chapter. The questions are all used as examples in the body of the chapter.

1. 33 L soln
2. 0.672 L H_2 gas
3. 68.9%
4. 0.4640 *M*
5. 3.3 L soln
6. 0.378 *M*
7. 88.4%

Most chemical reactions are run using solutions of the substances being reacted. There are several reasons for this. For example, reactions tend to be faster, handling of the materials is simpler, one can use small amounts of the substances, and so forth. Consequently, many stoichiometry problems that involve determining how much of one substance will react with or produce some other substance are expressed in terms of the concentrations of solutions of the various substances.

As you learned in Chapter 10, the way to solve stoichiometry problems is first to convert the substance in the information given to moles. Next you use the bridge conversion factor from the balanced equation to get to moles of the substance in the question asked. Then you convert its moles to the particular dimension asked for. You follow exactly the same procedure when you are dealing with solutions. And, if the concentration of the solution is expressed in molarity (or formality), it will give you the moles you need for the bridge. We'll see how this goes.

SECTION A   Problems with a Single
Solution of Known Molarity

Once you have a balanced equation for the reaction, you start by noting the question asked and the information given in the usual dimensional-analysis procedure. Question 1 on the pretest is a typical example. It is "How many liters of a 0.30 *M* solution of HCl do you need for the complete reaction of 500.0 g of $CaCO_3$ according to the reaction

$2HCl + CaCO_3 = CaCl_2 + H_2O + CO_2$?" The question asked is "How many liters of HCl solution?" and the information given is "500.0 g $CaCO_3$." The conversion factor given in the problem is that the HCl

solution is 0.30 M. Therefore you have _____. Also, since weight of $CaCO_3$ is in the information given, you will undoubtedly

need its _____ weight, _____.
  You start as usual:

  ? _____ = 500.0 g $CaCO_3$

Then, as in any stoichiometry problem, you convert to _____.

  ? L soln = 500.0 g $CaCO_3$ × _____

Now you can use the bridge from the balanced equation that relates the

$CaCO_3$ in the information given to the _____ in the solution. The

bridge is always, of course, in _____ of the substances.

  ? L soln = 500.0 g ~~CaCO₃~~ × $\dfrac{mol\ CaCO_3}{100.1\ g\ \text{~~CaCO₃~~}}$ × _____

The unwanted dimension is now moles HCl. The conversion factor that

is the _____ of the solution will get you directly to the units in the question asked:

  ? L soln = 500.0 g ~~CaCO₃~~ × $\dfrac{\text{~~mol CaCO₃~~}}{100.1\ \text{~~g CaCO₃~~}}$ × $\dfrac{2\ mol\ HCl}{\text{~~mol CaCO₃~~}}$

  × _____

The only dimension left uncanceled is _____, so the answer must be

  ? L soln = 500.0 × $\dfrac{1}{100.1}$ × 2 × $\dfrac{L\ soln}{0.30}$ = _____ L soln

  You could have the same type of problem, but the other way around. For example, "How many grams of $BaSO_4$ are formed by the reaction of 25.0 mL 0.100 M H_2SO_4 with an excess of $BaCl_2$?" The first thing you

need in order to do any stoichiometry problem is a _____

_____:

  $H_2SO_4 + BaCl_2 =$ _____ + _____

The bridge conversion factor you need contains the _____

mentioned in the problem. It is _____. Undoubtedly, since you must find a weight of $BaSO_4$, you will need the

  _____ of $BaSO_4$, $\dfrac{233.4\ g\ BaSO_4}{mol\ BaSO_4}$

Right column answers:

$\dfrac{0.30\ mol\ HCl}{L\ soln}$

molecular; $\dfrac{100.1\ g\ CaCO_3}{mol\ CaCO_3}$

L soln

moles
$\dfrac{mol\ CaCO_3}{100.1\ g\ CaCO_3}$

HCl

moles

$\dfrac{2\ mol\ HCl}{mol\ CaCO_3}$

molarity

$\dfrac{L\ soln}{0.30\ mol\ HCl}$
liters soln

33

balanced equation

$BaSO_4$; 2HCl

two substances
$\dfrac{mol\ H_2SO_4}{mol\ BaCl_2}$

molecular weight

$$\frac{0.100 \text{ mol } H_2SO_4}{L \text{ soln}}$$

And, of course, the molarity of the H_2SO_4 solution gives you another conversion factor, _____. (Note that you do not need the molecular weight of H_2SO_4, since the problem never mentions its weight.) Now setting up the solution to the problem is easy. You start with the question asked and the information given.

g $BaSO_4$; 25.0 mL soln

? _____ = _____

liter

Since the molarity of the solution is in moles per _____ of solution,

liters solution

you will have to convert from milliliters solution to _____.

$$? \text{ g } BaSO_4 = 25.0 \text{ mL soln} \times \frac{10^{-3} \text{ L soln}}{\text{mL soln}}$$

[You could have saved yourself one conversion factor if you had written

molarity

25.0 mL soln as 0.0250 L soln.] Now you can use the _____ of the solution to get you toward the bridge.

$$\frac{0.100 \text{ mol } H_2SO_4}{L \text{ soln}}$$

$$? \text{ g } BaSO_4 = 25.0 \text{ mL soln} \times \frac{10^{-3} \text{ L soln}}{\text{mL soln}} \times \text{_____}$$

Now you can use the bridge:

$$? \text{ g } BaSO_4 = 25.0 \text{ mL soln} \times \frac{10^{-3} \text{ L soln}}{\text{mL soln}} \times \frac{0.100 \text{ mol } H_2SO_4}{L \text{ soln}}$$

$$\frac{\text{mol } BaSO_4}{\text{mol } H_2SO_4}$$

$$\times \text{_____}$$

mol $BaSO_4$

The only dimension left is _____, which you can easily

molecular

convert to the dimension of the answer by using the _____ weight.

$$? \text{ g } BaSO_4 = 25.0 \text{ mL soln} \times \frac{10^{-3} \text{ L soln}}{\text{mL soln}} \times \frac{0.100 \text{ mol } H_2SO_4}{L \text{ soln}}$$

$$\frac{233.4 \text{ g } BaSO_4}{\text{mol } BaSO_4}$$

$$\times \frac{\text{mol } BaSO_4}{\text{mol } H_2SO_4} \times \text{_____}$$

g $BaSO_4$

You've got the answer. The only dimension left uncanceled is _____, the dimension of the answer. So

$233.4 \text{ g } BaSO_4$

$$? \text{ g } BaSO_4 = 25.0 \times 10^{-3} \times 0.100 \times 1 \times \text{_____}$$

g $BaSO_4$

$$= 0.584 \text{ _____}$$

(We rounded off 0.5835 g $BaSO_4$ to 0.584, using the rules for significant figures in Chapter 1, Section C.)

In these problems with a single solution of known molarity, you'll run into complications only if the other substance is expressed in units other than grams or moles. Question 2 in the pretest is an example of this. "How many liters of H_2 gas at STP can be produced by the reaction of 10.0 mL of 6.00 *M* HCl with an excess of Mg metal? Recall that there are 22.4 L/mol for an ideal gas at STP."

Before you start, the first thing you need is a _____ equation. (Your work in Chapter 8 should help you now.)

	balanced

$Mg +$ _____ $=$ _____ $+ H_2$

	$2HCl$; $MgCl_2$

It's a good idea also before you start to write down all the conversion factors that you think you will need. Certainly you will need the bridge from the balanced equation. In this problem, you are asked about H_2 and given information about _____, so the bridge you will use is _____. Also, since you want the volume of H_2 gas at STP, you need the conversion factor _____. And finally, you will use the molarity of the HCl solution, _____. Now do the problem. The question asked and the information given are:

	HCl; $\dfrac{2 \text{ mol HCl}}{\text{mol } H_2}$
	$\dfrac{22.4 \text{ L } H_2}{\text{mol } H_2}$
	$\dfrac{6.00 \text{ mol HCl}}{\text{L soln}}$

? _____ = _____

	$L\ H_2$; 0.0100 L soln

[Notice that we converted milliliters solution directly into liters solution.]

The unwanted dimension liters solution can be canceled using the

_____ of the solution.

$? \text{ L } H_2 = 0.0100 \text{ L soln} \times$ _____

	molarity
	$\dfrac{6.00 \text{ mol HCl}}{\text{L soln}}$

You are now ready for the bridge:

$? \text{ L } H_2 = 0.0100 \cancel{\text{ L soln}} \times \dfrac{6.00 \text{ mol HCl}}{\cancel{\text{L soln}}} \times$ _____

	$\dfrac{\text{mol } H_2}{2 \text{ mol HCl}}$

[Note that the bridge was inverted in order to cancel the unwanted dimension moles HCl.] Now that you have moles H_2, you can go directly to liters H_2 at STP.

$? \text{ L } H_2 = 0.0100 \cancel{\text{ L soln}} \times \dfrac{6.00 \cancel{\text{ mol HCl}}}{\cancel{\text{L soln}}} \times \dfrac{\text{mol } H_2}{2 \cancel{\text{ mol HCl}}} \times$ _____

	$\dfrac{22.4 \text{ L } H_2}{\text{mol } H_2}$

The only dimension left uncanceled is the dimension of the question asked. So the answer must be:

$? \text{ L } H_2 = 0.0100 \times 6.00 \times \dfrac{1}{2} \times 22.4 \text{ L } H_2 = 0.672 \text{ L } H_2$

Chemists often run a *titration* to analyze a substance for its percentage purity. The procedure involves adding a solution (the *titrant*) to a known amount of impure substance. The titrant contains a known molarity of some material that reacts stoichiometrically with the substance being analyzed for. They use some method to indicate when exactly the right amount of titrant needed for complete reaction has been added. This is called the *equivalence point*. The addition (titration) is stopped and the amount of titrant added is measured. They then know the amount of solution that is equivalent to a known weight of impure substance. This is a conversion factor.

Chemists use many ways to determine the equivalence point in a titration. For the titration of an acid with a base, there are indicators that change color as you go from a basic to an acidic solution. (This is covered in considerable detail in Chapter 21.) In some reactions, a precipitate appears as soon as the equivalence point is reached. In another type, I_2 is formed at the equivalence point. The I_2, in the presence of starch, gives a deep blue color that you can see. You can even observe the change in electrical voltage when a current is passed through the solution as an indication of the equivalence point.

Regardless of what method is used to determine the equivalence point, the method of solving the problems is the same. Question 3 on the pretest is a typical example: "What is the percent purity of an impure sample of Na_2CO_3, given that a 6.00 g sample of the impure material titrates to an equivalence point with 52.0 mL of 1.50 *M* HCl solution? The reaction produces NaCl, CO_2, and H_2O."

$$\frac{52.0 \text{ mL soln}}{6.00 \text{ g imp}}$$

The conversion factor given in the problem is the ratio of HCl solution to the weight of impure material, _____. You are also told the molarity of the HCl solution. So you know another conversion

$$\frac{1.50 \text{ mol HCl}}{\text{L soln}}$$

factor, _____. Since there are weights involved in the

$$\frac{\text{molecular}}{106.0 \text{ g } Na_2CO_3}$$
$$\frac{}{\text{mol } Na_2CO_3}$$

problem, you will also need the _____ weight of the Na_2CO_3 _____. And before you can do any stoichiometry

balanced equation

problem, you must have a _____.

Na_2CO_3 ; HCl

The reactants are _____ and _____. The problem tells you that the products are NaCl, CO_2, and H_2O. Therefore the balanced equation must be

2HCl; 2NaCl

$$Na_2CO_3 + \underline{} = \underline{} + CO_2 + H_2O$$

Na_2CO_3
HCl; $\frac{\text{mol } Na_2CO_3}{2 \text{ mol HCl}}$

Since the two substances involved in the stoichiometry are _____ and _____, the bridge conversion factor you need is _____.

You can see all the conversion factors you might need in the answer column on the left. All you have to do is to pick the one you need from among these, then cancel unwanted dimensions. The only question is how to start the problem. Recall that we reworded problems that asked for a conversion factor (percent, in this case). We can reword "What is

100

the percent purity?" to ask "How many grams of Na_2CO_3 = _____ g imp?" We now have both the question asked and the information given:

$$? \text{ g } Na_2CO_3 = 100 \text{ g imp}$$

Next, in order to cancel the unwanted dimension, we choose from the

grams imp
$$\frac{0.0520 \text{ L soln}}{6.00 \text{ g imp}}$$

conversion factors the one that has _____:

$$? \text{ g } Na_2CO_3 = 100 \text{ g imp} \times \underline{}$$

[Once again we change milliliters to liters.]

liters solution

The unwanted dimension is now _____ and the molarity conversion factor will take care of that:

$$? \text{ g Na}_2\text{CO}_3 = 100 \text{ g imp} \times \frac{0.0520 \text{ L soln}}{6.00 \text{ g imp}} \times \underline{\hspace{3cm}}$$

1.50 mol HCl
L soln

Now that you are at moles HCl, you can use the _____ conversion factor:

bridge

$$? \text{ g Na}_2\text{CO}_3 = 100 \text{ g imp} \times \frac{0.0520 \text{ L soln}}{6.00 \text{ g imp}} \times \frac{1.50 \text{ mol HCl}}{\text{L soln}}$$

$$\times \underline{\hspace{4cm}}$$

mol Na$_2$CO$_3$
2 mol HCl

It now remains only to get from _____ to grams Na_2CO_3. For this you use the _____ of Na_2CO_3.

moles Na$_2$CO$_3$
molecular weight

$$? \text{ g Na}_2\text{CO}_3 = 100 \text{ g imp} \times \frac{0.0520 \text{ L soln}}{6.00 \text{ g imp}} \times \frac{1.50 \text{ mol HCl}}{\text{L soln}}$$

$$\times \frac{\text{mol Na}_2\text{CO}_3}{2 \text{ mol HCl}} \times \underline{\hspace{3cm}}$$

106.0 g Na$_2$CO$_3$
mol Na$_2$CO$_3$

The only dimension left is the dimension of the answer, so the answer must be

$$? \text{ g Na}_2\text{CO}_3 = 100 \times \frac{0.0520}{6.00} \times 1.50 \times \frac{1}{2} \times 106.0 \text{ g Na}_2\text{CO}_3$$

$$= 68.9 \text{ g Na}_2\text{CO}_3 \text{ in } 100 \text{ g imp}$$

Therefore the material is _____ % pure.

68.9

Don't get the idea that *titration* is used only for determining percent purities. Any time you add a solution slowly to something, stop the addition when the reaction is complete, and measure the amount (generally the volume) of the added titrant, you say that you have run a *titration.*

SECTION B   Standardization of Solutions

Up to this point, we have assumed that the molarity of the solution is known. But how is it known? The determination of the molarity of a solution is called the *standardization* of the solution. You can do this by determining the volume of the solution required to completely react with a known amount of another substance. The choice of the substance used depends on the ease of getting a pure sample of the substance (called a *primary standard*) or the availability of a solution of the substance of very precisely known molarity (a *secondary standard*).

A commonly used primary standard for solutions of bases is potassium hydrogen phthalate, $C_6H_4(\text{COOK})(\text{COOH})$. It has one H that reacts with the OH^- ion from a base:

$$C_6H_4(\text{COOK})(\text{COOH}) + \text{NaOH} = C_6H_4(\text{COOK})(\text{COONa}) + H_2O$$

It is too much trouble to write out the formula for potassium hydrogen phthalate, so we'll just abbreviate it: *KHP*. Thus the balanced equation would be

$$KHP + NaOH = KNaP + H_2O$$

Pretest question 4 shows how you use KHP to find the molarity of a solution of a base. Suppose that you want to standardize a NaOH solution. You weigh out 2.114 g KHP, dissolve it in water, and titrate it to the equivalence point with the NaOH solution of unknown molarity. You find that it takes 22.31 mL of the solution of NaOH for complete reaction. You can then calculate the molarity of the NaOH solution. You ask: What is the molarity? That is,

$$\frac{mol\ NaOH}{L\ soln}$$

The information given is a conversion factor. There are 22.31 mL soln

per _____, and since the weight of KHP is in the problem, | **2.114 g KHP**

you need its _____. There are 204.2 g KHP per mol | **molecular weight**
KHP. And finally, for any stoichiometry problem, you need the

_____ conversion factor, _____. | **bridge;** $\dfrac{mol\ KHP}{mol\ NaOH}$

So you write the question:

$$\frac{?\ mol\ NaOH}{L\ soln} =$$

The best conversion factor to start with will have either moles NaOH in its numerator or liters solution in its denominator. In this way you are sure to get the right answer. The inverted bridge isn't bad:

$$\frac{?\ mol\ NaOH}{L\ soln} = \underline{\qquad\qquad}$$ | $\dfrac{mol\ NaOH}{mol\ KHP}$

The unwanted dimension is in the denominator this time. To remove it,

you multiply by _____. | $\dfrac{mol\ KHP}{204.2\ g\ KHP}$

$$\frac{?\ mol\ NaOH}{L\ soln} = \frac{mol\ NaOH}{\cancel{mol\ KHP}} \times \frac{\cancel{mol\ KHP}}{204.2\ g\ KHP}$$

You can cancel the unwanted dimension grams KHP by using the conversion factor given in the problem (converting milliliters immediately to liters):

$$\frac{?\ mol\ NaOH}{L\ soln} = \frac{mol\ NaOH}{\cancel{mol\ KHP}} \times \frac{\cancel{mol\ KHP}}{204.2\ g\ KHP} \times \underline{\qquad\qquad}$$ | $\dfrac{2.114\ g\ KHP}{0.02231\ L\ soln}$

And now the only dimensions left are those you want for your answer. Therefore

$$\frac{?\ mol\ NaOH}{L\ soln} = mol\ NaOH \times \frac{1}{204.2} \times \frac{2.114}{0.02231\ L\ soln}$$

$$= \frac{0.4649\ mol\ NaOH}{\underline{\qquad}} = 0.4640\ \underline{\qquad}$$ | $\dfrac{M}{L\ soln}$

When you do problems that have a conversion factor for the answer, you may prefer to reword the question so that the conversion factor provides both the question asked and the information given. Thus in this last problem, instead of asking "What is the molarity?" you could reword as follows: "How many moles NaOH are in 1.00 L?" This gives you a start in the right direction. You don't have to decide which conversion factor to use first. (Remember that the 1.00 L is a defined amount and has no effect on the significant figures.)

? mol NaOH = 1.00 L soln

You automatically use the only conversion factor that has liters solution in it.

? mol NaOH = 1.00 L soln × _____

$$\frac{2.114 \text{ g KHP}}{0.02231 \text{ L soln}}$$
molecular weight

Then, to cancel the unwanted grams KHP, you use the _____

? mol NaOH = 1.00 L̶ ̶s̶o̶l̶n̶ × $\dfrac{2.114 \text{ g KHP}}{0.02231 \text{ L̶ ̶s̶o̶l̶n̶}}$ × _____

$$\frac{\text{mol KHP}}{204.2 \text{ g KHP}}$$

And now you can use the bridge conversion factor:

? mol NaOH = 1.00 L soln × $\dfrac{2.114 \text{ g KHP}}{0.02231 \text{ L soln}}$ × $\dfrac{\text{mol KHP}}{204.2 \text{ g KHP}}$

× _____

$$\frac{\text{mol NaOH}}{\text{mol KHP}}$$
moles NaOH

The only dimension left is _____, which is the dimension of the question asked. So

? mol NaOH = 1.00 × $\dfrac{2.114}{0.02231}$ × $\dfrac{1}{204.2}$ × mol NaOH

= 0.4640 mol NaOH in 1.00 L soln = 0.4640 _____

M

[Notice that the 1.00 L has no effect on the number of significant figures. It is a defined quantity.]

If—instead of using a weighed amount of primary standard for your standardization—you use a solution of the standard substance whose molarity is known (a secondary standard), your problem will involve the reaction occurring when there are two solutions. We will cover this in the next section.

SECTION C Problems with Two Solutions Both in Molarity

Stoichiometry problems in which the two reactants are in separate solutions—except for the fact that you must be careful to label the dimensions of the solutions so that you can distinguish between them—are very easy. Question 5 in the pretest is a typical example. "How many liters of a 0.30 *M* solution of HCl are required for complete reaction with 5.0 L of a 0.10 *M* solution of Na_2CO_3? The reaction is
$Na_2CO_3 + 2HCl = 2NaCl + CO_2 + H_2O$."

Since there are two different solutions, you have to find some way to distinguish between liters of HCl solution and liters of Na_2CO_3 solution.

One way is to put the formula for the substance in the solution just after the L. Thus 5.0 L of the HCl solution would be 5.0 L HCl and 5.0 L of

5.0 L Na_2CO_3

the Na_2CO_3 solution would be _____. Perhaps a better way would be to put just the *non*spectator ion after the L. The net ionic equation is $2H^+ + CO_3^{2-} = CO_2 + H_2O$. (If you've forgotten this, see Chapter 8, Section B.) Thus 5.0 L of HCl solution would be 5.0 L H^+.

5.0 L CO_3^{2-}

And then 5.0 L of Na_2CO_3 solution would be _____. You could even put these as subscripts after the L. Thus you'd have 5.0 L_{H^+}

CO_3^{2-}

and 5.0 L _____. Whichever way you prefer is fine, as long as you are consistent. We'll use this last method here, since it eliminates mistakes as to what dimension must be canceled.

So, back to our problem. The question asked was "How many liters of a 0.30 *M* solution of HCl?" The information given is "5.0 L of a 0.10 *M* solution of Na_2CO_3." The conversion factors given in the problem are the two molarities

$\dfrac{0.30 \text{ mol HCl}}{L_{H^+}}$; $\dfrac{0.10 \text{ mol } Na_2CO_3}{L_{CO_3^{2-}}}$

_____ and _____

You will, of course, need the bridge between the two substances in the

$\dfrac{2 \text{ mol HCl}}{\text{mol } Na_2CO_3}$

question. This is _____.
   You start with

L_{H^+}; 5.0 $L_{CO_3^{2-}}$

? _____ = _____

and cancel the $L_{CO_3^{2-}}$ with the molarity of the Na_2CO_3 solution.

$\dfrac{0.10 \text{ mol } Na_2CO_3}{L_{CO_3^{2-}}}$

? $L_{H^+} = 5.0 \, L_{CO_3^{2-}} \times$ _____

bridge

Now that you are at mol Na_2CO_3, you can use the _____ conversion factor,

$\dfrac{2 \text{ mol HCl}}{\text{mol } Na_2CO_3}$

? $L_{H^+} = 5.0 \, \cancel{L_{CO_3^{2-}}} \times \dfrac{0.10 \text{ mol } Na_2CO_3}{\cancel{L_{CO_3^{2-}}}} \times$ _____

The molarity of the HCl solution takes you directly to the dimensions of the question asked:

? $L_{H^+} = 5.0 \, \cancel{L_{CO_3^{2-}}} \times \dfrac{0.10 \cancel{\text{ mol } Na_2CO_3}}{\cancel{L_{CO_3^{2-}}}} \times \dfrac{2 \text{ mol HCl}}{\cancel{\text{mol } Na_2CO_3}}$

$\dfrac{L_{H^+}}{0.30 \text{ mol HCl}}$

$\times$ _____

So, since the only dimension left is the dimension of the question asked, the answer must be

3.3 L_{H^+}

? $L_{H^+} = 5.0 \times 0.10 \times 2 \times \dfrac{L_{H^+}}{0.30} =$ _____

The 5.0 L of Na_2CO_3 solution thus require 3.3 L of the HCl solution.
   In Section B we mentioned the use of a solution as a secondary standard for the standardization of another solution. Question 6 in the pretest is an example of this. "It is found that 10.00 mL of an NaCl solution titrates to an equivalence point with 18.35 mL of 0.103 *M*

$Hg_2(NO_3)_2$. What is the molarity of the NaCl solution? The reaction considered is $Hg_2(NO_3)_2 + 2NaCl = Hg_2Cl_2 (s) + 2NaNO_3$." The net ionic equation is $Hg_2^{2+} + 2Cl^- = Hg_2Cl_2$.

The conversion factor given in the problem is the ratio of the two

solutions used, _____. You also have the molarity of

the $Hg_2(NO_3)_2$ solution,

$$\frac{0.01835 \ L_{Hg_2^{2+}}}{0.01000 \ L_{Cl^-}}$$

_____   and the bridge, $\dfrac{mol \ Hg_2(NO_3)_2}{2 \ mol \ NaCl}$

$$\frac{0.103 \ mol \ Hg_2(NO_3)_2}{L_{Hg_2^{2+}}}$$

You can solve for moles NaCl per liter$_{Cl^-}$ or you can reword the problem to ask "How many moles of NaCl are there in 1.00 L of solution?" Let's do it the second way.

? moles NaCl = _____

$$1.00 \ L_{Cl^-}$$

So, to cancel the unwanted L_{Cl^-}, you multiply by

? mol NaCl = 1.00 L_{Cl^-} × _____

$$\frac{0.01835 \ L_{Hg_2^{2+}}}{0.01000 \ L_{Cl^-}}$$

As in any stoichiometry, you head toward moles. So the next conversion is

? mol NaCl = 1.00 $\cancel{L_{Cl^-}}$ × $\dfrac{0.01835 \ L_{Hg_2^{2+}}}{0.01000 \ \cancel{L_{Cl^-}}}$ × _____

$$\frac{0.103 \ mol \ Hg_2(NO_3)_2}{L_{Hg_2^{2+}}}$$

The bridge now gets you directly to the dimension of the answer:

? mol NaCl = 1.00 $\cancel{L_{Cl^-}}$ × $\dfrac{0.01835 \ \cancel{L_{Hg_2^{2+}}}}{0.01000 \ \cancel{L_{Cl^-}}}$ × $\dfrac{0.103 \ mol \ Hg_2(NO_3)_2}{\cancel{L_{Hg_2^{2+}}}}$

$$\frac{2 \ mol \ NaCl}{mol \ Hg_2(NO_3)_2}$$

× _____

So

? mol NaCl = 1.00 × $\dfrac{0.01835}{0.01000}$ × 0.103 × 2 mol NaCl

= 0.378 _____ in 1.00 L or 0.378 __ NaCl

mol NaCl; *M*

If you had solved this problem without rewording, that is, if you had solved for moles NaCl per liter$_{Cl^-}$, your setup would have looked virtually the same, although the sequence of factors may have been different.

SECTION D   Back-titration

When running titrations, chemists may pass the end point, either accidentally or intentionally. One reason that they do this intentionally is to approach the equivalence point in a more convenient way. For example, you can recognize the change in color of some indicators more easily when going in one direction than in the other. Another reason is that an excess of titrant assures a rapid and complete reaction. Whatever the reason, the chemist has to *back-titrate* the reaction mixture to determine how much excess titrant has been added. The excess is then subtracted from the amount of titrant initially added, and the difference is then used for any calculations. What must be known is the relationship between the solution used for back-titrating and the amount of titrant initially added that is in excess.

The usual way to do this is to titrate the initial solution with the back-titrant directly. Question 7 in the pretest is an example of this method. "An impure sample of $CaCO_3$ is reacted with an excess of 1.50 M HCl solution. The excess HCl solution is then back-titrated with an NaOH solution to its equivalence point. The data are:

Weight impure $CaCO_3$ = 2.20 g
Volume 1.50 M HCl = 45.0 mL
Volume NaOH for back-titration = 14.0 mL
30.0 mL HCl titrates to an equivalence point with 22.0 mL NaOH.

What is the percent purity of the $CaCO_3$ sample?
The balanced equation for the reaction is

$$CaCO_3 + 2HCl = CaCl_2 + CO_2 + H_2O"$$

The net ionic equation is $CO_3^{2-} + 2H^+ = CO_2 + 2H_2O$.
The titration of the HCl solution with NaOH solution, $OH^- + H^+ = H_2O$, gives you the relation between the two solutions. It is a conversion factor,

$$\frac{30.0 \text{ mL}_{H^+}}{\rule{3cm}{0.4pt}}$$

22.0 mL$_{OH^-}$	

To determine the excess HCl solution, you say that it is equivalent to 14.0 mL of the NaOH solution:

14.0 mL$_{OH^-}$	? mL$_{H^+}$ (excess) = _____

Using the conversion factor that gives the relation between the two solutions,

$\dfrac{30.0 \text{ mL}_{H^+}}{22.0 \text{ mL}_{OH^-}}$; 19.1 mL$_{H^+}$? mL$_{H^+}$ (excess) = 14.0 mL$_{OH^-}$ × _____ = _____

The amount of the HCl solution that was actually required was the initial

minus   amount _____ the amount in excess:

− 19.1 mL$_{H^+}$; 25.9   ? mL$_{H^+}$ (required) = 45.0 mL$_{H^+}$ _____ = _____ mL$_{H^+}$

Now you solve the problem in the same way you did in the example in Section A. The weight of impure substance is 2.20 g and the amount of

25.9 mL soln   1.50 M HCl required is _____. (You can now call mL$_{H^+}$ "mL soln.") Starting with the reworded question, you have

g $CaCO_3$; 100 g imp   ? _____ = _____

You now have a conversion factor containing grams imp and the amount of HCl solution actually required, so

$\dfrac{0.0259 \text{ L soln}}{2.20 \text{ g imp}}$? g $CaCO_3$ = 100 g imp × _____

The molarity of the HCl solution gets you to moles HCl:

$$? \text{ g CaCO}_3 = 100 \text{ g imp} \times \frac{0.0259 \text{ L soln}}{2.20 \text{ g imp}} \times \underline{\hspace{2cm}}$$

Now that you are at moles HCl, you can use the _____ conversion factor from the balanced equation:

$$? \text{ g CaCO}_3 = 100 \text{ g imp} \times \frac{0.0259 \text{ L soln}}{2.20 \text{ g imp}} \times \frac{1.50 \text{ mol HCl}}{\text{L soln}}$$

$$\times \underline{\hspace{2cm}}$$

Since the dimension of the answer is grams CaCO$_3$, you multiply by the

_____ of CaCO$_3$:

$$? \text{ g CaCO}_3 = 100 \text{ g imp} \times \frac{0.0259 \text{ L soln}}{2.20 \text{ g imp}} \times \frac{1.50 \text{ mol HCl}}{\text{L soln}}$$

$$\times \frac{\text{mol CaCO}_3}{2 \text{ mol HCl}} \times \underline{\hspace{2cm}}$$

The only dimension left uncanceled is _____, which is the dimension of the question asked:

$$? \text{ g CaCO}_3 = 100 \times \frac{0.0259}{2.20} \times 1.50 \times \frac{1}{2} \times 100.1 \text{ g CaCO}_3$$

$$= 88.4 \text{ g CaCO}_3 \text{ in } \underline{\hspace{1cm}} \text{ g imp, so the sample is } \underline{\hspace{1cm}} \% \text{ pure}$$

Sometimes you are not given the relationship between titrant and back-titrant in milliliters of one solution per milliliter of the other. Instead you may be given the molarity of both solutions. In this case, you determine the number of moles of the solute in the titrant that were really required for the reaction by subtracting the moles in excess from the total number of moles put in initially. The following example brings this out.

The acid content of fats and oils is expressed commercially as the *saponification number.* This is defined as the number of milligrams of KOH that react with 1.0 g of the fat or oil. In order to assure complete reaction of the fat or oil, you always add an excess of KOH solution to the sample. Then, after reaction, you back-titrate the excess KOH with a standardized HCl solution.

The following data were obtained for an oil sample.

Weight of oil = 3.60 g
Volume 0.350 M KOH = 50.00 mL
Volume 0.550 M HCl for back-titration = 7.10 mL

The reaction between titrant and back-titrant is

$$\text{HCl} + \text{KOH} = \text{KCl} + \text{H}_2\text{O} \quad \text{or} \quad \text{H}^+ + \text{OH}^- = \text{H}_2\text{O}$$

First you calculate the number of *moles* of excess KOH:

Right margin notes:

$\dfrac{1.50 \text{ mol HCl}}{\text{L soln}}$

bridge

$\dfrac{\text{mol CaCO}_3}{2 \text{ mol HCl}}$

molecular weight

$\dfrac{100.1 \text{ g CaCO}_3}{\text{mol CaCO}_3}$

grams CaCO$_3$

100; 88.4

$\dfrac{\text{mol KOH}}{\text{mol HCl}}$

$$? \text{ mol KOH (excess)} = 0.00710 \text{ } \cancel{L_{H^+}} \times \dfrac{0.550 \text{ mol HCl}}{\cancel{L_{H^+}}} \times \underline{\hspace{2cm}}$$

$$= 0.00390 \text{ mol KOH (excess)}$$

Then you calculate the number of moles of KOH initially put in:

$\dfrac{0.350 \text{ mol KOH}}{L_{OH^-}}$

$$? \text{ mol KOH (initial)} = 0.05000 \text{ } L_{OH^-} \times \underline{\hspace{3cm}}$$

$$= 0.0175 \text{ mol KOH (initial)}$$

subtracting

You get the moles KOH actually required by $\underline{\hspace{3cm}}$ the

initial

moles KOH (excess) from the moles KOH ($\underline{\hspace{2cm}}$):

0.00390

$$\text{mol KOH (required)} = 0.0175 \text{ mol KOH} - \underline{\hspace{2cm}} \text{ mol KOH}$$

$$= 0.0136 \text{ mol KOH}$$

This gives you the conversion factor,

$$\dfrac{0.0136 \text{ mol KOH}}{3.60 \text{ g oil}}$$

and you can now solve the problem easily. You are asked "How many milligrams of KOH are required to titrate 1.0 g of oil?" You start as usual:

mg KOH

$$? \underline{\hspace{2cm}} = 1.0 \text{ g oil}$$

Then you use the conversion factor from the result of the back-titration:

$\dfrac{0.0136 \text{ mol KOH}}{3.60 \text{ g oil}}$

$$? \text{ mg KOH} = 1.0 \text{ g oil} \times \underline{\hspace{3cm}}$$

Now all you have to do is convert moles KOH to milligrams KOH:

$\dfrac{56.1 \text{ g KOH}}{\text{mol KOH}}$; $\dfrac{\text{mg KOH}}{10^{-3} \text{ g KOH}}$

$$? \text{ mg KOH} = 1.0 \text{ } \cancel{\text{g oil}} \times \dfrac{0.0136 \text{ mol KOH}}{3.60 \text{ } \cancel{\text{g oil}}} \times \underline{\hspace{2cm}} \times \underline{\hspace{2cm}}$$

$$= 1.0 \times \dfrac{0.0136}{3.60} \times 56.1 \times \dfrac{\text{mg KOH}}{10^{-3}}$$

$$= 212 \text{ mg KOH per 1.0-g sample}$$

[The 1.0 g is a defined quantity and does not enter into the significant figures.]

So the saponification number of the oil is 212.

PROBLEM SET

The first part of this problem set is divided into groups according to the sections in the chapter in which the type of problem is covered. Following this is a group of miscellaneous problems relating to all parts of the chapter. You shouldn't have any difficulty figuring them out after you have done the first set.

The first 17 problems are based on the balanced equations shown below. The pertinent equation is indicated in each problem. Also, to save you time, a listing of molecular weights of the compounds you will use is given after the list of balanced equations. In some of the later problems, you will have to balance the equation yourself. Problems that are more difficult are marked with an asterisk (*).

Balanced Equations

(a) $CaO + 2HCl = CaCl_2 + H_2O$

(b) $KMnO_4 + 5FeCl_2 + 8HCl = MnCl_2 + 5FeCl_3 + 4H_2O + KCl$

(c) $K_2Cr_2O_7 + 3H_2C_2O_4 + 8HCl = 2CrCl_3 + 6CO_2 + 7H_2O + 2KCl$

(d) $H_3PO_4 + 3KOH = K_3PO_4 + 3H_2O$

(e) $2H_3BO_3 + Ca(OH)_2 = Ca(H_2BO_3)_2 + 2H_2O$

Molecular Weights

AgCl	143.4 g/mol	Fe	55.8 g/mol	KCl	74.6 g/mol
$Ba(OH)_2$	171.3 g/mol	$FeCl_2$	126.8 g/mol	$K_2Cr_2O_7$	294.2 g/mol
Br_2	159.8 g/mol	H_3AsO_4	141.9 g/mol	KOH	56.1 g/mol
CH_3COOH	60.0 g/mol	$H_2C_2O_4$	90.0 g/mol	$KMnO_4$	158.0 g/mol
$CaCO_3$	100.1 g/mol	HCl	36.5 g/mol	$Na_2C_2O_4$	134.0 g/mol
$Ca(OH)_2$	74.1 g/mol	H_3PO_4	98.0 g/mol	NaOH	40.0 g/mol

Section A

1. How many liters of 0.10 *M* HCl solution are needed to react completely with 4.0 mol CaO according to reaction (a)?

2. How many liters of 0.50 *M* HCl are consumed by the complete reaction of 0.561 g of CaO according to reaction (a)?

3. How many grams of $KMnO_4$ are needed for complete reaction (b) with 150.0 mL of a 0.30 *M* $FeCl_2$ solution?

4. Suppose that you want to convert all the H_3BO_3 in 750 mL of a 0.50 *M* solution of H_3BO_3 to $Ca(H_2BO_3)_2$ according to reaction (e). How many grams of $Ca(OH)_2$ do you need? (You don't want to use more of the $Ca(OH)_2$ than is absolutely required.)

5. Potassium permanganate, $KMnO_4$, is added to an empty Erlenmeyer flask weighing 162.37 g. The flask and contents weigh 163.45. The $KMnO_4$ is dissolved in water and titrated with a 1.10 *M* solution of $FeCl_2$. How many milliliters of the $FeCl_2$ solution are added when the equivalence point for reaction (b) is reached?

*6. How many liters of 0.150 M KOH are required to titrate 10.0 mL of a 42.0 wt % solution of H_3PO_4 to its equivalence point? The density of the H_3PO_4 solution is 1.23 g/mL. (d)

7. What is the percent purity of an anhydrous (without water) sample of $H_2C_2O_4$ if 2.777 g of the impure material reacts to equivalence with 38.21 mL of 0.250 M $K_2Cr_2O_7$ according to reaction (c)?

*8. A solution is prepared by dissolving 6.34 g $FeCl_2$ in enough water to have 50.00 mL of solution. A 10.00 mL sample of this solution is then titrated with 0.110 M $KMnO_4$. How many milliliters of the $KMnO_4$ solution are needed for complete reaction according to equation (b)?

Section B

9. It is found that 35.0 mL of an H_3PO_4 solution titrates 0.250 g KOH to its equivalence point. What is the molarity of the H_3PO_4 solution? [This is reaction (d).]

10. The molarity of an $H_2C_2O_4$ solution is determined by titrating it against a weighed amount of $K_2Cr_2O_7$. Calculate the molarity if 0.580 g $K_2Cr_2O_7$ titrates with exactly 35.00 mL of the solution. [This is reaction (c).]

*11. What is the molarity of a $KMnO_4$ solution if 36.37 mL of it are required to react completely with 2.304 g of a 90.00% pure sample of $FeCl_2$ according to reaction (b)?

Section C

12. According to reaction (c), 4.5 L of a 0.10 M $K_2Cr_2O_7$ solution react with how many liters of a 0.50 M solution of $H_2C_2O_4$?

13. How many milliliters of a 0.020 M solution of $KMnO_4$ are required to titrate 60.0 mL of a 0.010 M solution of $FeCl_2$ according to reaction (b)?

14. Suppose that 35.0 mL of an H_3PO_4 solution react with exactly 50.0 mL of 2.5 M KOH according to reaction (d). What is the molarity of the H_3PO_4 solution?

15. It is found that 650 mL of a solution of H_3BO_3 react with exactly 1.30 L of 0.0250 M $Ca(OH)_2$ solution according to equation (e). From these data, calculate the molarity of the H_3BO_3 solution.

*16. What is the maximum amount (in grams) of KCl that can be formed by the reaction of 3.5 L of a 0.15 M $KMnO_4$ solution with 7.0 L of a 0.25 M $FeCl_2$ solution according to reaction (b)? [Be careful. You must determine which reactant is the controlling one (Chapter 11).]

Section D

*17. When a 10.0 g sample of impure $FeCl_2$ is reacted with 60.0 mL of a 0.20 M $KMnO_4$ solution, it is found that an excess of $KMnO_4$ solution has been used. The excess is back-titrated with 18.0 mL of $Na_2C_2O_4$ solution. In the absence of the $FeCl_2$, 10.0 mL of the $KMnO_4$ solution required 22.0 mL of the $Na_2C_2O_4$ to reach an end point. What is the percent purity of the $FeCl_2$ sample? [This is reaction (b).]

*18. A student was standardizing an NaOH solution by titrating it against a weighed sample of potassium acid phthalate (KHP). Unfortunately the student passed the equivalence point and had to back-titrate the excess NaOH solution, using a previously standardized HCl (0.1050 M) solution. The two reactions involved are

$$NaOH + KHP = KNaP + H_2O \quad (\text{MW of KHP} = 204.2 \text{ g/mol})$$

and

$$NaOH + HCl = NaCl + H_2O$$

The data are shown below.

Weight KHP   =   2.758 g
Volume NaOH = 28.69 mL
Volume 0.1050 M HCl for back-titration = 2.37 mL

What is the molarity of the NaOH solution?

*19. The "bromine number" of a lubricating oil is a measure of the number of double bonds in the oil. It is expressed in milligrams of Br_2 per gram of oil. To ensure complete reaction with the oil, an excess of Br_2 is used. When the reaction is completed, the excess Br_2 is back-titrated with a standardized solution of $Na_2S_2O_3$. The reaction between Br_2 and $Na_2S_2O_3$ in a basic solution is

$$Na_2S_2O_3 + 4Br_2 + 10NaOH = 2Na_2SO_4 + 8NaBr + 5H_2O$$

Using the following data, determine the bromine number for an oil sample.

Weight of oil = 16.0 g
Volume 0.100 M Br_2 = 50.0 mL
Volume 0.0125 M $Na_2S_2O_3$ for back-titration = 8.00 mL

Miscellaneous Problems

20. Suppose that it requires 2.00 L of an H_2SO_4 solution to completely react with 300.0 g of an 80% pure sample of NaOH. What is the molarity of the H_2SO_4 solution? Determine the balanced equation. The products are Na_2SO_4 and H_2O.

21. When an excess of NaCl solution was added to 75 mL of an $AgNO_3$ solution, it was found that 0.43 g of AgCl precipitated. Write the equation for the reaction; then determine the molarity of the $AgNO_3$ solution.

*22. A 0.500 g sample of impure $Ba(OH)_2$ requires 35.3 mL of an HCl solution to reach an equivalence point. What is the percent purity of the $Ba(OH)_2$ sample? The HCl solution was previously standardized by titration against a pure sample of $CaCO_3$, yielding the following data: Weight $CaCO_3$ = 0.240 g, volume HCl solution = 45.0 mL. Balance the equations.

*23. How many milliliters of a 0.20 *M* NaOH solution are needed to exactly neutralize (bring to the equivalence point) 5.0 mL of vinegar that is a 5.0 wt % solution of CH_3COOH? The density of the solution is 1.04 g/mL and the reaction is

$$CH_3COOH + NaOH = CH_3COONa + H_2O$$

24. It is found that 40.4 mL of a KOH solution require 25.0 mL of a 20.2 wt% solution of HCl (density = 1.1 g/mL) for complete neutralization. What is the molarity of the KOH solution? The reaction yields KCl and H_2O.

*25. You add 75.0 mL of a 0.20 *M* HCl solution to 25.0 mL of a 0.50 *M* NaOH solution. What is the molarity of the HCl solution that remains after the following reaction has taken place?

$$HCl + NaOH = NaCl + H_2O$$

(Don't forget to add the volumes.)

26. How many milliliters of a 0.30 *M* solution of $Fe(NO_3)_2$ must you add to 100.0 mL of 0.50 *M* $KMnO_4$ to completely reduce the $KMnO_4$ to $Mn(NO_3)_2$? The reaction is

$$KMnO_4 + 5Fe(NO_3)_2 + 8HNO_3 = Mn(NO_3)_2 + 5Fe(NO_3)_3 + 4H_2O$$
$$+ KNO_3$$

27. A 19.75 mL sample of vinegar (dilute CH_3COOH) whose density is 1.06 g/mL requires 43.24 mL of 0.400 *M* NaOH to reach the equivalence point. What is the weight percent CH_3COOH in the vinegar? The reaction is

$$CH_3COOH + NaOH = CH_3COONa + H_2O$$

28. An excess of HCl solution is added to a 0.500 g sample of impure $CaCO_3$. When the reaction is complete, the excess HCl is back-titrated with a standardized NaOH solution. From the following data, calculate the percent purity of the $CaCO_3$ sample.

Weight impure sample = 0.500 g
Volume 0.100 *M* HCl = 50.0 mL
Volume 0.105 *M* NaOH used for back-titration = 6.00 mL

29. An unknown acid is known to have one replaceable H atom. A 0.610 g sample of the acid is exactly neutralized (reaches its equivalence point) with 48.2 mL of a 0.104 *M* NaOH solution. What is the molecular weight of the acid? [*Hint:* The reaction must be HX + NaOH = NaX + H_2O, where HX is the unknown acid.]

*30. It is possible to determine the percent nitrogen in a protein sample by reacting the protein with H_2SO_4 (with HgO catalyst) to produce $(NH_4)_2SO_4$. Concentrated NaOH is then reacted with the $(NH_4)_2SO_4$ and NH_3 gas is formed. The NH_3 is led into a known volume of HCl solution of known molarity. The NH_3 consumes some of the HCl. The excess HCl is then back-titrated with a standardized NaOH solution. The reactions are

$$NH_3 + HCl = NH_4Cl \quad \text{and} \quad HCl + NaOH = NaCl + H_2O \quad \text{(back-titration)}$$

From the following data, determine the percent N in the protein sample:

Weight protein sample = 1.75 g
Volume 0.105 M HCl = 25.00 mL
Volume 0.130 M NaOH for the back-titration = 3.85 mL

PROBLEM SET ANSWERS

1. ? L soln = 4.0 ~~mol CaO~~ $\times \dfrac{2 \text{ mol HCl}}{\text{mol CaO}} \times \dfrac{\text{L soln}}{0.10 \text{ mol HCl}}$ = 80 L soln

2. ? L soln = 0.561 ~~g CaO~~ $\times \dfrac{\text{mol CaO}}{56.1 \text{ g CaO}} \times \dfrac{2 \text{ mol HCl}}{\text{mol CaO}} \times \dfrac{\text{L soln}}{0.50 \text{ mol HCl}}$

    = 4.0×10^{-2} L soln

3. ? g KMnO$_4$ = 0.1500 ~~L soln~~ $\times \dfrac{0.30 \text{ mol FeCl}_2}{\text{L soln}} \times \dfrac{\text{mol KMnO}_4}{5 \text{ mol FeCl}_2}$

    $\times \dfrac{158.0 \text{ g KMnO}_4}{\text{mol KMnO}_4}$ = 1.4 g KMnO$_4$

4. ? g Ca(OH)$_2$ = 0.75 ~~L soln~~ $\times \dfrac{0.50 \text{ mol H}_3\text{BO}_3}{\text{L soln}} \times \dfrac{\text{mol Ca(OH)}_2}{2 \text{ mol H}_3\text{BO}_3}$

    $\times \dfrac{74.1 \text{ g Ca(OH)}_2}{\text{mol Ca(OH)}_2}$ = 14 g Ca(OH)$_2$

5. First determine the weight of KMnO$_4$ in the flask:

    163.45 g – 162.37 g = 1.08 g KMnO$_4$

    ? mL soln = 1.08 ~~g KMnO$_4$~~ $\times \dfrac{\text{mol KMnO}_4}{158.0 \text{ g KMnO}_4} \times \dfrac{5 \text{ mol FeCl}_2}{\text{mol KMnO}_4}$

    $\times \dfrac{\text{L soln}}{1.10 \text{ mol FeCl}_2} \times \dfrac{\text{mL soln}}{10^{-3} \text{ L soln}}$ = 31.1 mL soln

6. ? L$_{OH^-}$ = 10.0 ~~mL$_{H^+}$~~ $\times \dfrac{1.23 \text{ g}_{H^+}}{\text{mL}_{H^+}} \times \dfrac{42.0 \text{ g H}_3\text{PO}_4}{100 \text{ g}_{H^+}}$

    $\times \dfrac{\text{mol H}_3\text{PO}_4}{98.0 \text{ g H}_3\text{PO}_4} \times \dfrac{3 \text{ mol KOH}}{\text{mol H}_3\text{PO}_4} \times \dfrac{\text{L}_{OH^-}}{0.150 \text{ mol KOH}}$

    = 1.05 L$_{OH^-}$

7. ? g H$_2$C$_2$O$_4$ = 100 ~~g imp~~ $\times \dfrac{0.03821 \text{ L}_{Cr_2O_7^{2-}}}{2.777 \text{ g imp}} \times \dfrac{0.250 \text{ mol K}_2\text{Cr}_2\text{O}_7}{\text{L}_{Cr_2O_7^{2-}}}$

    $\times \dfrac{3 \text{ mol H}_2\text{C}_2\text{O}_4}{\text{mol K}_2\text{Cr}_2\text{O}_7} \times \dfrac{90.0 \text{ g H}_2\text{C}_2\text{O}_4}{\text{mol H}_2\text{C}_2\text{O}_4}$

    = 92.9 g H$_2$C$_2$O$_4$ in 100 g imp or 92.9% pure

8. $? mL_{MnO_4^-} = 10.00 \text{ mL}_{Fe^{2+}} \times \dfrac{6.34 \text{ g FeCl}_2}{50.00 \text{ mL}_{Fe^{2+}}} \times \dfrac{\text{mol FeCl}_2}{126.8 \text{ g FeCl}_2}$

$\times \dfrac{\text{mol KMnO}_4}{5 \text{ mol FeCl}_2} \times \dfrac{L_{MnO_4^-}}{0.110 \text{ mol KMnO}_4} \times \dfrac{mL_{MnO_4^-}}{10^{-3} L_{MnO_4^-}}$

$= 18.2 \text{ mL}_{MnO_4^-}$

9. $? M = \dfrac{? \text{ mol H}_3PO_4}{\text{L soln}} = \dfrac{0.250 \text{ g KOH}}{0.0350 \text{ L soln}} \times \dfrac{\text{mol KOH}}{56.1 \text{ g KOH}} \times \dfrac{\text{mol H}_3PO_4}{3 \text{ mol KOH}}$

$= 4.24 \times 10^{-3} \ M$

10. [We will show this answer using the reworded-question method. You ask "How many moles of $H_2C_2O_4$ are in 1.00 L of solution?"]

$? \text{ mol H}_2C_2O_4 = 1.00 \text{ L soln} \times \dfrac{0.580 \text{ g K}_2Cr_2O_7}{0.03500 \text{ L soln}}$

$\times \dfrac{\text{mol K}_2Cr_2O_7}{294.2 \text{ g K}_2Cr_2O_7} \times \dfrac{3 \text{ mol H}_2C_2O_4}{\text{mol K}_2Cr_2O_7}$

$= 0.169 \text{ mol H}_2C_2O_4 \text{ in } 1.00 \text{ L soln or } 0.169 \ M \ H_2C_2O_4$

11. $? M = \dfrac{? \text{ mol KMnO}_4}{\text{L soln}} = \dfrac{2.304 \text{ g 90\%}}{0.03637 \text{ L soln}} \times \dfrac{90.00 \text{ g FeCl}_2}{100 \text{ g 90\%}} \times \dfrac{\text{mol FeCl}_2}{126.8 \text{ g FeCl}_2}$

$\times \dfrac{\text{mol KMnO}_4}{5 \text{ mol FeCl}_2} = 0.08993 \ M \ KMnO_4$

12. $? L_{C_2O_4^{2-}} = 4.5 \text{ L}_{Cr_2O_7^{2-}} \times \dfrac{0.10 \text{ mol K}_2Cr_2O_7}{L_{Cr_2O_7^{2-}}} \times \dfrac{3 \text{ mol H}_2C_2O_4}{\text{mol K}_2Cr_2O_7}$

$\times \dfrac{L_{C_2O_4^{2-}}}{0.50 \text{ mol H}_2C_2O_4} = 2.7 \ L_{C_2O_4^{2-}}$

13. $? mL_{MnO_4^-} = 0.0600 \text{ L}_{Fe^{2+}} \times \dfrac{0.010 \text{ mol FeCl}_2}{L_{Fe^{2+}}} \times \dfrac{\text{mol KMnO}_4}{5 \text{ mol FeCl}_2}$

$\times \dfrac{L_{MnO_4^-}}{0.020 \text{ mol KMnO}_4} \times \dfrac{mL_{MnO_4^-}}{10^{-3} L_{MnO_4^-}} = 6.0 \text{ mL}_{MnO_4^-}$

14. [We'll do this as a reworded problem.]

$? \text{ mol H}_3PO_4 = 1.00 \text{ L}_{H^+} \times \dfrac{0.0500 \text{ L}_{OH^-}}{0.0350 \text{ L}_{H^+}} \times \dfrac{2.5 \text{ mol KOH}}{L_{OH^-}}$

$\times \dfrac{\text{mol H}_3PO_4}{3 \text{ mol KOH}} = 1.2 \text{ mol H}_3PO_4 \text{ in } 1.00 \text{ L}_{H^+} \text{ or } 1.2 \ M \ H_3PO_4$

15. $? M = \dfrac{? \text{ mol H}_3BO_3}{L_{H^+}} = \dfrac{1.30 \text{ L}_{OH^-}}{0.650 \text{ L}_{H^+}} \times \dfrac{0.0250 \text{ mol Ca(OH)}_2}{L_{OH^-}} \times \dfrac{2 \text{ mol H}_3BO_3}{\text{mol Ca(OH)}_2}$

$= 0.100 \ M \ H_3BO_3$

16. First you determine the number of moles of each reactant:

$? \text{ mol KMnO}_4 = 3.5 \text{ L}_{MnO_4^-} \times \dfrac{0.15 \text{ mol KMnO}_4}{L_{MnO_4^-}} = 0.525 \text{ mol KMnO}_4$

$$? \text{ mol FeCl}_2 = 7.0\,\cancel{\text{L}_{Fe^{2+}}} \times \frac{0.25 \text{ mol FeCl}_2}{\cancel{\text{L}_{Fe^{2+}}}} = 1.75 \text{ mol FeCl}_2$$

From the balanced equation, 1 mol $KMnO_4$ would require 5 mol $FeCl_2$. Therefore 0.525 mol $KMnO_4$ would need 2.625 mol $FeCl_2$, which is not available. The $KMnO_4$ is the controlling reactant. You can now solve the problem using 0.525 mol $KMnO_4$ as the information given:

$$? \text{ g KCl} = 0.525 \,\cancel{\text{mol KMnO}_4} \times \frac{\cancel{\text{mol KCl}}}{\cancel{\text{mol KMnO}_4}} \times \frac{74.6 \text{ g KCl}}{\cancel{\text{mol KCl}}}$$

$$= 39 \text{ g KCl (to the correct significant figures)}$$

17. First you determine the $mL_{MnO_4^-}$ (excess).

$$mL_{MnO_4^-} \text{ (excess)} = 18.0\,\cancel{\text{mL}_{C_2O_4{}^{2-}}} \times \frac{10.0 \text{ mL}_{MnO_4^-}}{22.0\,\cancel{\text{mL}_{C_2O_4{}^{2-}}}}$$

$$= 8.2 \text{ mL}_{MnO_4^-} \text{ (excess)}$$

Therefore the amount of MnO_4^- actually needed is

$$mL_{MnO_4^-} \text{ (required)} = 60.0 \text{ mL}_{MnO_4^-} - 8.2 \text{ mL}_{MnO_4^-}$$

$$= 51.8 \text{ mL}_{MnO_4^-} \text{ (required)}$$

$$? \text{ g FeCl}_2 = 100\,\cancel{\text{g imp}} \times \frac{0.0518\,\cancel{\text{L}_{MnO_4^-}}}{10.0\,\cancel{\text{g imp}}} \times \frac{0.20\,\cancel{\text{mol KMnO}_4}}{\cancel{\text{L}_{MnO_4^-}}} \times \frac{5\,\cancel{\text{mol FeCl}_2}}{\cancel{\text{mol KMnO}_4}}$$

$$\times \frac{126.8 \text{ g FeCl}_2}{\cancel{\text{mol FeCl}_2}} = 66 \text{ g FeCl}_2 \text{ in 100 g imp} = 66\% \text{ pure}$$

18. In this problem, you have to say that the moles of HCl used in back-titration are the same as the moles of KHP present all the time. Thus you add the moles of HCl to the moles of KHP. The net ionic equation is $H^+ + OH^- = H_2O$. It doesn't matter where the H^+ comes from.

$$? \text{ mol HCl} = 0.00237\,\cancel{\text{L}_{HCl}} \times \frac{0.1050 \text{ mol HCl}}{\cancel{\text{L}_{HCl}}} = 0.000249 \text{ mol HCl}$$

You then calculate the number of moles of KHP:

$$? \text{ mol KHP} = 2.758\,\cancel{\text{g KHP}} \times \frac{\text{mol KHP}}{204.2\,\cancel{\text{g KHP}}} = 0.01351 \text{ mol KHP}$$

Each of these is contributing one H^+ per molecule. Thus

$$? \text{ mol H}^+ = 0.000249 \text{ mol H}^+ + 0.01351 \text{ mol H}^+ = 0.01376 \text{ mol H}^+$$

$$? \text{ mol NaOH} = 1.00\,\cancel{\text{L}_{OH^-}} \times \frac{0.01376\,\cancel{\text{mol H}^+}}{0.02869\,\cancel{\text{L}_{OH^-}}} \times \frac{\text{mol NaOH}}{\cancel{\text{mol H}^+}}$$

$$= 0.4796 \text{ mol NaOH in 1.00 L or } 0.4796\,M \text{ NaOH}$$

This is a good example of how the net ionic equation can help you.

19. This is a typical back-titration problem. First, from the back-titration data, you calculate the number of moles of Br_2 that were in excess.

$$? \text{ mol Br}_2 \text{ (excess)} = 0.00800 \text{ } \cancel{L_{S_2O_3{}^{2-}}} \times \frac{0.0125 \text{ } \cancel{\text{mol Na}_2\text{S}_2\text{O}_3}}{\cancel{L_{S_2O_3{}^{2-}}}} \times \frac{4 \text{ mol Br}_2}{\cancel{\text{mol Na}_2\text{S}_2\text{O}_3}}$$

$$= 0.000400 \text{ mol Br}_2 \text{ (excess)}$$

Then you determine the number of moles of Br_2 put in initially.

$$? \text{ mol Br}_2 \text{ (init)} = 0.0500 \text{ } \cancel{L_{Br_2}} \times \frac{0.100 \text{ mol Br}_2}{\cancel{L_{Br_2}}} = 0.00500 \text{ mol Br}_2 \text{ (init)}$$

You get the number of moles of Br_2 actually needed by subtracting:

$$\text{mol Br}_2 \text{ (required)} = 0.00500 \text{ mol Br}_2 - 0.000400 \text{ mol Br}_2$$

$$= 0.00460 \text{ mol Br}_2 \text{ (required)}$$

To get the bromine number,

$$? \text{ mg Br}_2 = 1.00 \text{ } \cancel{\text{g oil}} \times \frac{0.00460 \text{ } \cancel{\text{mol Br}_2}}{16.0 \text{ } \cancel{\text{g oil}}} \times \frac{159.8 \text{ } \cancel{\text{g Br}_2}}{\cancel{\text{mol Br}_2}} \times \frac{\text{mg Br}_2}{10^{-3} \text{ } \cancel{\text{g Br}_2}}$$

$$= 45.9 \text{ mg Br}_2 \text{ for } 1.0 \text{ g oil}$$

The bromine number is therefore 45.9

20. The balanced equation is

$$H_2SO_4 + 2NaOH = Na_2SO_4 + 2H_2O$$

$$? M = \frac{? \text{ mol H}_2\text{SO}_4}{L_{H^+}} = \frac{300.0 \text{ } \cancel{\text{g 80\%}}}{2.00 \text{ } L_{H^+}} \times \frac{80 \text{ } \cancel{\text{g NaOH}}}{100 \text{ } \cancel{\text{g 80\%}}} \times \frac{\cancel{\text{mol NaOH}}}{40.0 \text{ } \cancel{\text{g NaOH}}}$$

$$\times \frac{\text{mol H}_2\text{SO}_4}{2 \text{ } \cancel{\text{mol NaOH}}} = 1.5 \text{ } M \text{ H}_2\text{SO}_4$$

21. The balanced equation is

$$AgNO_3 + NaCl = AgCl + NaNO_3$$

[We'll do this one by rewording the question. How many moles of $AgNO_3$ are in 1.00 L ?]

$$? \text{ mol AgNO}_3 = 1.00 \text{ } \cancel{L_{Ag^+}} \times \frac{0.43 \text{ } \cancel{\text{g AgCl}}}{0.075 \text{ } \cancel{L_{Ag^+}}} \times \frac{\cancel{\text{mol AgCl}}}{143.4 \text{ } \cancel{\text{g AgCl}}} \times \frac{\text{mol AgNO}_3}{\cancel{\text{mol AgCl}}}$$

$$= 0.040 \text{ mol AgNO}_3 \text{ in } 1.00 \text{ } L_{Ag^+} = 0.040 \text{ } M \text{ AgNO}_3$$

22. First you determine the molarity of the HCl solution. The balanced equation for its standardization is

$$CaCO_3 + 2HCl = CaCl_2 + H_2O + CO_2$$

$$? M = \frac{? \text{ mol HCl}}{L_{H^+}} = \frac{0.240 \text{ } \cancel{\text{g CaCO}_3}}{0.0450 \text{ } L_{H^+}} \times \frac{\cancel{\text{mol CaCO}_3}}{100.1 \text{ } \cancel{\text{g CaCO}_3}} \times \frac{2 \text{ mol HCl}}{\cancel{\text{mol CaCO}_3}}$$

$$= 0.107 \text{ } M \text{ HCl}$$

The balanced equation for the reaction is

$$Ba(OH)_2 + 2HCl = BaCl_2 + 2H_2O$$

$$? \text{ g Ba(OH)}_2 = 100 \, \cancel{\text{g imp}} \times \frac{0.0353 \, \cancel{\text{L}_{H^+}}}{0.500 \, \cancel{\text{g imp}}} \times \frac{0.107 \, \cancel{\text{mol HCl}}}{\cancel{\text{L}_{H^+}}} \times \frac{\cancel{\text{mol Ba(OH)}_2}}{2 \, \cancel{\text{mol HCl}}}$$

$$\times \frac{171.3 \text{ g Ba(OH)}_2}{\cancel{\text{mol Ba(OH)}_2}} = 64.7 \text{ g Ba(OH)}_2 \text{ in 100 g imp}$$

or 64.7% pure

23. $? \text{ mL}_{OH^-} = 5.0 \, \cancel{\text{mL vinegar}} \times \dfrac{1.04 \, \cancel{\text{g vinegar}}}{\cancel{\text{mL vinegar}}} \times \dfrac{5.0 \, \cancel{\text{g CH}_3\text{COOH}}}{100 \, \cancel{\text{g vinegar}}}$

$$\times \frac{\cancel{\text{mol CH}_3\text{COOH}}}{60.0 \, \cancel{\text{g CH}_3\text{COOH}}} \times \frac{\cancel{\text{mol NaOH}}}{\cancel{\text{mol CH}_3\text{COOH}}} \times \frac{\cancel{\text{L}_{OH^-}}}{0.20 \, \cancel{\text{mol NaOH}}}$$

$$\times \frac{\text{mL}_{OH^-}}{10^{-3} \, \cancel{\text{L}_{OH^-}}} = 22 \text{ mL}_{OH^-} \text{ (to the correct number of significant figures)}$$

24. The balanced equation is $KOH + HCl = KCl + H_2O$. We'll call the KOH solution OH^- and the 20.2% HCl solution H^+.

$$? M = \frac{? \text{ mol KOH}}{\text{L}_{OH^-}} = \frac{25.0 \, \cancel{\text{mL}_{H^+}}}{0.0404 \, \text{L}_{OH^-}} \times \frac{1.1 \, \cancel{\text{g}_{H^+}}}{\cancel{\text{mL}_{H^+}}} \times \frac{20.2 \, \cancel{\text{g HCl}}}{100 \, \cancel{\text{g}_{H^+}}} \times \frac{\cancel{\text{mol HCl}}}{36.5 \, \cancel{\text{g HCl}}}$$

$$\times \frac{\text{mol KOH}}{\cancel{\text{mol HCl}}} = \frac{3.8 \text{ mol KOH}}{\text{L}_{OH^-}} = 3.8 \, M \text{ KOH}$$

25. First you determine the milliliters of the HCl solution consumed by the reaction

$$HCl + NaOH = NaCl + H_2O$$

$$? \text{ mL}_{H^+} = 0.0250 \, \cancel{\text{L}_{OH^-}} \times \frac{0.50 \, \cancel{\text{mol NaOH}}}{\cancel{\text{L}_{OH^-}}} \times \frac{\cancel{\text{mol HCl}}}{\cancel{\text{mol NaOH}}}$$

$$\times \frac{\cancel{\text{L}_{H^+}}}{0.20 \, \cancel{\text{mol HCl}}} \times \frac{\text{mL}_{H^+}}{10^{-3} \, \cancel{\text{L}_{H^+}}} = 62.5 \text{ mL}_{H^+}$$

You find the amount of the HCl solution remaining by subtracting the amount consumed from the initial amount:

$$75.0 \text{ mL}_{H^+} - 62.5 \text{ mL}_{H^+} = 12.5 \text{ mL}_{H^+} \text{ (remaining)}$$

The total volume is

$$75.0 \text{ mL} + 25.0 \text{ mL} = 100.0 \text{ mL (total volume)}$$

The molarity of the HCl solution remaining is

$$M = \frac{? \text{ mol HCl}}{\text{L}_{total}} = \frac{0.0125 \, \cancel{\text{L}_{H^+}}}{0.1000 \, \text{L}_{total}} \times \frac{0.20 \text{ mol HCl}}{\cancel{\text{L}_{H^+}}} = \frac{0.025 \, M \text{ HCl}}{\text{L}_{total}}$$

$$= 0.025 \, M \text{ HCl (to the correct significant figures)}$$

26. $? \text{ mL}_{Fe^{2+}} = 0.1000 \, \cancel{\text{L}_{MnO_4^-}} \times \dfrac{0.50 \, \cancel{\text{mol KMnO}_4}}{\cancel{\text{L}_{MnO_4^-}}} \times \dfrac{5 \, \cancel{\text{mol Fe(NO}_3)_3}}{\cancel{\text{mol KMnO}_4}}$

$$\times \frac{\cancel{\text{L}_{Fe^{2+}}}}{0.30 \, \cancel{\text{mol Fe(NO}_3)_3}} \times \frac{\text{mL}_{Fe^{2+}}}{10^{-3} \, \cancel{\text{L}_{Fe^{2+}}}}$$

$$= 8.3 \times 10^2 \text{ mL}_{Fe^{2+}} \text{ (to the correct significant figures)}$$

27. $? \text{ g CH}_3\text{COOH} = 100 \text{ g vinegar} \times \dfrac{\text{mL vinegar}}{1.06 \text{ g vinegar}} \times \dfrac{0.04324 \text{ mol OH}^-}{19.75 \text{ mL vinegar}}$

    $\times \dfrac{0.400 \text{ mol NaOH}}{\text{mol OH}^-} \times \dfrac{\text{mol CH}_3\text{COOH}}{\text{mol NaOH}} \times \dfrac{60.0 \text{ g CH}_3\text{COOH}}{\text{mol CH}_3\text{COOH}}$

    $= 4.96 \text{ g CH}_3\text{COOH in 100 g vinegar or 4.96\% CH}_3\text{COOH}$

28. You need two balanced equations:

    $\text{CaCO}_3 + 2\text{HCl} = \text{CaCl}_2 + \text{CO}_2 + \text{H}_2\text{O}, \quad \text{NaOH} + \text{HCl} = \text{NaCl} + \text{H}_2\text{O}$

    First find how many moles of HCl were in excess from the back-titration:

    $? \text{ mol HCl (excess)} = 0.00600 \text{ mol OH}^- \times \dfrac{0.105 \text{ mol NaOH}}{\text{mol OH}^-} \times \dfrac{\text{mol HCl}}{\text{mol NaOH}}$

    $= 0.000630 \text{ mol HCl (excess)}$

    Then you determine the initial number of moles of HCl.

    $? \text{ mol HCl (init)} = 0.0500 \text{ mol H}^+ \times \dfrac{0.100 \text{ mol HCl}}{\text{mol H}^+} = 0.00500 \text{ mol HCl (init)}$

    The number of moles of HCl required is the difference:

    $\text{mol HCl (required)} = 0.00500 \text{ mol HCl} - 0.000630 \text{ mol HCl}$

    $= 0.00437 \text{ mol HCl (required)}$

    $? \text{ g CaCO}_3 = 100 \text{ g imp} \times \dfrac{0.00437 \text{ mol HCl}}{0.500 \text{ g imp}} \times \dfrac{\text{mol CaCO}_3}{2 \text{ mol HCl}} \times \dfrac{100.1 \text{ g CaCO}_3}{\text{mol CaCO}_3}$

    $= 43.7 \text{ g CaCO}_3 \text{ in 100 g imp or 43.7\% pure}$

29. Since only one H atom in the acid can react with NaOH, you know that there is one mole of acid (HX) per one mole of NaOH. You can determine the molecular weight by rewording the question to: "How many grams HX equals a mole of HX?"

    $? \text{ g HX} = \text{mol HX} \times \dfrac{\text{mol NaOH}}{\text{mol HX}} \times \dfrac{\text{mol OH}^-}{0.104 \text{ mol NaOH}} \times \dfrac{0.610 \text{ g HX}}{0.0482 \text{ mol OH}^-}$

    $= 122 \text{ g HX in one mol HX} = 122 \text{ g/mol}$

30. First you determine the moles of HCl that were in excess:

    $? \text{ mol HCl (excess)} = 0.00385 \text{ mol OH}^- \times \dfrac{0.130 \text{ mol NaOH}}{\text{mol OH}^-} \times \dfrac{\text{mol HCl}}{\text{mol NaOH}}$

    $= 0.000500 \text{ mol HCl (excess)}$

    Then you determine the initial moles of HCl:

    $? \text{ mol HCl (init)} = 0.02500 \text{ mol H}^+ \times \dfrac{0.105 \text{ mol HCl}}{\text{mol H}^+} = 0.00262 \text{ mol HCl (init)}$

    You determine the number of moles of HCl required by subtracting:

    $? \text{ mol HCl (req)} = (0.00262 - 0.000500) \text{ mol HCl} = 0.00212 \text{ mol HCl (req)}$

    $? \text{ g N} = 100 \text{ g protein} \times \dfrac{0.002125 \text{ mol HCl}}{1.75 \text{ g protein}} \times \dfrac{\text{mol NH}_3}{\text{mol HCl}} \times \dfrac{14.0 \text{ g N}}{\text{mol NH}_3}$

    $= 1.70 \text{ g N in 100 g protein} \quad \text{or 1.70\% N}$

CHAPTER 17

Colligative Properties of Solutions

Boiling Point, Freezing Point, and Osmotic Pressure

There is no pretest for this chapter. The material is completely different from what has come before. In fact, we will have to start with a brief discussion of what happens to certain properties of a solvent when you dissolve a nonvolatile solute in it. The description here will be minimal. Your textbook is your best guide for the theory of what is going on. We will be concerned here almost entirely with the calculations.

SECTION A   Colligative Properties

There are several properties of a liquid solution that depend on the relative numbers of particles of solute and solvent, the so-called *colligative properties*: Among these are

1. The lowering of the vapor pressure
2. The lowering of the freezing point
3. The raising of the boiling point
4. The raising of the osmotic pressure

   Changes in the freezing and boiling points are both due to the lowering of the vapor pressure. It is easy to show how this occurs. A *phase diagram* (next page) shows vapor pressure versus temperature for a pure solvent (————) and for a solution of a nonvolatile solute (— — — —). The lines represent equilibria between liquid–gas, liquid–solid, and solid–gas.

   The *boiling point* is defined as the temperature at which the liquid is in equilibrium with gas at 1.0 atm pressure. (To be exact, this is called the *standard boiling point*. We will consider no other.) The *freezing point* is the temperature at which liquid and solid are at equilibrium at 1.0 atm pressure. On the following diagram, they are the temperatures at which the two equilibrium lines cross 1.0 atm pressure.

   As you see, because the liquid solution's vapor-pressure line is lower than that for the pure solvent, the crossing of the 1.0-atm line is at a higher temperature for the boiling points and at a lower temperature for

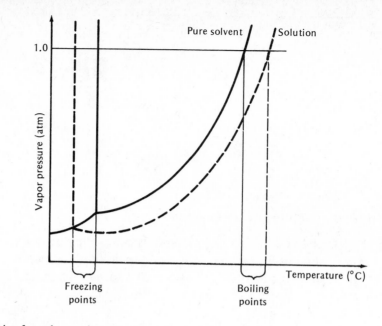

the freezing point. The number of degrees Celsius it changes is determined by three factors:

1. The number of solute particles
2. The number of solvent particles
3. The nature of the solvent

The amount the vapor pressure is lowered is independent of what the solute is so long as the vapor pressure of the solute is much, much lower than the vapor pressure of the solvent; that is, so long as it is *nonvolatile*.

It is most convenient to express the number of solute particles in moles and always to refer to the number of solvent particles present in 1000 g of solvent. This concentration unit,

$$\frac{\text{mol solute}}{\text{1000 g solvent}} \quad \text{or} \quad \frac{\text{mol solute}}{\text{kg solvent}}$$

molality   should look familiar to you. It is called _____ and abbreviated

m   ___. In Chapter 14, we said that this concentration unit was useful in only one type of problem. These colligative-property problems are that type.

Now we are going to set up a conversion factor which is the number of degrees Celsius that the boiling point is raised or the freezing point is lowered in a 1 molal solution of a nonvolatile solute. Each solvent has its own numerical value for this conversion factor. Thus, in much the same way that we can say that there are 18.0 g H_2O per mole of H_2O,

$$\frac{18.0 \text{ g } H_2O}{\text{mol } H_2O} \quad \text{or 72.0 g of } C_6H_6 \text{ per mole } C_6H_6, \quad \frac{72.0 \text{ g } C_6H_6}{\underline{\hspace{1.5cm}} \ C_6H_6}$$

mol

we can say that the boiling-point elevation is 0.51°C per molal solution in water,

$$\frac{0.51°C}{m\ H_2O}$$ and 2.5°C per molal solution in benzene (C_6H_6), $$\frac{2.5°C}{m\ C_6H_6}$$

It is customary to write the symbol for the number of degrees Celsius that the boiling point is raised as ΔT_B and the number of degrees Celsius that the freezing point is lowered as ΔT_F. They are always positive numbers and their units are degrees Celsius. We can therefore write our conversion factor as $(\Delta T)°C/m$. The numerical value of ΔT depends on the solvent. Of course, molality (m) is itself a conversion factor,

$$m = \frac{\text{mol solute}}{\text{kg solvent}}$$

Thus, rewriting our conversion factor for the boiling-point elevation or the freezing-point lowering, we have

$$\frac{\Delta T°C}{\frac{\text{mol solute}}{\text{kg solvent}}} \quad \text{or} \quad \frac{(\Delta T°C)(\text{kg solvent})}{\text{mol solute}}$$

The second conversion factor is simply an algebraic transposition of the first. In fact, you may later find it more convenient to write these conversion factors as

$$\frac{(\Delta T°C)(10^3\ \text{g solvent})}{\text{mol solute}}$$

These conversion factors have a name. If you are referring to the number of degrees Celsius that 1.0 mol of solute raises the boiling point of 1.0 kg of solvent, it is called the *molal boiling-point-elevation constant* and abbreviated K_B. If you are talking about the number of degrees Celsius that the freezing point is lowered by the presence of 1.0 mol of solute in 1.0 kg of solvent, it is called the *molal freezing-point-depression constant*, K_F. The important thing to remember is that K_B and K_F are conversion factors with the very strange units

$$\frac{(°C)(\text{kg solvent})}{\text{mol solute}}$$

This isn't the first time that you have seen a conversion factor with several different dimensions in it. Recall the way we initially set up a conversion factor from an equality (in Chapter 2). There is no reason why the equality could not have more than two dimensions in it. Take the expression "man-hours," a term usually used in reference to work. For example, "It took 1500 man-hours to build a tunnel." You could have one person working 1500 hours or 1500 persons each working one hour. The product of the number of persons times the number of hours is still the same. Similarly, our dimension (°C)(kg solvent), is just such a double dimension.

When you were working with gas laws in Chapter 12, you had a conversion factor with two of these double dimensions, a total of four different units in one conversion factor. The important thing to remember is to make sure when you solve a problem that all the unwanted dimensions cancel.

SECTION B Determining K_F or K_B

All you need to know in order to experimentally determine K_F or K_B is the freezing point of a solution of known amount of solute and known amount of solvent, and the freezing point of pure solvent (or for K_B the boiling points respectively). Here is an example.

What is the molal boiling-point-elevation constant, K_B, for the solvent cyclohexane, if it was found that dissolving 0.20 mol of a nonvolatile solute in 4.0 kg of cyclohexane increased the boiling point 0.42°C over that of pure cyclohexane?

K_B

The question asked is "What is _____ for cyclohexane?" You must, of course, write down the dimensions in order to solve the problem.

°C; kg cyclohexane

$$? \, K_B = \frac{(? \underline{\hspace{1cm}})(\underline{\hspace{3cm}})}{\underline{\hspace{2cm}} \text{ solute}}$$

mol

There is one conversion factor given in the problem,

0.20 mol

$$\underline{\hspace{3cm}} \text{ solute}$$

4.0 kg cyclohexane

$$\underline{\hspace{3cm}}$$

0.42°C

The information given in the problem is ΔT_B, which is _____. You have everything you need to get the dimensions of the answer.

$$? \, K_B = \frac{(? \, °C)(\text{kg cyclohexane})}{\text{mol solute}} = 0.42°C \times \frac{4.0 \text{ kg cyclohexane}}{0.20 \text{ mol solute}}$$

$$= \frac{(8.4°C)(\text{kg cyclohexane})}{\text{mol solute}}$$

Our old trick of rewording problems when the answer itself is a conversion factor doesn't work very well when the conversion factor has more than two dimensions. However, it can be done. You ask "How many degrees Celsius does the boiling point change when you have one mole of solute per kilogram of solvent?"

$$? \, °C = \frac{1 \text{ mol solute}}{\text{kg cyclohexane}}$$

The problem states that the temperature changes 0.42°C when you have 0.20 mol of solute in 4.0 kg of cyclohexane. This would be the double fraction

$$\frac{0.42°C}{\dfrac{0.20 \text{ mol solute}}{4.0 \text{ kg cyclohexane}}}$$

which can be rearranged to

$$\frac{0.42°C(4.0 \text{ kg cyclohexane})}{0.20 \text{ mol solute}}$$

If you multiply by this conversion factor, all the unwanted dimensions cancel:

$$? °C = \frac{1 \text{ mol solute}}{\text{kg cyclohexane}} \times \frac{0.42°C(4.0 \text{ kg cyclohexane})}{0.20 \text{ mol solute}}$$

When you are determining K_B or K_F, it is simpler to write in all the dimensions you want in the answer in their correct positions. Then you just insert the information given to match these dimensions.

Here is a more complicated example that gives the amount of solute in grams, not moles. A solution of 50.0 g $C_2H_6O_2$ in 700 g of water freezes at $-2.14°C$. Pure water freezes at $0.00°C$. What is K_F for water? Since

the question asked is "What is _____ for water?", you put this in dimensions and have

K_F

$$? K_F = \frac{(? \text{\_\_\_\_\_})(\text{_____})}{\text{\_\_\_\_\_ } C_2H_6O_2}$$

°C; kg water

mol

The information given is the lowering of the freezing point of water:

$$(\Delta T_F)°C = 0.00°C - (-2.14°C) = \text{\_\_\_\_\_}$$

2.14°C

[*Note:* ΔT is always a + number.] The problem gives one conversion factor,

$$\frac{\text{\_\_\_\_\_ } C_2H_6O_2}{0.700 \text{_____}}$$

50.0 g

kg water

(We changed grams water to kilograms water immediately to save one step.) Since the answer will have moles $C_2H_6O_2$ and the conversion factor has grams $C_2H_6O_2$, you need a conversion factor that relates these

two. It is _____, the molecular weight. You now have everything you need to solve the problem. You start with

$\frac{62.0 \text{ g } C_2H_6O_2}{\text{mol } C_2H_6O_2}$

$$? K_F = \frac{(? °C)(\text{kg water})}{\text{mol } C_2H_6O_2} = \text{\_\_\_\_\_}$$

2.14°C

Insert the information given first. Then multiply by the conversion

factor given in the problem, _____. You now have two of the dimensions of the answer, degrees Celsius and kilograms water.

$\frac{0.700 \text{ kg water}}{50.0 \text{ g } C_2H_6O_2}$

But you still have the unwanted dimension _____ and

grams $C_2H_6O_2$

you lack the dimension _____ $C_2H_6O_2$. But if you multiply by the

moles

conversion factor _____, you have only the dimensions of the answer left. The answer is therefore

$\frac{62.0 \text{ g } C_2H_6O_2}{\text{mol } C_2H_6O_2}$

$$? K_F = \frac{(?\ ^\circ C)(\text{kg water})}{\text{mol } C_2H_6O_2} = 2.14^\circ C \times \frac{0.700 \text{ kg water}}{50.0 \text{ g } C_2H_6O_2} \times \frac{62.0 \text{ g } C_2H_6O_2}{\text{mol } C_2H_6O_2}$$

$$= \frac{1.86^\circ C \text{ (kg water)}}{\text{mol solute}}$$

Actually, chemists rarely calculate the values for K_B and K_F. They are well known. They look them up in a text or in the *Handbook of Chemistry and Physics* [The Chemical Rubber Co., Boca Raton, FL, 33431]. Table 16.1 shows the values of K_B and K_F for a few solvents.

TABLE 16.1 Molal Freezing and Boiling Point Constants

Solvent	Boiling point	K_B	Freezing point	K_F
Acetic acid	118.5°C	3.1°/m	16.6°C	3.9°/m
Benzene	80.1°C	2.5°/m	5.5°C	5.1°/m
Camphor			78.4°C	38.0°/m
Carbon disulfide	46.3°C	2.3°/m		
Carbon tetrachloride	76.8°C	5.0°/m		
Naphthalene			80.2°C	6.9°/m
Water	100°C	0.51°/m	0.0°C	1.86°/m

SECTION C Determining Molecular Weight

A chemist often wants to know the molecular weight of an unknown nonvolatile compound. The easiest way to find it is to determine either the freezing-point lowering or boiling-point elevation caused by a known weight of the compound in a known weight of some solvent. The problem always gives one conversion factor (the relative weights of unknown and solvent) and one conversion factor, the molal freezing-point or boiling-point constant, which you look up in the table. It also gives ΔT in some form.

Here is an example. What is the molecular weight of an unknown compound, X, if a solution prepared by dissolving 1.0 g of X in 10.0 g benzene freezes at 2.3°C? Pure benzene freezes at 5.5°C.

The conversion factor given in the problem is

$$\frac{1.0 \text{ g } X}{10.0 \text{ g benzene}}$$

The table lists the molal freezing-point-depression constant for benzene as

$$(5.1^\circ C)\left(\frac{\text{kg benzene}}{\text{mol } X}\right)$$

You calculate the value for ΔT_F, the number of degrees that the freezing point of pure benzene has been lowered:

$$\Delta T_F = 5.5°C - \underline{\hspace{1cm}}°C = \underline{\hspace{1cm}}°C$$

2.3; 3.2

You now have everything you need. The question asked is "What is the molecular weight of X?" You must reword this using units, so you ask

"How many grams of X are in a _____ of X?" (In this case, rewording the problem is a big help.) You start, then, with the question

mole

$$? \text{ g } X = \text{mol } X$$

The unwanted dimension is _____ and you can use K_F to cancel it. You'll save yourself a step if you immediately write K_F in terms of grams of benzene rather than kilograms of benzene:

moles X

$$K_F = \frac{(5.1°C)(\underline{\hspace{1cm}} \text{ g benzene})}{\text{mol } X}$$

10^3 (or 1000)

So you have

$$? \text{ g } X = \cancel{\text{mol } X} \times \frac{(5.1°C)(\underline{\hspace{2cm}})}{\cancel{\text{mol } X}}$$

10^3 g benzene

The unwanted dimensions are now _____ and

_____. But the problem tells you that there is _____ per _____ g benzene. You can use this as a conversion factor to get rid of the unwanted grams benzene and also give you the dimension you want for the answer:

degrees Celsius

grams benzene; 1.0 g X

10.0

$$? \text{ g } X = \cancel{\text{mol } X} \times \frac{(5.1°C)(10^3 \text{ g benzene})}{\cancel{\text{mol } X}} \times \underline{\hspace{2cm}}$$

$\dfrac{1.0 \text{ g } X}{10.0 \text{ g benzene}}$

The only remaining unwanted dimension is now _____. It appears in the numerator. The only way to cancel it is to multiply by

degrees Celsius

_____. (It may strike you as strange to multiply by the reciprocal of the information given in the problem. But this isn't really stranger than multiplying by the number itself. So long as the unwanted dimensions cancel, everything is all right.)

$\dfrac{1}{\Delta T_F}$

$$? \text{ g } X = \cancel{\text{mol } X} \times \frac{(5.1\cancel{°C})(10^3 \cancel{\text{ g benzene}})}{\cancel{\text{mol } X}} \times \frac{1.0 \text{ g } X}{10.0 \cancel{\text{ g benzene}}} \times \frac{1}{3.2\cancel{°C}}$$

$$= 1 \times (5.1)(10^3) \times \frac{1.0 \text{ g } X}{10.0} \times \frac{1}{3.2}$$

$$= 1.6 \times 10^2 \text{ g } X \text{ in one mole } X$$

Thus the molecular weight is

$$\frac{1.6 \times 10^2 \text{ g } X}{\text{mol } X}$$

These problems are simple if you always express K_F or K_B as a conversion factor with all its dimensions. Sometimes, however, the relation between weight of solute and weight of solvent is more complicated, as in the next example.

What is the molecular weight of an unknown compound, X, if a 10.0 wt % solution of X in acetic acid boils at 122.5°C? Here the weight of solute per weight of solvent is in the 10.0 wt %. If you take 100.0 g of

10.0; 90.0

solution, it contains _____ g of X and _____ g of acetic acid. So you can write the conversion factor

10.0

$$\frac{\text{_____ g } X}{}$$

90.0

$$\text{_____ g acetic acid}$$

The other conversion factor, given in the table, is the K_B for acetic acid:

g acetic acid

$$\frac{(3.1°C)(10^3 \text{ _____})}{}$$

mol X

$$\text{_____}$$

Finally, you calculate the number of degrees Celsius that the boiling point of pure benzene has been raised:

122.5; 4.0

$$\Delta T_B = \text{_____} °C - 118.5°C = \text{_____} °C$$

You now have all the information you need. To get molecular weight, you ask the question

g; mol X

$$? \text{_____} X = \text{_____}$$

moles X

You start with K_B to cancel the unwanted dimension, _____. Then you multiply by the conversion factor given in the problem relating grams of

$\dfrac{1}{4.0°C}$

X to grams of benzene. Finally you multiply by _____ to remove the unwanted °C. Your answer is then

$$? \text{ g } X = \cancel{\text{mol } X} \times \frac{(3.1\cancel{°C})(10^3 \cancel{\text{ g benzene}})}{\cancel{\text{mol } X}} \times \frac{10.0 \text{ g } X}{90.0 \cancel{\text{ g benzene}}} \times \frac{1}{4.0\cancel{°C}}$$

$$= 86 \text{ g } X \text{ in one mole } X \quad \text{or} \quad \frac{86 \text{ g } X}{\text{mol } X}$$

SECTION D Determining Freezing and Boiling Points of Solutions

Often chemists want to know at what temperature a solution of a substance of known molecular weight will boil or freeze. What they do is determine ΔT_B or ΔT_F and then add to or subtract this number of degrees from the boiling or freezing point of pure solvent. Here is a simple example. What is the boiling point of a 6.0 m solution of anthracene in benzene? The problem tells you that the solution is 6.0 m. Written as a conversion factor, this is

mol

$$6.0 \text{ _____ anthracene}$$

kg

$$\text{_____ benzene}$$

From the table, you know another conversion factor, the molal boiling-

point-elevation constant, which is _____.

$$\frac{(2.5°C)(kg\ benzene)}{mol\ anthracene}$$

The question asked is really "How many degrees Celsius is the boiling point of pure benzene raised?" You start, then, by writing ? $°C =$

The problem gives you the conversion factor that the solution is 6.0 *m*. So you can start there:

$$?\,°C = \frac{6.0\ mol\ anthracene}{kg\ benzene}$$

Then you multiply by K_B:

$$?\,°C = \frac{6.0\ \cancel{mol\ anthracene}}{\cancel{kg\ benzene}} \times \frac{(2.5°C)(\cancel{kg\ benzene})}{\cancel{mol\ anthracene}}$$

The only dimension left is the dimension of the answer:

$$?\,°C = 6.0 \times 2.5°C = 15°C = \Delta T_B$$

To get the boiling point of the solution, you have to add ΔT_B to the boiling point of pure benzene.

$$\text{Boiling point, } °C = 80.1°C + 15°C = 95°C$$

(Notice how you add to get the correct number of significant figures.)

The last example was easy because the problem gave you the concentration of solute in molality. Such problems are not always so straightforward. But with your experience at converting from one dimension to another, a problem like the following shouldn't be too difficult.

At what temperature does a solution of ethyl alcohol in water freeze if you add 200 mL of ethyl alcohol, C_2H_5OH (density 0.79 g/mL) to 1000 mL of water (density 1.00 g/mL) to prepare the solution?

The problem gives three conversion factors:

200 mL ethyl alcohol
_____

1000 mL water

0.79 g _____,   1.00 g _____

ethyl alcohol;   water

_____   _____

mL ethyl alcohol; mL water

From the table you can get the molal freezing-point-depression constant for water, which is

$$\frac{(1.86°C)(10^3\ _____)}{mol\ ethyl\ alcohol}$$

g water

Since you have moles ethyl alcohol in one conversion factor and grams ethyl alcohol in another, you need the conversion factor

_____, which will get you from one to the other. Now you have all the conversion factors you need.

$$\frac{46.0\ g\ ethyl\ alcohol}{mol\ ethyl\ alcohol}$$

You start, then, with the question

? _____ =

°C

$$\frac{200 \text{ mL ethyl alcohol}}{1000 \text{ mL water}}$$

$$\frac{0.79 \text{ g ethyl alcohol}}{\text{mL ethyl alcohol}}$$

$$\frac{\text{mL water}}{1.00 \text{ g water}}$$

$$\frac{(1.86°C)(10^3 \text{ g water})}{\text{mol ethyl alcohol}}$$

$$\frac{\text{mol ethyl alcohol}}{46.0 \text{ g ethyl alcohol}}$$

degrees Celsius

First, the conversion factor given in the problem is _____ To cancel the unwanted dimension milliliters ethyl alcohol, you multiply by _____. To cancel the unwanted dimension milliliters water, you multiply by _____. You can cancel the unwanted dimension grams water by using the conversion factor from K_F, _____. Finally, to cancel both the unwanted dimensions mole ethyl alcohol and grams ethyl alcohol, you multiply by _____. The only dimension left now is

_____. So the number of degrees the freezing point has been lowered that will be in your worked-out setup must be

$$\Delta T_F = ? °C = \frac{200}{1000} \times 0.79 \times \frac{1}{1.00} \times 1.86°C \times 10^3 \times \frac{1}{46.0} = 6.4°C$$

subtract

To get the freezing point of the solution, you must now _____ ΔT_F from the freezing point of pure water:

Freezing point $= 0.0°C - 6.4°C = -6.4°C$

SECTION E Preparing Solutions with Known Freezing and Boiling Points

These problems are simply the problems in Section D turned the other way around. For example, how many grams of C_6H_5Cl would you have to add to 100 g of carbon tetrachloride solvent to have a solution that boils at 78.3°C?

You can't get the boiling point of the solution directly from a colliga-

greater

tive calculation. You can only determine how much _____ the boiling point is than that of pure solvent. You can determine ΔT_B:

78.3; 1.5

$$(\Delta T_B)°C = \underline{\hspace{1cm}} °C - 76.8°C = \underline{\hspace{1cm}} °C$$

grams of C_6H_5Cl

The question asked is "How many _____?" The

100 g

information given is the amount of solvent: _____ carbon tetrachloride. You also know K_B for the solvent and the calculated ΔT_B.

You start in the normal way:

g C_6H_5Cl

$$? \underline{\hspace{2cm}} = 100 \text{ g carbon tetrachloride}$$

Multiplying by the inverted K_B cancels the unwanted dimension, grams carbon tetrachloride:

$$? \text{ g } C_6H_5Cl = 100 \text{ g carbon tetrachloride}$$
$$\times \frac{\text{mol } C_6H_5Cl}{(5.0°C)(10^3 \text{ g carbon tetrachloride})}$$

Next you can cancel the unwanted dimension, moles C_6H_5Cl, and get

directly to the dimension of the answer by multiplying by _____.
You now have the dimension of the answer. But there remains one

unwanted dimension, _____. So you multiply by the

other piece of information given, _____, which is ΔT_B. The answer in
your worked-out setup is therefore

$$? \text{ g } C_6H_5Cl = 100 \times \frac{1}{(5.0)(10^3)} \times 112.5 \text{ g } C_6H_5Cl \times 1.5$$

$$= 3.4 \text{ g } C_6H_5Cl$$

The problem is a little strange because there are two pieces of "information given." You could have put them both in initially, but it is better
to wait with degrees Celsius until the end. The next example will show
why. It is the sort of problem that asks for the solvent instead of the
solute. What's more, the units are not metric.

How many gallons of water could you add to 1.0 quart of ethylene
glycol antifreeze ($C_2H_6O_2$; density = 0.80 g/mL) in order to have a
solution in your car's radiator that will not freeze until the temperature
drops below 23.0°F? The density of water is 1.0 g/mL.

First you determine ΔT_F in degrees Celsius. To do this, determine the
temperature in degrees Celsius below which it will freeze.

$$°C = \frac{5}{9}(\underline{\hspace{3cm}}) = \underline{\hspace{2cm}}$$

(If you have forgotten how to do this, check the last section of Chapter
3.) Now you calculate ΔT_F:

$$\Delta T_F = 0.0°C - (\underline{\hspace{1.5cm}}) = \underline{\hspace{1.5cm}}$$

The amount of $C_2H_6O_2$ is the information given in the problem. It is

_____. We'll pick up the other conversion factors given
in the problem as we go along. You also need the conversion factor, K_F,
for water. (Get it from the table.) But first let's start the problem:

$$? \underline{\hspace{4cm}} =$$

The information given is _____. So, as usual, you start
there:

$$? \text{ gal water} = \underline{\hspace{3cm}}$$

You can now cancel the unwanted dimension and work your way toward
moles $C_2H_6O_2$ by using the following conversion factors:

$$\frac{\underline{\hspace{1cm}} \text{ mL } C_2H_6O_2}{\text{qt } C_2H_6O_2} \times \frac{0.80 \text{ g } C_2H_6O_2}{\underline{\hspace{1cm}} C_2H_6O_2}$$

To get from grams $C_2H_6O_2$ to moles $C_2H_6O_2$, you multiply by
_____. The unwanted dimension is now _____,
but this is in K_F for water,

$\dfrac{112.5 \text{ g } C_6H_5Cl}{\text{mol } C_6H_5Cl}$

degrees Celsius

1.5°C

23.0°F − 32.0; −5.0°C

−5.0°C; 5.0°C

1.0 qt $C_2H_6O_2$

gal water

1.0 qt $C_2H_6O_2$

1.0 qt $C_2H_6O_2$

946

mL

$\dfrac{\text{mol } C_2H_6O_2}{62.0 \text{ g } C_2H_6O_2}$; moles $C_2H_6O_2$

10^3 g water

$$(1.86°C)(\underline{\hspace{3cm}})$$

mol

$$\underline{\hspace{2cm}} C_2H_6O_2$$

You can convert from grams water to gallons water in three steps:

qt water

1.0 g water

$$\frac{\text{mL water}}{\underline{\hspace{2cm}}} \times \frac{\underline{\hspace{2cm}}}{946 \text{ mL water}} \times \frac{\text{gal water}}{4 \text{ qt water}}$$

You now have the dimensions of the answer, but still have not canceled

degrees Celsius

$\dfrac{1}{5.0°C}$

$\underline{\hspace{4cm}}$. However, you have a value in degrees Celsius for ΔT_F that you can use. You multiply by $\underline{\hspace{1.5cm}}$. (Notice that once again you have used the reciprocal.) The worked-out answer, therefore, is

$$? \text{ gal water} = 1.0 \times 946 \times 0.80 \times \frac{1}{62.0} \times (1.86)(10^3)$$

$$\times \frac{1}{1.0} \times \frac{1}{946} \times \frac{\text{gal water}}{4} \times \frac{1}{5.0}$$

$$= 1.2 \text{ gal water}$$

That problem was especially complicated. If you were able to do it, congratulate yourself.

SECTION F   What Happens when the Solute Breaks Up into Several Particles

[Do not do this complex section unless instructed to do so.]

You may have noticed all along that the molecules of solute have not been mentioned, while the term *particles of solute* has been. The reason for this distinction is that some solute molecules *dissociate* (break up) into several charged particles (ions) when they are dissolved in a very polar solvent like water. The solutes that do this are either inorganic salts or acids and bases. These are the electrolytes covered in Chapter 8.

If these compounds are strong electrolytes (that is, if they dissociate 100% so that *all* the molecules break up into ions), then determining the colligative properties wouldn't be much more difficult. Thus, if you had a 1-molal NaCl solution and you knew that the NaCl broke up into Na^+ and Cl^- ions, then the solution would be 2 molal in particle concentration.

In general, dilute solutions (less than 0.1 *m*) of salts and of strong acids and bases are 100% dissociated. As the concentration of the solute increases, the percentage dissociation falls off. Weak acids and bases— even in the most dilute solution—dissociate only very slightly. But since even compounds that are capable of complete dissociation are not 100% dissociated at all concentrations, we will figure out how to count the total number of particles, both ions and molecules, present in the solution.

To see how you count particles when you know the percentage dissociation, let's try a sample problem. What is the total number of particles present in 5.0 mol of NaCl which is only 90% dissociated into Na^+ and Cl^- ions?

$$NaCl \longrightarrow Na^+ + Cl^-$$

Well, 90% of 5.0 mol is 4.5 mol. Each mole dissociated gives 2 mol of ions (1 Na^+ and 1 Cl^-). Thus you have 9.0 mol of mixed ions. Ten percent of the initial NaCl are still molecules, so you have 0.5 mol of molecules left. Thus the total number of moles of particles is 9.5.

Now let's see if we can make this more systematic. Instead of talking about percentage dissociation (the number per 100 total molecules), we'll talk about *fraction* dissociated (the number per 1 molecule). This fraction, f, equals the percentage dissociation divided by 100. Thus, if a salt is 80% dissociated, then the fraction dissociated, f, equals _____. If | 0.80

an acid is 0.15% dissociated, then the value for f is _____. | 0.0015

Therefore, if you have 5.0 mol of NaCl initially and it is 90% dissociated, you can count the number of particles this way:

$$(5.0)(f) = \text{Number dissociated} \quad = 5.0 \times 0.90 = 4.5$$
$$(5.0)(1 - f) = \text{Number undissociated} = 5.0 \times 0.10 = 0.5$$

Now we can count particles.

$$\text{Mol } Na^+ = (5.0)(f) \quad = \text{Number dissociated}$$
$$\text{Mol } Cl^- = (5.0)(f) \quad = \text{Number dissociated}$$
$$\text{Mol } NaCl = (5.0)(1 - f) = \text{Number undissociated}$$

Adding all these, you have

$$\text{Moles particles} = (5.0)(f) + (5.0)(f) + (5.0)(1 - f) = (5.0)(1 + f)$$

Since f in this case is 0.90,

$$\text{Moles particles} = (5.0)(1.90) = 9.5$$

We can make this still more general if we let the number of moles of molecules initially present be n. Then in the case in which the molecule breaks up into two ions, the total number of particles can be expressed as moles particles = $n(1 + f)$.

Now you try one. How many total moles of particles are present in a KI solution in which 5.0 mol of KI have dissolved but only 98% of the KI has dissociated into K^+ and I^- ions?

$f =$ _____ | 0.98

Moles particles = __ (1 _____) = _____ mol particles | 5.0; +0.98; 9.9

Try an example involving weights of solute rather than moles. How many moles of particles are present in a sample containing 18 g of CH_3COOH which is 2% dissociated into H^+ and $(CH_3COO)^-$ ions?

$$\frac{0.02}{\text{mol } CH_3COOH}$$
$$\frac{}{60.0 \text{ g } CH_3COOH}$$
0.30 mol

n; 0.306

$f =$ _____

Mole CH_3COOH init $= n = 18$ g CH_3COOH × _____

= _____

Moles particles $=$ __ $(1 + f) =$ _____ mol

Remember, n must be expressed in moles of solute initially put in the solution.

What happens when you have a molecule that dissociates into 3 ions, as, for example, $CaCl_2$?

$$CaCl_2 \longrightarrow Ca^{2+} + 2Cl^-$$

We will count as we did before. Suppose you start with n mol of $CaCl_2$ and f is the fraction dissociated.

$$\text{Number moles dissociated} = n(f)$$
$$\text{Number moles undissociated} = n(1 - f)$$
$$\text{So: Number moles } Ca^{2+} = n(f)$$
$$\text{Number moles } Cl^- = 2n(f)$$
$$\text{Number undissociated moles } CaCl_2 = n(1 - f)$$

Adding these, you have

$$\text{Total number of moles of particles} = n(f) + 2n(f) + n(1 - f) = n(1 + 2f)$$

Here's another problem. What is the total number of moles of particles when 8.0 mol of $CaCl_2$ are 94% dissociated?

0.94; 8.0 mol

1 + 2f; 2.88

$f =$ _____ $n =$ _____

Total number of moles of particles $= n($_____$) = 8.0($_____$)$

$= 23$ mol particles

Now that you can count the number of moles of particles, you will be able to do colligative-property problems like the following.

What is the boiling point of a solution prepared by dissolving 11.1 g $CaCl_2$ in 250 g water? The $CaCl_2$ is 97% dissociated at this concentration.

First you calculate the number of moles of particles, Ca^{2+}, Cl^-, and $CaCl_2$. To do this you need n, the number of moles of $CaCl_2$ put in initially.

$$\frac{\text{mol } CaCl_2}{111.1 \text{ g } CaCl_2}$$
0.100

0.97; 3

0.294

$n = ?$ init mol $CaCl_2 = 11.1$ g $CaCl_2$ × _____

= _____ mol $CaCl_2$

Since $f =$ _____ and $CaCl_2$ dissociates into _____ ions, the total

number of particles present is _____ mol, that is, $n(1 + 2f)$.

Now you can solve the problem exactly as you did in Section D. The conversion factor given in the problem has become

$$\frac{0.294 \text{ mol particles}}{}$$

0.250

_____ kg water

The question asked is, "How many degrees above the boiling point of pure water does the solution have to be to boil?" or, "What is the value for _____?" Starting with the conversion factor

$$(?\ \Delta T_B)°C = \underline{\hspace{3cm}}$$

You next multiply by the conversion factor, K_B, _____. Everything except the dimension of the answer cancels, so the answer is

$\Delta T_B = 0.60°C$ (You *should* be working these on scratch paper.)

To get the boiling point of the solution, you _____ ΔT_B to the boiling point of pure water. The boiling point of the solution is therefore

_____.

It is also possible to reverse this procedure. You can use an experimentally determined value of ΔT to get a percentage dissociation for a molecule. What you do is determine the number of moles of particles present in the first step. Using this information, you can then calculate the value for f.

Here is an example of this type of problem. What is the percentage dissociation of HIO_4 in a solution that contains 3.84 g HIO_4 in 250 g water, given that the solution freezes at $-0.15°C$?

First you calculate ΔT_F. Since water freezes at $0.00°C$, ΔT_F is

$$\Delta T_F = \underline{\hspace{1.5cm}}$$

Then you solve for the total number of moles of particles present using K_F:

? mol particles = 0.250 kg water $\times$ _____ $\times$ _____

= _____ mol particles

Now the acid HIO_4 dissociates into __ ions, H^+ and IO_4^-, so you know that the total number of particles is represented by the equation

Number of particles = _____

You can calculate n from the initial weight of HIO_4 used.

$$n = \text{mol } HIO_4 = \underline{\hspace{1cm}} \text{ g } HIO_4 \times \frac{\text{mol } HIO_4}{191.9 \text{ g } HIO_4}$$

$$= \underline{\hspace{3cm}}$$

Now you solve for f:

$0.0202 = 0.0200(1 + f)$ $f = \underline{\hspace{1cm}}$

Since f is the fraction and the question asks for percent, you must multiply f by _____. Therefore the percentage dissociation is _____.

Recall:
If the molecule splits into 2 ions: Total particles $= n(1 + f)$.
If the molecule splits into 3 ions: Total particles $= n(1 + 2f)$.

Margin answers:

ΔT_B
$\dfrac{0.294 \text{ mol particles}}{0.250 \text{ kg water}}$
$\dfrac{(0.51°C)(\text{kg water})}{\text{mol particles}}$

add

100.60°C

0.15°C

$\dfrac{\text{mol particles}}{(1.86°C)(\text{kg water})}$; 0.15°C
0.0202

2

$n(1 + f)$

3.84

0.0200 mol HIO_4

0.01

100; 1%

3

$1 + 4f$

If the molecule splits into 4 ions: Total particles $= n(1 + \underline{}f)$.

And (although there are not many) if it splits into 5 ions:

Total particles $= n(\underline{})$

SECTION G   Osmotic Pressure

This last colligative property comes about when you have a solution separated from pure solvent by a membrane that allows only the solvent molecules to go through. It doesn't allow the solute particles to pass through. This type of membrane is called *semipermeable.* The liquids on the two sides of the membrane try to reach the same vapor pressure. That is, they try to have the same concentration of solute on both sides of the membrane. Since the solute particles cannot cross the membrane to increase the concentration of solute in pure solvent, the solvent molecules leave the pure solvent and try to dilute the solution until the concentration of solute becomes zero. This, of course, is not possible.

Osmotic pressure is the pressure that must be applied to stop the flow of solvent molecules into the solution. It is dependent not only on the concentration of solute in the solution but also on the temperature. The relationship is

$$\pi = MRT$$

where π = osmotic pressure
  M = molarity
  R = gas constant for liters of solution and moles of solute
  T = temperature in K

This equation looks like the Ideal Gas Law, and you know that if

π is expressed in atmospheres (atm)

M is expressed in moles solute per liter solution, and

T is expressed in Kelvins (K), which equals ($^{\circ}$C+273), then

R is $0.0820 \dfrac{\text{(liter solution)(atmospheres)}}{\text{(mole solute)(Kelvins)}}$

(At higher concentrations, it is best to express the concentration in molality. That would be kilograms of solvent rather than liters of solution. However, almost all measurements of osmotic pressure are made at very low concentrations.)

A major reason for determining osmotic pressure is to find the molecular weights of very large molecules. An advantage is that it is simple to read very low pressures. A pressure of 10^{-2} atm is shown by a column of mercury 7.6 mm high, which is easily read. To read a temperature difference of $10^{-2}\,^{\circ}$C for a freezing-point lowering is hard to do in the lab. You can get good results with osmotic pressure even at very low concentrations of solute.

The expression you use for molecular weights is easily calculated from the expression

$$\pi = MRT$$

We will do it by dimensional analysis. We'll start with an example. What is the molecular weight of polyethylene plastic if the osmotic pressure of a solution prepared by dissolving 1.00 g of the polyethylene in 500 mL of tetralin solvent is 2.2×10^{-2} atm at $27°C$?

First you must convert the temperature from $°C$ to K:

$$K = (273 + °C) = \underline{\hspace{1.5cm}}$$

300 K

Since the solution is so dilute, you can assume that the volume of the solution is equal to the volume of the solvent. You can therefore write the conversion factor given in the problem as

1.00 g polyethylene
_____
_____ L soln

0.500

The *gas constant*, R, is a conversion factor with four dimensions,

$$\frac{0.0820 \text{ (L)(atm)}}{\text{(mol solute)(K)}}$$

The problem gives two pieces of information. The osmotic pressure is

_____, and the temperature is 300 K.

2.2×10^{-2} atm

Now we can start. You want to know the molecular weight of polyethylene, which is a conversion factor having the dimensions grams per mole. Consequently, you ask

? g polyethylene = mol polyethylene

We determine this exactly as we determined molecular weights of gases in Chapter 12, Section D. You cancel the unwanted dimension, moles polyethylene, using the gas-law conversion factor,

? g polyethylene = mol polyethylene $\times$ _____

$$\frac{0.0820(\text{L soln})(\text{atm})}{(\text{mol polyethylene})(\text{K})}$$

You can cancel the unwanted atm and K using the information given for the osmotic pressure and temperature:

? g polyethylene = ~~mol polyethylene~~ $\times \dfrac{0.0820(\text{L soln})(\text{atm})}{(\text{mol polyethylene})(\text{K})}$

$$\times \underline{\hspace{3cm}}$$

$$\frac{300 \text{ K}}{2.2 \times 10^{-2} \text{ atm}}$$
liter solution

The remaining unwanted dimension is _____. You can cancel this by using the conversion factor given in the problem, _____. This also gives you the dimension you want for the answer:

$$\frac{1.00 \text{ g polyethylene}}{0.500 \text{ L soln}}$$

$$? \text{ g polyethylene} = \cancel{\text{mol polyethylene}} \times \frac{0.0820 \cancel{(\text{L soln})}\cancel{(\text{atm})}}{\cancel{(\text{mol polyethylene})}\cancel{(\text{K})}}$$

$$\times \frac{300 \cancel{\text{K}}}{2.2 \times 10^{-2} \cancel{\text{atm}}} \times \frac{1.00 \text{ g polyethylene}}{0.500 \cancel{\text{L soln}}}$$

$$= 2.2 \times 10^3 \text{ g polyethylene in one mole}$$

Therefore the molecular weight is

$$\frac{2.2 \times 10^3 \text{ g}}{\text{mol}}$$

You could even use the short-cut formula we developed for molecular weights of gases at the end of Section D in Chapter 12:

$$\text{Molecular weight} = \frac{(\text{g}) RT}{PV}$$

Except in this case the pressure P becomes the osmotic pressure π and the V is the volume of the solution:

$$\text{Molecular weight} = \frac{(\text{g}) RT}{\pi(\text{volume soln})}$$

PROBLEM SET

The first seventeen questions are divided into sections corresponding to the section in the chapter that pertains to them. Problems with an asterisk are optional. The last questions come from all the sections. Use Table 16.1 for values of K_B and K_F.

Section B

1. The boiling point of pure tetralin is $207.3°C$. When 26.4 g of stearic acid, $C_{17}H_{35}COOH$ (MW = 284 g/mol), are dissolved in 225 g of tetralin, the solution boils at $210.1°C$. What is the value of K_B for tetralin?

2. A 10.0 wt % alloy of Cu in Ag melts at $876°C$. Pure Ag melts at $961°C$. What is the molal freezing-point-depression constant for Ag?

Section C

3. A solution that contains 10.0 g of an unknown substance in 150 g carbon disulfide boils at $48.3°C$. What is the molecular weight of the unknown substance?

4. The freezing point of a solution containing 2.0 g of an unknown compound in 20 g benzene is $3.4°C$ lower than the freezing point of pure benzene. Calculate the molecular weight of the unknown.

Section D

5. What is the freezing point of a solution that contains 3.42 g sugar (MW = 342 g/mol) in 10.0 g water?

6. What is the boiling point of a solution that contains 3.58 g C_6Cl_6 in 10.0 g carbon tetrachloride? The molecular weight of C_6Cl_6 is 285.0 g/mol.

*7. Determine the boiling point of a 0.600 M solution of sugar in benzene. The molecular weight of sugar is 342 g/mol, and the density of the solution is 0.90 g/mL.

Section E

8. How many grams of ethyl alcohol, C_2H_5OH, must be added to 250 g of water to have a solution that does not freeze until the temperature goes below $-3.72°C$?

9. How many grams of phenol, C_6H_5OH, must you add to 500 mL carbon tetrachloride (density = 1.59 g/mL) to increase the boiling point to 82.8°C?

*10. How many quarts of ethylene glycol, $C_2H_6O_2$, must you add to 3.0 gal of water (density = 1.00 g/mL) so that the solution does not freeze until the temperature gets below 14°F? There are 946 mL/qt, 4 qt/gal, and the density of ethylene glycol is 1.1 g/mL.

*Section F (Optional)

11. Determine the freezing point of a 0.10 m solution of $CaCl_2$ in water, given that the $CaCl_2$ is 100% dissociated at this concentration into Ca^{2+} and Cl^- ions.

12. A solution is prepared by dissolving 5.85 g NaCl in 100 g water and the NaCl is 90% dissociated at this concentration. What is the boiling point of of the solution?

13. What is the freezing point of an 18.75 wt % solution of $Cu(NO_3)_2$ in water? The salt is 80% dissociated into Cu^{2+} and NO_3^- ions.

14. A solution prepared by dissolving 0.50 mol CH_3COOH in 1.0 kg water freezes at $-0.95°C$. What is the percentage dissociation into $(CH_3COO)^-$ and H^+ ions?

15. A solution prepared by dissolving 292.5 g NaCl in 1.0 kg water boils at 103.5°C. What is the percentage dissociation of the NaCl?

Section G

16. Determine the molecular weight of a polypropylene sample if a solution prepared by dissolving 1.0 g of the polypropylene in 1.25 L of hexane has an osmotic pressure π of 0.00020 atm at 350 K.

17. A solution containing 2.00 g protein in 200 g water has an osmotic pressure π of 7.6 mmHg at 27°C. There are 760 mmHg/1.0 atm, and the density of water is 1.0 g/mL. What is the molecular weight of the protein fraction?

Miscellaneous Problems

18. Calculate the expected boiling-point elevation for a solution of 0.504 g of anthracene, $C_{14}H_{10}$, in 42.0 g of benzene.

19. A solution is prepared by dissolving 0.150 g of a protein in 100 mL of water. The solution is found to have an osmotic pressure of 1.16 mmHg at 25°C. From these data, determine the molecular weight of the protein.

20. It is found that when 0.218 g sulfur is dissolved in 17.79 g of carbon disulfide, the solution boils at 46.41°C. What is the molecular formula for sulfur? You may use 46.30°C as the boiling point of pure carbon disulfide.

21. Determine the osmotic pressure at 20°C for a solution containing 500 g of water and 17.1 g of sucrose, $C_{12}H_{22}O_{11}$ (MW = 342.0 g/mol). You may use molality in this case and so

$$R = \frac{0.0820 \, (\text{kg soln})(\text{atm})}{(\text{mol solute})(\text{K})}$$

22. When 1.00 g $CaCl_2$ is dissolved in 250.0 g of water, the $CaCl_2$ is 100% dissociated. What is the freezing point of the solution?

23. Dissolving 0.15 mol of squalene in 300.0 g of toluene produces a solution that boils at 112.1°C. What is the molal boiling-point-elevation constant for toluene? The standard boiling point for toluene is 110.6°C.

*24. A 5.00 *m* solution of NaCl boils at 103.5°C. What is the percent dissociation of the NaCl?

25. Determine the molality of a solution of sugar in camphor that freezes at 102.4°C.

PROBLEM SET ANSWERS

If you have understood this chapter, you should be able to do Problems 1–6, 8, 9, and either 16 or 17. If you run into trouble, check the section in the chapter that pertains to the problems you cannot do. The ΔT calculations are shown in parentheses at the end of each problem for which they are needed.

1. $? K_B = \dfrac{(? \,°C)(\text{kg tetralin})}{\text{mol stearic acid}}$

$= \dfrac{225 \, \text{g tetralin}}{26.4 \, \text{g stearic acid}} \times \dfrac{284 \, \text{g stearic acid}}{\text{mol stearic acid}} \times \dfrac{\text{kg tetralin}}{10^3 \, \text{g tetralin}} \times 2.8°C$

$= \dfrac{6.8°C(\text{kg tetralin})}{\text{mol solute}} \qquad (\Delta T_B = 210.1°C - 207.3°C = 2.8°C)$

2. $? K_F = \dfrac{(?\,°C)(kg\ Ag)}{mol\ Cu} = \dfrac{90.0\ \cancel{g\ Ag}}{10.0\ \cancel{g\ Cu}} \times \dfrac{kg\ Ag}{10^3\ \cancel{g\ Ag}} \times \dfrac{63.5\ \cancel{g\ Cu}}{mol\ Cu} \times 85°C$

$\qquad = \dfrac{49°C(kg\ Ag)}{mol\ solute} \qquad (\Delta T_F = 961°C - 876°C = 85°C)$

3. $? g = \cancel{mol\ X} \times \dfrac{2.3°\cancel{C}(kg\ \cancel{carbon\ disulfide})}{\cancel{mol\ X}} \times \dfrac{10.0\ g\ X}{0.150\ \cancel{kg\ carbon\ disulfide}} \times \dfrac{1}{2.0°\cancel{C}}$

$\qquad = 77\ g\ X$ in one mole $X = \dfrac{77\ g\ X}{mol\ X} \quad (\Delta T_B = 48.3°C - 46.3°C = 2.0°C)$

4. $? g = \cancel{mol\ X} \times \dfrac{5.1°\cancel{C}(10^3\ \cancel{g\ benzene})}{\cancel{mol\ X}} \times \dfrac{2.0\ g\ X}{20\ \cancel{g\ benzene}} \times \dfrac{1}{3.4°\cancel{C}}$

$\qquad = 1.5 \times 10^2\ g\ X$ in one mole $X = \dfrac{1.5 \times 10^2\ g\ X}{mol\ X}$

(Notice the correct way to give the number of significant figures.)

5. $? \Delta T_F °C = \dfrac{3.42\ \cancel{g\ sugar}}{10.0\ \cancel{g\ water}} \times \dfrac{1.86°C(10^3\ \cancel{g\ water})}{\cancel{mol\ sugar}} \times \dfrac{\cancel{mol\ sugar}}{342\ \cancel{g\ sugar}} = 1.86°C$

Freezing point $= 0.00°C - 1.86°C = -1.86°C$

6. $? \Delta T_B °C = \dfrac{3.58\ \cancel{g\ C_6Cl_6}}{10.0\ \cancel{g\ carbon\ tetrachloride}} \times \dfrac{5.0°C(10^3\ \cancel{g\ carbon\ tetrachloride})}{\cancel{mol\ C_6Cl_6}}$

$\qquad \times \dfrac{\cancel{mol\ C_6Cl_6}}{285.0\ \cancel{g\ C_6Cl_6}} = 6.3°C$

Boiling point $= 76.8°C + 6.3°C = 83.1°C$

7. First calculate the weight of 1.0 L of solution.

$? g\ soln = 1.0\ \cancel{L\ soln} \times \dfrac{\cancel{mL\ soln}}{10^{-3}\ \cancel{L\ soln}} \times \dfrac{0.90\ g\ soln}{\cancel{mL\ soln}} = 900\ g\ soln$

Then calculate the weight of sugar in 1.0 L of solution.

$? g\ sugar = 1.0\ \cancel{L\ soln} \times \dfrac{0.600\ \cancel{mol\ sugar}}{\cancel{L\ soln}} \times \dfrac{342\ g\ sugar}{\cancel{mol\ sugar}} = 205\ g\ sugar$

You then get the weight of solvent by subtracting:

$? g\ benzene = 900\ g - 205\ g = 695\ g\ benzene = 0.695\ kg\ benzene$

Now you know that you have 0.600 mol sugar/0.695 kg benzene

$? \Delta T_B °C = \dfrac{0.600\ \cancel{mol\ sugar}}{0.695\ \cancel{kg\ benzene}} \times \dfrac{2.5°C(kg\ \cancel{benzene})}{\cancel{mol\ sugar}} = 2.2°C$

Boiling point $= 80.1°C + 2.2°C = 82.3°C$

8. $\Delta T_F = 0.00°C - (-3.72°C) = 3.72°C$

$? g\ C_2H_5OH = 0.250\ \cancel{kg\ water} \times \dfrac{\cancel{mol\ C_2H_5OH}}{1.86°\cancel{C}(kg\ \cancel{water})} \times \dfrac{46.0\ g\ C_2H_5OH}{\cancel{mol\ C_2H_5OH}} \times 3.72°\cancel{C}$

$\qquad = 23.0\ g\ C_2H_5OH$

9. $\Delta T_B = 82.8°C - 76.8°C = 6.0°C$

? g C_6H_5OH = 500 ~~mL carbon tetrachloride~~ $\times \dfrac{1.59 \text{ g carbon tetrachloride}}{\text{mL carbon tetrachloride}}$

$\times \dfrac{\text{mol } C_6H_5OH}{5.0°C(10^3 \text{ g carbon tetrachloride})} \times \dfrac{94.0 \text{ g } C_6H_5OH}{\text{mol } C_6H_5OH} \times 6.00°C$

$= 90 \text{ g } C_6H_5OH$

10. First determine the freezing point in degrees Celsius:

$$°C = \frac{5}{9}(°F - 32) = \frac{5}{9}(-18) = -10°C$$

$$\Delta T_F = 0.0°C - (-10°C) = 10°C$$

? qt eth. glyc. = 3.0 ~~gal water~~ $\times \dfrac{4 \text{ qt water}}{\text{gal water}} \times \dfrac{946 \text{ mL water}}{\text{qt water}} \times \dfrac{1.00 \text{ g water}}{\text{mL water}}$

$\times \dfrac{\text{mol ethylene glycol}}{1.86°C \, (10^3 \text{ g water})} \times \dfrac{62.0 \text{ g ethylene glycol}}{\text{mol ethylene glycol}}$

$\times \dfrac{\text{mL ethylene glycol}}{1.1 \text{ g ethylene glycol}} \times \dfrac{\text{qt ethylene glycol}}{946 \text{ mL ethylene glycol}} \times 10°C$

$= 3.6$ qt ethylene glycol

11. Since the salt is 100% dissociated and a molecule of $CaCl_2$ produces three ions, one Ca^{2+} and two Cl^-'s, the solution will be 3 $\times$ 0.10 m in total particles.

? $\Delta T_F \, °C = \dfrac{0.30 \text{ mol particles}}{\text{kg water}} \times \dfrac{1.86°C(\text{kg water})}{\text{mol particles}} = 0.56°C$

Freezing point = $0.00°C - 0.56°C = -0.56°C$

12. Mol particles = $n(1 + f) = 5.85$ ~~g NaCl~~ $\times \dfrac{\text{mol NaCl}}{58.5 \text{ g NaCl}}(1 + 0.90)$

$= 0.19$ mol particles

? $\Delta T_B \, °C = \dfrac{0.19 \text{ mol particles}}{0.100 \text{ kg water}} \times \dfrac{0.51°C(\text{kg water})}{\text{mol particles}} = 0.97°C$

Boiling point = $100.00°C + 0.97°C = 100.97°C$

13. Mol particles = $n(1 + 2f) = 18.75$ ~~g Cu(NO₃)₂~~ $\times \dfrac{\text{mol Cu(NO}_3)_2}{187.5 \text{ g Cu(NO}_3)_2}(1 + 1.6)$

$= 0.26$ mol particles

Weight of water = 100 g - 18.75 g = 81.25 g water = 0.08125 kg water

? $\Delta T_F \, °C = \dfrac{0.26 \text{ mol particles}}{0.08125 \text{ kg water}} \times \dfrac{1.86°C(\text{kg water})}{\text{mol particles}} = 6.0°C$

Freezing point = $0.00°C - 6.0°C = -6.0°C$

14. First calculate the number of moles of particles from the freezing-point lowering, $\Delta T_F = 0.95°C$:

? mol particles $= 1.0 \,\text{kg water} \times \dfrac{\text{mol particles}}{1.86°C(\text{kg water})} \times 0.95°C$

$= 0.51$ mol particles

Then note that the moles CH_3COOH initially $= 0.50$ mol $= n$. Then calculate f:

Mol particles $= n(1 + f)$ so $0.51 = 0.50(1 + f)$ and $f = 0.02$

Percent dissociation $= 100(f) = 2\%$ dissociated

15. This problem is the same type as the one above.

? $\Delta T_B °C = 103.5°C - 100.0°C = 3.5°C$

? mol particles $= 1.0 \,\text{kg water} \times \dfrac{\text{mol particles}}{0.51°C(\text{kg water})} \times 3.5°C$

$= 6.9$ mol particles

? mol NaCl init $= 292.5 \,\text{g NaCl} \times \dfrac{\text{mol NaCl}}{58.5 \,\text{g NaCl}}$

$= 5.00$ mol NaCl $= n$

Mol particles $= n(1 + f)$ so $6.9 = 5.00(1 + f)$ and $f = 0.38$

Percent dissociation $= 100(f) = 38\%$ dissociated

(This is not a very realistic example; it is only to show you how to do these calculations.)

16. Since the solution is so dilute, you can assume that the volume of the solution equals the volume of the solvent.

? g polypropylene $= \text{mol polypropylene} \times \dfrac{0.0820(\text{L soln})(\text{atm})}{(\text{mol polypropylene})(\text{K})}$

$\times \dfrac{350 \,\text{K}}{0.00020 \,\text{atm}} \times \dfrac{1.0 \text{ g polypropylene}}{1.25 \,\text{L soln}}$

$= 1.1 \times 10^5$ g polypropylene in a mole $= \dfrac{1.1 \times 10^5 \text{ g polypropylene}}{\text{mol polypropylene}}$

17. This problem is similar to the one above, except that you convert the pressure to atmospheres, the temperature to Kelvins, and the volume of solution to liters.

? atm $= 7.6 \,\text{mmHg} \times \dfrac{\text{atm}}{760 \,\text{mmHg}} = 0.010$ atm

? K $= 27°C + 273 = 300$ K

? L soln $= 200 \,\text{g water} \times \dfrac{\text{mL water}}{1.0 \,\text{g water}} \times \dfrac{10^{-3} \,\text{L water}}{\text{mL water}} \times \dfrac{\text{L soln}}{\text{L water}}$

$= 0.20$ L soln

(Once again we have assumed that volume of solution equals volume of solvent.)

$$? \text{ g protein} = \cancel{\text{mol protein}} \times \frac{0.0820 (\cancel{\text{L soln}})(\cancel{\text{atm}})}{(\cancel{\text{mol protein}})(\cancel{\text{K}})} \times \frac{300 \cancel{\text{ K}}}{0.010 \cancel{\text{ atm}}} \times \frac{2.00 \text{ g protein}}{0.20 \cancel{\text{ L soln}}}$$

$$= 2.5 \times 10^4 \text{ g protein in one mole} = \frac{2.5 \times 10^4 \text{ g protein}}{\text{mol protein}}$$

18. You can start by determining the molecular weight of anthracene. It is 178.0 g/mol. Then you calculate ΔT_B.

$$? \text{ }^\circ\cancel{C} = \frac{0.504 \cancel{\text{ g anthracene}}}{42.0 \cancel{\text{ g benzene}}} \times \frac{2.5^\circ C (1000 \cancel{\text{ g benzene}})}{\cancel{\text{mol anthracene}}} \times \frac{\cancel{\text{mol anthracene}}}{178.0 \cancel{\text{ g anthracene}}}$$

$$= 0.17^\circ C$$

19. $? \text{ g protein} = \cancel{\text{mol protein}} \times \dfrac{0.0820 (\cancel{\text{L soln}})(\cancel{\text{atm}})}{(\cancel{\text{mol protein}})(\cancel{\text{K}})} \times \dfrac{0.150 \text{ g protein}}{0.100 \cancel{\text{ L soln}}}$

$$\times \frac{298 \cancel{\text{ K}}}{1.16 \cancel{\text{ mmHg}}} \times \frac{760 \cancel{\text{ mmHg}}}{\cancel{\text{atm}}}$$

$$= 2.40 \times 10^4 \text{ g in one mol}$$

So it is 2.40×10^4 g/mol. Or you could use the short cut:

$$MW = \frac{(g) RT}{\pi V} = \frac{(0.150)(0.0820)(298)}{\dfrac{1.16}{760} (0.100)} = 2.40 \times 10^4 \text{ g/mol}$$

20. $\Delta T_B = 46.41^\circ C - 46.30^\circ C = 0.11^\circ C$

$$? \text{ g sulfur} = \cancel{\text{mol sulfur}} \times \frac{2.3^\circ \cancel{C} (1000 \cancel{\text{ g carbon disulfide}})}{\cancel{\text{mol sulfur}}}$$

$$\times \frac{0.218 \text{ g sulfur}}{17.79 \cancel{\text{ g carbon disulfide}}} \times \frac{1}{0.11^\circ \cancel{C}}$$

$$= 2.6 \times 10^2 \text{ g of S in one mole S or } 2.6 \times 10^2 \text{ g/mol}$$

Since the atomic weight of S is 32.1 g/mol, it takes

$$\frac{2.6 \times 10^2}{32.1} = 8 \text{ atoms/molecule (that is, } S_8)$$

21. $? \text{ atm} = 17.1 \cancel{\text{ g sucrose}} \times \dfrac{\cancel{\text{mol sucrose}}}{342.0 \cancel{\text{ g sucrose}}} \times \dfrac{0.0820 (\cancel{\text{kg water}})(\cancel{\text{atm}})}{(\cancel{\text{mol sucrose}})(\cancel{\text{K}})}$

$$\times \frac{10^3 \cancel{\text{ g water}}}{\cancel{\text{kg water}}} \times \frac{293 \cancel{\text{ K}}}{500 \cancel{\text{ g water}}} = 2.40 \text{ atm}$$

22. You determine ΔT_F:

$$? \text{ }^\circ C = \frac{1.00 \cancel{\text{ g CaCl}_2}}{250.0 \cancel{\text{ g H}_2\text{O}}} \times \frac{\cancel{\text{mol CaCl}_2}}{111.1 \cancel{\text{ g CaCl}_2}} \times \frac{3 \cancel{\text{ mol particles}}}{\cancel{\text{mol CaCl}_2}}$$

$$\times \frac{1.86^\circ C (1000 \cancel{\text{ g H}_2\text{O}})}{\cancel{\text{mol particles}}} = 0.201^\circ C$$

Then subtract it from the freezing point of pure water:

Freezing point = 0.000°C – 0.201°C = –0.201°C

23. $\dfrac{(?\ ^\circ C)(\text{kg toluene})}{\text{mol squalene}} = \dfrac{(1.5^\circ C)(300.0\ \text{g toluene})}{0.15\ \text{mol squalene}} \times \dfrac{\text{kg toluene}}{10^3\ \text{g toluene}}$

$\qquad\qquad = \dfrac{3.0^\circ C(\text{kg toluene})}{\text{mol squalene}}$

$(\Delta T_B = 112.1^\circ C - 110.6^\circ C = 1.5^\circ C)$

24. First determine the number of moles of particles in one kilogram water:

$\Delta T_B = 103.5^\circ C - 100.0^\circ C = 3.5^\circ C$

? mol particles = $3.5^\circ C \times \dfrac{\text{mol particles}}{0.51^\circ C(\text{kg water})} \times \text{kg water}$

$\qquad\qquad = 6.86\ \text{mol particles}$

Since NaCl dissociates to give two particles,

Total particles = $n(1 + f) = 5.00(1 + f) = 6.86$ ($n = 5.00\ \text{mol/kg}$)

$\qquad\qquad 5.00 + 5.00f = 6.86$

$\qquad\qquad\qquad 5.00f = 1.86$,   so $f = 0.37$ (to correct significant figures)

Percent = $100f = 37\%$

25. $\Delta T_F = 178.4^\circ C - 102.4^\circ C = 76.0^\circ C$

It is easiest to use K_F in the form

$K_F = \dfrac{38.0^\circ C}{m}$

So

? $m = 76^\circ C \times \dfrac{m}{38.0^\circ C} = 2.00\ m$

CHAPTER 18

Calculations Involving Heat

PRETEST

1. How many joules are needed to raise the temperature of 50.0 g of methyl alcohol 15.0°C? The specific heat of methyl alcohol is 2.510 J/(g)(°C).

2. How many kilojoules are needed to heat 11.0 mol of water from 20.0°C to 100.0°C? The specific heat of water is 4.184 J/(g)(°C).

3. You know that 60.2 J are given off when a 25.0 g sample of lead cools from 98.0°C to 80.0°C. What is the specific heat of lead metal?

4. How many joules are required to vaporize 35.0 g of methyl alcohol? The heat of vaporization of alcohol is 1.10 kJ/gram.

5. How many calories are needed to convert 45.0 mL of ether at 0.0°C to vapor at its boiling point (34.6°C)? Ether's density is 0.714 g/mL. The specific heat of ether is 0.547 cal/(g)(°C), and its heat of vaporization is 89.3 cal/gram.

6. How much heat can you get by burning 4.80 kg of carbon with all substances at 25°C? The reaction is $C(s) + O_2(g) \rightarrow CO_2(g)$. The standard heat of reaction is -394 kJ/mol.

7. What is the heat of reaction for the process

    $$FeS(s) + 2\,HCl(aq) = FeCl_2(aq) + H_2S(g)$$

    The notation (aq) means "in dilute solution in water." The standard heats of formation are:

$FeS(s)$	$\Delta H_f^0 = -95.0$ kJ/mol
$HCl(aq)$	$\Delta H_f^0 = -167$ kJ/mol
$FeCl_2(aq)$	$\Delta H_f^0 = -422.6$ kJ/mol
$H_2S(g)$	$\Delta H_f^0 = -20.2$ kJ/mol

8. A 10.00 g piece of aluminum is heated to 102.25°C and placed in a calorimeter containing 25.0 g of water at 20.50°C. The temperature of the water rises to 27.25°C. What is the specific heat of aluminum? Assume that the body of the calorimeter does not absorb any heat.

9. Suppose that you put 25.0 g of ice at 0.00°C into a calorimeter that contains 300.0 g of water at 21.16°C. The ice melts and the temperature of the water in the calorimeter finally drops to 13.40°C. What is the heat of fusion for water, in kilojoules per mole? The heat capacity K_c of the calorimeter equals 3.3 J/°C.

PRETEST ANSWERS

All the questions are worked out as examples in the body of the chapter. If you were able to do questions 1, 2, and 3, you can skip Section A in this chapter. If you were able to do question 4 but not 5, you need to cover only the last half of Section B. If you missed both, do all of Section B. You should be able to do both question 6 and question 7. If you missed either, do Section C. You should be able to do question 8. You can find it worked out in the first part of Section D. Question 9 is rough. It is covered at the end of Section D. Your instructor will tell you whether you need to be able to do this type of calculation.

1. 1.88×10^3 J

2. 66.3 kJ

3. 0.134 J/(g)(°C)

4. 38.5 kJ

5. 3.48×10^3 cal

6. -1.58×10^5 kJ

7. −14 kJ/mol

8. 0.941 J/(g)(°C)

9. 6.0 kJ/mol

In this chapter, we'll examine some of the types of calculations that you use when you are dealing with heat. We will consider only three types of processes that can either consume or give off heat:

1. Changing the temperature of a substance
2. Changing the phase of a substance
3. Having the substance undergo a chemical reaction: *thermochemistry*

Finally, we will consider *calorimetry*, that is, how this heat is measured in the laboratory.

   As we go along, you will see that occasionally a simple, one-step dimensional-analysis solution is not possible. You may have to add or subtract several solutions to arrive at the final answer. However, this shouldn't pose any real difficulty for you because the methods are very straightforward.

SECTION A Specific Heat

To raise the temperature of a substance, you must put heat into it. When the substance cools, it gives off heat to its surroundings. The amount of heat it gives off depends on two things: the amount of the substance and the change in its temperature. When you express this heat per gram of substance and also per degree Celsius change, you have a conversion factor that has the units

$$\frac{\text{joule}}{\text{(gram)(degree Celsius)}} \quad \text{or} \quad \frac{\text{J}}{\text{(g)(°C)}}$$

The *joule*, abbreviated J, is the currently used unit of heat. It is one of the few SI units (see Chapter 3) that has gained wide acceptance. In terms of SI base units, energy is equal to *work*, which can be expressed as the force used to move an object times the distance it is moved. Work equals the mass times the distance squared per second squared. The joule is therefore defined as the energy needed to move (1 kg mass)(1 meter)^2 per sec^2:

$$1 \text{ J} = \frac{1 \text{ kg m}^2}{\text{sec}^2}$$

Don't worry about this, since chemists have little use for the units that make up the joule. We will simply consider the joule as a prime unit of heat.

The unit formerly used for heat is the *calorie*. Many older books, in fact, still refer to calories (abbreviated cal). The calorie, like many of our units, is defined on the basis of the properties of water. (The Celsius temperature scale is based on the freezing point, 0°C, and the boiling point, 100°C. The liter is based on the volume occupied by 1 kg water.) The *calorie* is defined as the heat needed to increase the temperature of 1.00 g of water from 14.5°C to 15.5°C (that is, 1°C):

   1 cal = (1.00 g water)(1.00°C)

We will work mostly in joules. However, it is easy to convert from joules to calories. The conversion factor(s) is

$$\frac{4.18400 \text{ J}}{\text{cal}} \quad \text{or} \quad \frac{0.2390057 \text{ cal}}{\text{J}}$$

The definition of calorie actually expresses what is called the *specific heat* of water, that is, the number of calories per (g)(°C). In the case of water,

$$\text{Specific heat} = \frac{1.00 \text{ cal}}{\text{(g)(°C)}}$$

In SI units, the specific heat of water would be

$$\text{Specific heat} = \frac{\underline{\qquad} \text{ J}}{\text{(g)(°C)}}$$

4.184

Every substance has its own specific heat, which varies somewhat with the temperature at which you consider it. The same thing holds true for the boiling point. Each substance has its own boiling point, which varies with the pressure at which you consider it. There are tables with specific heats for various substances at various temperatures. For example, the specific heat of methyl alcohol at 0°C is

$$\frac{2.37 \text{ J}}{\text{(g alcohol)(°C)}} \quad \text{but at 20°C it is} \quad \frac{2.51 \text{ J}}{\text{(g alcohol)(°C)}}$$

For copper metal at 0°C, the specific heat is

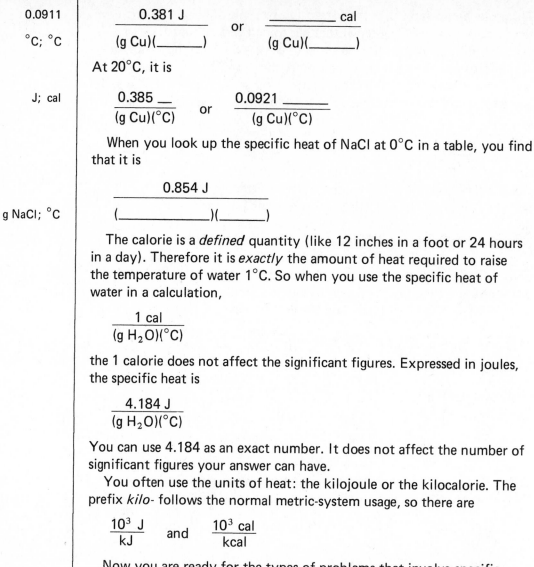

0.0911

°C; °C

$$\frac{0.381 \text{ J}}{(\text{g Cu})(\underline{\hspace{1cm}})} \quad \text{or} \quad \frac{\underline{\hspace{1cm}} \text{ cal}}{(\text{g Cu})(\underline{\hspace{1cm}})}$$

At 20°C, it is

J; cal

$$\frac{0.385 \underline{\hspace{0.5cm}}}{(\text{g Cu})(°\text{C})} \quad \text{or} \quad \frac{0.0921 \underline{\hspace{1cm}}}{(\text{g Cu})(°\text{C})}$$

When you look up the specific heat of NaCl at 0°C in a table, you find that it is

g NaCl; °C

$$\frac{0.854 \text{ J}}{(\underline{\hspace{2cm}})(\underline{\hspace{1cm}})}$$

The calorie is a *defined* quantity (like 12 inches in a foot or 24 hours in a day). Therefore it is *exactly* the amount of heat required to raise the temperature of water 1°C. So when you use the specific heat of water in a calculation,

$$\frac{1 \text{ cal}}{(\text{g H}_2\text{O})(°\text{C})}$$

the 1 calorie does not affect the significant figures. Expressed in joules, the specific heat is

$$\frac{4.184 \text{ J}}{(\text{g H}_2\text{O})(°\text{C})}$$

You can use 4.184 as an exact number. It does not affect the number of significant figures your answer can have.

You often use the units of heat: the kilojoule or the kilocalorie. The prefix *kilo-* follows the normal metric-system usage, so there are

$$\frac{10^3 \text{ J}}{\text{kJ}} \quad \text{and} \quad \frac{10^3 \text{ cal}}{\text{kcal}}$$

Now you are ready for the types of problems that involve specific heat. Here is question 1 on the pretest. How many joules are needed to raise the temperature of 50.0 g of methyl alcohol 15.0°C? The specific heat of the alcohol is 2.510 J/(g)(°C).

joules

The question asked is "How many _____?" The problem gives you two pieces of information. It may be best, in general, to start with the amount of the substance. So you write

J

? ___ = 50.0 g alcohol

Now you can use the conversion factor, the specific heat, to cancel the unwanted dimension, grams alcohol:

$$\frac{2.510 \text{ J}}{(\text{g alcohol})(°\text{C})}$$

? J = 50.0 g alcohol × _____

degrees Celsius

The unwanted dimension that remains is _____. You

can get rid of this by multiplying by the other piece of information

given, _____. Thus you have as the complete solution

$$? \, J = 50.0 \, \cancel{\text{g alcohol}} \times \frac{2.510 \, J}{(\cancel{\text{g alcohol}})(\cancel{^\circ C})} \times 15.0 \cancel{^\circ C}$$

$$= 1882.5 \, J \quad \text{or} \quad 1.88 \times 10^3 \, J$$

in scientific notation and to the correct significant figures.

Some problems may not give you the weight of the substance in grams. Then you have to convert whatever units it may have into grams. Here is an example. How many joules are given off when 25.0 mL of mercury cool from 25.5°C to 20.4°C? The density of mercury is 13.5 g/mL and its specific heat is 0.139 J/(g)(°C).

Starting in the usual way, you write

$$? \, \_\_ =$$

The information given is the volume of Hg, _____, and its initial

and final temperatures. You need to know the _____ of Hg and its *change* in temperature. The density gets you from the volume to the

weight of Hg. You can get the change in temperature by _____ the initial temperature from the final temperature. The change is abbreviated ΔT:

$$\Delta T = T_{\text{final}} - T_{\text{init}} = \_\_\_\_ {}^\circ C - \_\_\_\_ {}^\circ C = -5.1 {}^\circ C$$

Notice that, since the temperature dropped, the sign on ΔT is negative.

Now we can set up the problem. The amount of Hg given in the prob-

lem was _____ mL Hg.

$$? \, J = 25.0 \, \text{mL Hg}$$

You can get from milliliters Hg to grams Hg by using the conversion

factor _____

$$? \, J = 25.0 \, \cancel{\text{mL Hg}} \times \frac{13.5 \, g \, Hg}{\cancel{\text{mL Hg}}}$$

Now you can use the _____,

$$\frac{0.139 \, J}{(g \, Hg)({}^\circ C)}$$

to cancel the unwanted dimension grams Hg:

$$? \, J = 25.0 \, \cancel{\text{mL Hg}} \times \frac{13.5 \, \cancel{g \, Hg}}{\cancel{\text{mL Hg}}} \times \frac{0.139 \, J}{(\cancel{g \, Hg})({}^\circ C)}$$

The only remaining unwanted dimension is _____. You

can cancel this by multiplying by _____:

15.0°C

J

25.0 mL

weight

subtracting

20.4; 25.5

25.0

$\dfrac{13.5 \, g \, Hg}{mL \, Hg}$

specific heat

degrees Celsius

ΔT

-5.1°C

$? J = 25.0 \, \text{mL Hg} \times \dfrac{13.5 \, \text{g Hg}}{\text{mL Hg}} \times \dfrac{0.139 \, J}{(\text{g Hg})(°C)} \times \underline{\hspace{1cm}}$

$= -239 \, J \quad (-2.4 \times 10^2 \, J$ to the correct significant figures)

Notice that "joules" has a negative sign. This shows that the process has given off heat. Such a process is called *exothermic*. A process that takes up heat is called *endothermic*, and the sign of the heat is then positive. As you know, + signs are generally not written; they are understood.

Some problems ask you to find the heat required for a change in temperature and give you the amount of the substance in moles. You can easily convert moles to grams by using as a conversion factor the

molecular weight

_____ of the substance. Here is question 2 on the pretest. How many kilojoules are needed to heat 11.0 mol of water from 20.0°C to 100.0°C? The specific heat of water is 4.184 J/(g)(°C).

molecular weight

The _____ of water takes you from moles of water to grams of water. To find the temperature change, you subtract the

initial; final

_____ temperature from the _____ temperature:

final; init; 100.0; 20.0; 80.0

$\Delta T = T\underline{\hspace{1cm}} - T\underline{\hspace{1cm}} = \underline{\hspace{0.7cm}}°C - \underline{\hspace{0.7cm}}°C = \underline{\hspace{0.7cm}}°C$

endothermic

Notice that the sign on ΔT is positive because this is an _____ process. Heat must be put in.

Now, starting in the usual way, you write

$? \, kJ = 11.0 \, \text{mol } H_2O$

Using the molecular weight of water to shift to grams of water, you have

$\dfrac{18.0 \, \text{g water}}{\text{mol water}}$

$? \, kJ = 11.0 \, \text{mol } H_2O \times \underline{\hspace{2cm}}$

specific heat

Now you can use the conversion factor, the _____ of water:

$\dfrac{4.184 \, J}{(\text{g water})(°C)}$

$? \, kJ = 11.0 \, \text{mol water} \times \dfrac{18.0 \, \text{g water}}{\text{mol water}} \times \underline{\hspace{2cm}}$

ΔT

You can cancel the unwanted °C by multiplying by _____, which

80.0

equals _____°C:

$? \, kJ = 11.0 \, \text{mol water} \times \dfrac{18.0 \, \text{g water}}{\text{mol water}} \times \dfrac{4.184 \, J}{(\text{g water})(°C)} \times 80.0°C$

The only dimension remaining is joules, but the dimension of the answer

kilojoules

must be _____. Therefore you multiply by the conversion

$\dfrac{kJ}{10^3 \, J}$

factor _____

$? \, kJ = 11.0 \, \text{mol water} \times \dfrac{18.0 \, \text{g water}}{\text{mol water}} \times \dfrac{4.184 \, J}{(\text{g water})(°C)} \times 80.0°C$

$\times \dfrac{kJ}{10^3 \, J} = 66.3 \, kJ$

Calculating the specific heat of a substance is simple if you know the number of joules involved in changing the temperature of a given weight of the substance by a known number of degrees Celsius. Since you want to calculate a conversion factor, the specific heat, you could reword the problem as we have done before for determining density, percent, molal freezing-point-depression constant, molecular weight, and so on. However, you may find it easier just to take the values for joules, for ΔT, and for the weight, and then set them into a fraction that gives you all the correct dimensions in their correct places. Question 3 on the pretest is an example of this type of calculation. You know that 60.2 J are given off when a 25.0 g sample of lead cools from 98.0°C to 80.0°C. What is the specific heat of lead metal?

First you determine ΔT:

$$\Delta T = T\underline{\quad\quad} - T\underline{\quad\quad} = \underline{\quad\quad}°C - \underline{\quad\quad}°C = \underline{\quad\quad}°C$$

final; init; 80.0; 98.0; −18.0

The problem tells you that 60.2 J are given off. Since heat is given off, this is an _____ process and the sign on the number of joules is

exothermic

_____. The number of joules is _____. You are also told that this heat is given off by a 25.0 g sample. You now have all the data that you need to set up the specific-heat conversion factor. The units of this

− (negative); −60.2 J

factor are _____. When you put in the values −18.0°C, −60.2 J, and 25.0 g, so that the dimensions are as desired for specific heat, you have

$$\frac{J}{(g)(°C)}$$

$$\text{Specific heat} = \frac{\overline{\underline{\quad\quad\quad}}}{(\underline{\quad\quad})(\underline{\quad\quad})}$$

−60.2 J

25.0 g; −18.0°C

Dividing the numbers in the numerator by those in the denominator gives

$$\text{Specific heat} = \frac{\overline{\underline{\quad\quad}}}{(g)(°C)}$$

0.134 J

Notice that the two negative signs cancel. Specific heat always has a + sign.

You may prefer to reword problems in which the answer is itself a conversion factor so that the question asked and the information given both come from the desired conversion factor. If so, you can reword specific-heat problems so that the question is "How many joules?" and the information given is "1.0 g substance" and "1.0°C." Both the 1.0's are defined terms and do not affect the number of significant figures. In the example we have just done, it would go like this:

? J = 1.0 g lead × 1.0°C

The remaining information given in the problem is that −60.2 J are required per 25.0 g lead per −18.0°C. Using this information to cancel the unwanted dimensions gives you

−60.2 J

25.0 g lead

0.134 J

$$? J = 1.0 \text{ g lead} \times 1.0°\cancel{C} \times \frac{\underline{\hspace{3cm}}}{(\underline{\hspace{2cm}})(-18.0°\cancel{C})}$$

$$= \underline{\hspace{1.5cm}}$$

Since this is the number of joules needed to change the temperature of

specific heat

1 g of lead 1°C, it is, of course, equal to the _____.
    Another type of problem that doesn't come up often involves determining the change in temperature that results from the gain or loss of a given number of joules. With dimensional analysis, it is no harder to determine the change in temperature when you know the number of joules than it is to determine the number of joules when you know the change in temperature. In both cases, you use the specific heat of the substance as your main conversion factor. Here's an example of this type of problem: If you add exactly 750 joules to a 50.0 g sample of water at 11.0°C, what does the temperature become?
    First you must realize that you cannot calculate the final temperature

change

directly. The specific heat tells you only about the _____ in temperature. You can calculate ΔT. Once you know ΔT, you can get the

adding

final temperature by _____ ΔT to the initial temperature.

+

$$T_{final} = T_{init} - \Delta T$$

Thus the question changes to "How many degrees Celsius is ΔT?" and the information given is +750 J and 50.0 g water. The joules have a plus

endothermic

sign because heat is being added. This is an _____ process.
    You will find it best to start with 750 J as the information given:

$$? °C = 750 \text{ J}$$

You can cancel the unwanted dimension joules by using the specific heat

$\dfrac{4.184 \text{ J}}{(\text{g water})(°\text{C})}$

of water, _____, as your conversion factor.

$$? °C = 750 \cancel{J} \times \frac{(\text{g water})(°\text{C})}{4.184 \cancel{J}}$$

inverted

Notice that you had to use the specific heat in its _____ form to cancel the unwanted joules.

grams water

    The unwanted dimension that remains is _____, and you have this in the information given. But be careful. You must put it in as 1/g water in order to cancel the unwanted dimension. This may seem strange, but we did the same thing in Chapter 16 when we were dealing with colligative properties. If you can multiply by the information given, you can also divide by it, since dividing is the same as multiplying by one over something. So you finish the problem by writing

$\dfrac{1}{50.0 \text{ g water}}$; 3.59°C

$$? °C = 750 \cancel{J} \times \frac{(\text{g water})(°\text{C})}{4.184 \cancel{J}} \times \underline{\hspace{2cm}} = \underline{\hspace{1.5cm}}$$

This 3.59°C represents ΔT. To find the final temperature of the water, you must _____ ΔT to the starting temperature of the water.

$$T_{final} = T_{init} + \underline{\hspace{1.5cm}} = \underline{\hspace{1.5cm}}°C + 3.59°C$$

$$= \underline{\hspace{1.5cm}}°C \quad \text{(Notice the correct significant figures.)}$$

add

ΔT; 11.0

14.6

SECTION B   Heat Involved in Changes of Phase

When ice melts, its temperature remains constant at 0°C throughout the melting process. Nevertheless, heat must be added to make the melting continue. Similarly, when water is boiling at standard pressure, its temperature remains at 100°C. Nonetheless, heat must be added in order to convert liquid water at 100°C to gaseous water at 100°C. The heat required to change the phase of a substance without changing its temperature is called the *latent heat.* Generally a more exact name is used to indicate which change of phase is occurring. Thus when you have a phase change between liquid and gas, you speak of the *heat of vaporization.* When the phase change is between solid and liquid, you speak of the *heat of fusion.* If the phase change is directly from solid to gas, you use the term *heat of sublimation.*

These latent heats are all conversion factors, since the more substance you have, the more heat is involved. However, since the temperature does not change when the phase changes, degrees Celsius doesn't appear as it did in specific heat. So the dimensions of these latent heats are

$$\frac{\text{joules}}{\text{gram}} \quad \text{or} \quad \frac{J}{g} \quad \text{or often in older books} \quad \frac{\text{calories}}{\text{gram}} \quad \text{or} \quad \frac{\text{cal}}{g}$$

Thus the heat of vaporization of water is given as

$$\frac{2.26 \text{ kJ}}{\text{g water}} \quad \text{or} \quad \frac{540 \text{ cal}}{\text{g water}}$$

and the heat of fusion of water is given as

$$\frac{333 \text{ J}}{\underline{\hspace{1.5cm}}} \quad \text{or} \quad \frac{79.7 \text{ cal}}{\underline{\hspace{1.5cm}}}$$

g water; g water

The heat of fusion for Cu metal is given as

$$\frac{49.0 \text{ cal}}{\underline{\hspace{1.5cm}}} \quad \text{or} \quad \frac{\underline{\hspace{1.5cm}} \text{ kJ}}{\underline{\hspace{1.5cm}}} \quad \text{(there are 4.184 J/cal)}$$

0.205

g Cu; g Cu

Latent heats for these changes in phase are often given with the dimensions

$$\frac{\text{kJ}}{\text{mol}} \quad \text{or} \quad \frac{\text{kcal}}{\text{mol}}$$

You can easily calculate the value in one set of dimensions from the value in the other. Here is an example of that calculation in calories. What is the heat of vaporization for water in kilocalories per mole if its value is 540.0 cal/g?

In this case, it is best to reword the question as "How many kilocalories?" and have as the information given "one mole." Once again, the mole is a defined number and doesn't affect the significant figures.

1 mol

? kcal = _____ water

To convert the unwanted moles water to grams water, you can use the

molecular weight

_____ of water as your conversion factor.

$\dfrac{18.0 \text{ g water}}{\text{mol water}}$

? kcal = 1 mol water × _____

Now you may cancel the unwanted dimension "g water" with the conversion factor given in the problem, the heat of _____ of water.

vaporization

$\dfrac{540 \text{ cal}}{\text{g water}}$

? kcal = 1 ~~mol water~~ × $\dfrac{18.0 \text{ g water}}{\text{mol water}}$ × _____

Now all you must do is convert calories to _____.

kilocalories

$\dfrac{\text{kcal}}{10^3 \text{ cal}}$

? kcal = 1 ~~mol water~~ × $\dfrac{18.0 \text{ g water}}{\text{mol water}}$ × $\dfrac{540.0 \text{ cal}}{\text{g water}}$ × _____

9.72 kcal

= 1 × 18.0 × 540.0 × $\dfrac{\text{kcal}}{10^3}$ = _____

Thus the vaporization of 1 mol of water at 100°C requires 9.72 kcal. This is the latent heat of vaporization in kilocalories per mole. Just as specific heats vary slightly with the temperature at which they are considered, so do heats of vaporization. We won't consider this problem. Instead, we will always consider the heat of vaporization at the standard boiling point.

Question 4 on the pretest is a good example of the type of problem that you will face. How many joules are required to vaporize 35.0 g of methyl alcohol? The heat of vaporization for alcohol is 1.10 kJ/g.

joules

The question asked is "How many _____?" and the information given is "_____." So start by writing

35.0 g alcohol

J

? __ = 35.0 g alcohol

You can cancel the unwanted dimension by multiplying by the conversion factor given in the problem, the _____ of alcohol:

heat of vaporization

$\dfrac{1.10 \text{ kJ}}{\text{g alcohol}}$; 38.5 × 10³

? J = 35.0 g alcohol × _____ × $\dfrac{10^3 \text{ J}}{\text{kJ}}$ = _____ J

This is such a large number that it would be best to express it in kilojoules:

? kJ = 38.5 × 10³ J × _____ = 38.5 kJ

$$\frac{kJ}{10^3 \ J}$$

Since converting a liquid to a gas or converting a solid to a liquid

always requires heat, they are _____ processes and the   endothermic

sign on the number of joules or calories is always _____. On the   + (positive)
other hand, converting a gas to a liquid or converting a liquid to a solid

gives off heat. These are _____ processes, and the number of   exothermic

joules or calories is _____.   – (negative)

   We can write the processes as a reaction

   liquid + heat ⟶ gas   gas – heat ⟶ liquid
                          and
   solid + heat ⟶ liquid   liquid – heat ⟶ solid

If you consider the same amount of substance, then the same number of
joules are consumed in changing a liquid to the gas as are given off by
changing the gas to liquid. And the same number of joules are needed to
change the solid to the liquid as are given off by changing the

_____ to the _____. This is always true. The same   liquid; solid
amount of heat is involved in both a process and its reverse. Only the
sign on the heat is opposite. You know that it requires 38.5 kJ to convert 1 mol of water to 1 mol of steam. The heat you get when 1 mol of

steam condenses to liquid water is _____. It has a __ sign   38.5 kJ; –

because it is an _____ process.   exothermic
   Problems with latent heats are almost always simple. The possible
difficulties occur only when the amount of substance is not given in
grams and you have to convert to grams, or when the substance is not at
the temperature at which the phase change occurs. In this last case, you
calculate the amount of heat needed to get the substance to the constant
temperature of the phase change, and then add this amount to the heat
needed for the phase change. Question 5 on the pretest has both these
complications. For a change, we did this one in calories.
   How many calories are needed to convert 45.0 mL of ether at 0.0°C
to its vapor at its boiling point, 34.6°C? Ether's density is 0.714 g/mL.
The specific heat of ether is 0.547 cal/(g)(°C), and its heat of vaporization is 89.3 cal/g.
   The first thing that you must determine is the heat needed to bring
the 45.0 mL of ether from 0.0°C to its boiling point, 34.6°C. The ΔT

for this process is _____°C. Proceeding as you did in Section A for   34.6
specific heat, you write

   ? _____ = _____   cal; 45.0 mL ether

Then you must convert milliliters to _____ so that you can use the   grams
specific-heat conversion factor. You can do this by multiplying by the

_____.   density

$\dfrac{0.714 \text{ g ether}}{\text{mL ether}}$

? cal = 45.0 mL ether × _____

Then you can use the specific heat of ether:

$\dfrac{0.547 \text{ cal}}{(\text{g ether})(^{\circ}\text{C})}$

$$? \text{ cal} = 45.0 \text{ mL ether} \times \dfrac{0.714 \text{ g ether}}{\text{mL ether}} \times \underline{\qquad}$$

degrees Celsius

The remaining unwanted dimension is _____, and you

34.6°C

can cancel this by multiplying by ΔT, which is _____.

$$? \text{ cal} = 45.0 \text{ mL ether} \times \dfrac{0.714 \text{ g ether}}{\text{mL ether}} \times \dfrac{0.547 \text{ cal}}{(\text{g ether})(^{\circ}\text{C})} \times 34.6^{\circ}\text{C}$$

$$= 608 \text{ cal}$$

This is the amount of heat needed to get the ether to the temperature at which it changes to gas. Now you determine the heat needed for the phase change that occurs at the boiling point.

? cal = 45.0 mL ether

Once again, you must get to grams to use the heat of vaporization:

$\dfrac{0.714 \text{ g ether}}{\text{mL ether}}$

? cal = 45.0 mL ether × _____

Now you can multiply by the heat of vaporization,

$\dfrac{89.3 \text{ cal}}{\text{g ether}}$

$$? \text{ cal} = 45.0 \text{ mL ether} \times \dfrac{0.714 \text{ g ether}}{\text{mL ether}} \times \underline{\qquad}$$

and the only dimension left is the desired dimension, calories.

$$? \text{ cal} = 45.0 \text{ mL ether} \times \dfrac{0.714 \text{ g ether}}{\text{mL ether}} \times \dfrac{89.3 \text{ cal}}{\text{g ether}} = 2869 \text{ cal}$$

The total heat needed to warm the ether and to vaporize it is thus

608 cal; 2869 cal

_____ + _____ = 3477 cal

Written to the correct number of significant figures, this is

3.48×10^{3}; 3.48

_____ cal or _____ kcal.

You can see that here we took a process that occurs in several steps, and added the heats of the individual steps to get the total heat of the process. If we write out the steps, we have

Step 1: $\text{ether}_{(\text{liq. } 0^{\circ}\text{C})} + \text{heat}_1 = \text{ether}_{(\text{liq. } 34.6^{\circ}\text{C})}$

Step 2: $\text{ether}_{(\text{liq. } 34.6^{\circ}\text{C})} + \text{heat}_2 = \text{ether}_{(\text{gas } 34.6^{\circ}\text{C})}$

If you add these two equations by adding all the things on the left side of the equals sign and setting these equal to the sum of all the things on the right side of the equals sign, you have

$\text{ether}_{(\text{liq. } 0^{\circ}\text{C})} + \text{heat}_1 + \text{ether}_{(\text{liq. } 34.6^{\circ}\text{C})} + \text{heat}_2$

$= \text{ether}_{(\text{liq. } 34.6^{\circ}\text{C})} + \text{ether}_{(\text{gas } 34.6^{\circ})}$

Since ether$_{(liq. 34.6°C)}$ appears on both sides of the equal sign, it cancels and the sum of the two steps is really

ether$_{(liq. 0°C)}$ + heat$_1$ + heat$_2$ = ether$_{(gas\ 34.6°C)}$

The fact that you can determine the heat of a process by adding the heats of the steps that get you from the starting conditions to the final conditions is called *Hess's law.* Your process may not actually go through these steps, but if you can write a series of steps that, by addition, result in the process in which you are interested, then you can simply add the heats of these steps.

Here is another example of this. Suppose that you want to know the heat needed to convert 1 mol of ice at 0°C to water vapor at the same temperature. The process of going from the solid directly to the gas is called *sublimation*. You can write this process as two steps.

Step 1:   Ice + heat$_1$ = liquid water

Step 2:   Liquid water + heat$_2$ = water vapor

Adding both steps, you have

Ice + heat$_1$ + liquid water + heat$_2$ = liquid water + water vapor

Since liquid water appears on both sides of the equals sign, it cancels, and you have for the total process

Ice + _____ + _____ = water vapor   heat$_1$; heat$_2$

This is the sublimation of ice. The values for heat$_1$ and heat$_2$ can be found in tables. They are 5.98 kJ/mol and 44.85 kJ/mol, respectively. Therefore you can write

Ice + _____ + _____ = water vapor   5.98 kJ; 44.85 kJ

Ice + _____ kJ = water vapor   50.83

You need 50.83 kJ to sublime one mole of water.

The use of *Hess's law* is going to be much more important in the next section, in which we consider the heat involved in chemical reactions.

SECTION C   Heat Involved in Chemical Reactions

Practically every chemical reaction either needs heat to occur or gives off heat as it proceeds. The amount of heat depends on how much material is reacting. It is most convenient to express this amount of heat as a conversion factor that has the dimensions

$$\frac{kilojoules}{mole} \quad \left(\frac{kJ}{mol}\right) \quad or \quad \frac{kilocalories}{mole} \quad \left(\frac{kcal}{mol}\right)$$

If the reaction requires heat, it is an _____ reaction and the sign   endothermic

+ (positive)

exothermic; – (negative)

on the heat is _____. If the reaction gives off heat, it is an

_____ reaction and the sign on the heat is _____.

As your text tells you, the change in energy ΔE of a process equals the difference between the internal energy of the products and that of the reactants. That is,

$$\Delta E = E_{products} - E_{reactants}$$

You can measure ΔE from the heat either given off or consumed by the process *so long as the process is carried out at constant volume.* But in practice most chemical reactions are carried out at constant (atmospheric) pressure and the volume may change a great deal. So a new property had to be defined: the *enthalpy H.* This includes the energy used in changing the volume:

$$H = E + P\Delta V$$

For a process occurring at constant pressure, you can write an expression for the change in enthalpy ΔH (that is, $H_{products} - H_{reactants}$) as

$$\Delta H = \Delta E + P\Delta V$$

The $P\Delta V$ term represents the work done by the process. If this is the only work, you can write

$$\Delta E = q - P\Delta V$$

where q is the heat involved. When you substitute this value for ΔE in our equation for ΔH, it gives you

$$\Delta H = q - P\Delta V + P\Delta V \quad \text{or} \quad \Delta H = q$$

The change in enthalpy equals the heat given off or consumed by the process.

Because the two terms are equivalent, we often speak of "the heat of reaction" when in fact we really mean "the enthalpy of reaction."

For example, the burning of coal is an exothermic reaction that gives off 394 kJ heat per mole of carbon burned. You would write its heat of reaction (change in enthalpy) as

$$\Delta H^0 = \frac{394 \text{ kJ}}{\text{mol C}}$$

The ^{0} that follows the H indicates that the reaction is occurring at 25°C, and that, if gases are present, they are at 1.0 atm pressure. When you write the equation for the reaction, you must note the phases, solid (*s*), liquid (*l*) or gas (*g*), since the enthalpy is different if a change of phase is required. Thus, when you write the equation for the burning of coal (carbon) with oxygen, you have

$$C(s) + O_2(g) = CO_2(g), \quad \Delta H^0 = -394 \text{ kJ/mol}$$

You can actually write the heat in the equation as if it were a reactant or a product:

$$C(s) + O_2(g) = CO_2(g) + 394 \text{ kJ}$$

This shows that 394 kJ are produced. However, the convention is that you always show the heat term on the reactant side of the equation. You can get it there by subtracting 394 kJ from both sides of the equation:

$$C(s) + O_2(g) - 394 \text{ kJ} = CO_2(g) + 394 \text{ kJ} - 394 \text{ kJ}$$

or

$$C(s) + O_2(g) - 394 \text{ kJ} = CO_2(g)$$

This convention is what determines that the heat term in an exothermic process always has a _____ sign and that the heat in an

endothermic process has a _____ sign.

– (negative)

+ (positive)

By writing the heat of reaction in the equation, you can easily solve problems that ask you to find the amount of heat involved in a reaction. You consider the heat as if it were one of the substances in the reaction, and you set up a bridge conversion factor relating heat to moles of the other components. Thus, in the foregoing example, you can write

$$\frac{-394 \text{ kJ}}{\text{mol C}} \quad \text{or} \quad \frac{-394 \text{ kJ}}{\underline{\hspace{1cm}} O_2} \quad \text{or} \quad \frac{\underline{\hspace{1cm}}}{\text{mol CO}_2}$$

$\dfrac{394 \text{ kJ}}{\text{mol}}$

Question 6 on the pretest is an example of the type of problem that you may face. How much heat can you get from burning 4.80 kg of carbon, with all substances at 25°C? The reaction is $C(s) + O_2(g) \rightarrow CO_2(g)$. The standard heat of reaction is –394 kJ/mol.

As with normal stoichiometry problems, the first thing that you must

have is a _____ equation.

balanced

$$C(s) + O_2(g) - 394 \text{ kJ} = CO_2(g)$$

The question asked is "How many kilojoules (the most convenient unit

for heat)?" and the information given is "_____."

4.80 kg C

? kJ = _____

4.80 kg C

To use the bridge, you must convert kilograms C to _____. You can do this in two steps.

moles C

? kJ = 4.80 kg C × _____ × _____

$\dfrac{10^3 \text{ g C}}{\text{kg C}} ; \dfrac{\text{mol C}}{12.0 \text{ g C}}$

Now you can use the bridge conversion factor that relates the heat to moles C:

$$? \text{ kJ} = 4.80 \text{ kg C} \times \frac{10^3 \text{ g C}}{\text{kg C}} \times \frac{\text{mol C}}{12.0 \text{ g C}} \times \underline{\hspace{1cm}} = \underline{\hspace{1cm}}$$

$\dfrac{-394 \text{ kJ}}{\text{mol C}} ; -157{,}600 \text{ kJ}$

Written in scientific notation to the correct number of significant figures,

this is _____.

-1.58×10^5 kJ

You may have a problem in which this entire procedure is reversed. You know the amount of heat involved when a given weight of a reactant is used, and you are asked to determine ΔH^0. Since ΔH^0 is a con-

version factor, _____, you reword and ask "How many _____?" and

$\dfrac{\text{kJ}}{\text{mol}} ;$ kJ (kilojoules)

1 mol	take as your information given "_____." This is our usual way to solve for conversion factors. Here is an example of this type of problem: What is ΔH^0 for the reaction

$$PCl_5(g) = PCl_3(g) + Cl_2(g)$$

if 1.94 kJ are needed to decompose 4.17 g of PCl_5? Since the problem

| + (positive) | states that the heat is "required," the sign on the heat is _____. |

Starting with the reworded question and the information given, you have

| kJ; 1 mol PCl_5 | ? _____ = _____ |

The problem tells you that you need 1.94 kJ for 4.17 g PCl_5. Therefore

| grams | you must convert moles PCl_5 to _____ PCl_5 by using the molecular weight, 208.5 g PCl_5 per mole PCl_5: |

| $\dfrac{208.5 \text{ g } PCl_5}{\text{mol } PCl_5}$ | ? kJ = 1 mol PCl_5 × _____ |

Now you can use the conversion factor given in the problem, the number of kilojoules per gram PCl_5:

| $\dfrac{+1.94 \text{ kJ}}{4.17 \text{ g } PCl_5}$ | $? \text{ kJ} = 1 \cancel{\text{mol } PCl_5} \times \dfrac{208.5 \text{ g } PCl_5}{\cancel{\text{mol } PCl_5}} \times \underline{\qquad\qquad}$ |
| +97.0 kJ | = _____ (to the correct number of significant figures) |

And this is ΔH^0 for the reaction.

Adding Heats (Enthalpies) of Stepwise Reactions

In Section B, we introduced Hess's law for summing the heats of the steps that lead from starting conditions to final conditions. All the cases shown there were for only a single substance that changed temperature or phase. But Hess's law can be applied equally well to chemical reactions. In a way, this is the same thing as doing stoichiometry for consecutive reactions. If you don't remember this, refer to Section F, Chapter 11. There we covered two methods of doing this. In the first, we used two separate bridges from the two equations. In the second, we added the two consecutive reactions and then got the bridge from this composite equation. Only this last method works for thermodynamic equations because the heats (enthalpies) are additive and are not multiplied together.

Here is an example that will make this clear. Suppose that you want to find the amount of heat that you get from burning 4.80 kg of carbon in oxygen, to yield CO_2, but you know only the heats of reaction for the two reactions.

$$C(s) + \tfrac{1}{2}O_2(g) - 111 \text{ kJ} = CO(g)$$

$$CO(g) + \tfrac{1}{2}O_2(g) - 283 \text{ kJ} = CO_2(g)$$

You can add these two steps by adding all the terms on the left side of the equals sign and setting this equal to the sum of everything on the right side of the equals sign. This gives you

$$C(s) + \tfrac{1}{2}O_2(g) - 111 \text{ kJ} + \cancel{CO(g)} + \tfrac{1}{2}O_2(g) - 283 \text{ kJ}$$

$$= \cancel{CO(g)} + CO_2(g)$$

Since $CO(g)$ appears on both sides of the equals sign, it cancels, and two $\tfrac{1}{2}O_2(g)$ is the same as just one $O_2(g)$, so you have as the sum

$$C(s) + O_2(g) - 394 \text{ kJ} = CO_2(g)$$

This is exactly the same thermodynamic equation that you had before when you were given the heat of reaction. The sum of the steps that make up a reaction gives you the heat according to _____ law.

Now you can solve the problem exactly as before.

$$? \underline{\hspace{2cm}} = 4.80 \text{ kg C} \times \underline{\hspace{3cm}} \times \underline{\hspace{3cm}}$$

$$= 4.80 \,\cancel{\text{kg C}} \times \frac{10^3 \,\cancel{\text{g C}}}{\cancel{\text{kg C}}} \times \frac{\text{mol C}}{12.0 \,\cancel{\text{g C}}} \times \underline{\hspace{2cm}}$$

$$= \underline{\hspace{3cm}}$$

It may look strange to you to have balanced equations with $\tfrac{1}{2}O_2$. Up to this time, we have never used fractional coefficients to balance chemical equations. However, in equations that show the heat (*thermodynamic* equations), it's more convenient to have the heat expressed per one mole of the substance. For example, if you want to know the amount of heat given off when $HCl(g)$ is formed from its elements, you write the thermodynamic equation as

$$\tfrac{1}{2}H_2(g) + \tfrac{1}{2}Cl_2(g) - 92.3 \text{ kJ} = HCl(g)$$

The value -92.3 kJ represents the molar heat of formation for $HCl(g)$ at 25°C. You use the symbol ΔH_f^0 for *molar heats of formation*. We will soon see how very useful the values for ΔH_f^0 can be.

Sometimes when you use a series of steps to get the total reaction in which you are interested, you may find that you have to write one of the steps in reverse. For example, suppose that you are interested in the reaction

$$2Al(s) + Fe_2O_3(s) = 2Fe(s) + Al_2O_3(s)$$

The steps whose heats of reaction you know are:

Step 1: $2Al(s) + \tfrac{3}{2}O_2(g) - 400.5 \text{ kcal} = Al_2O_3(s)$

Step 2: $2Fe(s) + \tfrac{3}{2}O_2(g) - 196.5 \text{ kcal} = Fe_2O_3(s)$

(This time we'll work in calories rather than joules—just for practice.)

These two steps have all the substances that appear in the equation for the reaction. But notice that the $Fe_2O_3(s)$ appears on the left side in the statement of the equation, and on the right side in Step 2. If you are going to add Step 1 to Step 2 to get the equation that you want, you have to write Step 2 in reverse. Once you have done this, you can add them:

Step 1: $2Al(s) + \frac{3}{2}O_2(g) - 400.5 \text{ kcal} = Al_2O_3(s)$

Step 2: $Fe_2O_3(s) = -196.5 \text{ kcal} + \frac{3}{2}O_2(g) + 2Fe(s)$

Adding everything on the left, you have

$2Al(s) + \frac{3}{2}O_2(g) - 400.5 \text{ kcal} + Fe_2O_3(s)$

and adding everything on the right, you have

$Al_2O_3(s) - 196.5 \text{ kcal} + \frac{3}{2}O_2(g) + 2Fe(s)$

Since the $\frac{3}{2}O_2(g)$ appears on both the left and the right sides, it cancels, and you can write

$2Al(s) - 400.5 \text{ kcal} + Fe_2O_3(s) = Al_2O_3(s) - 196.5 \text{ kcal} + Fe(s)$

To get all the kilocalories on the left, you add +196.5 kcal to both sides, and you have

$2Al(s) + Fe_2O_3(s) - 400.5 \text{ kcal} + 196.5 \text{ kcal} = Al_2O_3(s) + Fe(s)$

If you algebraically add the −400.5 kcal and the +196.5 kcal, you have the final thermodynamic equation:

−204.0

$2Al(s) + Fe_2O_3(s) \rule{2cm}{0.4pt} \text{ kcal} = Al_2O_3(s) + 2Fe(s)$

You now know the heat (enthalpy) of the reaction for the initial reaction in which you were interested.

Let's try another of these with the reactions of C with O_2. Suppose that you want to know the heat of reaction for

$C(s) + \frac{1}{2}O_2(g) = CO(g)$

and the steps for which you have values are:

Step 1: $C(s) + O_2(g) - 394 \text{ kJ} = CO_2(g)$

Step 2: $CO(g) + \frac{1}{2}O_2(g) - 283 \text{ kJ} = CO_2(g)$

In the reaction in which you are interested, the $CO(g)$ is on the

right; left

$\rule{2cm}{0.4pt}$ side of the equals sign. In Step 2, the $CO(g)$ is on the $\rule{2cm}{0.4pt}$ side of the equals sign. Therefore, before you add the steps, you must

reverse

write Step 2 in its $\rule{2cm}{0.4pt}$ form.

Step 1: $C(s) + O_2(g) - 394 \text{ kJ} = CO_2(g)$

Step 2: $CO_2(g) = -283 \text{ kJ} + \frac{1}{2}O_2(g) + CO(g)$

Now you can add these steps:

$$C(s) + O_2(g) - 394 \text{ kJ} + CO_2(g)$$
$$= CO_2(g) - 283 \text{ kJ} + \underline{\hspace{2cm}} + CO(g)$$

$\frac{1}{2}O_2(g)$

The $CO_2(g)$ appears on both sides, so it \underline{\hspace{2cm}}. If you subtract

cancels

$\frac{1}{2}O_2(g)$ from both sides and add \underline{\hspace{1cm}} to both sides, what remains is

283 kJ

$$C(s) + \frac{1}{2}O_2(g) \underline{\hspace{2cm}} = CO(g)$$

−111 kJ

This is the thermodynamic equation in which you are interested.

From all this, you can see that to get the thermodynamic equations that allow you to solve a numerical problem, you may have to add several steps whose thermodynamic equations you know. And you can reverse any steps to get to the reaction of interest. You may even have to multiply some of the steps to give you the correct balanced equation. You may even need more than just two steps. Consider the following example.

The equation of interest is the burning of CH_4, methane:

$$CH_4(g) + 2O_2(g) = CO_2(g) + 2H_2O(l)$$

The only thermodynamic equations that you have are the heats of formation of the various compounds from their elements. Recall that we said earlier that these ΔH_f^0 values would be very useful.

Step 1: $C(s) + 2H_2(g) - 74.9 \text{ kJ} = CH_4(g)$

Step 2: $C(s) + O_2(g) - 394 \text{ kJ} = CO_2(g)$

Step 3: $H_2(g) + \frac{1}{2}O_2(g) - 286 \text{ kJ} = H_2O(l)$

When you compare these steps with the statement of the equation, you see that the CH_4 is on the left side of the equals sign in the equation of

interest, but on the \underline{\hspace{1cm}} side in Step 1. Therefore, when you add the

right

steps, you must \underline{\hspace{2cm}} the form of Step 1. Also, in the balanced

reverse

equation of interest there are \underline{\hspace{1cm}} H_2O's, but in Step 3, there is only

two

one. Consequently, you must multiply Step 3 by \underline{\hspace{0.3cm}} all the way across.

2

This means that the heat term will be \underline{\hspace{1cm}} kJ for two times Step 3.

−572

Now you can write the steps in the form that adds up to the equation of interest.

Step 1: $CH_4(g) = \underline{\hspace{2cm}} + 2H_2(g) + C(s)$

−74.9 kJ

Step 2: $C(s) + O_2(g) - 394 \text{ kJ} = CO_2(g)$

Step 3: $2H_2(g) + O_2(g) \underline{\hspace{1cm}} \text{ kJ} = 2H_2O(l)$

−572

Now you add all terms on the left side of the equals sign.

	Step 1: $CH_4(g) =$
-394	*Step* 2: $C(s) + O_2(g)$ _____ kJ $=$
-572	*Step* 3: $2H_2(g) + O_2(g)$ _____ kJ $=$
-966	$\quad CH_4(g) + C(s) + 2H_2(g) + 2O_2(g)$ _____ kJ $=$

Next you add all the terms on the right side of the equals sign.

-74.9	*Step* 1: $=$ _____ kJ $+ 2H_2(g) + C(s)$
	Step 2: $= CO_2(g)$
$2H_2O(l)$	*Step* 3: $=$ _____

$2H_2O(l)$	$= -74.9$ kJ $+ 2H_2(g) + C(s) + CO_2(g) +$ _____

As you can see, you have one $C(s)$ and two $H_2(g)$ on both the left and

cancel	the right side. Therefore they _____. Thus the total equation is

$$CH_4(g) + 2O_2(g) - 966 \text{ kJ}$$

-74.9	$=$ _____ kJ $+ CO_2(g) + 2H_2O(l)$

Adding $+74.9$ kJ to both sides of the equation brings all the kJ terms together. So as the final equation you have

-891	$CH_4(g) + 2O_2(g)$ _____ kJ $= CO_2(g) + 2H_2O(l)$

This is your equation of interest.

Using Heats (Enthalpies) of Formation, ΔH_f^0

As mentioned at the beginning of the last problem, each of the steps used to build up the total equation is simply the heat of formation of the compound from its elements. These standard heats (enthalpies) of formation have the abbreviation ΔH_f^0. You used the thermodynamic equation for the formation of the compounds on the left side of the equals sign, the *reactants*, as it stands. But for the compounds on the right side of the equals sign, the *products*, you had to reverse the equation for the heat of formation. The mathematical effect of reversing the thermodynamic equation is to change the sign of the heat. Also, when the compound in Step 3 appeared with a coefficient of 2 in the equation of interest [it was $2H_2O(l)$], you had to double the thermodynamic equation. This resulted in the heat being doubled for that step.

What all this means is that a very neat short cut is available if you know the heats of formation of all the compounds in an equation. All you have to do is write the equation and place the heats of formation beneath the compound. If there is a coefficient on a substance, you multiply the heat of formation by that coefficient. Then you add all the heats of formation of the products on the right side of the equals sign, and you subtract from this the sum of all the heats on the left side of the equals sign.

Let's look at this last example with the heats of formation written below each substance.

$$CH_4(g) \quad + \quad O_2(g) \quad = \quad CO_2(g) \quad + \quad 2H_2O(l)$$
$$(-74.9 \text{ kJ}) \qquad (0) \qquad (-394 \text{ kJ}) \qquad 2(-286 \text{ kJ})$$

Notice that the element oxygen has a heat of formation of zero. This entire procedure is based on the fact that the standard enthalpy of formation of all elements is zero.

Adding all the heats on the products side of the equation algebraically, you have

$-394 \text{ kJ} + (\underline{\hspace{1cm}}) \text{ kJ} = \underline{\hspace{1cm}} \text{ kJ}$ −572; −966

On the reactant side of the equation, you have simply

$-74.9 \text{ kJ} + (\underline{\hspace{0.5cm}}) \text{ kJ} = \underline{\hspace{1cm}} \text{ kJ}$ 0; −74.9

If you subtract the reactant sum from the product sum, you have

$-966 \text{ kJ} - (-74.9 \text{ kJ}) = \underline{\hspace{1cm}} \text{ kJ}$ −891

which is exactly the same value that we got before.

To give you a little more practice, let's look again at an earlier example.

$$2Al(s) + Fe_2O_3(s) = 2Fe(s) + Al_2O_3(s)$$

The heats of formation that we used were (recall that we worked this one in calories):

$$Fe_2O_3(s) \qquad \Delta H_f^0 = -196.5 \text{ kcal/mol}$$
$$Al_2O_3(s) \qquad \Delta H_f^0 = -400.5 \text{ kcal/mol}$$

You don't consider Al(s) or Fe(s) because they are _____ and   elements

their heats of formation are _____. Writing these values beneath the   zero
compounds in the equation, you have

$$2Al(s) \quad + \quad Fe_2O_3(s) \quad = \quad 2Fe(s) \quad + \quad Al_2O_3(s)$$
$$2(0) \text{ kcal} \quad (-196.5 \text{ kcal}) \quad 2(0) \text{ kcal} \quad (-400.5 \text{ kcal})$$

If you subtract the sum of the heats of formation of the reactants from that of the products, you have

$\underline{\hspace{1cm}} \text{ kcal} - (\underline{\hspace{1cm}}) \text{ kcal} = -204.0 \text{ kcal}$ −400.5; −196.5

This is the same value we got before. When you write the thermochemical

equation, be careful to put the heat on the reactant, or _____, side. So   left
you have

$$2Al(s) + Fe_2O_3(s) - 204.0 \text{ kcal} = 2Fe(s) + Al_2O_3(s)$$

Here is question 7 on the pretest. What is the heat of reaction for the process

$$FeS(s) + 2HCl(aq) = FeCl_2(aq) + H_2S(g)$$

The notation (*aq*) is an abbreviation for aqueous, which means "as a dilute solution in water." The heats of formation are FeS(*s*) = -95.1 kJ, HCl(*aq*) = -167 kJ/mol, FeCl$_2$(*aq*) = -422.6 kJ, and H$_2$S(*g*) = -20.1 kJ.

To get the heat of reaction for the process, you write the various heats

balanced

of formation underneath the compounds in the _____ equation:

$$\text{FeS}(s) \quad + \quad 2\text{HCl}(aq) \quad = \quad \text{FeCl}_2(aq) \quad + \quad \text{H}_2\text{S}(g)$$

2(-167 kJ)

(-95.1 kJ) _____ (-422.6 kJ) (-20.2 kJ)

add

Then you _____ the heats of formation on the right (product) side:

-442.8 kJ

-422.6 kJ + (-20.2 kJ) = _____

left

Next you add the heats of formation of the reactants on the _____ side of the equals sign:

-334; -429

-95.1 kJ + (_____) kJ = _____ kJ

subtract

Then you must _____ the sum of the heats of the reactants

products

from the sum of the heats of the _____.

-14

-442.8 kJ - (-429 kJ) = _____ kJ

You now know that the heat of this reaction is -14 kJ.

There are tables that give the standard heats (enthalpies) of formation of many compounds. Your textbook probably has a relatively short list containing most of the values that you will need for problems in this text. Nonetheless, you will be given the values in the problem set.

SECTION D Calorimetry

So far we have been talking about the heats that are given off and the heats that are consumed. But we never mentioned just how these heats were measured in the laboratory. Actually, it is quite simple. You just need to have the process in which you are interested occur in something whose specific heat is known. Water is used most often for this, but other liquids can be used just as well. If you know the amount of water and can measure how much the temperature changes as a result of the process occurring in the water, you can calculate the heat change of the water. Since heat is conserved in the same way that matter is conserved, the heat that the water has gained or lost must have come from your process having lost or gained an equivalent amount of heat. Thus:

Heat gained by water = heat lost by process

or

gained

Heat lost by water = heat _____ by process

The apparatus for measuring this change in temperature is called a *calorimeter*. It is important for the calorimeter body to be well insulated so that very little heat escapes from the calorimeter. There are many types of calorimeter, but the simplest version is just a styrofoam cup in which a thermometer is suspended.

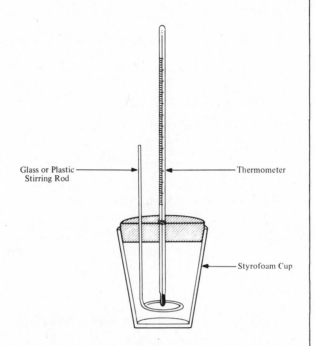

Glass or Plastic Stirring Rod

Thermometer

Styrofoam Cup

Figure 18.1   Calorimeter

To measure the amount of heat gained or lost, you place a weighed amount (or a known volume) of water in the calorimeter and record its temperature. Then you allow the process in which you are interested to take place in the water. After the process is complete (indicated by no further change in the temperature of the water), you record the final

temperature of the water. Subtracting the _____ temperature   initial

of the water from the _____ temperature gives you ΔT. Knowing the   final
weight of the water, its specific heat, and ΔT lets you calculate the
amount of heat either lost or gained by the calorimeter. We covered this
in Section A of this chapter.

The symbol Q is frequently used to represent the heat. The idea of balancing the change in the heat of the process with that in the calorimeter can be shown in an equation,

$$Q_{calor.} = -Q_{process}$$

where $Q_{calor.}$ is the heat change in the calorimeter and $Q_{process}$ is the heat change of the process. $Q_{process}$ has a negative sign because, if the process

loses heat, the calorimeter _____ an equivalent amount of heat.   gains

Question 8 on the pretest is a good example of how this works. A 10.0 g sample of aluminum is heated to 102.25°C and placed in a calorimeter containing 25.0 g water at 20.50°C. The temperature of the water rises to 27.25°C. What is the specific heat of aluminum? Assume that the body of the calorimeter does not absorb any heat.

To determine the heat change in the calorimeter, you must know three things: the weight of water, the specific heat of water, and ΔT, the temperature change of the water. The weight of water is given in the problem as _____. The specific heat of water is known to be

_____. So all you must do is calculate ΔT. You do this by subtracting _____ T_{init} from T_{final}:

$$\Delta T = T_____ - T_____$$

In this problem, the calculation is

$$\Delta T = _____°C - _____°C = _____°C$$

If you follow the methods you used in Section A of this chapter for determining the heat involved in a temperature change, you start by asking "How many _____?" and take as the information given the _____ of the substance. So in this problem you would start with

$$? J = _____$$

To cancel the unwanted dimension grams water, you use the

_____ of water:

$$? J = 25.0 \text{ g water} \times _____$$

Finally, to cancel the unwanted degrees Celsius, you multiply by _____, which is 6.75°C:

$$? J = 25.0 \text{ g water} \times \frac{4.184 \text{ J}}{\text{(g water)}(°C)} \times _____ = _____$$

The heat gained in the calorimeter, $Q_{calor.}$, must equal the heat lost by the aluminum, $Q_{process}$. But since the calorimeter gains heat, its joules have a __ sign. The aluminum must lose heat, so its joules have a __ sign:

$$Q_{calor.} = \_\_Q_{process}$$

So from the temperature change in the calorimeter, you have found that the number of joules the aluminum has lost is _____. To answer the question "What is the _____ heat of Al?" you must know three things: the number of joules, the _____ of the aluminum, and its _____. As in the preceding example, you can get ΔT by subtracting the _____ temperature from the _____ temperature. The initial temperature of the aluminum is given in the problem as _____°C, and

Margin answers (left column):

25.0 g

$\dfrac{4.184 \text{ J}}{\text{(g water)}(°C)}$

subtracting

final; init

27.25; 20.50; 6.75

joules

weight

25.0 g water

specific heat

$\dfrac{4.184 \text{ J}}{\text{(g water)}(°C)}$

ΔT

6.75°C; 706 J

+; −

−

−706 J

specific

weight

ΔT

initial; final

102.25

its final temperature is the same as that of the water in the calorimeter,

_____°C.

$$\Delta T = T\_\_\_\_ - T\_\_\_\_ = \_\_\_\_°C - \_\_\_\_°C$$

$$= _____°C$$

(Notice the minus sign, which shows that the temperature has decreased.)
   You now have all the information you need to get the specific heat of
the aluminum:

_____ J,   _____ g Al,   _____°C = ΔT

We can use the short-cut way of getting the specific heat, as we did in
Section A. We simply noted that the dimensions of specific heat are

$$\frac{\_\_\_\_}{(\_\_)(\_\_\_\_)}$$

and put in our values in such a way as to match these dimensions.
   Thus we have

$$\text{Specific heat} = \frac{_____}{(\_\_\_\_)(_____)} = _____$$

(Notice that the two minus signs canceled. The specific heat is always a
positive number.)

Calculating the Heat Capacity of the Calorimeter, K_c

Now we are going to introduce a new term, *heat capacity*. This is the
general term for the heat required to raise the temperature of a substance
1°C. If the heat capacity is expressed per gram of substance, it is called
specific heat. If it is expressed per mole of substance, it is called the
molar heat capacity. If the substance has only one possible weight, you
just say *heat capacity*. Since each calorimeter has a weight that doesn't
change (and also loses the same amount of heat to the surroundings), you
can simply say *heat capacity of the calorimeter*. Its value is a constant,
and the symbol K_c is used. It has as its dimensions

$$\frac{\text{joules}}{\text{degree Celsius}} \text{ or } \frac{J}{°C} \quad \text{or} \quad \frac{\text{calories}}{\text{degree Celsius}} \text{ or } \frac{\text{cal}}{°C}$$

   Therefore, if you want to know how much heat the calorimeter
absorbs and also loses to its surroundings, you multiply the value of K_c
by ΔT, the change in temperature of the calorimeter. This amount of
heat must be added to the heat gained or lost by the water in the
calorimeter to give the total change in heat of the calorimeter.

$$Q_{\text{calor.}} = Q_w + K_c \times \Delta T = Q_w + Q_c$$

The abbreviation Q_c represents the heat gained by the calorimeter itself,
and lost to the surroundings.

Margin answers

27.25

final; init; 27.25; 102.25

−75.00

−706; 10.0; −75.00

J

g; °C

−706 J

10.0 g; −75.00°C ; $\frac{0.941\ J}{(g)(°C)}$

Question 8 on the pretest told you to assume that the body of the calorimeter holding the water didn't absorb or give off any heat itself. Also, it implied that there was no heat lost to or gained from the surroundings. If you are working in the laboratory with a thermometer that allows you to read temperatures only to ±0.3°C, and you are using a styrofoam cup to hold the water, you can probably assume this. The styrofoam absorbs or loses heat very slowly, so it provides excellent insulation from the surroundings. Any differences caused by these changes would be too small to read on your thermometer. However, if you are doing precision work using a thermometer that reads to ±0.02°C and you are using a calorimeter made of metal, you have to worry about the heat gained or lost by the calorimeter itself. Then you'll have to use the equation shown at the bottom of the preceding page:

K_c; Q_w; Q_c

$$Q_{calor.} = Q_w + \underline{\quad\quad} \times \Delta T = \underline{\quad\quad} + \underline{\quad\quad}$$

Then, when you write the equation used in calorimetry,

$$Q_{calor.} = -Q_{process} \qquad \text{you have to write} \qquad Q_w + Q_c = -Q_{process}$$

Or, expressing Q_c in terms of the calorimeter constant K_c, you have

$K_c \times \Delta T$

$$Q_w + (\underline{\quad\quad\quad\quad}) = -Q_{process}$$

This last equation gives you a way to determine the K_c for a calorimeter. What you do is perform some process in the calorimeter whose heat is known. Thus you know $Q_{process}$. You measure the ΔT for the water in the calorimeter, and from the weight of the water and the specific heat of water, you calculate Q_w. Then, if you subtract Q_w from both sides of the equation,

Q_c

$$Q_w + Q_c = -Q_{process} \qquad \text{you have} \qquad \underline{\quad\quad} = -Q_{process} - Q_w$$

ΔT; dividing

Since $Q_c = \underline{\quad\quad} \times K_c$, you can calculate K_c by $\underline{\quad\quad\quad\quad}$ both

ΔT

sides of this equation by $\underline{\quad\quad}$:

$$\frac{Q_c}{\underline{\quad\quad}} = K_c$$

ΔT

It is adequate to assume that the ΔT for the calorimeter is identical to the ΔT for the water that it contains. Therefore you get the values for ΔT by subtracting the initial temperature of the water in the calorimeter from its final temperature.

Here is an example that shows how this is done experimentally. The process that produces a known amount of heat is the cooling down of a piece of aluminum. You can calculate the heat that this process produces

weight

if you know three things: the $\underline{\quad\quad\quad\quad}$ of aluminum, the specific

change

heat of aluminum, and the $\underline{\quad\quad\quad\quad}$ in its temperature.

In this example, the Al weighs 9.575 g, and its specific heat is now

$\dfrac{J}{(g\ Al)(^\circ C)}$

known to be 0.9412 $\underline{\quad\quad\quad\quad}$. Its initial temperature is 102.25°C.

It cools in the calorimeter until the temperature of the water in the calorimeter is 27.05°C. This last temperature is the final temperature of the aluminum, the water in the calorimeter, and the calorimeter body.

To get ΔT for the aluminum, you subtract its _____ temperature from its final temperature:

$$\Delta T = T_{final} - T\_\_\_\_\_ = \_\_\_\_\_°C - \_\_\_\_\_°C$$

$$= \_\_\_\_\_°C$$

From these three pieces of information, you can calculate the heat $Q_{process}$ given off by the aluminum.

$$? \, Q_{process} = ? \, J = \_\_\_\_\_ = 9.575 \text{ g Al} \times \_\_\_\_\_$$

$$= 9.575 \, \text{g Al} \times \frac{0.9412 \text{ J}}{(\text{g Al})(°C)} \times \_\_\_\_\_$$

$$= \_\_\_\_\_$$

Next you must determine Q_w, the heat absorbed by the water in the calorimeter. You need to know three things: the weight of the water, the _____ of water, and the ΔT of the water. In this example, the water weighs 25.00 g, and its specific heat is _____. The initial temperature is 20.79°C, and its final temperature is 27.05°C. The ΔT for water is

$$\Delta T = \_\_\_\_\_°C - \_\_\_\_\_°C = \_\_\_\_\_°C$$

With these three pieces of information, you can calculate Q_w:

$$? \, Q_w = ? \, J = 25.00 \text{ g water} \times \_\_\_\_\_$$

$$= 25.00 \, \text{g water} \times \frac{4.184 \text{ J}}{(\text{g water})(°C)} \times \_\_\_\_\_$$

$$= \_\_\_\_\_$$

You now have $Q_{process}$ and Q_w, and you can calculate Q_c from the equation

$$Q_c = -(Q_{process}) - Q_w = \_\_\_\_\_ \, J - \_\_\_\_\_ \, J = \_\_\_\_\_ \, J$$

[If you have forgotten the sign convention, it is that $-(-)$ and $+(+)$ both give +. But $-(+)$ or $+(-)$ both give $-$.]
  It is now possible to calculate K_c from the equation

$$K_c = \frac{Q_c}{\_\_\_\_\_}$$

The value for ΔT is the same as it was for the water in the calorimeter, which was _____. Putting in the values for Q_c and ΔT gives you

---

initial

initial; 27.05; 102.25

−75.20

10.00 g Al; $\dfrac{0.9012 \text{ J}}{(\text{g Al})(°C)}$

−75.20°C

−677.7 J

specific heat

$\dfrac{4.184 \text{ J}}{(\text{g water})(°C)}$

27.05; 20.79; 6.26

$\dfrac{4.184 \text{ J}}{(\text{g water})(°C)}$

6.26°C

654.8 J

−(−677.7); 654.8; +22.9

ΔT

6.26°C

$\dfrac{22.9 \text{ J}}{6.26°\text{C}}$; 3.66

$$K_c = \dfrac{\underline{\quad\quad}}{\underline{\quad\quad}} = \underline{\quad\quad}$$

(to the correct number of significant figures). Notice that the sign of K_c is +. Just as specific heat is always positive, any heat capacity is positive. Once you have determined K_c for a calorimeter, you can use it any time you are using that particular calorimeter.

Using the Heat Capacity of the Calorimeter in Calculations

Question 9 of the pretest is an example of how you can calculate the heat of a process using a calorimeter whose heat capacity is known: Suppose that you put 25.0 g of ice at 0.00°C into a calorimeter that contains 300.0 g of water at 21.16°C. The ice melts and lowers the temperature of the water in the calorimeter to 13.40°C. What is the heat of fusion for water, in kilojoules/mole? The heat capacity of the calorimeter, K_c, equals 3.3 J/°C.

As you did in pretest Question 8, you must determine the heat change

Q_c

of the calorimeter. But in this case, it will be the sum of Q_w and _____. In question 8, we assumed that Q_c was equal to zero. In this case, it is not.

Q_c

$$Q_{\text{calor.}} = Q_w + \underline{\quad\quad}$$

You have all the information that you need to calculate Q_w. You have

300.0

$\dfrac{4.184 \text{ J}}{\text{(g water)}(°\text{C})}$

initial; final

the weight of the water, _____ g, you know the specific heat of water,

_____, and you can calculate ΔT of the water by subtracting

the _____ temperature of the water from the _____ temperature of the water.

13.40; 21.16; −7.76

$$\Delta T = \underline{\quad\quad}°\text{C} - \underline{\quad\quad}°\text{C} = \underline{\quad\quad}°\text{C}$$

Thus

$\dfrac{4.184 \text{ J}}{\text{(g water)}(°\text{C})}$

−7.76°C

$$? Q_w = ? \text{ J} = 300.0 \text{ g water} \times \underline{\quad\quad}$$

$$= 300.0 \text{ g water} \times \dfrac{4.184 \text{ J}}{\text{(g water)}(°\text{C})} \times \underline{\quad}$$

−9740

$$= \underline{\quad\quad} \text{ J}$$

(We'll keep an extra figure until the end.) To get Q_c, you use the heat capacity of the calorimeter

$$? Q_c = ? \text{ J} = \dfrac{3.3 \text{ J}}{°\text{C}}$$

To cancel the unwanted degrees Celsius, you use the ΔT for the calorimeter, which is the same as the ΔT for the water that it contains,

−7.76

_____°C.

$$Q_c = \frac{3.3\ J}{^\circ C} \times \underline{\hspace{3cm}} = \underline{\hspace{2cm}}$$

You now have Q_w and Q_c, so you can get $Q_{calor.}$.

$$Q_{calor.} = Q_w - Q_c = \underline{\hspace{2cm}} J + \underline{\hspace{1.5cm}} J$$

$$= \underline{\hspace{2cm}} J$$

Once you know the heat given off by the calorimeter, you can tell the heat picked up by the process.

$$Q_{calor.} = \underline{\hspace{3cm}} \quad or \quad \underline{\hspace{3cm}} = Q_{process}$$

The process here is one of the two-step processes that we discussed at the end of Section B. Here you first change the phase of the substance (ice to water), and then you raise its temperature. By Hess's law, the heat for the total process is the _____ of the heats of the individual steps.

$$Q_{process} = Q_{fusion} + Q_{temp.\ change}$$

You already know $Q_{process}$, and it's easy to calculate $Q_{temp.\ change}$ for water. So to get Q_{fusion}, you subtract _____ from $Q_{process}$:

$$Q_{process} - \underline{\hspace{2.5cm}} = Q_{fusion}$$

To calculate the heat involved in the temperature change, you need three pieces of information: the weight of the water that has come from the ice, _____ g, the specific heat of water, and the _____ of the water. The ice, which has liquefied, is at _____°C. Its temperature rises to the final temperature of the calorimeter, _____°C. So

$$\Delta T = \underline{\hspace{1.5cm}}^\circ C - \underline{\hspace{1.5cm}}^\circ C = \underline{\hspace{1.5cm}}^\circ C$$

Therefore

$$? Q_{temp.\ change} = ?\ J = 25.0\ g\ water \times \underline{\hspace{2.5cm}}$$

$$= 25.0\ g\ water \times \frac{4.184\ J}{(g\ water)(^\circ C)} \times \underline{\hspace{1.5cm}}$$

$$= \underline{\hspace{1.5cm}} J$$

Now you can calculate Q_{fusion}, the heat from the melting of ice at 0°.

$$Q_{fusion} = Q_{process} - \underline{\hspace{2.5cm}}$$

$$= \underline{\hspace{1.5cm}} J - \underline{\hspace{1.5cm}} J = \underline{\hspace{2cm}}$$

But the value you have calculated is the heat required to melt 25.0 g of ice at 0.00°C. The question asked for the value per mole. This is a simple dimensional-analysis problem.

$$? J = \underline{\hspace{1.5cm}}$$

Margin answers:

-7.76°C; -26 J

+; -9740; -26

-9766

-$Q_{process}$; +9766 J

sum

$Q_{temp.\ change}$

$Q_{temp.\ change}$

25.0; ΔT

0.00

13.40

13.40; 0.00; 13.40

$\frac{4.184\ J}{(g\ water)(^\circ C)}$

13.40°C

1402

$Q_{temp.\ change}$

9766; 1402; 8364 J

1 mol

molecular	You can use the _____ weight to get from moles to grams:
$\dfrac{18.0 \text{ g water}}{\text{mol water}}$	? J = 1 mol water X _____
	And the preceding calculation has given you a conversion factor that
$\dfrac{8364 \text{ J}}{25.0 \text{ g water}}$	relates joules to grams of water, _____.
$\dfrac{8364 \text{ J}}{25.0 \text{ g water}}$	? J = 1 ~~mol water~~ X $\dfrac{18.0 \text{ g water}}{\text{~~mol water~~}}$ X _____
6.022; 6.02	= _____ J = _____ kJ for 1 mol or 6.02 kJ/mol

(It is best to use kilojoules and get the correct number of significant figures.)

One word of caution about calorimetry. We have had only pure water as the liquid in our calorimeter, and we know its specific heat. However, if you have some process that occurs in the calorimeter that involves a substance that dissolves in water, the specific heat of the solution formed will be different from that of pure water. When this comes up, you will be told the specific heat of the solution.

PROBLEM SET

The first 14 problems are divided into groups on the basis of the section of the chapter in which the material is covered. The last 10 problems are a miscellaneous collection and you will have to decide into what category they fall. The more difficult problems are marked with an asterisk (*). The values for the standard enthalpies of formation are given in the problems.

Section A

1. How many calories are needed to raise the temperature of exactly 500 g of water from 22.15°C to 24.70°C?

2. How many joules are released when 30.0 mL of chloroform cool 18.0°C? The specific heat of chloroform is 0.971 J/(g)(°C) and its density is 1.50 g/mL.

3. The specific heat of benzene is 1.83 J/(g)(°C). Suppose that you have a 20.0 g sample of benzene at 45.0°C and remove 1.463 kJ. What does its temperature become?

4. Determine the specific heat of Cu from the fact that 64.0 J are needed to raise the temperature of 15.0 g of Cu metal from 22.0°C to 33.0°C.

Section B

5. How many kilocalories are used to vaporize 85.0 g of liquid water at 100°C to steam, also at 100°C? The molar heat of vaporization of water is 9.71 kcal/mole.

6. Liquid water at 25.0°C is converted to steam at 100.0°C. How many kilojoules are needed if the sample of water weighs 5.00 g and the molar heat of vaporization of water is 40.6 kJ/mole?

*7. How many joules are required to convert 10.0 g of solid ethyl alcohol at −180.3°C to its vapor at its boiling point, 78.3°C? The specific heat for solid ethyl alcohol is 0.971 J/(g)(°C). For liquid ethyl alcohol, the specific heat is 2.30 J/(g)(°C). The melting point of the alcohol is −117.3°C. The heat of fusion is 218 J/g and the heat vaporization is 854 J/g.

Section C

8. The heat of combustion for ethene, C_2H_4, is −1413 kJ/mol for the reaction

$$C_2H_4(g) + 3O_2(g) = 2CO_2(g) + 2H_2O(l)$$

How many kilojoules are supplied by burning 450 g of C_2H_4?

9. Use the heat of combustion given in Problem 8 and the ΔH_f^0 values $CO_2(g) = -394$ kJ/mol and $H_2O(l) = -286$ kJ/mol to determine the standard enthalpy of formation for ethene, C_2H_4.

10. Determine the heat of reaction for

$$CuCl(s) + \tfrac{1}{2}Cl_2(g) = CuCl_2(s)$$

by using the following two reactions:

$$Cu(s) + Cl_2(g) - 206 \text{ kJ} = CuCl_2(s) \qquad Cu(s) + \tfrac{1}{2}Cl_2(g) - 136 \text{ kJ} = CuCl(s)$$

*11. Determine the heat (enthalpy) of reaction for

$$3NO_2(g) + H_2O(l) = 2HNO_3(aq) + NO(g)$$

The standard enthalpies of formation, ΔH_f^0, that you will need are $H_2O(l) = -286$ kJ/mol, $NO(g) = +90.4$ kJ/mol, $NO_2(g) = +33.9$ kJ/mol, and $HNO_3(aq) = -207$ kJ/mol.

Section D

12. How much heat is generated by a process occurring in a calorimeter when the calorimeter contains 50.0 g of water at 22.5°C before the process occurs and the temperature rises to 31.0°C? The heat capacity of the calorimeter is 6.3 J/°C. (Be careful with the sign of your answer.)

13. Suppose that you added 35.0 g of water at 99.50°C to a calorimeter containing 150.0 g of water at 23.00°C and the temperature of the calorimeter and its contents rose to 35.60°C. What is the heat capacity of the calorimeter, expressed in joules/°C?

14. A 50.3 g piece of a metal at 100.0°C is put into a calorimeter containing 150.0 g of water at 23.0°C. The temperature of the water in the calorimeter rises to 24.8°C. What is the specific heat of the metal if the heat capacity of the calorimeter K_c is 6.5 cal/°C? (For a change, do this problem in calories.)

Miscellaneous

15. The body consumes sucrose (cane sugar) to produce energy in the form of heat. The reaction is

$$C_{12}H_{22}O_{11}(s) + 12\,O_2(g) - 5640\text{ kJ} = 12CO_2(g) + 11H_2O(l)$$

How many kilojoules can be gotten from a teaspoon of sugar? There are 80 teaspoons per pound and 454 g per pound. To save you a little time, the molecular weight of sucrose is 342 g/mol.

16. Calculate ΔH_f^0 for sucrose using the data from Problem 15 and the standard enthalpies of formation -394 kJ/mol for $CO_2(g)$ and -286 kJ/mol for $H_2O(l)$.

*17. A chemist determines the heat capacity K_c for a calorimeter by using, as the process that produces a known amount of heat, the reaction

$$HCl(aq) + NaOH(aq) - 57.3\text{ kJ} = NaCl(aq) + H_2O(l)$$

What the chemist does is to add 50.0 mL of a 2.00 M HCl solution at 22.0°C to the calorimeter, then add 50.0 mL of a 2.00 M NaOH solution, also at 22.0°C. The temperature of the 100.0 mL of NaCl solution in the calorimeter rises to 35.0°C. The density of the NaCl solution is 1.04 g/mL and its specific heat is 3.93 J/(g)(°C). What is K_c for the calorimeter?

18. Calculate the quantity of heat required to change 4.13×10^{-3} mol of solid aluminum (specific heat = 0.224 cal/g°C) from 600.0°C to the liquid state at its melting point, 660.2°C. The heat of fusion for Al is 95.2 cal/g. (Work in calories for a change.)

19. What is the heat of combustion (burning with oxygen) for liquid ethyl alcohol, C_2H_5OH? The products are $CO_2(g)$ and $H_2O(l)$. The standard enthalpies of formation are -235 kJ/mol for $C_2H_5OH(l)$, -394 kJ/mol for $CO_2(g)$, and -286 kJ/mol for $H_2O(l)$.

20. Sulfur occurs in two crystalline forms: rhombic sulfur, S_r, and monoclinic sulfur, S_m. What is the standard enthalpy (heat) of transition from S_r to S_m if the heat of combustion of S_r is -296.6 kJ/mol and for S_m is 296.9 kJ/mol? Both forms produce $SO_2(g)$.

*21. Determine the enthalpy of reaction for the partial oxidation of ethyl alcohol

$$2C_2H_5OH(l) + O_2(g) = 2C_2H_4O(l) + 2H_2O(l)$$

from the two reactions

$$C_2H_5OH(l) + 3\,O_2(g) - 3168\text{ kJ} = 2CO_2(g) + 3H_2O(l)$$

and

$$C_2H_4O(l) + \tfrac{5}{2}O_2(g) - 1167\text{ kJ} = 2CO_2(g) + 2H_2O(l)$$

22. The specific heat of liquid ammonia is 4.38 J/(g)(°C). Calculate the amount of heat needed to raise the temperature of 3.50 mol NH_3 from -78.0°C to its boiling point, -33.5°C.

23. The usual method of determining the standard enthalpy of formation for organic compounds is to determine their *heat of combustion* (reaction

with oxygen). The products are CO_2, H_2O, and—if N is in the organic compound—NO_2. What is the ΔH_f^0 for nitrobenzene, $C_6H_5NO_2$, if its heat of combustion is -3093 kJ/mole? The balanced equation for the combustion reaction is

$$4C_6H_5NO_2\,(l) + 29\,O_2\,(g) = 24CO_2\,(g) + 4NO_2\,(g) + 10H_2O\,(l)$$

and the ΔH_f^0 values you will need are $CO_2\,(g) = -394$ kJ/mol, $NO_2\,(g) = +33.9$ kJ/mol, and $H_2O(l) = -286$ kJ/mol.

24. A student determines the specific heat of a metal using a calorimeter like that shown in Figure 18.1. The student heats a weighed sample of the metal and then places it in the calorimeter containing a known amount of water at a known temperature. The student records the rise in temperature of the water in the calorimeter, and from this determines the specific heat of the metal. The K_c value for the calorimeter is known to be 3.5 J/°C. The data are as follows.

Weight of metal sample	19.32 g
Initial temperature of metal	92.0°C
Weight of water in calorimeter	50.50 g
Initial temperature of water	24.6°C
Final temperature of water and metal	27.3°C
K_c for calorimeter	3.5 J/°C

What is the specific heat of the metal in joules/(g)(°C)?

25. How many joules are released when one mole of acetylene, C_2H_2, is *hydrogenated* (adds H_2)? The reaction is

$$C_2H_2\,(g) + 2H_2\,(g) = C_2H_6\,(g)$$

and the heats of combustion for all compounds are known.

$$C_2H_2\,(g) + \tfrac{5}{2}O_2\,(g) - 1300 \text{ kJ} = 2CO_2\,(g) + H_2O(l)$$

$$H_2\,(g) + \tfrac{1}{2}O_2\,(g) - 286 \text{ kJ} = H_2O(l)$$

$$C_2H_6\,(g) + \tfrac{7}{2}O_2\,(g) - 1560 \text{ kJ} = 2CO_2\,(g) + 3H_2O(l)$$

PROBLEM SET ANSWERS

1. ? cal $= 500$ ~~g water~~ $\times \dfrac{1 \text{ cal}}{\text{(g water)}(°C)} \times 2.55°\cancel{C} = 1275$ cal $= 1.28 \times 10^3$ cal

   $[\Delta T = 24.70°C - 22.15°C = 2.55°C]$

2. ? J $= 30.0$ ~~mL $CHCl_3$~~ $\times \dfrac{1.50 \text{ g } \cancel{CHCl_3}}{\text{mL } \cancel{CHCl_3}} \times \dfrac{0.971 \text{ J}}{(\text{g } \cancel{CHCl_3})(°C)} \times (-18.0°C) = -786$ J

3. ? ΔT $=$? °C $= -1463$ J $\times \dfrac{(\text{g } \cancel{benzene})(°C)}{1.83 \text{ J}} \times \dfrac{1}{20.0 \text{ g } \cancel{benzene}} = -40.0°C$

   $T_{final} = T_{init} + \Delta T = 45°C - 40.0°C = 5.0°C$

4. Specific heat $= \dfrac{?\text{ cal}}{(\text{gram Cu})(^{\circ}\text{C})} = \dfrac{64.0\text{ J}}{(15.0\text{ g Cu})(11.0^{\circ}\text{C})} = \dfrac{0.388\text{ J}}{(\text{gram Cu})(^{\circ}\text{C})}$

$[\Delta T = 33.0^{\circ}\text{C} - 22.0^{\circ}\text{C} = 11.0^{\circ}\text{C}]$

5. $?\text{ kcal} = 85.0 \text{ g water} \times \dfrac{\text{mole water}}{18.0 \text{ g water}} \times \dfrac{9.71\text{ kcal}}{\text{mole water}} = 45.9\text{ kcal}$

6. This is a two-step process. First water must go from 25.0°C to 100.0°C, so

$\Delta T = 75.0^{\circ}\text{C}$, 5.00 g water, $\dfrac{4.184\text{ J}}{(\text{gram water})(^{\circ}\text{C})}$

$?\text{ J} = 5.00 \text{ g water} \times \dfrac{4.184\text{ J}}{(\text{g water})(^{\circ}\text{C})} \times 75.0^{\circ}\text{C} = 1569\text{ J}$

Then the water must be vaporized.

$?\text{ J} = 5.00 \text{ g water} \times \dfrac{\text{mole water}}{18.0 \text{ g water}} \times \dfrac{40.6 \times 10^{3}\text{ J}}{\text{mole water}} = 11{,}278\text{ J}$

The total heat is the sum of the steps.

$?\text{ J} = 1569\text{ J} + 11{,}278\text{ J} = 12{,}847\text{ J} = 12.8\text{ kJ}$

(to the correct number of significant figures)

7. This is a four-step process. Step 1: The solid must be warmed to its melting point. Step 2: The solid must be converted to a liquid. Step 3: The liquid must be warmed to its boiling point. Step 4: The liquid must be converted to a gas.

Step 1.　　10.0 g alc; $\dfrac{0.971\text{ J}}{(\text{gram alc})(^{\circ}\text{C})}$;

$\Delta T = -180.3^{\circ}\text{C} - (-117.3^{\circ}\text{C}) = 63.0^{\circ}\text{C}$

$?\text{ J} = 10.0 \text{ g alc} \times \dfrac{0.971\text{ J}}{(\text{gram alc})(^{\circ}\text{C})} \times 63.0^{\circ}\text{C} = 612\text{ J}$

Step 2.　　$?\text{ J} = 10.0 \text{ g alc} \times \dfrac{218\text{ J}}{\text{gram alc}} = 2180\text{ J}$

Step 3.　　10.0 g alc; $\dfrac{2.30\text{ J}}{(\text{gram alc})(^{\circ}\text{C})}$;

$\Delta T = 78.3^{\circ}\text{C} - (-117.3^{\circ}\text{C}) = 195.6^{\circ}\text{C}$

$?\text{ J} = 10.0 \text{ g alc} \times \dfrac{2.30\text{ J}}{(\text{gram alc})(^{\circ}\text{C})} \times 195.6^{\circ}\text{C} = 4499\text{ J}$

Step 4.　　$?\text{ J} = 10.0 \text{ g alc} \times \dfrac{854\text{ J}}{\text{gram alc}} = 8540\text{ J}$

The total heat is the sum of the heats of the individual steps.

$?\text{ J} = 612\text{ J} + 2180\text{ J} + 4499\text{ J} + 8540\text{ J} = 15{,}831\text{ J}$

$= 15.8\text{ kJ}$ (to the correct significant figures)

8. $? \text{ kJ} = 450 \, \cancel{\text{g } C_2H_4} \times \dfrac{\cancel{\text{mole } C_2H_4}}{28.0 \, \cancel{\text{g } C_2H_4}} \times \dfrac{(-1413 \text{ kJ})}{\cancel{\text{mole } C_2H_4}} = -2.27 \times 10^4 \text{ kJ}$

9. $C_2H_4 \ + \quad 3\,O_2 \quad - 1413 \text{ kJ} = \quad 2CO_2 \quad + \quad 2H_2O$

    (? kJ)   3(0.00 kJ)   2(−394 kJ)   2(−286 kJ)

Sum of heats of formation of products (right side):

    2(−394 kJ) + 2(−286) = −1360 kJ

Sum of heats of formation of reactants (left side):

    ? kJ + 3(0.00 kJ) − 1413 kJ = (? kJ) − 1413 kJ

The sums on both sides of the equation must be equal:

    ? kJ − 1413 kJ = −1360 kJ

So

    ? kJ = −1360 kJ + 1413 kJ = +53 kJ

10. You must reverse the second equation so that the sum of the equation gives you the reaction you are interested in:

    $Cu(s) + Cl_2(g) - 206 \text{ kJ} = CuCl_2(s)$

                 $CuCl(s) = -136 \text{ kJ} + \frac{1}{2}Cl_2(g) + Cu(s)$

Adding these gives

    $Cu(s) + Cl_2(g) - 206 \text{ kJ} + CuCl(s) = CuCl_2(s) - 136 \text{ kJ} + \frac{1}{2}Cl_2(g) + Cu(s)$

The Cu(s) cancels from both sides. If you subtract $\frac{1}{2}Cl_2(g)$ and −136 kJ from both sides, you have

    $CuCl(s) + \frac{1}{2}Cl_2(g) - 70 \text{ kJ} = CuCl_2(s)$

The heat of the reaction is −70 kJ.

11. You put all the standard heats of formation in the equation for the reaction:

    $3NO_2(g) \ + \quad H_2O(l) \quad = \quad 2HNO_3(aq) \ + \quad NO(g)$

    3(33.9 kJ)   (−286 kJ)   2(−207 kJ)   (90.4 kJ)

The sum of heats of formation for the products is −324 kJ. The sum of heats of formation for the reactants is −184 kJ. Subtracting the sum for the reactants from the sum for the products gives the heat of reaction:

    −324 kJ − (−184 kJ) = −140 kJ

12. The data you need are

$$50.0 \text{ g water, } \frac{4.184 \text{ J}}{(\text{g water})(°\text{C})} \text{ and } \Delta T = 31.0°\text{C} - 22.5°\text{C} = 8.5°\text{C}$$

The heat gained by the calorimeter is the sum of the heat gained by the water Q_w and the heat gained by the body of the calorimeter Q_c:

$$? \, Q_w = 50.0 \text{ g water} \times \frac{4.184 \text{ J}}{(\text{g water})(°\text{C})} \times 8.5°\text{C} = 1778 \text{ J}$$

$$? \, Q_c = 8.5°\text{C} \times \frac{6.3 \text{ J}}{°\text{C}} = 54 \text{ J}$$

The total heat gained by the calorimeter = 1778 J + 54 J = 1832 J. Or to the correct significant figures, 1.8 kJ. Since $Q_{calor.} = -Q_{process}$, the $Q_{process}$ equals −1.8 kJ.

13. The process that is occurring in the calorimeter is the cooling of 35.0 g of water from 99.50°C to 35.60°C. Therefore

$$\Delta T = 35.60°\text{C} - 99.50°\text{C} = -63.90°\text{C}$$

$$? \, Q_{process} = ? \text{ J} = 35.0 \text{ g water} \times \frac{4.184 \text{ J}}{(\text{g water})(°\text{C})} \times (-63.90°\text{C}) = -9358 \text{ J}$$

You calculate the heat picked up by the water in the calorimeter, Q_w, from its weight, 150.0 g, and its $\Delta T = 35.60°\text{C} - 23.0°\text{C} = 12.60°\text{C}$. Thus

$$Q_w = ? \text{ J} = 150.0 \text{ g water} \times \frac{4.184 \text{ J}}{(\text{g water})(°\text{C})} \times 12.60°\text{C} = 7908 \text{ J}$$

$$Q_w + Q_c = -Q_{process}, \text{ so } 7908 \text{ J} + Q_c = 9358 \text{ J}$$

Or

$$Q_c = 1450 \text{ J} \qquad K_c = \frac{? \text{ J}}{°\text{C}} = \frac{1450 \text{ J}}{12.60°\text{C}} = 115 \text{ J/}°\text{C}$$

14. First you must determine the heat gained by the calorimeter. This is the sum of Q_w and Q_c. The ΔT is 24.8°C − 23.0°C = 1.8°C.

$$? \, Q_w = ? \text{ cal} = 150.0 \text{ g water} \times \frac{1 \text{ cal}}{(\text{gram water})(°\text{C})} \times 1.8°\text{C} = 270 \text{ cal}$$

$$? \, Q_c = ? \text{ cal} = 1.8°\text{C} \times \frac{6.5 \text{ cal}}{°\text{C}} = 12 \text{ cal}$$

$$Q_{calor.} = Q_w + Q_c = 270 \text{ cal} + 12 \text{ cal} = 282 \text{ cal}$$

Then you know that $Q_{calor.} = -Q_{process}$, so −282 cal = $Q_{process}$. To get the specific heat, you must know the weight of the substance, the heat of the process, and $\Delta T = 24.8°\text{C} - 100.0°\text{C} = -75.2°\text{C}$.

$$\text{Specific heat} = \frac{? \text{ cal}}{(\text{gram metal})(°\text{C})} = \frac{-282 \text{ cal}}{(50.3 \text{ g metal})(-75.2°\text{C})}$$

$$= \frac{0.0746 \text{ cal}}{(\text{gram metal})(°\text{C})}$$

15. $? kJ = 1 \text{ tsp} \times \dfrac{\text{pound sucrose}}{80 \text{ tsp}} \times \dfrac{454 \text{ g sucrose}}{\text{pound sucrose}} \times \dfrac{\text{mole sucrose}}{342 \text{ g sucrose}}$

$\times \dfrac{5640 \text{ kJ}}{\text{mole sucrose}} = 93.6 \text{ kJ}$

16. $C_{12}H_{22}O_{11}(s) + \quad 12\,O_2(g) \quad - 5640 \text{ kJ} = \quad 12CO_2(g) \quad + \quad 11H_2O(l)$

$\quad$ (? kJ) $\qquad\quad$ 12(0.00 kJ) $\qquad\qquad\qquad\quad$ 12(-394 kJ) $\quad$ 11(-286 kJ)

The sum of the heats of formation on the product side is

$\qquad$ 12(-394) + 11(-286) = -7874 kJ

The sum of the heats on the left side is (? kJ) - 5640 kJ, so

$\qquad$? kJ - 5640 kJ = -7874 kJ $\quad$ or $\quad$? kJ = -7874 kJ + 5640 kJ = 2234 kJ

17. First you must determine the number of moles that reacted:

$\qquad ? \text{ mol HCl} = 0.0500 \text{ L}_{H^+} \times \dfrac{2.00 \text{ mol HCl}}{\text{L}_{H^+}} = 0.100 \text{ mol HCl}$

$\qquad ? \text{ mol NaOH} = 0.0500 \text{ L}_{OH^-} \times \dfrac{2.00 \text{ mol NaOH}}{\text{L}_{OH^-}} = 0.100 \text{ mol NaOH}$

Since the number of moles of both reactants is the same, you can use either to calculate the heat given off by the process, $Q_{process}$:

$\qquad ? Q_{process} = ? \text{ kJ} = 0.100 \text{ mol HCl} \times \dfrac{-57.3 \text{ kJ}}{\text{mol HCl}} \times \dfrac{10^3 \text{ J}}{\text{kJ}} = -5730 \text{ J}$

Then you determine the heat picked up by the NaCl solution in the calorimeter, $Q_{NaCl \, soln}$. The ΔT is $35.0°C - 22.0°C = 13.0°C$.

$\qquad ? Q_{NaCl \, soln} = ? \text{ J} = 100.0 \text{ mL}_{NaCl \, soln} \times \dfrac{1.04 \text{ g}_{NaCl \, soln}}{\text{mL}_{NaCl \, soln}} \times \dfrac{3.93 \text{ J}}{(\text{g}_{NaCl \, soln})(°C)}$

$\qquad\qquad \times 13.0°C = 5313 \text{ J}$

Then

$\qquad Q_{NaCl \, soln} + Q_c = -Q_{process} \quad$ and $\quad 5313 \text{ J} + Q_c = 5730 \text{ J} \quad$ or $\quad Q_c = 417 \text{ J}$

$\qquad K_c = \dfrac{? \text{ J}}{°C} = \dfrac{417 \text{ J}}{13.0°C} = 32 \text{ J/°C}$

18. This is a two-step process. First the solid Al must be heated to its melting point. The $\Delta T = 660.2°C - 600.0°C = 60.2°C$.

$\qquad ? \text{ cal} = 4.13 \times 10^{-3} \text{ mol Al} \times \dfrac{27.0 \text{ g Al}}{\text{mol Al}} \times \dfrac{0.224 \text{ cal}}{(\text{g Al})(°C)} \times 60.2°C = 1.50 \text{ cal}$

The second step is to determine the heat required to melt the Al.

$\qquad ? \text{ cal} = 4.13 \times 10^{-3} \text{ mol Al} \times \dfrac{27.0 \text{ g Al}}{\text{mol Al}} \times \dfrac{95.2 \text{ cal}}{\text{g Al}} = 10.62 \text{ cal}$

The heat required is the sum of the two steps:

$\qquad ? \text{ cal} = 1.50 \text{ cal} + 10.62 \text{ cal} = 12.12 \text{ cal}$

19. The balanced equation for the combustion of C_2H_5OH (with the ΔH_f^0 values shown below the compounds) is

$$C_2H_5OH(l) + 3O_2(g) = 2CO_2(g) + 3H_2O(l)$$
$$(-278 \text{ kJ}) \quad 3(0.00 \text{ kJ}) \quad 2(-394 \text{ kJ}) \quad 3(-286 \text{ kJ})$$

The sum of the ΔH_f^0 values for the products is

$$2(-394 \text{ kJ}) + 3(-286 \text{ kJ}) = -1646 \text{ kJ}$$

The sum for the reactants is

$$(-278 \text{ kJ}) + 0.00 \text{ kJ} = -278 \text{ kJ}$$

The heat of reaction is the $\Delta H_{products} - \Delta H_{reactants}$.

$$-1646 \text{ kJ} - (-278 \text{ kJ}) = -1368 \text{ kJ}$$

20. The reaction you are interested in is $S_r \rightarrow S_m$, so you can write the two combustion reactions with the one relating to S_m backward:

$$S_r + O_2(g) - 296.6 \text{ kJ} = SO_2(g)$$
$$SO_2(g) = -296.9 \text{ kJ} + O_2(g) + S_m$$

Adding these, you have

$$S_r - 296.6 \text{ kJ} = -296.9 \text{ kJ} + S_m \quad \text{or} \quad S_r + 0.3 \text{ kJ} = S_m$$

21. You have to double the first reaction,

$$2C_2H_5OH + 6O_2 + 2(-3168 \text{ kJ}) = 4CO_2 + 6H_2O$$

and double and reverse the direction of the second reaction:

$$4H_2O + 4CO_2 = 2(-1167 \text{ kJ}) + 5O_2 + 2C_2H_4O$$

so that adding will give the desired reaction. Written out completely, the added reactions are

$$2C_2H_5OH + 6O_2 - 6336 \text{ kJ} + 4H_2O + 4CO_2$$
$$= 4CO_2 + 6H_2O - 2334 \text{ kJ} + 5O_2 + 2C_2H_4O$$

Eliminating and simplifying the compounds that appear on both sides gives

$$2C_2H_5OH + O_2 - 6336 \text{ kJ} = 2H_2O + 2C_2H_4O - 2334 \text{ kJ}$$

And getting all the kilojoule terms together (using the order of reactants and products in the problem):

$$2C_2H_5OH + O_2 - 4002 \text{ kJ} = 2C_2H_4O + 2H_2O$$

Thus the enthalpy for the reaction is -4002 kJ, or 4.00×10^3 kJ to the correct significant figures.

22. $? \text{ J} = 3.50 \,\cancel{\text{mol NH}_3} \times \dfrac{17.0 \,\cancel{\text{g NH}_3}}{\cancel{\text{mol NH}_3}} \times \dfrac{4.38 \text{ J}}{(\cancel{\text{g NH}_3})(^\circ\cancel{C})} \times 44.5\,^\circ\cancel{C}$

$= 11.6$ kJ (to the correct significant figures)

$\Delta T = -33.5^\circ C - (-78.0^\circ C) = 44.5^\circ C$

23. First it is best to write the reaction for one mole of $C_6H_5NO_2$ and then put the ΔH_f^0 values for all other compounds beneath the formulas:

$$C_6H_5NO_2 + \tfrac{29}{4}O_2 - 3093 \text{ kJ} = 6CO_2 + NO_2 + 2\tfrac{1}{2}H_2O$$

$$\text{(? kJ)} \quad \tfrac{29}{4}(0.00 \text{ kJ}) \qquad 6(-394 \text{ kJ}) \quad (+33.9 \text{ kJ}) \quad 2\tfrac{1}{2}(-286 \text{ kJ})$$

The sum of the heats on the product side is

$$6(-394 \text{ kJ}) + 33.9 \text{ kJ} + 2\tfrac{1}{2}(-286 \text{ kJ}) = -3045 \text{ kJ}$$

The sum of the heats on the reactant side is

$$\text{(? kJ)} + 0.00 \text{ kJ} - 3093 \text{ kJ} = \text{(? kJ)} - 2093 \text{ kJ}$$

Since the heats on both sides must be equal,

$$\text{(? kJ)} - 3093 \text{ kJ} = -3045 \text{ kJ} \quad \text{or} \quad \text{? kJ} = \Delta H_f^0 = 48 \text{ kJ}$$

Therefore ΔH_f^0 for C_6H_5 (*l*) = +48 kJ

24. First determine the heat gained by the calorimeter, $Q_{calor.}$

$$\Delta T = 27.3°C - 24.6°C = 2.7°C$$

$$Q_{calor.} = Q_w + Q_c = 50.50 \text{ g water} \times \frac{4.184 \text{ J}}{\text{(g water)}(°C)} \times 2.7°C + \frac{3.5 \text{ J}}{°C} \times 2.7°C$$

$$= 570.5 \text{ J} + 9.5 \text{ J} = 580 \text{ J}$$

The $Q_{calor.} = -Q_{process}$, so $Q_{process} = -580$ J. For the metal, $\Delta T = 27.3°C - 92.0°C = -64.7°C.$

$$\text{Specific heat} = \frac{-580 \text{ J}}{(19.32 \text{ g})(-64.7°C)} = 0.46 \text{ J/(g)}(°C)$$

25. The first equation gives you the C_2H_2 on the correct side, but the second equation must be doubled to have $2H_2(g)$ on the left. The third equation must be reversed to have the C_2H_6 on the right.

$$C_2H_2(g) + \tfrac{5}{2}O_2(g) - 1300 \text{ kJ} = 2CO_2(g) + H_2O(l)$$

$$2H_2(g) + O_2(g) - 2(286 \text{ kJ}) = 2H_2O(l)$$

$$3H_2O(l) + 2CO_2(g) = -1560 \text{ kJ} + \tfrac{7}{2}O_2(g) + C_2H_6(g)$$

There are $\tfrac{7}{2}O_2(g)$ on both the left and the right. There are $2CO_2(g)$ on both sides and $3H_2O(l)$ on both sides. These all cancel. This leaves

$$C_2H_2(g) + 2H_2(g) - 1300 \text{ kJ} - 572 \text{ kJ} = C_2H_6(g) - 1560 \text{ kJ}$$

Combining all the heats, you have

$$C_2H_2(g) + 2H_2(g) - 312 \text{ kJ} = C_2H_6(g)$$

The hydrogenation releases 312 kJ per mole of acetylene.

Appendixes

Four-place Logarithms

Four-Place Logarithms

	0	1	2	3	4	5	6	7	8	9
1.0	.0000	.0043	.0086	.0128	.0170	.0212	.0253	.0294	.0334	.0374
1.1	.0414	.0453	.0492	.0531	.0569	.0607	.0645	.0682	.0719	.0755
1.2	.0792	.0828	.0864	.0899	.0934	.0969	.1004	.1038	.1072	.1106
1.3	.1139	.1173	.1206	.1239	.1271	.1303	.1335	.1367	.1399	.1430
1.4	.1461	.1492	.1523	.1553	.1584	.1614	.1644	.1673	.1703	.1732
1.5	.1761	.1790	.1818	.1847	.1875	.1903	.1931	.1959	.1987	.2014
1.6	.2041	.2068	.2095	.2122	.2148	.2175	.2201	.2227	.2253	.2279
1.7	.2304	.2330	.2355	.2380	.2405	.2430	.2455	.2480	.2504	.2529
1.8	.2553	.2577	.2601	.2625	.2648	.2672	.2695	.2718	.2742	.2765
1.9	.2788	.2810	.2833	.2856	.2878	.2900	.2923	.2945	.2967	.2989
2.0	.3010	.3032	.3054	.3075	.3096	.3118	.3159	.3160	.3181	.3201
2.1	.3222	.3243	.3263	.3284	.3304	.3324	.3345	.3365	.3385	.3404
2.2	.3424	.3444	.3464	.3483	.3502	.3522	.3541	.3560	.3579	.3598
2.3	.3617	.3636	.3655	.3674	.3692	.3711	.3729	.3747	.3766	.3784
2.4	.3802	.3820	.3838	.3856	.3874	.3892	.3909	.3927	.3945	.3962
2.5	.3979	.3997	.4014	.4031	.4048	.4065	.4082	.4099	.4116	.4133
2.6	.4150	.4166	.4183	.4200	.4216	.4232	.4249	.4265	.4281	.4298
2.7	.4314	.4330	.4346	.4362	.4378	.4393	.4409	.4425	.4440	.4456
2.8	.4472	.4487	.4502	.4518	.4533	.4548	.4564	.4579	.4594	.4609
2.9	.4624	.4639	.4654	.4669	.4683	.4698	.4713	.4728	.4742	.4757
3.0	.4771	.4786	.4800	.4814	.4829	.4843	.4857	.4871	.4886	.4900
3.1	.4914	.4928	.4942	.4955	.4969	.4983	.4997	.5011	.5024	.5038
3.2	.5051	.5065	.5079	.5092	.5105	.5119	.5132	.5145	.5159	.5172
3.3	.5185	.5198	.5211	.5224	.5237	.5250	.5263	.5276	.5289	.5302
3.4	.5315	.5328	.5340	.5353	.5366	.5378	.5391	.5403	.5416	.5428
3.5	.5441	.5453	.5465	.5478	.5490	.5502	.5514	.5527	.5539	.5551
3.6	.5563	.5575	.5587	.5599	.5611	.5623	.5635	.5647	.5658	.5670
3.7	.5682	.5694	.5705	.5717	.5729	.5740	.5752	.5763	.5775	.5786
3.8	.5798	.5809	.5821	.5832	.5843	.5855	.5866	.5877	.5888	.5899
3.9	.5911	.5922	.5933	.5944	.5955	.5966	.5977	.5988	.5999	.6010
4.0	.6021	.6031	.6042	.6053	.6064	.6075	.6085	.6096	.6107	.6117
4.1	.6128	.6138	.6149	.6160	.6170	.6180	.6191	.6201	.6212	.6222
4.2	.6232	.6243	.6253	.6263	.6274	.6284	.6294	.6304	.6314	.6325
4.3	.6335	.6345	.6355	.6365	.6375	.6385	.6395	.6405	.6415	.6425
4.4	.6435	.6444	.6454	.6464	.6474	.6484	.6493	.6503	.6513	.6522
4.5	.6532	.6542	.6551	.6561	.6571	.6580	.6590	.6599	.6609	.6618
4.6	.6628	.6637	.6646	.6656	.6665	.6675	.6684	.6693	.6702	.6712
4.7	.6721	.6730	.6739	.6749	.6758	.6767	.6776	.6785	.6794	.6803
4.8	.6812	.6821	.6830	.6839	.6848	.6857	.6866	.6875	.6884	.6893
4.9	.6902	.6911	.6920	.6928	.6937	.6946	.6955	.6964	.6972	.6981
5.0	.6990	.6998	.7007	.7016	.7024	.7033	.7042	.7050	.7059	.7067
5.1	.7076	.7084	.7093	.7101	.7110	.7118	.7126	.7135	.7143	.7152
5.2	.7160	.7168	.7177	.7185	.7193	.7202	.7210	.7218	.7226	.7235
5.3	.7243	.7251	.7259	.7267	.7275	.7284	.7292	.7300	.7308	.7316
5.4	.7324	.7332	.7340	.7348	.7356	.7364	.7372	.7380	.7388	.7396

	0	1	2	3	4	5	6	7	8	9
5.5	.7404	.7412	.7419	.7427	.7435	.7443	.7451	.7459	.7466	.7474
5.6	.7482	.7490	.7497	.7505	.7513	.7520	.7528	.7536	.7543	.7551
5.7	.7559	.7566	.7574	.7582	.7589	.7597	.7604	.7612	.7619	.7627
5.8	.7634	.7642	.7649	.7657	.7664	.7672	.7679	.7686	.7694	.7701
5.9	.7709	.7716	.7723	.7731	.7738	.7745	.7752	.7760	.7767	.7774
6.0	.7782	.7789	.7796	.7803	.7810	.7818	.7825	.7832	.7839	.7846
6.1	.7853	.7860	.7868	.7875	.7882	.7889	.7896	.7903	.7910	.7917
6.2	.7924	.7931	.7938	.7945	.7952	.7959	.7966	.7973	.7980	.7987
6.3	.7993	.8000	.8007	.8014	.8021	.8028	.8035	.8041	.8048	.8055
6.4	.8062	.8069	.8075	.8082	.8089	.8096	.8102	.8109	.8116	.8122
6.5	.8129	.8136	.8142	.8149	.8156	.8162	.8169	.8176	.8182	.8189
6.6	.8195	.8202	.8209	.8215	.8222	.8228	.8235	.8241	.8248	.8254
6.7	.8261	.8267	.8274	.8280	.8287	.8293	.8299	.8306	.8312	.8319
6.8	.8325	.8331	.8338	.8344	.8351	.8357	.8363	.8370	.8376	.8382
6.9	.8388	.8395	.8401	.8407	.8414	.8420	.8426	.8432	.8439	.8445
7.0	.8451	.8457	.8463	.8470	.8476	.8482	.8488	.8494	.8500	.8506
7.1	.8513	.8519	.8525	.8531	.8537	.8543	.8549	.8555	.8561	.8567
7.2	.8573	.8579	.8585	.8591	.8597	.8602	.8609	.8615	.8621	.8627
7.3	.8633	.8639	.8645	.8651	.8657	.8663	.8669	.8675	.8681	.8686
7.4	.8692	.8698	.8704	.8710	.8716	.8722	.8727	.8733	.8739	.8745
7.5	.8751	.8756	.8762	.8768	.8774	.8779	.8785	.8791	.8797	.8802
7.6	.8808	.8814	.8820	.8825	.8831	.8837	.8842	.8848	.8854	.8859
7.7	.8865	.8871	.8876	.8882	.8887	.8893	.8899	.8904	.8910	.8915
7.8	.8921	.8927	.8932	.8938	.8943	.8949	.8954	.8960	.8965	.8971
7.9	.8976	.8982	.8987	.8993	.8998	.9004	.9009	.9015	.9020	.9026
8.0	.9031	.9036	.9042	.9047	.9053	.9058	.9063	.9069	.9074	.9079
8.1	.9085	.9090	.9096	.9101	.9106	.9112	.9117	.9122	.9128	.9133
8.2	.9138	.9143	.9149	.9154	.9159	.9165	.9170	.9175	.9180	.9186
8.3	.9191	.9196	.9201	.9206	.9212	.9217	.9222	.9227	.9232	.9238
8.4	.9243	.9248	.9253	.9258	.9263	.9269	.9274	.9279	.9284	.9289
8.5	.9294	.9299	.9304	.9309	.9315	.9320	.9325	.9330	.9335	.9340
8.6	.9345	.9350	.9355	.9360	.9365	.9370	.9375	.9380	.9385	.9390
8.7	.9395	.9400	.9405	.9410	.9415	.9420	.9425	.9430	.9435	.9440
8.8	.9445	.9450	.9455	.9460	.9465	.9469	.9474	.9479	.9484	.9489
8.9	.9494	.9499	.9504	.9509	.9513	.9518	.9523	.9528	.9533	.9538
9.0	.9542	.9547	.9552	.9557	.9562	.9566	.9571	.9576	.9581	.9586
9.1	.9590	.9595	.9600	.9605	.9609	.9614	.9619	.9624	.9628	.9633
9.2	.9638	.9643	.9647	.9652	.9657	.9661	.9666	.9671	.9675	.9680
9.3	.9685	.9689	.9694	.9699	.9703	.9708	.9713	.9717	.9722	.9727
9.4	.9731	.9736	.9741	.9745	.9750	.9754	.9759	.9763	.9768	.9773
9.5	.9777	.9782	.9786	.9791	.9795	.9800	.9805	.9809	.9814	.9818
9.6	.9823	.9827	.9832	.9836	.9841	.9845	.9850	.9854	.9859	.9863
9.7	.9868	.9872	.9877	.9881	.9886	.9890	.9894	.9899	.9903	.9908
9.8	.9912	.9917	.9921	.9926	.9930	.9934	.9939	.9943	.9948	.9952
9.9	.9956	.9961	.9965	.9969	.9974	.9978	.9983	.9987	.9991	.9996

APPENDIX II

Conversion Factors

Metric Units

Length		Mass		Volume	
Angstrom	$\dfrac{\text{Å}}{10^{-10}\ \text{m}}$	Microgram	$\dfrac{\mu\text{g}}{10^{-6}\ \text{g}}$	Milliliter	$\dfrac{\text{mL}}{10^{-3}\ \text{L}}$
*Micron	$\dfrac{\mu}{10^{-6}\ \text{m}}$	Milligram	$\dfrac{\text{mg}}{10^{-3}\ \text{g}}$	Cubic centimeter	$\dfrac{1\ \text{mL}}{\text{cm}^3}$
Millimeter	$\dfrac{\text{mm}}{10^{-3}\ \text{m}}$	Kilogram	$\dfrac{\text{kg}}{10^3\ \text{g}}$	Microliter	$\dfrac{\mu\text{L}}{10^{-6}\ \text{L}}$
Centimeter	$\dfrac{\text{cm}}{10^{-2}\ \text{m}}$				
Kilometer	$\dfrac{\text{km}}{10^3\ \text{m}}$				

*Now called a micrometer, with the symbol μm

American (English)

Length	Mass	Volume
$\dfrac{12\ \text{in.}}{\text{ft}}$	$\dfrac{16\ \text{oz (avoir)}}{\text{lb (avoir)}}$	$\dfrac{16\ \text{oz (fluid)}}{\text{pt}}$
$\dfrac{3\ \text{ft}}{\text{yd}}$	$\dfrac{2000\ \text{lb}}{\text{ton}}$	$\dfrac{2\ \text{pt}}{\text{qt}}$
$\dfrac{5280\ \text{ft}}{\text{mi}}$		$\dfrac{4\ \text{qt}}{\text{gal}}$

Metric to American (English)

Length	Mass	Volume
$\dfrac{2.54\ \text{cm}}{\text{in.}}$	$\dfrac{454\ \text{g}}{\text{lb}}$	$\dfrac{946\ \text{mL}}{\text{qt}}$
$\dfrac{1.61\ \text{km}}{\text{mi}}$	$\dfrac{2.20\ \text{lb}}{\text{kg}}$	

Pressure

$\dfrac{760\ \text{mmHg}}{\text{atm}}$	$\dfrac{760\ \text{torr}}{\text{atm}}$	$\dfrac{76.0\ \text{cmHg}}{\text{atm}}$	$\dfrac{14.7\ \text{psi}}{\text{atm}}$	$\dfrac{29.9\ \text{in. Hg}}{\text{atm}}$

Heat

Calorie, cal	$\dfrac{10^3\ \text{cal}}{\text{kcal}}$	$\dfrac{4.184\ \text{J}}{\text{cal}}$	*Specific heat of water*
Joule, J	$\dfrac{10^3\ \text{J}}{\text{kJ}}$	$\dfrac{0.2390\ \text{cal}}{\text{J}}$	$\dfrac{1\ \text{cal}}{(\text{g water})(^\circ\text{C})}$
			$\dfrac{4.184\ \text{J}}{(\text{g water})(\text{K})}$

Ideal Gas Constant

$\dfrac{0.0820\ (L)(atm)}{(mol)(K)}$	$\dfrac{22.4\ L\ (STP)}{mol}$	$\dfrac{1.987\ cal}{mol^\circ K}$	$\dfrac{8.314\ J}{mol\ K}$

Avogadro's Number (6.02×10^{23})

$\dfrac{6.02 \times 10^{23}\ molecules}{mol\ molecules}$	$\dfrac{6.02 \times 10^{23}\ atoms}{mol\ atoms}$	$\dfrac{6.02 \times 10^{23}\ electrons}{mol\ electrons}$	$\dfrac{6.02 \times 10^{23}\ electrons}{\mathscr{F}}$

Temperature (not conversion factors; formulas)

$$^\circ C = \frac{5}{9}(^\circ F - 32) \qquad ^\circ F = \frac{9}{5}(^\circ C) + 32 \qquad K = {}^\circ C + 273$$

Physical Properties Expressed as Conversion Factors

Property	*Units*		
Density	$\dfrac{g}{cm^3}$ (solids) $\quad \dfrac{g}{mL}$ (liquids) $\quad \dfrac{g}{L}$ (gases)		
Specific gravity	$\dfrac{g\ sample}{g\ water\ occupying\ same\ volume\ as\ sample}$	or	$\dfrac{density\ sample}{density\ water}$
Percent	$\dfrac{parts\ substance\ of\ interest}{100\ parts\ total\ mixture}$		
Weight percent	$\dfrac{g\ substance\ of\ interest}{100\ g\ total\ mixture}$		
Volume percent	$\dfrac{mL\ substance\ of\ interest}{100\ mL\ total\ mixture}$		
Percent purity	$\dfrac{g\ pure\ component}{100\ g\ impure\ substance}$		

Chemical Properties Expressed as Conversion Factors

Property	*Units*	*Property*	*Units*
Molecular weight	$\dfrac{g}{mol}$	Molarity (*M*)	$\dfrac{\text{moles of solute}}{\text{liter solution}}$
Atomic weight	$\dfrac{g}{mol}$	Formality (*F*)	$\dfrac{\text{formula weights of solute}}{\text{liter solution}}$
Formula weight	$\dfrac{g}{mol}$	Molality (*m*)	$\dfrac{\text{moles of solute}}{1000\ g\ solvent}$
Molal freezing-point-depression constant (K_f) Molal boiling-point-elevation constant (K_b)	$\dfrac{^\circ C}{m}$ or		$\dfrac{(^\circ C)(1000\ g\ solvent)}{mol\ solute}$

Thermodynamic Properties Expressed as Conversion Factors

Property	*Units*
Specific heat	$\dfrac{cal}{g\,^\circ C}$ or $\dfrac{J}{g\,^\circ C}$
Calorimeter constant (K_c)	$\dfrac{cal}{^\circ C}$ or $\dfrac{J}{^\circ C}$
Heats of fusion, vaporization, or sublimation	$\dfrac{cal}{g}$ or $\dfrac{kcal}{mol}$ or $\dfrac{J}{g}$ or $\dfrac{kJ}{mol}$
Molar heat of reaction (ΔH^0)	$\dfrac{kcal}{mol}$ or $\dfrac{kJ}{mol}$
Heat of formation (ΔH_f^0)	$\dfrac{kcal}{mol}$ or $\dfrac{kJ}{mol}$

APPENDIX III

International System of Units (SI)

Starting in about 1960, the International Bureau of Weights and Measures, in a series of international committee meetings and general conferences, worked out an agreement on a system of units that they propose be used by all scientists throughout the world. This Système Internationale d'Unites, abbreviated SI, is closely related to the metric system (meter-kilogram-second or mks). But the SI system reduces the number of different units by expressing everything in terms of just seven *base* units.

To see how this works, consider the metric unit of volume, the liter. As you know, you can express any volume as the product of three lengths. A liter represents the volume of a cube that is 10 cm on each edge (10 cm $\times$ 10 cm $\times$ 10 cm = 1000 cm^3 = 1000 mL = 1.000 L). Thus the SI unit equivalent to one liter is one decimeter cubed, or dm^3 (a dm is 10^{-1} m) or 10^{-3} m^3. Volume is thus expressed in terms of length, and there is no need for a special unit for volume.

The table below shows the seven base units. The last, the candela, is not normally of interest to chemists.

Very large and very small multiples of the base units can be expressed in the same way as in the metric system, by putting a prefix before the name to indicate by what power of 10 the base unit is multiplied. On the next page is a partial list of these prefixes and the power of 10 they represent.

Some units are in such common usage that it would be impractical to eliminate them. Therefore they have been retained. Thus, although the base unit for time is seconds, the units for minutes, hours, and days have been kept. This is also true for atmospheres (pressure), liters (volume), and tonne (10^3 kg). Some familiar units have, however, been removed. Thus the angstrom (Å) (10^{-10} m

SI Base Units

Quantity	Name of Unit	Abbreviation
Length	meter	m
Mass	kilogram	kg
Time	second	s
Electric current	ampere	A
Temperature (thermodynamic)	kelvin	K
Amount of substance	mole	mol
Luminous intensity	candela	cd

SI Prefixes

Power of 10	Prefix	Symbol
10^3	kilo	k
10^1	deca	da
10^{-1}	deci	d
10^{-2}	centi	c
10^{-3}	milli	m
10^{-6}	micro	μ
10^{-9}	nano	n

or 10^{-8} cm), the torr or mmHg, and the calorie have been deleted. The micron (μ) was first changed to micrometer (mμ), then finally eliminated altogether. Also the word and symbol for degree ($°$) have been removed from temperatures when they are expressed in Kelvin notation. Thus water boils at 373 kelvins (373 K). However, the degree, word and symbol, has been kept if you express the temperature in Celsius (centigrade). Water boils at 100 degrees Celsius (100°C).

Actually, there is not too much difference between the metric system and SI. The following tables show those units that are important to chemists and that vary from the metric system. If in the future you need a source of all the SI units, send for a pamphlet, National Bureau of Standards Special Publication 330, 1972 Edition, SD Catalog No. 13.10:330/2, from the U.S. Government Printing Office, Washington, D.C.

Conversion Factors, Metric to SI

Length			*Volume*		
Angstrom (Å)	$\dfrac{\text{Å}}{10^{-10}\ \text{m}}$ or	$\dfrac{\text{Å}}{0.1\ \text{nm}}$	Liter (L)	$\dfrac{\text{L}}{10^{-3}\ \text{m}^3}$ or	$\dfrac{\text{L}}{\text{dm}^3}$
Millimicron (mμ) [Micron (μ)]	$\dfrac{\text{m}\mu}{\text{nm}}$ or	$\dfrac{10^{-3}\mu}{\text{nm}}$	(The liter and its multiples can still be used, but for high precision the use of m^3 is recommended)		

Pressure		*Temperature*
The SI unit is the pascal (Pa)		The SI unit is the kelvin (K)
mmHg	$\dfrac{\text{mmHg}}{133.3\ \text{Pa}}$	$\dfrac{°\text{K}}{\text{K}}$
torricelli (torr)	$\dfrac{\text{torr}}{133.3\ \text{Pa}}$	
atmosphere (atm)	$\dfrac{\text{atm}}{101.3 \times 10^3\ \text{Pa}}$ (The atm can still be used)	

Conversion Factors, Metric to SI   (continued)

Heat

calorie (cal)	$\dfrac{0.2390 \text{ cal}}{\text{J}}$	or	$\dfrac{4.184 \text{ J}}{\text{cal}}$

Ideal Gas Constant

$$\frac{0.0820 \text{ (L)(atm)}}{\text{(mol)(K)}} = \frac{8.31 \text{ (m}^3\text{)(Pa)}}{\text{(mol)(K)}} = \frac{8.314 \text{ J}}{\text{(mol)(K)}}$$

Properties Expressed as Conversion Factors with SI Equivalent

	Metric SI

Density $\qquad\qquad \dfrac{\text{g}}{\text{mL}} = \dfrac{\text{kg}}{\text{m}^3}$

Molecular weight $\dfrac{\text{g}}{\text{mol}} = \dfrac{10^{-3} \text{ kg}}{\text{mol}}$

Molarity, the word, is not to be used in SI, but the concentration is still allowed, and for high-accuracy work it should be expressed as

$$\frac{\text{mol}}{\text{L}} = \frac{\text{mol}}{10^{-3} \text{ m}^3} = \frac{\text{mol}}{\text{dm}^3}$$